Endothelium-Dependent Hyperpolarizations

Endothelium-Dependent Hyperpolarizations

Edited by

Paul M. Vanhoutte

Institut de Recherches Internationales Servier
Courbevoie Cedex
France

harwood academic publishers

Australia • Canada • France • Germany • India • Japan
Luxembourg • Malaysia • The Netherlands • Russia • Singapore
Switzerland

Amsteldijk 166
1st Floor
1079 LH Amsterdam
The Netherlands

British Library Cataloguing in Publication Data

Endothelium-dependent hyperpolarizations
   1. Endothelins – Congresses 2. Vascular endothelium –
Congresses
I. Vanhoutte, P. M. (Paul M.)
612.1′3

ISBN: 90-5702-492-6

TABLE OF CONTENTS

PREFACE

This monograph is the second of a series devoted to endothelium-dependent hyperpolarizations. It consists of the Proceedings of the Second International Symposium discussing this topic, which was held in Vaux de Cernay from 5-6 June, 1998. Whereas the first meeting established the existence of endothelium-dependent hyperpolarization and of endothelium-derived hyperpolarizing factor (EDHF), this symposium constitutes a major step forward as it established the multiplicity of factors involved, and reinforced the concept that EDHF, as opposed to nitric oxide (NO) is of particular importance in smaller blood vessels. The symposium also stressed how EDHF has become an exploding field of interest world-wide.

The first chapters discuss the many candidates as EDHF, analysing the contribution of mediators proposed at the first meeting, and suggesting sometimes surprising new ones. A second series of chapters detail, almost at the molecular level, the action of EDHF on vascular smooth muscle cells. A third set of contributions continue to enlarge our knowledge of the physiological and pathological roles of endothelium-dependent hyperpolarizations. The final series of chapters reports on the, still limited but crucial, knowledge of the contribution of EDHF in human blood vessels, in health and disease.

As always with multi-authored texts, the responsibility for the scientific content rests with the individual authors. Hence the statements made are not necessarily endorsed by the editor. His task has been mainly to select the authors, to streamline their texts, and to achieve, as much as possible, uniformity of presentation.

Endothelium-Dependent Hyperpolarizations will be of interest not only to physiologists and pharmacologists puzzled by the complexity of the interactions between the endothelium and the underlying vascular smooth muscle cells but also to clinical researchers and to the physicians who treat patients with cardiovascular diseases. Indeed, the understanding of the nature and the role of EDHF already appears to be crucial in the quest for an improvement in the treatment of hypertension and other vascular disorders.

Paul M. Vanhoutte

ACKNOWLEDGEMENTS

The participants of the Second International Symposium on Endothelium-Derived Hyperpolarizing Factor will not soon forget the charm and dedication of Mrs Denise Maggi, Ms Susan Blanchard and Ms Rowena McKeown. This monograph would not have been possible without the total dedication of Mr Robert R. Lorenz who took the responsibility of the illustrations. The administrative help of Ms Pascale Olivier and Ms Marie Palumbo was also instrumental in keeping the project within reasonable deadlines. The editor would like to thank most sincerely Dr Laurent Perret of the Institut de Recherches Internationales Servier, who supported the endeavor to try to make this monograph into a reasonably uniform text, despite the many contributors. He also would like to thank the authors most sincerely for their collaboration and understanding when faced with extensive editing of their manuscripts. Last, but not least, the staff of Harwood Academic Publishers should be complimented for a most efficient handling of the manuscripts.

CONTRIBUTORS

Abe, I.
Second Department of Internal Medicine
Faculty of Medicine
Kyushu University
Maidashi 3-1-1, Higashi-ku
Fukuoka 812-8582
Japan

Alkayed, N.
Department of Anesthesiology and
  Critical Care Medicine
School of Medicine
Johns Hopkins University
Baltimore, MD 21205-2196
USA

Alvarez de Sotomayor, M.
Laboratoire de Pharmacologie et
  Physiologie Cellulaires
Faculté de Pharmacie
Université Louis Pasteur de Strasbourg
CNRS ERS 653, BP 24
67401 Illkirch Cedex
France

Andersson, D.A.
Department of Clinical Pharmacology
Institute of Laboratory Medicine
Lund University
S-221 85 Lund
Sweden

Andriantsitohaina, R.
Laboratoire de Pharmacologie et
  Physiologie Cellulaires
Faculté de Pharmacie
Université Louis Pasteur de Strasbourg
CNRS ERS 653, BP 24
67401 Illkirch Cedex
France

Ballard, K.
Center for Experimental Therapeutics
Baylor College of Medicine
One Baylor Plaza, Room 802E
Houston, TX 77030
USA

Belfort, M.A.
Department of Perinatology
Utah Valley Regional Medical Center
1034 North 500 West
Provo, UT 84604
USA

Bény, J.-L.
Department of Zoology and Animal
  Biology
Geneva University
Sciences III, 30 quai E. Ansermet
CH-1211 Geneva 4
Switzerland

Berg, G.
Department of Medicine and Therapeutics
Western Infirmary
Glasgow, G11 6NT
UK

Bidouard, J.-P.
Département de Pathologies Cardiaques
    et Vasculaires
Institut de Recherches Servier
11 rue des Moulineaux
92150 Suresnes
France

Bolotina, V.M.
Vascular Biology Unit
Department of Medicine
Boston University Medical Center
88 East Newton Street
Boston, MA 02118-2393
USA

Boyle, J.P.
Department of Cell Physiology and
    Pharmacology
Maurice Shock Medical Sciences Building
University of Leicester
PO Box 138
Leicester, LE1 9HN
UK

Bult, H.
Division of Pharmacology
University of Antwerp (UIA)
Universiteitsplein 1
B-2610 Wilrijk
Belgium

Burnette, E.
University of Ottawa Heart Institute
40 Ruskin Street
Ottawa, K1Y 4W7
Canada

Busse, R.
Institut für Kardiovaskuläre Physiologie
Klinikum der J.W. Goethe-Universität
Theodor-Stern-Kai 7
D-60590 Frankfurt am Main
Germany

Cambarrat, C.
Département de Pathologies Cardiaques
    et Vasculaires
Institut de Recherches Servier
11 rue des Moulineaux
92150 Suresnes
France

Campbell, W.B.
Department of Pharmacology and
    Toxicology
Medical College of Wisconsin
8701 Watertown Plank Road
Milwaukee, Wl 53226
USA

Challiss, R.A.J.
Department of Cell Physiology and
    Pharmacology
Maurice Shock Medical Sciences Building
University of Leicester
PO Box 138
Leicester, LE1 9HN
UK

Chataigneau, T.
Service de Pharmacologie
Hôpital Pitié Salpétriére
47-88 boulevard de l'Hôpital
75651 Paris Cedex 13
France

Chen, G.
University of Ottawa Heart Institute
40 Ruskin Street
Ottawa, K1Y 4W7
Canada

Cheung, D.W.
University of Ottawa Heart Institute
40 Ruskin Street
Ottawa, K1Y 4W7
Canada

Cohen, R.A.
Vascular Biology Unit
Department of Medicine
Boston University School of Medicine
80 E. Concord Street
Boston, MA 02118
USA

Corriu, C.
Département de Diabétologie
Institut de Recherches Servier
11 rue des Moulineaux
92150 Suresnes
France

Darker, I.T.
School of Biomedical Sciences
University of Nottingham Medical School
Queen's Medical Centre
Nottingham, NG7 2UH
UK

Delescluse, I.
Département de Pathologies Cardiaques
  et Vasculaires
Institut des Recherches Servier
11 rue des Moulineaux
92150 Suresnes
France

Dominiczak, A.F.
Department of Medicine and Therapeutics
Western Infirmary
Glasgow, G11 6NT
UK

Dora, K.A.
Department of Pharmacology
University of Bristol
University Walk
Bristol, BS8 1TD
UK

Duhault, J.
Département de Diabétologie
Institut de Recherches Servier
11 rue des Moulineaux
92150 Suresnes
France

Dunn, W.R.
School of Biomedical Science
University of Nottingham Medical School
Queen's Medical Centre
Nottingham, NG7 2UH
UK

Eckman, D.M.
Department of Pharmacology
Given Building, B326
School of Medicine
University of Vermont
Burlington, VT 05405
USA

Edgemond W.S.
Department of Pharmacology and
  Toxicology
Medical College of Wisconsin
8701 Watertown Plank Road
Milwaukee, Wl 53226
USA

Edwards, G.
School of Biological Sciences
University of Manchester
G.38 Stopford Building
Oxford Road
Manchester, M13 9PT
UK

Fachney, A.
Department of Medicine and Therapeutics
Western Infirmary
Glasgow, G11 6NT
UK

Félétou, M.
Département de Diabétologie
Institut de Recherches Servier
11 rue des Moulineaux
92150 Suresnes
France

Fishman, M.C.
Cardiovascular Research Center and
Department of Medicine
Massachusetts General Hospital and
Harvard Medical School
Charlestown, MA 02129
USA

Fleming, I.
Institut für Kardiovaskuläre Physiologie
Klinikum der J.W. Goethe-Universität
Theodor-Stern-Kai 7
D-60590 Frankfurt am Main
Germany

Fournet-Bourguignon, M.P.
Département de Pathologies Cardiaques
   et Vasculaires
Institut de Recherches Servier
11 rue des Moulineaux
92150 Suresnes
France

Frank, S.
Department of Medical Biochemistry
Karl-Franzens University of Graz
Harrachgasse 21/III
A-8010 Graz
Austria

Fujii, K.
Second Department of Internal Medicine
Faculty of Medicine
Kyushu University
Maidashi 3-1-1, Higashi-ku
Fukuoka 812-8582
Japan

Fujishima, M.
Second Department of Internal Medicine
Faculty of Medicine
Kyushu University
Maidashi 3-1-1, Higashi-ku
Fukuoka 812-8582
Japan

Gardener, M.J.
School of Biological Sciences
University of Manchester
G.38 Stopford Building
Oxford Road
Manchester, M13 9PT
UK

Garfield, R.E.
Reproductive Sciences
Department of Obstetrics and Gynecology
University of Texas Medical Branch
301 University Boulevard
Galveston, TX 77555-1062
USA

Garland, C.J.
Department of Pharmacology
University of Bristol
University Walk
Bristol, BS8 ITD
UK

Gebremedhin, D.
Cardiovascular Research Center
Medical College of Wisconsin
8701 Watertown Plank Road
Milwaukee, WI 53226
USA

Ghiadoni, L.
Department of Internal Medicine
University of Pisa
Via Roma 67
56100 Pisa
Italy

Graier, W.F.
Department of Medical Biochemistry
Karl-Franzens University of Graz
Harrachgasse 21/III
A-8010 Graz
Austria

Griswold, M.C.
Department of Medicine
Boston University School of Medicine
80 E. Concord Street
Boston, MA 02118
USA

Haefliger, I.O.
Laboratory of Ocular Pharmacology and
    Physiology
University Eye Clinic Basel
Mittlere Strasse 91
PO Box CH-4012 Basel
Switzerland

Hamilton, C.A.
Department of Medicine and Therapeutics
Western Infirmary
Glasgow, G11 6NT
UK

Harder, D.R.
Cardiovascular Research Center
Medical College of Wisconsin
8701 Watertown Plank Road
Milwaukee, Wl 53226
USA

Harris, D.
School of Biomedical Sciences
University of Nottingham Medical School
Queen's Medical Centre
Nottingham, NG7 2UH
UK

Herman, A.G.
Division of Pharmacology
University of Antwerp (UIA)
Universiteitsplein 1
B-2610 Wilrijk
Belgium

Hirooka, Y.
Research Institute of Angiocardiology
    and Cardiovascular Clinic
Kyushu University School of Medicine
3-1-1 Maidashi, Higashi-ku
Fukuoka 812-8582
Japan

Hoebel, B.G.
Department of Medical Biochemistry
Karl-Franzens University of Graz
Harrachgasse 21/III
A-8010 Graz
Austria

Högestätt, E.D.
Department of Clinical Pharmacology
Institute of Laboratory Medicine
Lund University
S-221 85 Lund
Sweden

Hopkins, N.
Department of Physiology and Cell
    Biology
University of Nevada
School of Medicine
352 Anderson Building
Reno, NV 89557
USA

Holland, M.
Department of Cell Physiology and
    Pharmacology
Maurice Shock Medical Sciences Building
University of Leicester
PO Box 138
Leicester, LEI 9HN
UK

Holzmann, S.
Department of Pharmacology and
    Toxicology
Karl-Franzens University of Graz
A-8010 Graz
Austria

Huang, P.L.
Cardiovascular Research Center and
Department of Medicine
Massachusetts General Hospital and
Harvard Medical School
Charlestown, MA 02129
USA

Ibayashi, S.
Second Department of Internal Medicine
Faculty of Medicine
Kyushu University
Maidashi 3-1-1, Higashi-ku
Fukuoka 812-8582
Japan

Jain, V.
Reproductive Sciences
Department of Obstetrics and Gynecology
University of Texas Medical Branch
301 University Boulevard
Galveston, TX 77555-1062
USA

Jardine, E.
Department of Medicine and Therapeutics
Western Infirmary
Glasgow, G11 6NT
UK

Kanno, M.
Department of Pharmacology
Hokkaido University School of Medicine
N-15, W-7, Kita-ku
Sapporo 060-8638
Japan

Keef, K.D.
Department of Physiology and Cell Biology
University of Nevada School of Medicine
352 Anderson Building
Reno, NV 89557
USA

Kendall, D.A.
School of Biomedical Sciences
University of Nottingham Medical School
Queen's Medical Centre
Nottingham, NG7 2UH
UK

Kitabatake, A.
Department of Cardiovascular Medicine
Hokkaido University School of Medicine
N-15, W-7, Kita-ku
Sapporo 060-8638
Japan

Kitazono, T.
Second Department of Internal Medicine
Faculty of Medicine
Kyushu University
Maidashi 3-1-1, Higashi-ku
Fukuoka 812-8582
Japan

Kleschyov, A.L.
Laboratoire de Pharmacologie et
    Physiologie Cellulaires
Faculté de Pharmacie
Université Louis Pasteur de Strasbourg
CNRS ERS 653, BP 24
67401 Illkirch Cedex
France

Koehler, R.C.
Department of Anesthesiology and
    Critical Care Medicine
School of Medicine
Johns Hopkins University
Baltimore, MD 21205-2196
USA

Komalavilas, P.
Department of Pathology
University of Alabama
Birmingharn, AL 35294
USA

Kostner, G.M.
Department of Medical Biochemistry
Karl-Franzens University of Graz
Harrachgasse 21/III
A-8010 Graz
Austria

Lagaud, G.J.L.
Department of Cardiovascular Research
St. Paul's Hospital, Room 292
University of British Columbia
1081 Burrard Street
Vancouver BC, V6T 1Z3
Canada

Laher, I.
Department of Pharmacology and
    Therapeutics
St. Paul's Hospital, Room 292
University of British Columbia
1081 Burrard Street
Vancouver BC, V6T 1Z3
Canada

Lawrence, R.N.
School of Biomedical Science
University of Nottingham Medical School
Queen's Medical Centre
Nottingham, NG7 2UH
UK

Li, G.
University of Ottawa Heart Institute
40 Ruskin Street
Ottawa, KIY 4W7
Canada

Li, P.-L.
Department of Pharmacology and
    Toxicology
Medical College of Wisconsin
8701 Watertown Plank Road
Milwaukee, W1 53226
USA

Lincoln, T.M.
Department of Pathology
University of Alabama
Birmingham, AL 35294
USA

Liu, M.-Y.
Department of Pharmacology
Hokkaido University School of Medicine
N-15, W-7, Kita-ku
Sapporo 060-8638
Japan

Lonigro, A.J.
Departments of Pharmacological and
    Physiological Sciences
Saint Louis University School of Medicine
1402 S. Grand Boulevard
St. Louis, MI 63104
USA

Malblanc, S.
Laboratoire de Pharmacologie et
    Physiologie Cellulaires
Faculté de Pharmacie
Université Louis Pasteur de Strasbourg
CNRS ERS 653, BP 24
67401 Illkirch Cedex
France

Malinski, T.
Department of Chemistry
Institute of Biotechnology
Oakland University
Rochester, MI 48309
USA

Matthys, K.E.
Division of Pharmacology
University of Antwerp (UIA)
Universiteitsplein 1
B-2610 Wilrijk
Belgium

McArthur, K.
Department of Medicine and Therapeutics
Western Infirmary
Glasgow, G11 6NT
UK

McBride, C.
Department of Physiology and Cell Biology
University of Nevada
School of Medicine
352 Anderson Building
Reno, NV 89557
USA

Medhora, M.
Cardiovascular Research Center
Medical College of Wisconsin
8701 Watertown Plank Road
Milwaukee, Wl 53226
USA

Millns, P.J.
School of Biomedical Sciences
University of Nottingham Medical School
Queen's Medical Centre
Nottingham, NG7 2UH
UK

Mombouli, J.-V.
Center for Experimental Therapeutics
Baylor College of Medicine
One Baylor Plaza, Room 802E
Houston, TX 77030
USA

Mosse, I.
Institute of Pharmacology and Toxicology
Academy of Medical Sciences
14 Eugene Pottier Street
252057 Kiev
Ukraine

Movahed, P.
Department of Clinical Pharmacology
Institute of Laboratory Medicine
Lund University Hospital
S-221 85 Lund
Sweden

Muller, B.
Laboratoire de Pharmacologie et
    Physiologie Cellulaires
Faculté de Pharmacie
Université Louis Pasteur de Strasbourg
CNRS ERS 653, BP 24
67401 Illkirch Cedex
France

Munzenmaier, D.
Cardiovascular Research Center
Medical College of Wisconsin
8701 Watertown Plank Road
Milwaukee, Wl 53226
USA

Mutafova, V.
Department of Physiology and Cell Biology
University of Nevada
School of Medicine
352 Anderson Building
Reno, NV 89557
USA

Nagao, T.
Second Department of Internal Medicine
Faculty of Medicine
Kyushu University
Maidashi 3-1-1, Higashi-ku
Fukuoka 812-8582
Japan

Nakashima, M.
Surgical Center
Saga Medical School
1-1 5-chome, Nabeshima
Saga 849
Japan

Narayanan, J.
Cardiovascular Research Center
Medical College of Wisconsin
8701 Watertown Plank Road
Milwaukee, WI 53226
USA

Nithipatikom, K.
Department of Pharmacology and
    Toxicology
Medical College of Wisconsin
8701 Watertown Plank Road
Milwaukee, W1 53226
USA

Ntsikoussalabongui, B.
ASTER CEPHAC
90 Avenue du haut de chaume
BP 28, 86281 Saint-Benoit Cedex
France

Oishi, H.
Department of Community Health Science
Saga Medical School
1-1 5-chome, Nabeshima
Saga 849
Japan

Onaka, U.
Second Department of Internal Medicine
Faculty of Medicine
Kyushu University
Maidashi 3-1-1, Higashi-ku
Fukuoka 812-8582
Japan

Orskiszewski, R.
Center for Experimental Therapeutics
Baylor College of Medicine
One Baylor Plaza, Room 802E
Houston, TX 77030
USA

Parshikov, A.
Institute of Pharmacology and Toxicology
Academy of Medical Sciences
14 Eugene Pottier Street
252057 Kiev
Ukraine

Perrault, L.P.
Research Centre and Department of
    Surgery
Montreal Heart Institute
5000 Belanger East
Montreal, Quebec H1T 1C8
Canada

Petersson J.
Department of Clinical Pharmacology
Institute of Laboratory Medicine
Lund University Hospital
S-221 85 Lund
Sweden

Pfister, S.L.
Department of Pharmacology and
    Toxicology
Medical College of Wisconsin
8701 Watertown Plank Road
Milwaukee, Wl 53226
USA

Plane, F.
Department of Pharmacology
University of Bristol
University Walk
Bristol, BS8 ITD
UK

Quignard, J.-F.
Laboratoire de Physiopathologie
et Pharmacologie Vasculaire
Université de Bordeaux II
146 rue Léo Saignat
33076 Bordeaux Cedex
France

Randall, M.D.
School of Biomedical Sciences
University of Nottingham Medical School
Queen's Medical Centre
Nottingham, NG7 2UH
UK

Reid, J.L.
Department of Medicine and Therapeutics
Western Infirmary
Glasgow, G11 6NT
UK

Roman, R.J.
Cardiovascular Research Center
Medical College of Wisconsin
8701 Watertown Plank Road
Milwaukee, Wl 53226
USA

Ruiz, E.
Department of Pharmacology
School of Medicine
Complutense University
28040 Madrid
Spain

Saade, G.R.
Department of Obstetrics and Gynecology
University of Texas Medical Branch
Reproductive Sciences
301 University Boulevard
Galveston, TX 77555-1062
USA

Saboureau, D.
Département de Pathologies Cardiaques
    et Vasculaires
Institut des Recherches Servier
11 rue des Moulineaux
92150 Suresnes
France

Sakuma, I.
Department of Cardiovascular Medicine
Hokkaido University School of Medicine
N-15, W-7, Kita-ku
Sapporo 060-8638
Japan

Salvetti, A.
Department of Internal Medicine
University of Pisa
Via Roma 67
56100 Pisa
Italy

Sato, A.
Department of Pharmacology
Hokkaido University School of Medicine
N-15, W-7, Kita-ku
Sapporo 060-8638
Japan

Scott-Burden, T.
Texas Heart Institute
University of Texas at Houston
Houston, TX 77030
USA

Shimokawa, H.
Research Institute of Angiocardiology
    and Cardiovascular Clinic
Kyushu University School of Medicine
3-1-1 Maidashi, Higashi-ku
Fukuoka 812-8582
Japan

Skarsgard, P.L.
Department of Pharmacology and
    Therapeutics
St. Paul's Hospital, Room 292
University of British Columbia
1081 Burrard Street
Vancouver BC, V6T lZ3
Canada

Soloviev, A.
Institute of Pharmacology and Toxicology
Academy of Medical Sciences
14 Eugene Pottier Street
252057 Kiev
Ukraine

Sørgård, M.
Department of Clinical Pharmacology
Institute of Laboratory Medicine
Lund University Hospital
S-221 85 Lund
Sweden

Spitzbarth, N.
Department of Pharmacology and
 Toxicology
Medical College of Wisconsin
8701 Watertown Plank Road
Milwaukee, W1 53226
USA

Sprague, R.S.
Department of Internal Medicine
Saint Louis University School of Medicine
1402 S. Grand Boulevard
St. Louis, MI 63104
USA

Standen, N.B.
Department of Cell Physiology and
 Pharmacology
Maurice Shock Medical Sciences Building
University of Leicester
PO Box 138
Leicester, LEI 9HN
UK

Stefanov, A.
Institute of Pharmacology and Toxicology
Academy of Medical Sciences
14 Eugene Pottier Street
252057 Kiev
Ukraine

Stephenson, A.H.
Departments of Pharmacological and
 Physiological Sciences
Saint Louis University School of Medicine
1402 S. Grand Boulevard
St. Louis, Ml 63104
USA

Stoclet, J.-C.
Laboratoire de Pharmacologie et
 Physiologie Cellulaires
Faculté de Pharmacie
Université Louis Pasteur de Strasbourg
CNRS ERS 653, BP 24
67401 Illkirch Cedex
France

Taddei, S.
Department of Internal Medicine
University of Pisa
Via Roma 67
56100 Pisa
Italy

Tagawa, H.
Research Institute of Angiocardiology
 and Cardiovascular Clinic
Kyushu University School of Medicine
3-1-1 Maidashi, Higashi-ku
Fukuoka 812-8582
Japan

Takeshita, A.
Research Institute of Angiocardiology
 and Cardiovascular Clinic
Kyushu University School of Medicine
3-1-1 Maidashi, Higashi-ku
Fukuoka 812-8582
Japan

Taylor, A.A.
Center for Experimental Therapeutics
Baylor College of Medicine
One Baylor Plaza, Room 802E
Houston, TX 77030
USA

Tejerina, T.
Department of Pharmacology
School of Medicine
Complutense University
28040 Madrid
Spain

Thollon, C.
Département de Pathologies Cardiaques
  et Vasculaires
Institut de Recherches Servier
11 rue des Moulineaux
92150 Suresnes
France

Tishkin, S.
Institute of Pharmacology and Toxicology
Academy of Medical Sciences
14 Eugene Pottier Street
252057 Kiev
Ukraine

Tomokuni, K.
Department of Community Health Science
Saga Medical School
1-1 5-chome, Nabeshima
Saga 849
Japan

Totoki, T.
Department of Anesthesiology and
  Critical Care Medicine
Saga Medical School
1-1 5-chome, Nabeshima
Saga 849
Japan

Urakami-Harasawa, L.
Research Institute of Angiocardiology
  and Cardiovascular Clinic
Kyushu University School of Medicine
3-1-1 Maidashi, Higashi-ku
Fukuoka 812-8582
Japan

van Breemen, C.
Department of Pharmacology and
  Therapeutics
St. Paul's Hospital, Room 292
University of British Columbia
1081 Burrard Street
Vancouver BC, V6T 1Z3
Canada

Van de Voorde, J.
Department of Physiology and
  Physiopathology
University of Ghent, U.Z.-Blok B
De Pintelaan 185
B-9000 Ghent
Belgium

Van Hove, C.E.
Division of Pharmacology
University of Antwerp (UIA)
Universiteitsplein 1
B-2610 Wilrijk
Belgium

Vanheel, B.
Department of Physiology and
  Physiopathology
University of Ghent, U.Z.-Blok B
De Pintelaan 185
B-9000 Ghent
Belgium

Vanhoutte, P.M.
Institut de Recherches Internationales
  Servier
6 Place des Pléiades
92410 Courbevoie cedex
France

Vedernikov, Y.P.
Reproductive Sciences
Department of Obstetrics and Gynecology
University of Texas Medical Branch
301 University Boulevard
Galveston, TX 77555-1062
USA

Vilaine, J.-P.
Département de Pathologies Cardiaques
  et Vasculaires
Institut de Recherches Servier
11 rue des Moulineaux
92150 Suresnes
France

Villeneuve, N.
Département de Pathologies Cardiaques
  et Vasculaires
Institut de Recherches Servier
11 rue des Moulineaux
92150 Suresnes
France

Virdis, A.
Department of Internal Medicine
University of Pisa
Via Roma 67
56100 Pisa
Italy

Weintraub, N.L.
Department of Internal Medicine
University of Iowa College of Medicine
200 Hawkins Drive
Iowa City, IA 52242
USA

Weisbrod, R.M.
Department of Medicine
Boston University School of Medicine
80 E. Concord Street
Boston, MA 02118
USA

Weston, A.H.
School of Biological Sciences
University of Manchester
G.38 Stopford Building
Oxford Road
Manchester, M13 9PT
UK

Wilson, V.G.
School of Biomedical Science
University of Nottingham Medical School
Queen's Medical Centre
Nottingham, NG7 2UH
UK

Zakharov, S.I.
Department of Medicine
Vascular Biology Unit
Boston University School of Medicine
80 E. Concord Street
Boston, MA 02118
USA

Zeldin, D.
Laboratory of Pulmonary Pathobiology
NIH/NIEHS
111 TW Alexander Drive
Building 101, D236
Research Triangle Park, NC 27709
USA

Zou, A-P.
Department of Pharmacology and
  Toxicology
Medical College of Wisconsin
8701 Watertown Plank Road
Milwaukee, W1 53226
USA

Zygmunt, P.M.
Department of Clinical Pharmacology
Institute of Laboratory Medicine
Lund University Hospital
S-221 85 Lund
Sweden

1 Endothelium-Dependent Hyperpolarization Cannot Be Explained by Electrical Coupling Between the Endothelial and the Smooth Muscle Cells in Muscular Arteries

Jean-Louis Bény[a] and Ivan O. Haefliger[b]

[a] *Department of Zoology and Animal Biology, Sciences III, Geneva University, 30 quai Ernest Ansermet, CH–1211 Genève 4, Switzerland, and*
[b] *Laboratory of Ocular Pharmacology and Physiology, University Eye Clinic Basel, Mittlere Strasse 91, PO Box CH–4012 Basel, Switzerland*
Address for correspondence: Prof. Jean-Louis Bény, Geneva University, Department of Zoology and Animal Biology, Sciences III, 30 quai E. Ansermet, CH–1211 Genève 4, Switzerland. Tel: +41–22–702–6766, Fax: +41–22–781–1747, E-mail: Jean-Louis.Beny@zoo.unige.ch

In the porcine coronary artery, endothelium-derived hyperpolarizing factor (EDHF) is responsible for 70% of the endothelium-dependent vasodilatation caused by bradykinin. This relaxation is associated with the simultaneous hyperpolarization of endothelial and smooth muscle cells. The synchronism of these two hyperpolarizations suggests that an electrical communication between the endothelial and smooth muscle cells may underly endothelium dependent hyperpolarizations of vascular smooth muscle, a phenomenon that has been attributed to EDHF. By contrast, in the porcine ciliary artery, EDHF is not implicated in the bradykinin-evoked endothelium-dependent relaxations. Experiments were designed to compare the behavior of arteries with and without EDHF -mediated responses. In a strip of ciliary artery incubated *in vitro*, dye coupling experiments demonstrated heterocellular coupling between the endothelial and smooth muscle cells. In addition, a 12 mV transient bradykinin-induced hyperpolarization was measured using microelectrodes in endothelial cells. Nevertheless, bradykinin evoked no hyperpolarization of smooth muscle cells in this artery, although a 4 mV hyperpolarization could be recorded In a few smooth muscle cells next to the endothelium. These observations are compatible with the concept that in arteries where the EDHF-component is absent, the current which causes the hyperpolarization of the endothelial cell is not strong enough to passively change the membrane potential of the coupled multiple layers of smooth muscle cells. As a corollary, a phenomenon other than passive electrical coupling alone must be responsible for the transmission of the hyperpolarization of the endothelial cells to the smooth muscle cells in arteries where endothelium-dependent hyperpolarizations occur.

KEYWORDS: EDHF, ciliary artery, coronary artery, hyperpolarization, bradykinin.

INTRODUCTION

In blood vessels, endothelium-dependent relaxation of underlying smooth muscle cells can involve various factors, such as nitric oxide (NO), prostacyclin, or the putative endothelium-derived hyperpolarizing factor (EDHF) (Mombouli and Vanhoutte, 1997). Endothelium-dependent relaxations can occur after activation of endothelial membrane receptors by different agonists, such as bradykinin. The EDHF-mediated relaxation is characterized by hyperpolarization of the vascular smooth muscle cell. Bradykinin, in common with other agonists inducing EDHF-mediated relaxations, can also evoke endothelial cell hyperpolarization (Busse *et al.*, 1988a, Brunet and Bény, 1989). The fact that during endothelium-dependent relaxations, attributed to EDHF, synchronous hyperpolarization of both endothelial and smooth muscle cells can be observed suggests the existence of a passive electrotonic conduction between the cells (Davies *et al.*, 1988; Brunet and Bény, 1989; Bény, 1990).

The present chapter summarizes results demonstrating that, despite electrical coupling between endothelial and smooth muscle cells, hyperpolarization of the former does not necessarily imply that of the latter. Thus, passive electrotonic conduction, alone, is not necessarily sufficient to account for the hyperpolarization of smooth muscle cells observed relaxations attributed to the release of EDHF.

METHODS

Preparation of Tissues

Pigs were killed by electrocution in a slaughterhouse. The eyes or pieces of the heart including left descending branch were dissected and immediately transported to the laboratory in cold (4°C), oxygenated (95% O_2, 5% CO_2) Krebs-Ringer solution (10^{-3} M: NaCI 118.7, KCI 4.7, $CaCI_2$ 2.5, KH_2PO_4 1.2, $NaHCO_3$ 24.8, $MgSO_4$ 1.2, glucose 10.1) (Bény, Brunet and Huggel, 1987; Zhu *et al.*, 1997; Bény, Zhu and Haefliger, 1997).

Dissection

Porcine coronary artery

Upper segments of the left descending branch of the coronary artery were cleaned of adherent tissue, and cut into rings of about 2 mm width. These rings were then cut longitudinally to give strips of about 5 mm length. In some experiments, the endothelium was removed by rubbing the luminal face of the strip with a cotton-tip. The absence of endothelium was confirmed by the lack of response to bradykinin (Bény, Brunet and Huggel, 1987).

Ciliary artery

A short segment of the ciliary arteries travelling on a horizontal plane to reach the bulbus was dissected free. For pharmacological experiments, this segment was cut into small rings (2 mm width). In a few blood vessels the endothelium was removed by rubbing the luminal surface of the vessels with a human hair. For electrophysiological experiments, the vessel was cut longitudinally with care to preserve a functional endothelium, giving a strip of about 5 mm in length and 1.5 mm in width (Zhu *et al.*, 1997; Bény, Zhu and Haefliger, 1997).

Pharmacology

Porcine coronary artery

To obtain concentration-relaxation curves, ligatures were attached to both ends of the strips, which were suspended in a 85 μl tissue bath with a resting isometric tension of about 10 mN. To establish the concentration-response curves of bradykinin, strips, contracted by 10^{-5} M prostaglandin. F2α, were superfused with Krebs solution containing increasing concentrations of bradykinin given in non-cumulative manner. Bradykinin was administered to the preparations by diluting it in the plastic beaker that contained the perfusion solution. Different inhibitors were used: indomethacin (10^{-5} M) to block cyclooxygenase and N$^{\omega}$-nitro-L-arginine (10^{-6} M) for inhibiting the synthesis of nitric oxide. These inhibitors were administered to the strips in the perfusion fluid for at least 25 minutes prior to the first application of bradykinin (Pacicca, von der Weid and Bény, 1992).

Ciliary artery

Two tungsten wires were passed through the lumen of the arterial rings. One wire was connected to an isometric force transducer and the other to a micro manipulator to adjust the muscle resting isometric tension. The rings were immersed into organ chambers filled with Krebs-Ringer bicarbonate solution with indomethacin (10^{-5} M). To obtain concentration-response curves to bradykinin, cumulative concentrations of the peptide were added to vessels contracted by U 46619 (10^{-7} M). Concentration-response curves to bradykinin were obtained in the absence or presence of the inhibitor of nitric oxide synthase, N$^{\omega}$-nitro-L-arginine methyl ester (10^{-5} M; $n = 6$). This inhibitor was administered to the strips in the perfusion fluid for at least 25 minutes prior to the first application of bradykinin (Zhu *et al.*, 1997).

Electrophysiology

Ciliary artery and coronary artery strips were incubated in a organ chamber (100 μl) which was continuously superfused (1.3 ml/min., 37°C) with oxygenated Krebs-Ringer solution. The bottom of the chamber was coated with Sylgard and each strip was pinned to it with the intima facing upwards. Membrane potentials of endothelial and smooth muscle cells were measured using glass microelectrodes with tips filled with 3M KCl. In order to identify the impaled cell during dye-coupling experiments, the cell membrane potential was measured with microelectrodes with the tip filled with ethidium bromide solution (5% in KCI 5.10^{-2} M) or lucifer yellow dilithium (5% in water) and backfilled with 0.15 M LiCI. The electrode resistance was 80–120 mΩ when filled with 3 M KCI. Criteria for accepting a record were a sharp drop in potential both at cell penetration and withdrawal of the electrode, as well as a stable measured potential. To impale an endothelial cell, the electrode was gently brought towards the intima until the sudden appearance of a negative potential. To confirm that the recorded cell was indeed an endothelial cell, lucifer yellow contained in the tip of the electrode was injected microiontophoretically by applying a direct hyperpolarizing current of 3.5 nA through the electrode. Alternatively, ethidium bromide was injected microiontophoretically by applying a direct depolarizing current of 0.35 nA for 2 to 5 minutes. After these experiments, tissues were examined directly and

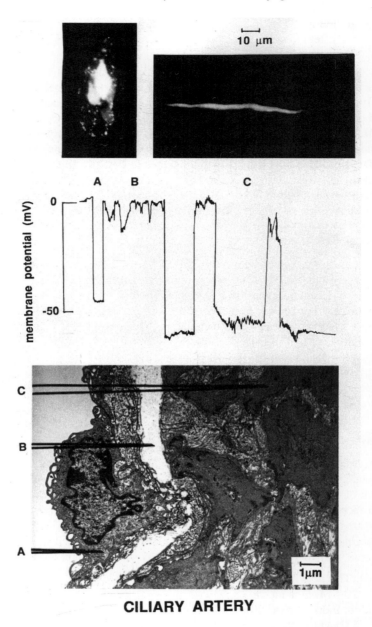

CILIARY ARTERY

Figure 1–1. Representation of the methods used. The electrode was gently brought towards the intima of a ciliary artery strip until the sudden appearance of a negative potential indicated the penetration of an endothelial cell (A). To confirm that this cell was within the endothelium, lucifer yellow was injected microiontophoretically into the impaled cell through the electrode and the intimal face of the arterial strip photographed with a fluorescence microscope (upper left panel). As the electrode penetrated deeper into the tissue, it passed through the endothelial cell and crossed the internal elastic membrane (B). The cells penetrated after that are smooth muscle cells (C) as shown by the injection of the dye followed by the examination of the strip from the intimal face (upper right panel).

photographed with a fluorescent microscope in physiological solution (Figure 1–1). The cells penetrated after having crossed the endothelium, and the internal elastic lamina were identified as smooth muscle cells (Figure 1–1). (Bény, 1990; Bény, Zhu and Haefliger, 1997).

Drugs

For coronary artery experiments, Bradykinin was purchased from Bachem Feinchemikalen (Bubendorf, Switzerland) and was prepared at a concentration of 1 mg/ml in 0.25% acetic acid, stored in 50 μl aliquots and frozen (–20°C) until used. For ciliary artery experiments, Bradykinin, N^{ω}-nitro-L-arginine methyl ester, N^{ω}-nitro-D-arginine methyl ester, indomethacin and U 46619 were purchased from Sigma (St. Louis, MO). Indomethacin was dissolved in Na_2CO_3 (10^{-5} M), and all other drugs in distilled water. For both arteries, concentrations are expressed as final molar concentrations in the organ chamber. Relaxations were expressed as percentage of contractions to U 46619 (10^{-7} M) (ciliary artery) or prostaglandin. F2α (10^{-5} M) (coronary artery). For each concentration-response curve, the maximal response was measured and the concentrations of bradykinin evoking 50% (IC_{50}) relaxation were calculated.

Statistical Analysis

Results are given as mean and standard error of the mean (mean ± SEM), and n equals the number of artery pieces studied (one ciliary and one coronary artery per animal). Statistical analysis of the data was carried out by Student's test. A difference was considered to be statistically significant, when the P was less than 0.05.

RESULTS

Pharmacology

Porcine coronary artery

Bradykinin transiently relaxed, tonically contracted strips in a concentration-dependent manner, in the presence of indomethacin (10^{-5} M). The IC_{50} was 3.2 ± 1 10^{-9} M Bradykinin ($n = 4$). Maximal relaxation reached $93 \pm 3\%$ of the prostaglandin. F2α-induced contraction ($n = 8$). The perfusion of strips with a Krebs-Ringer solution containing N^{ω}-nitro-L-arginine (10^{-4} M) plus indomethacin (10^{-5} M) shifted the concentration-response curve for bradykinin-induced relaxation to the right and diminished the maximal relaxation by 27%. In this situation, the IC_{50} was 21 ± 7 10^{-9} M bradykinin ($n = 4$) and the maximal relaxation was $68 \pm 18\%$ ($n = 4$) (Figure 1–2). After removal of the endothelium, the relaxing effect of bradykinin was abolished, confirming that relaxations to bradykinin are endothelium-dependent (Pacicca, von der Weid and Bény, 1992).

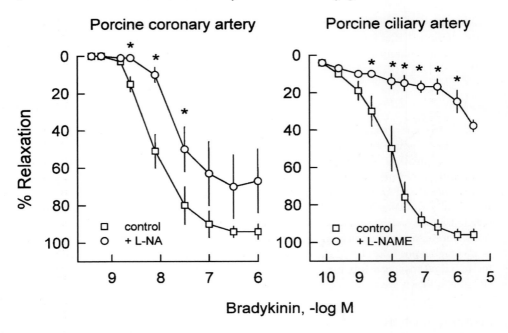

Figure 1–2. Effect of inhibition of nitric oxide synthesis on the endothelium-dependent relaxation to bradykinin. Relaxations were expressed as percentage of contractions induced by U4619 (10^{-7} M) (ciliary artery) or prostaglandin. F2α (PGF2α) (10^{-5} M) (coronary artery). Left panel: In the porcine coronary artery, bradykinin evoked an endothelium-dependent relaxation that was inhibited by up to 30% by N$^\omega$-nitro-L-arginine (L-NA) ($n = 4$–8). Right panel: in porcine ciliary artery the relaxations were blunted by at least 60% by N$^\omega$-nitro-L arginine methyl ester (L-NAME). ($n = 6$). All the experiments were done in the presence of indomethacin to inhibit the synthesis of prostacyclin. The asterisk indicates that the difference between the control value and the L-NA, respectively L-NAME value is statistically significant (P < 0.05).

Ciliary artery

In porcine ciliary arteries contracted with the thromboxane A$_2$ analog U 46619 (10^{-7} M), Bradykinin evoked concentration-dependent relaxations. Maximal relaxation reached $99 \pm 3\%$ and the IC$_{50}$ was $8.1 \pm 0.2\ 10^{-9}$ M ($n = 6$). After removal of the endothelium, the relaxing effect of bradykinin was abolished. In the presence of the nitric oxide synthase inhibitor, N$^\omega$-nitro-L-arginine methyl ester (10^{-4} M), relaxations to bradykinin were inhibited significantly. In this situation, the IC$_{50}$ exceeded $3\ 10^{-7}$ M bradykinin ($n = 6$) and the maximal relaxation averaged $39 \pm 4\%$ of the U 46619-induced contraction ($n = 6$) (Zhu *et al.*, 1997) (Figure 1–2).

Electrophysiology

Porcine coronary artery

A successful simultaneously recording of the membrane potentials of both an endothelial and a smooth muscle cell from the same artery during the application of bradykinin, was obtained nine times. The resting membrane potential of the endothelial cells averaged

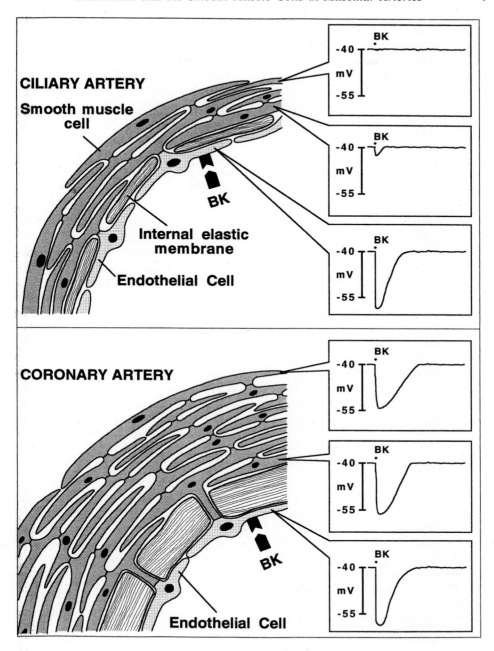

Figure 1–3. In ciliary artery, the hyperpolarization caused by bradykinin is electrotonically transmitted to the smooth muscle cells but this hyperpolarization dissipates within the big syncytium of smooth muscle cells that constitutes the media since the well coupled cells of the media cause a low input resistance. The same thing should happened in coronary artery, where the syncytium of coupled smooth muscles is even bigger but the hyperpolarization must be in some way "regenerated" to avoid this dissipation, leading to endothelium dependent hyperpolanzations.

-40.3 ± 3.6 mV ($n = 9$) and that of the smooth muscle cells -53.7 ± 3 mV ($n = 9$). Bradykinin caused a transient hyperpolarization of the endothelial cells of 13.8 ± 1.9 mV ($n = 9$) with the membrane potential at the peak of the response to the peptide reaching -54 ± 4.6 mV. Bradykinin caused a transient hyperpolarization of the smooth muscle cells of 11.1 ± 1.3 mV with the membrane potential at the peak of the response reaching -64.8 ± 3 mV ($n = 9$). It was not possible to detect any delay between the two responses (Bény, 1990) (Figure 1–3).

Ciliary artery

In the porcine ciliary arteries, the resting potential of the smooth muscle cells was -66 ± 4 mV ($n = 5$). The thromboxane A_2 analog U 46619 ($1.5 \ 10^{-7}$ M), contracted the strips but did not significantly change the cell membrane potential (-63 ± 3 mV, $n = 10$). In the strips incubated with U 46619, bradykinin ($2.5 \ 10^{-7}$ M) relaxed the strips but had no significant effect on the membrane potential of most of the smooth muscle cells (nine out of ten; mean change for these nine cells: 0.6 ± 1 mV). However, when smooth muscle cells were impaled close to the surface of the strip (next to the endothelium) a 4 mV hyperpolarization in response to the peptide was detected once (one out of ten observations). Because of the lack of effect of U 46619 and bradykinin on the cell membrane potential, it was important to demonstrate that the electrodes were inserted inta smooth muscle cells. This was done by identifying the recorded cells by dye injection and further examination with fluorescence (Figure 1–3).

The membrane potential of endothelial cells was -46 mV ± 2 mV ($n = 17$). Exposure to bradykinin ($2.25 \ 10^{-7}$ M) hyperpolarized the endothelial cells by 12 ± 1 mV ($n = 13$), in those cells where the recording electrode stayed in place after application of the drug. The cells were identified by injection of fluorescent dye followed by microscopic examination (Zhu *et al.*, 1997; Bény, Zhu and Haefliger, 1997) (Figure 1–3).

DISCUSSION

The present results demonstrate that, in porcine ciliary and coronary arteries, bradykinin evokes endothelium-dependent relaxations. Furthermore, endothelial cell hyperpolarization could be evoked by bradykinin and an electrical conduction between endothelial and smooth muscle cells could also be demonstrated in both types of artery. However, in contrast to the ciliary artery, in the coronary artery relaxations evoked by bradykinin were associated with hyperpolarization of smooth muscle cells deep in the media of the vascular wall. Therefore, these observations suggest that, a passive electrotonic conduction alone is not sufficient to account for hyperpolarization of smooth muscle cells observed during endothelium-dependent relaxations evoked by bradykinin.

The existence of a passive electrotonic conduction between endothelial and adjacent smooth muscle cells could be demonstrated in both ciliary and coronary arteries. Indeed, in the ciliary artery, a transfer of dye (positively and negatively charged) could be elicited from endothelial to smooth muscle cells, thus implying that these cells are coupled electrically (Bény, Zhu and Haefliger, 1997). In the coronary artery, electrical coupling directly has been demonstrated with electrophysiological recordings (Bény, 1997).

There exists a marked electrical coupling between smooth muscle cells of muscular arteries, implying an overall low resistance within the media of these blood vessels. Therefore, it is expected that a hyperpolarization arising in an endothelial cell and being electronically conducted to the underlying smooth muscle cells would quickly vanish in the wide syncytium formed by the smooth muscle cells of the media. This is what apparently happens in the ciliary artery, where despite electrical coupling between endothelial and smooth muscle cells, the endothelium-dependent relaxation evoked by bradykinin was only associated with hyperpolarization of endothelial cells but not of smooth muscle cells. Indeed the relaxing effect of bradykinin is mediated by nitric oxide in the ciliary artery (Zhu *et al.*, 1997; Bény, Zhu and Haefliger, 1997).

In the coronary artery, in which an electrical coupling between endothelial and smooth muscle cells occurs, the endothelium-dependent relaxation evoked by bradykinin is associated with a synchronous hyperpolarization of both endothelial and smooth muscle cells (Bény, 1990). Therefore, like in the ciliary artery, as it appears unlikely that, passive electrical conduction alone, can lead to the hyperpolarization of the syncytium of smooth muscle cells of the media. Thus, the observations made in the porcine coronary artery implies the existence of an additional phenomenon (beside electrical conduction) to account for the hyperpolarization of smooth muscle cells observed during endothelium-dependent relaxations of the coronary artery to bradykinin.

The most likely hypothesis to explain this additional phenomenon is the existence of a freely diffusible chemical factor released by the endothelial cells. The action of this factor on the smooth muscle cells would be to amplify (regenerate) the small hyperpolarization transmitted electrotonically from the endothelium. This hypothesis would explain the synchronicity of the hyperpolarizations recorded simultaneously in the endothelial and smooth muscle cells in the porcine coronary artery, as well as the difficulty in demonstrating the diffusible nature of EDHF (Kauser and Rubanyi, 1992). This interpretation is supported by the observations in rabbit conduit arteries that specifically inhibiting gap junction-coupling with a peptide homologue to a conserved portion of one extracellular loop of connexin inhibits phenomena attributed to EDHF (Chaytor, Evans and Griffiths, 1998). If this were true, endothelium-dependent hyperpolarizations would result from an electrotonic spread of the endothelial cell hyperpolarization, coupled to a regeneration of this hyperpolarization in the smooth muscle cells due to the release of an endothelial regenerating factor.

2 Epoxyeicosatrienoic Acids as Endothelium-Derived Hyperpolarization Factors in Coronary Arteries: Potassium Channel Activation Through Endogenous ADP-Ribosylation of Gs

William B. Campbell, Ai-Ping Zou and Pin-Lan Li

Departments of Pharmacology and Toxicology and Physiology, Medical College of Wisconsin, 8701 Watertown Plank Road, Milwaukee, Wisconsin 53226.
Phone: (414) 456-8267, Fax: (414) 456-6545, E-Mail: wbcamp@mcw.edu

Epoxyeicosatrienoic acids (EETs) function as endothelium-derived hyperpolarizing factors (EDHFs) in bovine coronary arteries. The molecular target for the EETs in vascular smooth muscle cells is the large conductance (256 pS) Ca^{2+}-activated potassium (K_{Ca}) channels. EETs increased the K_{Ca} channel activity in coronary smooth muscle cells through activation of Gs. However, the mechanism by which EETs activate Gs remains unknown. In the present study, the hypothesis was tested that EETs stimulate the endogenous ADP-ribosylation of Gs that regulates the K_{Ca} channel activity. Endogenous ADP-ribosylation was examined by incubating coronary arterial homogenates with [adenylate-^{32}P]NAD. Three ^{32}P-labeled proteins were observed at 52, 80 and 124 kDa in the homogenates. The 52 kDa acceptor protein of endogenous ADP-ribosylation, which comigrated with $G_S\alpha$, was removed by anti-$G_S\alpha$ immunoprecipitation. The ADP-ribosylation of G_S was enhanced by addition of 11,12-EET. Western blot analysis using a specific antibody against ADP-ribosyltransferase demonstrated a 42 kDa protein in the coronary arterial homogenates, which was primarily present in the cytosol of coronary smooth muscle cells. Using the patch clamp techniques, 11,12-EET activated the K^+ channels in coronary smooth muscle cells. Inhibitors of endogenous ADP-ribosylation, 3-aminobenzamide, blocked EET-induced K_{Ca} channel activation. These results indicate that an endogenous ADP-ribosyltransferase is present in coronary smooth muscle cells. This enzyme transfers ADP-ribose to $G_S\alpha$. Endogenous ADP-ribosylation mediates the activation of the K_{Ca} channels by EETs in coronary smooth muscle cells and may represent the mechanism of the action of these EDHFs.

Keywords: ADP-ribosylation, eicosanoids, endothelium-derived hyperpolarization factor, coronary artery, potassium channel

INTRODUCTION

Epoxyeicosatrienoic acids (EETs) activated K_{Ca} channels, hyperpolarize vascular smooth muscle and dilate blood vessels (Campbell *et al.*, 1996; Li and Campbell, 1997; Rosolowsky and Campbell, 1993; Zou *et al.*, 1996). EETs as endothelium-derived hyperpolarizating factors (EDHF) mediated the hyperpolarizing and vasodilator effect of acetylcholine, bradykinin and arachidonic acid (Campbell *et al.*, 1996). EETs activate the K_{Ca} channels in coronary smooth muscle cells through a guanine nucleotide binding protein, G_S (Li and Campbell, 1997). Like EETs, cholera toxin, an bacterial ADP-ribosyltransferase also activated this K_{Ca} channel (Li and Campbell, 1997). It seems that EETs act similarly as stimulators of ADP-ribosylation. In fact, EETs stimulate endogenous ADP-ribosylation in liver cells (Seki *et al.*, 1992). It remains to be determined whether endogenous ADP-ribosyltransferase is activated by EETs in coronary arterial smooth muscle and whether ADP-ribosylation is involved in the EETs-induced activation of the K_{Ca} channels.

The purpose of the present study was to identify a mono-ADP-ribosyltransferase in coronary smooth muscle cells and to determine the contribution of endogenous ADP-ribosylation of Gs to EETs-induced activation of the K_{Ca} channels.

MATERIALS AND METHODS

Assay of Endogenous ADP-Ribosylation

Small coronary arteries of bovine heart were microdissected, and the homogenates were prepared as described (Li and Campbell, 1997; Li *et al.*, 1994). An endogenous ADP-ribosylation assay was carried out (Seki *et al.*, 1992). Homogenates (140 μg) was incubated in 200 μl of Tris buffer with 5 μCi of [^{32}P]NAD$^+$. 11,12-Epoxyeicosatrienoic acid (11,12-EET) was added to the reaction mixture at a concentration of 10^{-5}M. The proteins in the reaction mixtures were fractioned by a 4–6% urea gradient/SDS-PAGE gel. The gel was dried, and the radioactivity detected by autoradiography on Kodak Omat film.

Western Blot Analysis

Fifty microgram of homogenates, microsomes or cytosols were subjected to 12% SDS-PAGE gel after heating at 100°C for 3 min. After the protein was electrophoretically transferred onto nitrocellulose membrane, the membrane was probed with polyclonal antibody against the synthetic peptide of human mono-ADP-ribosyltransferase-3. The membrane was washed three times with TBS-T, and then incubated for 1 h with 1:1000 hoseradish peroxidase-labeled goat anti-rabbit. 10 ml of detection solution 1 and 2 (1:1) (Amersham, IL) were added directly to the blots on the surface carrying the protein. After incubation for 1 min at room temperature, the membrane was wrapped in Saran wrap and then exposed to Kodak Omat film (Li *et al.*, 1994).

Patch Clamp Study

Small bovine coronary arterial smooth muscle cells were dissociated enzymatically, and K$^+$ channels were recorded and characterized (Li and Campbell, 1997; Li *et al.*, 1994). The effects of mono-ADP-ribosyltransferase inhibitor, 3-aminobenzamide on EETs-induced activation of the K$^+$ channels were examined in the cell-attached patch mode.

RESULTS

A typical autoradiograph of a gel shows endogenous ADP-ribosylation in the homogenates prepared from small bovine coronary arteries (Figure 2-1). Three proteins with 52, 80 and 124 kDa were ADP-ribosylated under control conditions in the presence of GTP (10^{-3}M).

A

kDa

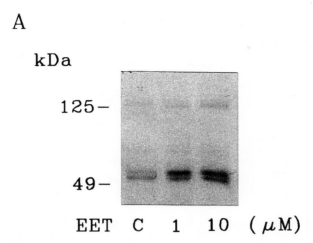

125 —

49 —

EET   C   1   10   (μM)

B

kDa

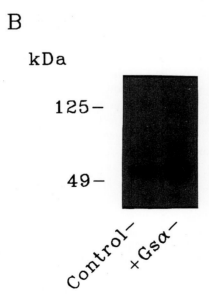

125 —

49 —

Figure 2-1. Autoradiographs of ADP-ribosylation in coronary arterial homogenates. A: Endogenous ADP-ribosyltransferase activity in the presence of GTP and 11,12-EET. B: Effect of anti-$G_S\alpha$ immunoprecipitation on endogenous ADP-ribosylation.

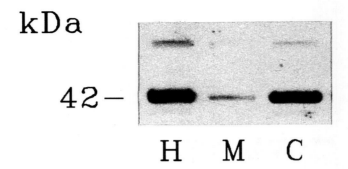

Figure 2-2. Western blot analysis of ADP-ribosyltransferase in homogenate, microsomes and cytosols prepared from small bovine coronary arteries. An antibody against a synthetic peptide derived from bovine ADP-ribosyltransferase-3 was used. H, C and M indicate homogenates, cytosols and microsomes prepared from bovine coronary arteries, respectively.

This ADP-ribosylation was specific for arginine residues (data not shown). When the coronary homogenates were treated with 11,12-EET, ADP-ribosylation of 52 kDa protein was significantly increased (panel A). By immunoblotting the same membrane after autoradiography using an anti-$G_S\alpha$ antibody, the 52 kDa immunoreactive protein was detected (data not shown). When $G_S\alpha$ was removed from the coronary homogenates by immunoprecipitation, the 52 kDa ADP-ribosylated protein band was decreased (panel B).

By Western blot, a 42 kDa protein was recognized by a specific antibody against human mono-ADP-ribosyltransferase-3 (Figure 2-2). This protein was detected in the coronary arterial homogenates, microsomes and cytosols, and the amount of this protein in the cytosolic fraction was much greater compared to the microsomal fraction.

Typical recordings of single K channel currents in a cell-attached patch depict the effects of 11,12-EET and 3-aminobenzamide on the K_{Ca} channel activity (Figure 2-3). Addition of 11,12-EET (10^{-7}M) to the bath solution markedly increased the activity of the K_{Ca} channels. It produced four-fold increase in the open probability (NP_O) of the K_{Ca} channels. In the presence of 3-aminobenzamide (10^{-3}M), the effects of 11,12-EET on the K_{Ca} channel activity were completely blocked. Similar results were obtained with other inhibitors or ADP-ribosylation, vitamin K, and novobiocin.

DISCUSSION

The present study provides the evidence that a NAD:arginine mono-ADP-ribosyltransferase with a molecular size of 42 kDa is present in bovine coronary arterial smooth muscle. This enzyme catalyzes an endogenous ADP-ribosylation of cellular proteins in the presence of NAD. Three proteins of vascular smooth muscle with molecular sizes of 52, 80, 124 kDa were ADP-ribosylated. By Western blot analysis, this mono-ADP-ribosyltransferase was primarily present in the cytosolic fraction. This result is consistent with previous findings indicating that endogenous ADP-ribosylation occurred in the cytosol of rat liver cells (Seki *et al.*, 1992) and human platelets (Brune, Molina, and Lapetina, 1990; Brune and Lapetina, 1989; Clancy *et al.*, 1993 and Molina *et al.*, 1989). However, several studies demonstrated

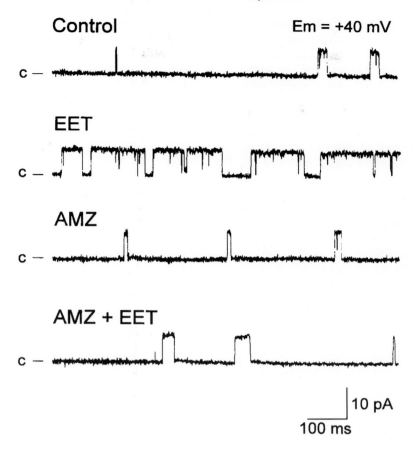

Figure 2-3. A representative recordings depicting the effects of 11, 12-EET (10^{-7}M) and 3-aminobenzamide (10^{-3}M) on K_{Ca} channel activity in cell-attached patches of coronary smooth muscle cells.

that arginine-specific mono-ADP-ribosyltransferases in rabbit or human skeletal and cardiac muscle, rodent lymphocytes, and human testis are glycosyl-phosphatidylinositol anchored proteins and present on the surface of cells (Terashima *et al.*, 1992; Tsuchiya *et al.*, 1994; Zolkiewska, Nightingale and Moss, 1992; Zolkiewska, Okazaki and Moss, 1994). This mono-ADP-ribosyltransferase in bovine coronary smooth muscle cells is also arginine specific.

The acceptor proteins for mono-ADP-ribosyltransferases possess a wide spectrum of molecular weights from 20–120 kDa and varied activities. These acceptor proteins include G-proteins, actin, calcium-dependent ATPase, cyclic AMP-independent protein kinase, glyceraldehyde 3-phosphate dehydrogenase, elongation factor-2 and some unknown cellular proteins (Li *et al.*, 1994; Molina *et al.*, 1989; Reilly *et al.*, 1981; Seki *et al.*, 1992; Terashima *et al.*, 1992). In the present study, the major ADP-ribosylated protein was 52 kDa. This protein was the same molecular size as the protein ADP-ribosylated by cholera toxin and was recognized by an antibody against $G_S\alpha$. Removal of $G_S\alpha$ from coronary arterial homogenates by immunoprecipitation abolished ADP-ribosylation of 52 kDa protein. These data suggest that the endogenous mono-ADP-ribosyltransferase may catalyze ADP-ribosylation of $G_S\alpha$ in coronary arterial smooth muscle. In addition, endogenous ADP-ribosylation also occurs on proteins with molecular weights of 80 and 124 kDa. The identity of these proteins remains to be determined.

Previous studies have indicated that endogenous ADP-ribosylation of cellular proteins may play an important role in the transmembrane and intracellular signal transduction. A number of hormones or bioactive substances such as nitric oxide, prostacyclin, adenosine and isopreterenol activated or modulated the activity of mono-ADP-ribosyltransferases in various animal tissues (Jacquemin *et al.*, 1986; Li *et al.*, 1994; Molina, Nolan and Lapetina, 1989; Reilly *et al.*, 1981). The present study determined the role of mono-ADP-ribosyltransferase in EET-induced activation of the $G_S\alpha$ and the K_{Ca} channels. Addition of 11,12-EET into the reaction mixture of coronary arterial homogenate and [^{32}P]-NAD markedly enhanced the ADP-ribosylation of G_S, suggesting that the mono-ADP-ribosyltransferase activity was stimulated by 11,12-EET. Using the patch clamp technique, a specific inhibitor of mono-ADP-ribosyltransferase, 3-amino-benzamide, blocked the 11,12-EET-induced increase in the K_{Ca} channel activity. These results indicate that mono-ADP-ribosyltransferase in coronary smooth muscle cells may play an important role in mediating the effects of EETs on the K_{Ca} channel activity.

Based on these results, EETs stimulate an mono-ADP-ribosyltransferase in coronary smooth muscle cells, which results in the ADP-ribosylation and activation of $G_S\alpha$. This ADP-ribosyltransferase-dependent activation of $G_S\alpha$ may function in concert with direct action of EETs on $G_S\alpha$ in mediating EETs-induced activation of the K_{Ca} channel and hyperpolarization in coronary smooth muscle cells. This may represent an important mechanism of endothelium-dependent relaxation in coronary circulation (Figure 2-4).

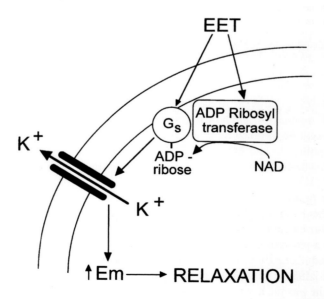

Figure 2-4.   A signaling pathway involving EETs, ADP-ribosyltransferase, G_S and K_{Ca} channels.

3 Endothelium-Derived Eicosanoids from Lipoxygenase Relax the Rabbit Aorta by Opening Potassium Channels

Sandra L. Pfister, Nancy Spitzbarth, Kasem Nithipatikom, William S. Edgemond and William B. Campbell

Department of Pharmacology and Toxicology, Medical College of Wisconsin, 8701 Watertown Plank Road, Milwaukee, Wisconsin 53226 U.S.A.
Address for Correspondence: Sandra L. Pfister, Ph.D. Department of Pharmacology and Toxicology, 8701 Watertown Plank Road, Milwaukee, Wisconsin 53226 U.S.A.
Tel: (414) 456-8285, Fax: (414) 456-6545, e-mail: spfister@mcw.edu

This study describes an endothelium-dependent, lipoxygenase metabolite of arachidonic acid that elicits relaxation of the rabbit aorta. In contracted rabbit aortas, arachidonic acid elicited a concentration-dependent relaxation in 4 mM extracellular potassium ($[K]_o$) If $[K]_o$ was increased to 20 mM, the relaxations to arachidonic acid were blocked. The aorta synthesized a previously unidentified [^{14}C]-labeled arachidonic acid metabolite. Both the production of this unknown factor and arachidonic acid-induced relaxations were dependent on an intact endothelium indicating that the cellular source of the unknown factor was the endothelial cell. This suggests that this factor represents a novel eicosanoid which produces relaxation via an activation of K channels of vascular smooth muscle. Further purification of the factor gave a major peak (peak 2) which relaxed contracted rabbit aortas. For structural analysis, peak 2 was derivatized to its pentafluorobenzyl (PFB) ester-trimethylsilyl (TMS) ether and subjected to negative ion chemical ionization gas chromatography/mass spectrometry analysis. This analysis indicated a major product eluting at 17.1 min. The most abundant ions in the spectra of this product were 569 (M-PFB) and 479 (M-PFB + $(CH_3)_3SiOH$). This indicates a molecular weight of 570 for the TMS ether of peak 2 and suggests a trihydroxyeicosatrienoic acid (THETA) structure. For the methyl ester-TMS ether derivative of peak 2, positive ion chemical ionization gas chromatography/mass spectrometry analysis indicated the presence of two compounds at 14.08 and 14.2 min. The two compounds differed in the intensities of the major ions indicating that they are regioisomeric THETAs. Thus, an endothelium-derived factor(s) has been identified that mediates arachidonic acid-induced relaxations of the rabbit aorta as two regioisomeric THETAs. At the present time, it is not known which isomer is the active metabolite. This arachidonic acid metabolite(s) appears to act through K channel activation and may represent an endothelium-derived hyperpolarizing factor(s).

KEYWORDS: lipoxygenase, arachidonic acid, vasorelaxation, rabbit aorta

INTRODUCTION

The endothelium is important in synthesizing and releasing compounds that are involved in the regulation of vascular tone (Furchgott and Vanhoutte, 1989). Endothelial cells from different vascular sources synthesize a large number of vasoactive compounds including prostacyclin, endothelium-derived relaxing factor or nitric oxide (NO), endothelium-derived hyperpolarizing factor (EDHF), endothelium-derived contracting factor and endothelin. Alterations in the production of these compounds may be associated with cardiovascular diseases, including atherosclerosis, coronary vasospasm and hypertension.

17

Arachidonic acid is metabolized by the vascular endothelium to a variety of products by cyclooxygenases, lipoxygenases and cytochrome P_{450} epoxygenases (Needleman *et al.*, 1986). The identity and potential biological activity of some of these arachidonic acid metabolites have been determined; however, many metabolites have not been well characterized, either structurally or biologically. The lipoxygenases are a family of enzymes that convert arachidonic acid into a series of monohydroperoxyeicosatetraenoic acids (HPETEs) (Pace-Asciak and Asotra, 1989). The hydroperoxy derivatives are metabolized by peroxidases to the monohydroxyeicosatetraenoic acids (HETEs). The HPETEs and to a lesser extent, the HETEs elicit relaxation in certain vascular tissues (Rosolowsky *et al.*, 1990b; d'Alarcao *et al.*, 1987). However, in the rabbit aorta, 12- and 15-HETE are inactive (Pfister, Falck, and Campbell, 1991). In various tissues and cells both the 12- and 15-HPETEs can undergo additional transformations to a variety of other compounds including leukotrienes, lipoxins, long chain aldehydes, trihydroxyeicosatrienoic acid and hydroxyepoxyeicosatrienoic acids (Pace-Asciak, Granstrom, and Samuelsson, 1983; Bryant and Bailey, 1981; Serhan, 1994). The rabbit aorta synthesizes 15(S)-HETE, the isomer produced by 15-lipoxygenase (Pfister, Falck, and Campbell, 1991). There is little information on the biological significance of the hydroxy-epoxy- and trihydroxy products derived from 15-lipoxygenase.

Arachidonic acid causes relaxation of blood vessels which is dependent on the presence of the endothelium (Pinto, Abraham, and Mullane, 1986; Singer and Peach, 1983b; DeMey, Claeys, and Vanhoutte, 1982; Rubanyi and Vanhoutte, 1985b). Vanhoutte and coworkers (DeMey, Claeys, and Vanhoutte, 1982) reported that arachidonic acid relaxed canine femoral arteries and provided evidence that this response was mediated in part by the endothelial production of the cyclooxygenase metabolite, prostacyclin. Based on the ability of inhibitors of arachidonic acid metabolism to block the relaxant response to arachidonic acid in rabbit aorta, Singer and Peach (Singer and Peach, 1983b) concluded that arachidonic acid-induced relaxations were mediated by a noncyclooxygenase factor. Arachidonic acid is able to release a lipoxygenase metabolite from endothelial cells which relaxes rabbit vascular smooth muscle (Pfister and Campbell, 1992). The identity of this metabolite was not determined.

Thus, there were two main objectives of the present study. First, to examine the hypothesis that in rabbit aorta, arachidonic acid-induced relaxations are mediated by a lipoxygenase metabolite via an activation of potassium channels. A second objective was to chemically identify the biologically active factor synthesized from arachidonic acid by the rabbit aorta.

METHODS

Metabolism of ^{14}C-Arachidonic Acid by Rabbit Aorta

Aortas (either freshly isolated from one to two month old New Zealand White rabbits or purchased from Pel-Frez Biologicals, Rogers, AR) were obtained and cleaned of adhering connective tissue and fat. The vessels were rinsed in Tris buffer (0.05 M, pH 7.5), and then cut into small pieces. They were placed in fresh buffer (500 mg of tissue/10 ml of buffer), centrifuged at $750 \times g$ for 15 min and the supernatant used. Aliquots (5 mg/ml) of the supernatant were incubated for 20 minutes in Tris-HCl buffer containing indomethacin (10^{-5} M) and [^{14}C(U)]-arachidonic acid (0.05 μCi, 10^{-7} M). In some incubations, 15-lipoxygenase (soybean lipoxidase, 0.2 mg/ml) was included. Alternatively, the homogenate was incubated with [^{14}C(U)]-15-HPETE instead of [^{14}C(U)]-arachidonic acid.

All reactions were stopped by adding ethanol to a final concentration of 15%. The samples were acidified (pH less than 3.5) and extracted using octadecylsilyl extraction columns (Pfister *et al.*, 1996). The extracted arachidonic acid metabolites were then resolved by reverse phase HPLC (Pfister *et al.*, 1996). Solvent system I consisted of solvent A which was water and solvent B which was acetonitrile containing 0.1% glacial acetic acid. The program was a 40 min linear gradient from 50% solvent B in A to 100% solvent B at a flow rate of 1 ml/min. Fraction 15–50, corresponding to the prostaglandin metabolites were pooled, acidified, extracted with cyclohexane/ethyl acetate (50/50) and rechromatographed on reverse phase HPLC using solvent system II. Solvent system II consisted of solvent A which was water containing 0.1% glacial acetic acid and solvent B which was acetonitrile. The program consisted of a 5 minute isocratic phase with 35% B in A, followed by a 35 minute linear gradient to 85% B at a flow rate of 1 ml/min. Radioactivity of the column eluate was collected in 0.2 ml aliquots and measured by liquid scintillation spectrometry.

Vascular Reactivity

Rabbits were sacrificed (pentobarbital 120 mg/kg, iv), the thoracic aorta removed and placed in Kreb's bicarbonate buffer (118 mM NaCl, 4 mM KCl, 3.3 mM $CaCl_2$, 24 mM $NaHCO_3$, 1.2 mM KH_2PO_4, 1.2 mM $MgSO_4$ and 11 mM glucose). The tissue was carefully cleaned of adhering fat and connective tissue and cut into rings (3 mm long) taking care not to damage the endothelium. Aortic rings were suspended in organ chambers containing 6 ml Kreb's-bicarbonate buffer maintained at 37°C and continuously bubbled with 95% O_2–5% CO_2. Isometric tension was measured with force-displacement transducers (Grass) and recorded with a polygraph (Grass model 7D). Resting tension was adjusted to its length-tension maximum of 2 grams, and vessels allowed to equilibrate for one hour. Contractions were produced by increasing the KCl concentration to 40 mM. Once the vessels reached peak contraction, the tissues were rinsed, and allowed to return to resting tension. After the aortic rings had reproducible, stable responses to KCl, they were contracted with norepinephrine (10^{-7} M). When the contraction stabilized, cumulative concentration-response curves to arachidonic acid (10^{-7} to 10^{-4} M) were obtained. The response to arachidonic acid was repeated in the presence of various inhibitors including the cyclooxygenase inhibitor, indomethacin (10^{-5} M), the lipoxygenase inhibitor, cinnamyl-3,4-dihydroxy-a-cyanocinnamate (CDC) (5×10^{-5} M), and in vessels in which the endothelium was removed by gently rubbing the intimal surface with a moist cotton swab. In some cases, arachidonic acid-induced relaxations were examined in the Kreb's-bicarbonate buffer containing 20 mM potassium. To further investigate the possible role of potassium channels in mediating arachidonic acid-induced relaxations, aortic rings were treated with inhibitors of potassium channels, including, charybdotoxin (2×10^{-8} M), apamin (5×10^{-9} M) or iberiotoxin (10^{-7} M) prior to the addition of norepinephrine. In the experiments with 20 mM KCl or the potassium channel inhibitors, indomethacin (10^{-5} M) was present.

Biological Activity of Arachidonic Acid Metabolites

Rabbit aortas (obtained from Pel-Frez) were incubated as before with indomethacin, ^{14}C-arachidonic acid and A23187. Identical control (cell free) incubations without tissue were carried out in parallel. Following incubation and extraction, both samples were chromatographed on reverse phase HPLC as described above. The major peaks labeled peaks 1–4 were collected from the solvent system II, extracted, and tested for biological activity. Although no radioactive peaks were observed in the cell-free incubations (data not shown), the fractions corresponding to peaks 1–4 in the aortic incubations were collected and also tested for biological activity. For the vascular reactivity experiment, rings of aorta from rabbit were suspended in organ chambers as described above. The blood vessels were contracted with norepinephrine (10^{-7} M) followed by peaks 1–4 obtained from the aorta or cell free incubations. The samples were suspended in a known volume of ethanol and 10 μl per 6 ml organ chamber was the maximal concentration administered. Aorta and cell free fractions were always analyzed in parallel.

Gas Chromatography/Mass Spectrometry

The fractions (88–94) which corresponded to the unknown factor previously described (Pfister *et al.*, 1996) were isolated from solvent system II, extracted with cyclohexane/ethyl acetate, evaporated to dryness under nitrogen and derivatized for gas chromatography/mass spectrometry (Pfister *et al.*, 1988; Pfister, Falck, and Campbell, 1991). For electron impact or positive ion chemical ionization mass spectrometry analysis, the sample was derivatized to the methyl ester, TMS ether and for negative ion chemical ionization mass spectrometry, the sample was derivatized to its PFB ester, TMS ether. Gas chromatography/mass spectrometry was performed with a Hewlett Packard 5989A Mass Spectrometer coupled with a 5890 Series 2 Gas Chromatograph. Ionization of the samples was done by electron impact at 65–70 eV or collisionally using methane as the reagent gas. The derivatized factor was resolved using a 14 m capillary DB-5 column with a linear gradient from 100° to 300°C. Standards were derivatized and analyzed by gas chromatography/mass spectrometry using the identical methods described for the biological samples.

Synthesis of 15-Lipoxygenase Standards

15-Lipoxygenase standards were obtained by incubating with [[14]C(U)]-arachidonic acid in buffer containing hematin (1 μM) and soybean lipoxidase (0.2 mg/ml) for 10 min at room temperature or chemically as described by Falck *et al.* (Falck *et al.*, 1983). 15-HPETE was synthesized by incubating arachidonic acid with soybean lipoxidase under a stream of oxygen for 30 min at room temperature. The reaction was acidified to pH 5 with 5% hydrochloric acid and extracted with diethyl ether. The combined extracts were washed with saturated NaCl, dried with $MgSO_4$, filtered and evaporated to dryness. The extract was purified by silica gel column chromatography using hexane-diethyl ether-acetic acid (2.5/10/0.25)) at 4°C. The yield is approximately 50%. A similar approach was used to synthesize [[14]C(U)]-15-HPETE.

Preparation of Microsomes

Rabbit aortas were isolated and homogenized in cold 0.25 M sucrose solution. Subcellular fractions were prepared by differential centrifugation (Capdevila *et al.*, 1990). The homogenate was centrifuged at $5000 \times g$ for 20 min to remove cellular debris, mitochondria and nuclei. The supernatant was further centrifuged at $100,000 \times g$ for 60 min. The microsomal pellet was resuspended in 1.15% KCl and homogenized. The homogenate was centrifuged at $100,000 \times g$ for 60 min. The pelleted microsomes were resuspended in microsome incubation buffer (Tris 50 mM, KCl 150 mM, $MgCl_2$ 10 mM, pH 7.5) and stored at $-80°C$. Protein concentrations were determined using the Bradford technique (Bio-Rad) with IgG as the standard. Aliquots of microsomal protein (2 mg/ml) were incubated with $[^{14}C(U)]$-arachidonic acid and 15-lipoxygenase as described above. Alternatively, incubations were performed with $[^{14}C(U)]$-15-HPETE for 10 min at 37°C. In a separate experiment, microsomes were treated with the cytochrome P_{450} inhibitor, miconazole (10^{-5} M), prior to the addition of $[^{14}C(U)]$-15-HPETE. Metabolites were extracted and chromatographed as described above using solvent system I then II.

Materials

The following drugs were used: norepinephrine, arachidonic acid (sodium salt), indomethacin and A23187 were all from Sigma Chemical Co., St. Louis, MO. CDC, charybdotoxin and apamin were from BIOMOL Research Laboratories, Plymouth Meeting, PA. Iberiotoxin was from RBI, Natick, MS. $[^{14}C(U)]$-Arachidonic acid (specific activity 920 mCi/mmol) was obtained from New England Nuclear, Boston, MA. All solvents were HPLC grade and purchased from Burdick and Jackson, Muskegan, MI.

Statistical Analysis

The vascular reactivity data were expressed as the mean ± SEM. Statistical evaluation of the data was performed by either using a one-way analysis of variance followed the Sidak multiple comparison test when significant differences were present, or data were analyzed by the Student's 't' test for paired observations. A value of $p < 0.05$ was considered statistically significant.

RESULTS

Arachidonic Acid-Induced Relaxations of Rabbit Aorta

In norepinephrine-contracted isolated rings of rabbit aorta, arachidonic acid elicited a concentration-related relaxation (Pfister and Campbell, 1992; Pfister *et al.*, 1996). Treatment with indomethacin, a cyclooxygenase inhibitor, potentiated the arachidonic acid-induced relaxations (Pfister and Campbell, 1992; Pfister *et al.*, 1996). However, treatment of vessels with CDC, a lipoxygenase inhibitor, blocked the arachidonic acid-induced relaxations (Figure 3-1, Pfister *et al.*, 1996)). Removal of the endothelium also inhibited the relaxations (Pfister and Campbell, 1992; Pfister *et al.*, 1996).

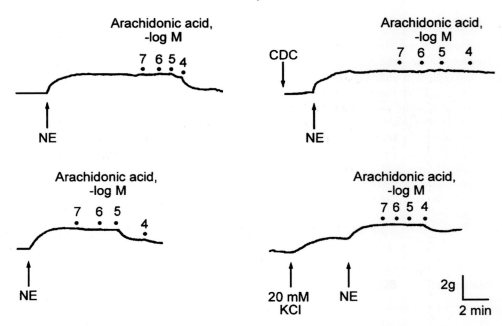

Figure 3-1. Tracing showing the effect of CDC (top panel) and elevated K+ (bottom panel) on AA-induced relaxations of rabbit aorta. *Panel A*, vessels were treated with vehicle or CDC (5×10^{-5} M) in normal Kreb's solution. Vessels were then contracted with norepinephrine (10^{-7} M). Arachidonic acid was added in increasing concentrations and changes in isometric tension continuously monitored. *Panel B*, arachidonic acid-induced contractions were measured in normal Kreb's containing 4.8 mM KCl and Kreb's containing 20 mM KCl. Indomethacin (10^{-5} M) was present throughout both experiments.

Increasing the extracellular potassium concentration to 20 mM significantly attenuated the arachidonic acid-induced relaxations (Figure 3-1, maximal response; $49.1 \pm 3.9\%$ vs $20.6 \pm 2.8\%$, 4.8 mM vs 20 mM). To further investigate the role of potassium channels, the effect of potassium channel inhibitors on arachidonic acid-induced relaxations was studied. Charybotoxin, an inhibitor of large conductance calcium-activated potassium channels had no effect on arachidonic acid-induced relaxations (maximal relaxation, $41.5 \pm 2.9\%$ vs $40.8 \pm 5.3\%$, control vs treated) whereas apamin, an inhibitor of small conductance calcium-activated potassium channels, significantly attenuated arachidonic acid-induced relaxations (maximal response, $41.5 \pm 2.9\%$ vs $25.7 \pm 6.9\%$, control vs treated). Treatment with iberiotoxin also significantly reduced the relaxations to arachidonic acid (maximal relaxation, $41.5 \pm 2.9\%$ vs $29.9 \pm 3.1\%$, control vs treated). The combination of iberiotoxin plus apamin caused the greatest inhibition of arachidonic acid-induced relaxations (maximal response, $49.1 \pm 3.9\%$ vs $20.1 \pm 11.6\%$, control vs treated).

Formation of Metabolites

When aortic homogenate was incubated with [^{14}C(U)]-arachidonic acid, extracted, and metabolites resolved by reverse-phase HPLC using solvent system I, there was the synthesis of radioactive products that migrated with the prostaglandins, the dihydroxyeico-

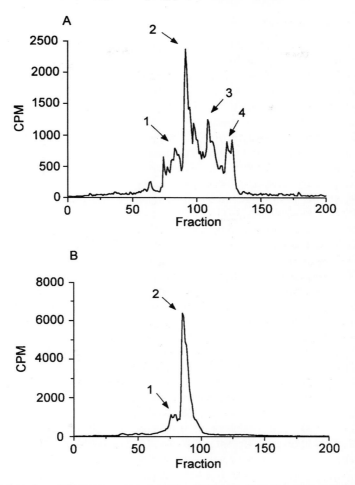

Figure 3-2. Metabolism of [^{14}C (U)]-arachidonic acid by rabbit aorta. Aortic homogenates were incubated with [^{14}C (U)]-arachidonic acid for 20 min, and metabolites were separated by reverse phase HPLC with use of solvent system I, and fractions 27–35 were collected, extracted and rechromatographed using solvent system II (Panel A). Aortic microsomes incubated with 15-lipoxygenase and [^{14}C (U)]-arachidonic acid, and metabolites were separated by reverse phase HPLC with use of solvent system I, and fractions 27–35 were collected, extracted and rechromatographed using solvent system II (Panel B).

satrienoic-dihydroxyeicosatetranoic acids and the HETEs ((Pfister *et al.*, 1996), data not shown). Fractions 27–35 (5–7.5 min) were analyzed further by reverse-phase HPLC using solvent system II (Figure 3-2). A number of radioactive peaks were observed and were labeled peaks 1–4. The major peak (peak 2) migrated at 17.5–18.5 minutes (fractions 88–94). The four peaks were isolated from aortic homogenates incubated with arachidonic acid and tested for activity on isolated rings of rabbit aorta. In contracted vessels, peak 2 elicited a concentration-dependent relaxation response (Pfister *et al.*, 1996). The other peaks were inactive. The cell free incubation with arachidonic acid did not relax the rabbit aorta (data not shown).

Gas Chromatography-Mass Spectrometry Analysis

In order to identify the metabolite associated with peak 2, gas chromatography/mass spectrometry analysis of the methyl ester-TMS ether derivative or PFB ester-TMS ether derivative was performed. To determine its molecular weight , peak 2 was derivatized to its PFB ester-TMS ether and subjected to negative ion chemical ionization gas chromatography/mass spectrometry analysis. The major product eluted at 17.1 min. The most abundant ions in the spectra of this peak were 569 (M-PFB) and 479 (M-PFB + $(CH_3)_3SiOH$). This indicates a molecular weight of 570 for the TMS ether of peak 2 and suggests a trihydroxy-eicosatrienoic acid structure. The positive chemical ionization gas chromatography/mass spectrometry analysis of the methyl ester-TMS ether derivative of peak 2 showed major products eluting at 14.08 and 14.20 min (Figure 3-3). The mass spectrum of the 14.08 minute product (compound A) revealed the presence of major ions at m/z 569 ($M^+ - 15$; loss of CH_3), 411 ($M^+ - 173$; loss of $[(CH_3)_3SiO]-(CH_2)_4-CH_3$), 405 ($M^+ - 179$; loss of $(CH_3)_3SiOH$ and $(CH_3)_3SiO$), 301 ($M^+ - 283$; loss of $[(CH_3)_3SiO]-CH-$ $(CH_2-CH=CH)_2-(CH_2)_3-COOCH_3$), 283 ($M^+ - 301$; loss of $[(CH_3)_3SiO]-CH-CH=CHCH$ $[(CH_3)_3SiO-](CH_2)_4-CH_3$), and 173 ($M^+ - 411$; loss of $CH=CH-CH[(CH_3)_3SiO-]CH(CH_3)_3$ $SiO-(CH_2-CH=CH)_2-(CH_2)_3-COOCH_3$) (Figure 3-3). The mass spectrum of compound B which had a gas chromatography retention time of 14.22 minutes (Figure 3-3) showed a similar mass spectrum (Figure 3-3). The two compounds (A and B) differed in the intensities of the 173 and 283 m/z ions with 283 being greater than 173 in compound A and the opposite in compound B (relative abundance 173 vs 283; 30.6% vs 60.0% for compound A and 100.0% vs 8.3% for compound B). The spectra indicated a molecular weight of 584 for the methyl ester-TMS ether of compound A and B. Based on these mass spectra, peak 2 contains two metabolites, identified as the methyl-ester TMS ether derivatives of 11,12,15-trihydroxyeicosatrienoic acid (11,12,15-THETA) and 11,14,15-trihydroxyeicosatrienoic acid (11,14,15-THETA). Chemical synthesis of 11,14,15-THETA as described by Falck *et al.* (Falck *et al.*, 1983), or by incubating 15-lipoxygenase and hematin with arachidonic acid (Bryant and Bailey, 1981) resulted in a derivatized 11,14,15-THETA standard which comigrated with the derivatized unknown by gas chromatography analysis (14.2 min). The 11,14,15-THETA standard also gave a similar mass spectrum (data not shown) when compared to the unknown. This structure is also consistent with chemical modification studies and synthesis of the unknown from 15-HPETE.

Biosynthesis of Trihydroxyeicosatrienoic Acids

Because these results suggested that peak 2 was derived from lipoxygenase-mediated metabolism of arachidonic acid, the effect of the addition of 15-lipoxygenase to the aortic homogenate incubations was investigated. ^{14}C-Arachidonic acid was incubated with 15-lipoxygenase plus the aortic homogenate or 15-lipoxygenase alone. The only product synthesized by 15-lipoxygenase alone comigrated with 15-HETE on HPLC using solvent system system I (data not shown). With the combination of 15-lipoxygenase and the homogenate, three major radioactive peaks were observed with solvent system I. One peak eluted with the unknown (fractions 27–35, 5–7.5 min), the second less polar peak eluted at 11.5–14 min (fractions 57.5–70), and the third major radioactive peak eluted with

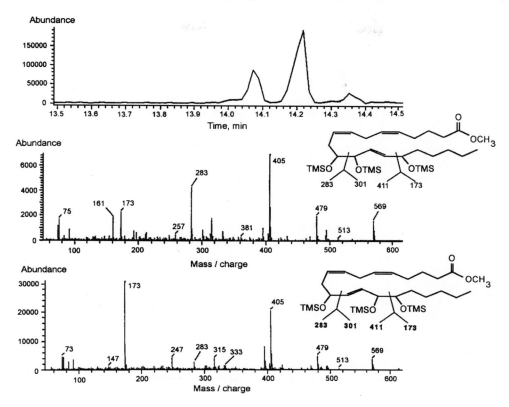

Figure 3-3. Mass spectrum of the methyl ester-TMS derivative of peak 2 (17.5–18.5 min) collected from reverse phase HPLC with use of solvent system II. The top panel is a representative tracing from the gas chromatography analysis. The proposed structure of compound A and B is given in the middle and top panel and its prominent ions are indicated.

15-HETE (fractions 100–120, 20–22 min). Analysis of the 5–7.5 min fraction using solvent system II indicated a pattern identical to the pattern observed with arachidonic acid and the homogenate, i.e., peak 2 was the major product. With equal amounts of tissue incubated, the synthesis of peak 2 was much greater when 15-lipoxygenase was added. The 11.5–14 min fraction comigrates with the hydroxyepoxyeicosatrienoic acids (HEETAs) according to previous reports (Weiss, Arnold, and Estabrook, 1987; Bryant and Bailey, 1981). In an additional experiment, the 11.5–14 minute fraction was tested for biological activity. This fraction relaxed the contracted rabbit aorta by 51 ± 5%.

Additionally, aortic microsomes were incubated with [^{14}C(U)]-arachidonic acid and 15-lipoxygenase and extracted metabolites analyzed by reverse phase HPLC using solvent system I followed by solvent system II. The major radioactive peak also migrated at 17.5–18.5 minutes with peak 2 (Figure 3-2). When aortic microsomes were incubated with [^{14}C(U)]-15-HPETE, the major metabolite again comigrated with peak 2 (17.5–18.5 min). Treatment with miconazole blocked the formation of peak 2 from ^{14}C-15-HPETE.

DISCUSSION

Many studies have documented the importance of the vascular endothelium in the regulation of vascular tone. A number of EDRF's have been identified including nitric oxide (Furchgott and Vanhoutte, 1989), prostacyclin (Moncada and Vane, 1979) and the epoxyeicosatrienoic acids (Campbell *et al.*, 1996). Arachidonic acid causes relaxation of blood vessels which is dependent on the presence of the endothelium (Pinto, Abraham, and Mullane, 1986; Singer and Peach, 1983b; DeMey, Claeys, and Vanhoutte, 1982; Rubanyi and Vanhoutte, 1985b). In the canine coronary artery, arachidonic acid-induced relaxations were mediated by a noncyclooxygenase metabolite and were blocked by ouabain (Rubanyi and Vanhoutte, 1985b). Previous work supported this observation and described an endothelium-dependent, lipoxygenase metabolite of arachidonic acid that elicited relaxation of the rabbit aorta (Pfister *et al.*, 1996). Arachidonic acid-induced relaxations were enhanced by treatment with the cyclooxygenase inhibitor, indomethacin and inhibited by lipoxygenase inhibitors, including nordihydroguaiaretic acid and CDC. The cytochrome P_{450} epoxygenase inhibitor metyrapone had no effect on the response. Arachidonic acid does not increase cyclic AMP or cyclic GMP in the rabbit aorta (Pfister and Campbell, 1992). Since EDHF is thought to act by increasing the efflux of potassium from the smooth muscle cell, most likely by activating a potassium channel (Brayden and Nelson, 1992; Garland *et al.*, 1995), the present study investigated whether arachidonic acid-induced relaxations were mediated by an EDHF. In the presence of an elevated extracellular potassium concentration, arachidonic acid failed to elicit relaxations. Furthermore, apamin and iberiotoxin, inhibitors of calcium activated potassium channels, blunted the relaxations to arachidonic acid. These data suggest that arachidonic acid releases a hyperpolarizing factor that relaxes the rabbit aorta.

Incubation of rabbit aorta with ^{14}C-arachidonic acid resulted in the synthesis of a previously unidentified ^{14}C-labeled metabolite. The production of the unknown factor was enhanced by indomethacin and decreased by inhibitors of lipoxygenase (Pfister *et al.*, 1996). Both the production of the unknown factor and arachidonic acid-induced relaxations were dependent on an intact endothelium, indicating that the cellular source of the unknown relaxant factor was the endothelial cell. The present study has further characterized this metabolite and found the presence of two different THETAs, 11,12,15- and 11,14-15-THETA. The 15-lipoxygenase product, 15-HPETE, can be converted to a series of hydroxy-epoxy and trihydroxy products by rat liver microsomes (Weiss, Arnold, and Estabrook, 1987) and rabbit peritoneal polymorphonuclear leukocytes (Narumiya *et al.*, 1981). Until the present report, there was no evidence that these metabolites were made in vascular tissue or endothelial cells. In the presence of soybean lipoxidase, arachidonic acid elicited concentration-dependent relaxations of a denuded preparation of canine coronary arteries (Rubanyi and Vanhoutte, 1987). In the absence of soybean lipoxidase and the endothelium, arachidonic acid contracted the vessels. These results support the conclusion that arachidonic acid produces a relaxing factor via 15-lipoxygenase.

Aortic homogenate incubated in the presence of a 15(S)-HPETE generating system or ^{14}C-15(S)-HPETE produced the unknown metabolite. In the presence of 15-lipoxygenase, the synthesis of the unknown factor was increased. The incubation of 15-lipoxygenase alone with arachidonic acid did not produce the unknown factor indicating that an additional component or enzyme was present in the homogenate. Similarly, incubation of

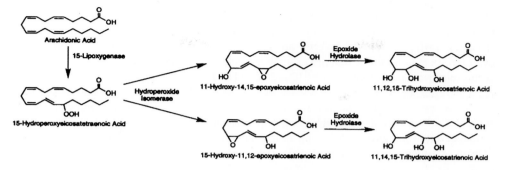

Figure 3-4. Proposed scheme depicting biosynthesis of 15-H-11, 12-EETA, 11-H-14, 15-EETA, 11, 12, 15-THETA and 11, 14, 15-THETA in the rabbit aorta.

[14]C-15-HPETE with aortic homogenates resulted in the formation of the unknown factor. Similar results were obtained when aortic microsomes were substituted for the aortic homogenate. Thus, the additional component or enzyme that converted 15-HPETE to the unknown factor was microsomal. These data indicate that the unknown factor represents a metabolite of 15(S)-HPETE and suggests the presence of a 15-hydroxyl group in the molecule. Results also showed that when arachidonic acid was incubated with aortic homogenate plus 15-lipoxygenase, a less polar metabolite was observed that eluted at 11.5–14 minutes on reverse phase HPLC using solvent system I. This fraction comigrates with the HEETAs, and gas chomatography/mass spectrometry analysis of the methyl ester-TMS ether derivative suggests that it is a mixture of 11-H-14,15-EETA and 15-H-11,12-EETA.

There is no information on the biological importance of the HEETAs and THETAs derived from 15-lipoxygenase. However, the present results indicated that peak 2 which contains both 11,12,15- and 11,14,15-THETA relaxed rabbit aorta. The corresponding HEETA also relaxed contracted rabbit aortas. It is not known if the HEETAs are vasodilators or whether HEETAs are metabolized by the aorta to 11,12,15- and 11,14,15-THETA which then mediate the relaxations.

If the biosynthetic pathway for HEETAs and THETAs in rabbit aorta is similar to the synthesis of hepoxilins and trioxilins (Pace-Asciak, Granstrom, and Samuelsson, 1983), the present results indicate that 15-HPETE is converted to 15-H-11,12-EETA and 11-H-14,15-EETA by a hydroperoxide isomerase (Figure 3-4). Little is known about this enzyme. Several heme containing enzymes, cytochrome P_{450}, hemoglobin and hematin, can function as hydroperoxide isomerases and catalyze this rearrangement (Weiss, Arnold, and Estabrook, 1987; Bryant and Bailey, 1981). In an analogous manner, prostacyclin synthase is a cytochrome P_{450} enzyme that catalyzes the rearrangement of prostaglandin H_2 to prostacyclin (Saez *et al.*, 1975). These rearrangements do not require cofactors such as O_2 or NADPH or reductases. The endothelial cell has several enzymes that contain heme and may serve as a hydroperoxide isomerase. These include cytochrome P_{450}, NO synthase, prostacyclin synthase and cyclooxygenase. Since the synthesis of 11, 12, 15- and 11,14,15-THETA was increased rather than decreased by inhibitors of cyclooxygenase (Pfister *et al.*, 1996), it is unlikely that cyclooxygenase is involved. Rat liver microsomal cytochrome P_{450} catalyzed the transformation of 15-HPETE into HEETAs and THETAs (Weiss, Arnold, and Estabrook,

1987). It is not known if cytochrome P_{450} catalyzes the reaction in the rabbit aorta; however, the experiments which demonstrated that microsomes derived from rabbit aortas synthesized the THETAs from 15-HPETE are supportive for this mechanism. In addition, the cytochrome P_{450} inhibitor, miconazole, attenuated the production of the ^{14}C-THETAs in aortic microsomes incubated with [14]C-15-HPETE. In aortas with endothelium of rabbits fed a 2% cholesterol diet for two weeks, there is an enhanced synthesis of the cytochrome P_{450} epoxygenase metabolites of arachidonic acid (Pfister, Falck, and Campbell, 1991). Removal of the endothelium suppressed the formation of cytochrome P_{450} metabolites suggesting that the enzyme is found in the endothelial, but not vascular smooth muscle, cells. The presence of the cytochrome P_{450} epoxygenase in endothelial cells has been confirmed (Pinto, Abraham, and Mullane, 1987; Pfister, Falck, and Campbell, 1991; Rosolowsky and Campbell, 1996). Since cholesterol-feeding increases the production of the unknown factor and the EETs (Pfister *et al.*, 1996; Pfister, Falck, and Campbell, 1991), these results support the role of cytochrome P_{450} in the synthesis of the THETAs. The mechanism whereby the THETAs cause relaxation is not known at the present time. While the effect of potassium on the THETA-induced relaxations was not determined, the results with arachidonic acid are supportive of these compounds acting as hyperpolarizing factors.

The factor(s) that mediate(s) arachidonic acid-induced relaxations of the rabbit aorta has been identified as metabolite(s) of 15-lipoxygenase, 11, 12, 15- and 11, 14, 15-THETA, using bioassay, HPLC, chemical modification and gas chromatography/mass spectrometry. While the physiological importance of these THETAs is not clearly known, previous work has shown that the calcium ionophore, A23187 stimulated their release from the rabbit aorta (Pfister *et al.*, 1996). Additionally, the activity and synthesis of the THETAs were enhanced in vessels obtained from cholesterol-fed rabbits (Pfister *et al.*, 1996). It is of interest that 15-lipoxygenase activity is increased in aortas obtained from hypercholesterolemic rabbits (Henricksson, Hamberg, and Diczfalusy, 1985). The present studies suggest that arachidonic acid is metabolized by 15-lipoxygenase to 15-HPETE which undergoes an enzymatic rearrangement to either 11-H-14, 15-EETA, 15-H-11, 12-EETA or both. Hydrolysis of the epoxy group results in the formation of 11, 12, 15- and 11, 14, 15-THETA. The HEETAs and THETAs relax the rabbit aorta and one or both may represent the active metabolite of arachidonic acid. At the present time, it is not known which isomer is the active metabolite. Thus, 15-series THETAs represents new members of the family of endothelium-derived relaxing factors which also includes prostacyclin, nitric oxide and epoxyeicosatrienoic acids.

ACKNOWLEDGMENTS

We thank Donna Kotulock for technical assistance and Gretchen Barg for secretarial assistance. These studies were supported by a grant from the National Heart, Lung and Blood Institute (HL-37981).

4 Phospholipid-Derived Epoxyeicosatrienoic Acids Mediate The Relaxations Attributed to Endothelium-Derived Hyperpolarizing Factor

Jean-Vivien Mombouli, §Bernard Ntsikoussalabongui, Kevin Ballard, Ralf Orskiszewski, Addisson A. Taylor and *Paul M. Vanhoutte

*Center for Experimental Therapeutics, Baylor College of Medicine, Houston, Texas, USA; §ASTER CEPHAC, Saint-Benoit, France and *Institut de Recherches Internationales Servier, Courbevoie, France.*
Address for correspondence: Jean-Vivien Mombouli, Ph. D., Center For Experimental Therapeutics, Baylor College of Medicine, One Baylor Plaza, Room 802 E, Houston, TX 77030, U.S.A. Tel: (1-713) 798 5187. Fax: (1-713) 799 2469.
E-mail: ombouli@bcm.tmc.edu
Bernard Ntsikoussalabongui, Ph. D., ASTER CEPHAC, 90 Avenue du haut de chaume, BP 28, 86281 Saint-Benoit-Cedex, France. Tel: (33-5) 4957 0404. Fax: (33-5) 4957 2239.
Kevin Ballard, Ph. D., Center for Experimental Therapeutics, Baylor College of Medicine, One Baylor Plaza, Room 802 E, Houston, TX 77030, U.S.A. Tel: (1-713) 798 4727. Fax: (1-713) 799 2469.
Ralph Orskiszewski, Center for Experimental Therapeutics, Baylor College of Medicine, One Baylor Plaza, Room 802 E, Houston, TX 77030, U.S.A. Tel: (1-713) 798 4727. Fax: (1-713) 799 2469.
Paul M. Vanhoutte, M.D., Ph. D., Institut de Recherches Internationales Servier, 6 Place des pléiades, 92415 Courbevoie-cedex, France.
Addisson A. Taylor, M.D., Ph. D., Center for Experimental Therapeutics, Baylor College of Medicine, One Baylor Plaza, Room 802 E, Houston, TX 77030, U.S.A. Tel: (1-713) 798 4727. Fax: (1-713) 799 2469.

Cytochrome P450 epoxygenases process arachidonic acid into epoxyeicosatrienoic acids (EETs) which, based on pharmacological evidence, may represent endothelium-derived hyperpolarizing factors. Endothelial cells can store EETs in phospholipids. Therefore, activation of endothelial phospholipases A_2 and/or C in response to agonists may trigger the release of EETs from this phospholipid pool, independently of their *de novo* synthesis. Hence, if EETs are endothelium-derived hyperpolarizing factors, inhibition of epoxygenases using cytochrome P450 inhibitors may not acutely impair the endothelium-dependent responses attributed to endothelium-dependent hyperpolarization. Organ bath studies were conducted using canine coronary arterial rings with endothelium. Inhibitors of both NO synthase and prostaglandin synthesis were present. Arachidonic acid induced endothelium dependent relaxations that were abolished by miconazole, an inhibitor of cytochrome P450. Endothelium-dependent relaxations to bradykinin were not affected by miconazole, but they were reduced by inhibitors of phospholipase A_2 and C. The combination of miconazole with a phospholipase A_2 inhibitor was used to restrict the pool of EETs that might be mobilized by bradykinin. Under these conditions, a progressive reduction of the relaxation to bradykinin attributed to endothelium-derived hyperpolarizing factor was obtained after each challenge. This progressive inhibition is consistent with the interpretation that phospholipid-derived (not newly synthesized) EETs account for endothelium-derived hyperpolarizing factor in isolated canine coronary arteries.

29

KEYWORDS: Cytochrome P450, phospholipases, arachidonic acid, bradykinin, vascular smooth muscle, potassium channels.

INTRODUCTION

In endothelial cells, cytochrome P450 epoxygenases process arachidonic acid into epoxyeicosatrienoic (EET) acids, yielding four regioisomers (5,6-EET, 8,9-EET, 11,12-EET, and 14,15-EET; Rosolowski *et al.*, 1990; Rosolowski, Falck & Campbell, 1991; Rosolowski, Falck and Campbell, 1993). EETs have been proposed to represent endothelium-derived hyperpolarizing factors (EDHF) for the following reasons: (a) EETs and EDHF activate charybdotoxin-inhibitable large conductance K^+ channels (BKCa) in vascular smooth muscle cells from various species (Gebremedhin *et al.*, 1992; Hu and Kim, 1993; Campbell *et al.*, 1996a; Zou *et al.*, 1996); (b) charybdotoxin attenuates the relaxation caused by either EETs or EDHF (Campbell *et al.*, 1996a); (c) depletors of cytochrome P450 monooxygenases (which include epoxygenases) impair; but inducers of these enzymes augment EDHF-mediated, endothelium-dependent relaxation of vascular smooth muscle (Pinto, Abraham and Mullane, 1986; Chen and Cheung, 1996); (d) inhibitors of cytochrome P450 monooxygenases impair EDHF-mediated relaxations (Pinto, Abraham and Mullane, 1986; Hecker *et al.*, 1994; Campbell *et al.*, 1996; Chen and Cheung, 1996; Popp *et al.*, 1996); and (e) during the inhibition of cyclooxygenase, arachidonic acid, the precursor of the EETs, elicits endothelium-dependent relaxations which are sensitive to cytochrome P450 inhibitors (Pinto, Abraham and Mullane, 1986; Rosolowski *et al.*, 1990; Rosolowski, Falck & Campbell, 1990; Rosolowski and Campbell, 1993; Campbell *et al.*, 1996a) and charybdotoxin (Campbell *et al.*, 1996a). The inhibitors of cytochrome P450 that are so far available are notoriously non-specific (Oyekan *et al.*, 1994; Corriu *et al.*, 1996a; Zygmunt *et al.*, 1996). This casts doubt on the validity of the EDHF/EET hypothesis (for review see Mombouli and Vanhoutte, 1997).

The impairment of EDHF by inhibitors of cytochrome P450 monooxygenases has been interpreted as evidence that endothelial agonists (e.g. acetylcholine, bradykinin) mobilize arachidonic acid through Ca^{2+} dependent and/or -independent mechanisms. The fatty acid is then converted by the microsomal epoxygenases into EETs which would thereafter be secreted in an 'on demand' fashion to act on vascular smooth muscle. As in hepatocytes and vascular smooth muscle cells (Karara *et al.*, 1989; Fang *et al.*, 1995), the metabolism of EETs in endothelial cells is characterized by the rapid incorporation of EETs into phospholipids, which prolongs their biological half-life by several hours (VanRollins *et al.*, 1993). Furthermore, acylation of the EETs provides a pool of the mediator which may be readily released upon activation of phospholipase A_2 and C by endothelial agonists such as thrombin or Ca^{2+} ionophores (VanRollins *et al.*, 1993). In this context (Figure 4-1), unless the inhibitors tested have secondary actions that are not related to inhibition of epoxygenase activity, inhibition of the *de novo* synthesis of EETs may cause only a partial acute inhibition of the secretion of EETs by endothelial cells. This chapter discusses results in support of the hypothesis that phospholipid-derived, not newly synthesized EETs account for the endothelium-dependent relaxation attributed to EDHF in isolated canine coronary arteries.

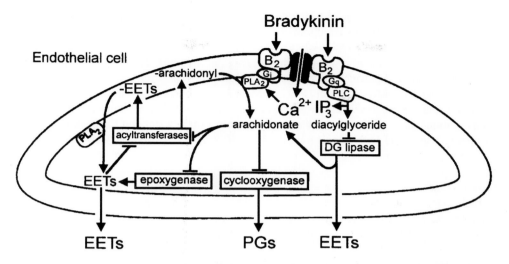

Figure 4-1. Shematic representation of the metabolism of EETs. EETs are synthesized from arachidonic acid by epoxygenases, which are members of the monooxygenase superfamilly of cytochrome P450 enzymes. Like arachidonic acid, but unlike prostaglandins (PGs), EETs are incorporated by acyltransferases into phospholipids (mostly phosphatidylcholine and phosphatidylinositol); therefore, EETs may be released by phospholipases (A_2 and/or C) following activation of Ca^{2+}-mobilizing G protein coupled receptors by e.g. bradykinin. By contrast, prostaglandins which are derived from cyclooxygenases are not acylated into phospholipids. The acylation process of EETs may represent a form of storage of the active mediator in endothelial cells or termination of its action in vascular smooth muscle cells.

MATERIAL AND METHODS

Mongrel dogs (15–25 kg of either sex) were anesthesized by intravenous injection of sodium pentobarbital (30 mg/mVkg weight) and exsanguinated in accordance with institutional guidelines. Hearts were collected, placed in cold modified Krebs-Ringer bicarbonate solution [composition in mM; NaCl 118.3; KCl 4.7; $CaCl_2$ 2.5; $MgSO_4$ 1.2; KH_2PO_2 1.2; $NaHCO_3$ 25; glucose 11.1; calcium disodium edetate 0.026, pH 7.4 (control solution)], and cleaned of extraneous tissue. The left circumflex coronary arteries were dissected and cleaned of connective tissue. The arteries were cut into rings (4 mm). Special care was taken not to damage the luminal surface of the preparations. Unless otherwise stated, the experiments were carried out in the presence of both N^{ω}-nitro-L-arginine (10^{-4} M, to block the production of nitric oxide) and indomethacin (10^{-5} M to avoid possible interference of endogenous prostanoids).

Coronary artery rings were suspended between two stirrups in organ chambers (10 ml) filled with control solution (gassed with 95% O_2–5% CO_2 and maintained at 37°C). One of the stirrups was anchored to the bottom of the organ chamber and the other was connected to a force transducer (UC2, Gould Inc., Cleveland, Ohio) to record changes in isometric tension. The rings were stretched to the optimal point of their length-active tension relationship as determined by the contraction to 60 mM K^+ at progressive levels of stretch. Relaxations to arachidonic acid or bradykinin were observed in rings contracted with prostaglandin $F_{2\alpha}$ (8×10^{-6} M). The inhibitor of angiotensin converting enzyme perindoprilat (10^{-6} M; 30 minute incubation) was present when experiments involved bradykinin, to optimize the effects of the nonapeptide (Mombouli *et al.*, 1992).

The following drugs were used; arachidonic acid sodium salt, bradykinin acetate, indomethacin, miconazole, prostaglandin $F_{2\alpha}$ (Sigma, St-Louis, MO, USA), AACOCF3, N^{ω}-nitro-L-arginine (Aldrich, Milwaukee, WI, USA), perindoprilat (S9490-3, a generous gift from Servier, Neuilly-sur-Seine, France). Most drugs were prepared in water. AACOCF3 and miconazole were dissolved in dimethyl sulfoxide (DMSO, less than 0.1% final bath concentration). Indomethacin was dissolved by sonication in water and Na_2CO3 which had no effect at the final bath concentration of 5×10^{-6} M.

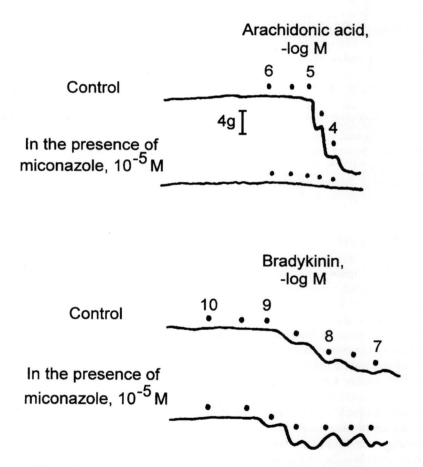

Figure 4-2. Miconazole and relaxations to arachidonic acid and bradykinin. Relaxations were obtained by cumulative additions of either arachidonic acid (top two traces) or bradykinin were obtained in canine coronary artery rings with endothelium contracted with prostaglandin $F_{2\alpha}$ (8×10^{-6} M). All of the rings were treated with both indomethacin (10^{-5} M) and N^{ω}-nitro-L-arginine (10^{-4} M). Half of the rings were treated with the P450-inhibitor miconazole. Experiments involving bradykinin were performed in the presence of perindoprilat (10^{-6} M). Dots represent additions of increasing concentrations of the relaxing agent in log(M) units. Experiments with bradykinin were carried at 5 times the speed of those with arachidonic acid. For sake of clarity the time scales were omitted.

RESULTS

Stable contractions to prostaglandin $F_{2\alpha}$ were obtained in rings of canine coronary arteries with endothelium, treated with both N^{ω}-nitro-L arginine and indomethacin. During the plateau of the contraction to prostaglandin $F_{2\alpha}$, cumulative additions of either arachidonic acid (Figure 4-2) or bradykinin (Figure 4-2, rings treated with 10^{-6} M perindoprilat) elicited concentration-dependent relaxations. Treatment of the arterial rings with the cytochrome P450 inhibitor miconazole abolished the relaxations induced by arachidonic acid (Figure 4-2). However, miconazole had no effect on the relaxations evoked by bradykinin (Figure 4-2). In a third series of experiments, repeated relaxations were induced using a maximally effective concentration of bradykinin (10^{-7} M, in the presence of perindoprilat; see also Mombouli *et al.*, 1992). Under such conditions the relaxation attributed to EDHF was reproducible (Figure 4-3, top row). After observation of an initial relaxation, treatment with miconazole had no effect on subsequent responses obtained during the course of these experiments (Figure 4-3, second row from top). The phospholipase A_2 antagonist AACOCF3 acutely reduced the relaxation to bradykinin. However, the magnitude of the component of the relaxation that was resistant to AACOCF3 was not changed in subsequent tests (Figure 4-3, third row from top). A combination of miconazole and AACOCF3 acutely reduced the relaxation to bradykinin. In addition, the component resistant to the first challenge with the combination of miconazole and AACOCF3 was progressively reduced in subsequent tests, which were performed in the continuous presence of these inhibitors (Figure 4-3, bottom row).

DISCUSSION

The results show that during inhibition of the synthesis of both NO and prostaglandins, the epoxygenase inhibitor miconazole abolished the endothelium-dependent relaxation to arachidonic acid, but had no effect on the relaxation induced by bradykinin. However, the latter were reduced after acute inhibition of phospholipase A_2. Furthermore, combination of the inhibitors of phospholipase A_2 and of epoxygenase provoked a progressive reduction of the relaxation to bradykinin. In conjunction with current knowledge on the metabolism of EETs in endothelial cells and their actions on vascular smooth muscle, these data are consistent with the hypothesis that phospholipid-derived EETs participate to the NO and prostacyclin-independent relaxations, usually ascribed to an EDHF.

Arachidonic acid caused concentration-dependent relaxation which were blocked by the epoxygenase inhibitor miconazole. Since indomethacin was present, these relaxation may be attributed to *de novo* synthesis of EETs, as demonstrated in canine and bovine coronary arteries (Rosolowski *et al.*, 1990; Rosolowski and Campbell, 1993). This observation is also consistent with the idea that 'on demand' synthesis of epoxygenase metabolites in response to endothelial stimulation could in principle sustain NO and prostanoid independent relaxations. Contrary to this view, miconazole did not affect the relaxation to bradykinin, which under the given experimental conditions is mediated by EDHF (Mombouli *et al.*, 1992 and 1996). Similar resistance to inhibitors of cytochrome P450 has been obtained in porcine, rat and guinea pig arteries (Zygmunt *et al.*, 1996; Corriu *et al.*, 1996a; Graier *et al.*, 1996; Fukao *et al.*, 1997a). This has been considered as evidence that EDHF is not

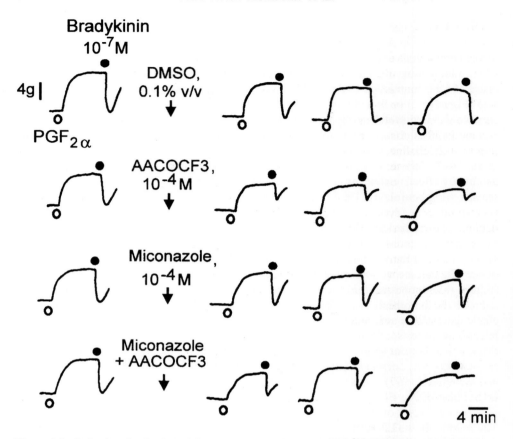

Figure 4-3. Induction of tachyphylaxis by combined use of inhibitors of phospholipase A$_2$ and epoxygenase. Canine coronary artery rings with endothelium were treated with both indomethacin (10^{-5} M) and N$^{\omega}$-nitro-L arginine (10^{-4} M). Open dots represent addition of prostaglandin F$_{2\alpha}$ (8×10^{-6} M) to induce contraction. Closed dots represent additions of bradykinin to induce relaxation. Perindoprilat (10^{-6} M) was present throughout the experiment. the first row of traces represents control runs. Afterwards, either vehicle (0.1% vol/vol DMSO, for both miconazole and AACOCF3), or AACOCF3 (10^{-4} M) alone or in combination with miconazole (10^{-5} M) was added. The rings were rinsed five times after each run. After each rinses the bath concentrations these inhibitors were renewed. Vertical bar represents development of isometric force expressed in g. Horizontal bar represents time.

an EET. A view that would be accurate only if newly synthesized EETs accounts for the EDHF-mediated relaxations to bradykinin, as they do for relaxations to arachidonic acid (Rosolowski *et al.*, 1990; Rosolowski and Campbell, 1993).

EDHF-mediated relaxations to bradykinin were reduced acutely by the inhibitor of phospholipase A$_2$ AACOCF3. In unpublished studies, the inhibitor of phospholipase C U73122 nearly abolished the relaxations to bradykinin, without having a significant effect on the relaxation evoked by arachidonic acid. Similar results have been obtained in porcine coronary and rat mesenteric arteries (Weintraub *et al.*, 1994; Fukao *et al.*, 1997b). Hence, activation of phospholipases appears to play an important role in the release of EDHF (see Mombouli and Vanhoutte, 1997 for review).

Results presented in the present study further show that combination of epoxygenase and phospholipase A_2 inhibitors induced tachyphylaxis of the component of the relaxation to bradykinin that remains after phospholipases A_2 inhibition. Whereas, miconazole alone did not have a significant effect. Taken in conjunction with previous reports (Weintraub *et al.*, 1994; Fukao *et al.*, 1997b), these results can be interpreted as follows. By inhibiting simultaneously epoxygenase and phospholipase A_2 activities, one eliminates the involvement of *de novo* synthesis EETs that follows the mobilization of endogenous arachidonic acid by agonists. Also, one eliminates the mobilization of EETs from epoxyeicosanotrienoyl phosphatidylcholine, which is the most abundant phospholipid pool of EETs (VanRollins *et al.*, 1993). Hence, it is conceivable that under such conditions only the pools of EETs associated with phosphatidylinositol (mobilized by phospholipase C-type cleavage) and phosphatidylethanolamine (mobilized by phospholipase D-type cleavage) can be generated in response to bradykinin. The tachyphylaxis would be a consequence of the progressive depletion after each stimulation of these relatively small pools, and the catabolism of EETs by epoxide hydrolases. The fact that miconazole alone does not induce tachyphylaxis suggests that in native endothelial cells of the canine coronary arteries the phospholipid pools of EETs are very large and difficult to deplete.

The observation that EETs can induce relaxation in canine coronary arteries further supports the hypothesis that EDHF is an EET in that artery (see Rosolowski and Campbell, 1993). EETs cause relaxation of vascular smooth following the activation of K^+ channels (Hecker *et al.*, 1994; Campbell *et al.*, 1996; Zou *et al.*, 1996). In other blood vessels, the electrophysiological actions of EETs do not always match those of EDHF (Zygmunt and Hogestatt, 1996; Corriu *et al.*, 1996b; Chataigneau *et al.*, 1998a; for review see Mombouli and Vanhoutte, 1997); this suggests that other endothelial products may act in concert with or in place of the EETs in these cases.

However, a set of unpublished and published observations from this laboratory are consistent with the EDHF/EETs hypothesis. In the left circumflex canine coronary arteries used in the present study, the relaxation attributed to EDHF is abolished by charybdotoxin. Remote detection of EDHF released by isolated canine arteries is made possible under superfusion cascade bioassay conditions by using the acyltransferase inhibitor thimerosal (Mombouli *et al.*, 1996). EDHF activates K^+ channels in vascular smooth muscle cells, as demonstrated by the inhibition of the relaxation by tetraethylammonium (Mombouli *et al.*, 1996). Levels of EETs above threshold can be measured in the perfusates during relaxations of the bioassay tissue by EDHF using gas chromatography/mass spectrometry. Moreover, the lack of effect of oxyhemoglobin, catalase and superoxide dismutase rules out the participation of CO and reactive oxygen species.

The present results are consistent with (but not proof of) the essential role phospholipid-derived EETs. This implies that acute inhibition of EDHF by cytochrome P450 inhibitors may result from an epoxygenase unrelated action of these compounds, such as inhibition of K^+ channels in vascular smooth muscle, or in some cases interference with Ca^{2+} signaling in endothelial cells. Thus, impairment of EDHF-mediated responses by cytochrome P450 inhibitors is an insufficient criteria to confirm the participation of EETs in vasomotion. Likewise, the absence of acute effect of epoxygenase inhibitors may simply reflect the predominant role of phospholipid-derived EETs; hence, it can not be a basis for the rejection of the EET/EDHF hypothesis. As done previously for NO, to definitively establish (or rule out) the role of EETs as EDHF, further consideration must be given to their chemistry, metabolism and pharmacology.

5 Lack of Evidence for the Involvement of Cytochrome P450 Mono-Oxygenase-Dependent Metabolites of Arachidonic Acid and Cannabinoids in Endothelium-Dependent Relaxations of the Guinea-Pig Basilar Artery

Jesper Petersson, Peter M. Zygmunt, Morten Sørgård, Pouya Movahed and Edward D. Högestätt

Department of Clinical Pharmacology, Institute of Laboratory Medicine, Lund University Hospital, S-221 85 Lund, Sweden
Address for correspondence: Jesper Petersson, Phone+4640331000; Fax+4640336228;
E-mail: Jesper.Petersson@neurolmas.lu.se
Peter M. Zygmunt, Phone+4646173809; Fax:+46462111987;
E-mail: Peter.Zygmunt@klinfarm.lu.se
Morten Sørgård, Phone+4646173679; Fax+46462111987; E-mail: LLV95MOS@student1.lu.se
Pouya Movahed, Phone+4646173679; Fax+46462111987; E-mail: MED95POM@student2.lu.se
Edward D Högestätt, Phone: +4646173359; Fax:+46462111987;
E-mail: Edward.Hogestatt@klinfarm.lu.se

In the guinea-pig basilar artery, both endothelium-derived hyperpolarizing factor (EDHF) and nitric oxide (NO) contribute to endothelium-dependent relaxations. In this artery, the EDHF-mediated relaxation is totally inhibited by a combination of charybdotoxin plus apamin, whereas the NO-mediated vasodilatation is abolished by the guanylate cyclase inhibitor ODQ or the NO synthase inhibitor N^{ω}-nitro-L-arginine. The identity of EDHF in the guinea-pig basilar artery, however, remains to be determined.

The epoxyeicosatrienoic acids, which are derivatives of arachidonic acid formed by cytochrome P450-dependent monoxygenase, act as endothelium-derived hyperpolarizing factors in some vascular tissues. In the guinea-pig basilar artery, 17-octadecynoic acid, which is considered a selective inhibitor of cytochrome P450 monoxygenase, and 5,8,11,14-eicosatetraynoic acid, a non-selective inhibitor of arachidonic acid metabolism, were unable to antagonize the EDHF-mediated relaxation induced by acetylcholine in the guinea-pig basilar artery. Furthermore, 11,12-epoxyeicosatrienoic acid did not produce a relaxing response under conditions when EDHF causes vasodilatation.

NO synthase and cytochrome P450 monoxygenase show similar sensitivity to oxygen. In the present study, relaxations mediated by endothelium-derived NO were abolished under hypoxic conditions (P_{O2} = 6 mmHg), whereas EDHF-mediated relaxations were almost intact. In contrast to the endogenous NO response, the NO donor S-nitroso-N-acetylpenicillamine induced a concentration-dependent relaxation during hypoxia.

Endocannabinoids such as anandamide (arachidonoylethanolamide) may participate in regulation of vascular tone. Anandamide, acting at cannabinoid receptors (CB1), has been proposed as a candidate for EDHF in the rat isolated perfused mesenteric vascular bed. In the guinea-pig basilar artery, anandamide induced concentration-dependent relaxations, which were partially inhibited by the selective CB1 receptor antagonist SR141716. In contrast, SR141716 had no effect on the EDHF-mediated response induced by acetylcholine. Furthermore, the combination of charybdotoxin and apamin did not affect anandamide-induced relaxations in the guinea-pig basilar artery, whereas EDHF-mediated responses were abolished. Two other endogenous cannabinoid receptor ligands, 2-arachidonoylglycerol and palmitoylethanolamide, were devoid of relaxant effect.

The results of the present study suggest that EDHF is not a cytochrome P450 monoxygenase-dependent arachidonic acid metabolite such as epoxyeicosatrienoic acid or the endocannabinoid anandamide in the guinea-pig basilar artery.

KEYWORDS: Cerebral arteries, cytochrome P450, cannabinoids, hyperpolarization, hypoxia, vascular endothelium.

INTRODUCTION

The endothelium generates another relaxing factor besides nitric oxide (NO) and prostacyclin, which activates potassium channels in vascular smooth muscle (Mombouli and Vanhoutte, 1997). This factor, termed endothelium-derived hyperpolarizing factor (EDHF), has been demonstrated in various blood vessels, including human pial (Petersson *et al.*, 1995), guinea-pig basilar (Nishiye *et al.*, 1989; Petersson, Zygmunt and Högestätt, 1997), and rabbit and cat middle cerebral (Brayden and Wellman, 1989; Brayden, 1990) arteries. While the arachidonic acid derivatives epoxyeicosatrienoic acids (EETs) and anandamide (arachidonoylethanolamide) have been proposed as candidates for EDHF in coronary and mesenteric arteries, respectively (Hecker *et al.*, 1994; Campbell *et al.*, 1996; Popp *et al.*, 1996; Randall *et al.*, 1996), the chemical nature of EDHF is still elusive in cerebral arteries.

Oxygen deprivation influences endothelium-dependent modulation of vascular tone in several ways (Wadsworth, 1994). The synthesis of NO in the vascular endothelium is oxygen-dependent and is impaired under hypoxic conditions (Wadsworth, 1994). The formation of EETs from arachidonic acid by cytochrome P450 monoxygenase also requires oxygen (McGiff, 1991). Since the K_mO_2 values for these two enzymatic pathways are similar (Jones, 1986; Rengasamy and Johns, 1996), the effects of hypoxia on EDHF- and NO-induced relaxations were examined in the guinea-pig basilar artery to determine whether cytochrome P450 monoxygenase is involved in the production of EDHF. The effects of 17-octadecynoic acid, a suicide substrate inhibitor of cytochrome P450 monoxygenase (Zou *et al.*, 1994), and 5,8,11,14-eicosatetraynoic acid (ETYA), a non-selective inhibitor of arachidonic acid metabolism (McGiff, 1991), on EDHF-mediated relaxations were also studied. Furthermore, relaxant effects of 11,12-EET, a representative of the EET family, and of the endogenous cannabinoid receptor ligands anandamide, 2-arachidonoylglycerol and palmitoylethanolamide, were investigated. Finally, the effects of SR141716, a selective cannabinoid-1 receptor (CB1) inhibitor, on relaxations induced by anandamide and EDHF were determined.

METHODS

Vascular Preparations

Guinea-pigs of male gender (300 g) were killed by CO_2 asphyxia followed by exsanguination. After removal of the brainstem, the basilar artery was dissected free and flushed with physiological salt solution to remove blood products. The physiological salt solution was composed of NaCl 119 mM, KCl 4.6 mM, $CaCl_2$ 1.5 mM, $MgCl_2$ 1.2 mM, $NaHCO_3$ 15 mM, NaH_2PO_4 1.2 mM and D-glucose 6 mM.

Tension Experiments

Tension experiments were performed in tissue baths containing physiological salt solution at 37°C. To provide oxygenation and a pH of 7.4, the solution in the tissue baths was continuously bubbled with carbogen (95% O_2, 5% CO_2). In brief, ring preparations (1–2 mm long) of the arteries were gently thread on two stainless steel wires (Högestätt, Andersson and Edvinsson, 1983). One of the wires was connected to a force-displacement transducer (model FT03 C, Grass Instruments, USA) for isometric tension recording. The output from the transducer was displayed on a polygraph (model 7D, Grass Instruments, USA).

The blood vessel segments were allowed to equilibrate for 60 min under a passive load of 2–4 mN (corresponding to a wall tension of 1–2 mN/mm). An isosmolar 60 mM K^+ solution (physiological salt solution with an equimolar amount of NaCl substituted with KCl) was then introduced to assess the contractile capacity of the preparation. Prostaglandin $F_{2\alpha}$ ($10^{-7} – 10^{-6}$M) was used to induce stable submaximal contractions. Inhibitors were generally added 30 min prior to induction of relaxant responses. To avoid interference by functional antagonism, the level of contraction was titrated in each experiment to 90 to 100% of the initial contraction obtained with 60 mM K^+. Acetylcholine ($10^{-8} – 10^{-4}$M) and S-nitroso-N-acetylpenicillamine (SNAP; $10^{-8} – 10^{-5}$M) were added cumulatively to elicit relaxations. Indomethacin (10^{-5}M) was present in all experiments. The EDHF candidates anandamide ($10^{-6} – 3 \times 10^{-5}$M), 11,12-EET ($3 \times 10^{-7} – 3 \times 10^{-6}$M), 2-arachidonoylglycerol ($10^{-6} – 10^{-5}$M) and palmitoylethanolamide ($10^{-6} – 10^{-5}$M) were also added cumulatively.

Induction of Hypoxia

Hypoxia was achieved by bubbling the physiological salt solution in the tissue bath with a mixture of N_2(95%) and CO_2 (5%). To ensure low and stable levels of O_2, a box was placed over the tissue baths. The box was continuously flushed with N_2/CO_2 to minimize oxygen exchange with the surrounding air. Drugs were added through elastic membranes situated above each organ bath. Relaxation experiments were performed after 30 min of hypoxia. Since prostaglandin $F_{2\alpha}$ failed to induce stable contractions under hypoxic conditions, endothelin-1 ($10^{-9} – 3 \times 10^{-8}$M) was used to contract the arteries in both controls (standard conditions with 95% O_2 and 5% CO_2) and during hypoxia. Acetylcholine-induced relaxations mediated by NO and EDHF do not differ whether prostaglandin $F_{2\alpha}$ or endothelin-1 is used to contract the arteries (Petersson *et al.*, 1998). The P_{O2} level was measured with a Clark-type electrode connected to an oximeter (Combi-Analysator, L. Eschweiler, Kiel, Germany). After 30 min of hypoxia, the P_{O2} level was 5.9 ± 1.0 mmHg (mean of 7 experiments), corresponding to an O_2 concentration of 7.6×10^{-6}M at 37°C.

Calculations and Statistics

The negative logarithm of the drug concentration eliciting half maximal relaxation (pEC_{50}) was calculated by linear regression, using the values immediately above and below half maximal response. E_{max} refers to the maximal relaxation achieved (100% denotes a complete reversal of the contraction). Results are given as arithmetic mean values ± s.e.

mean. Two-tailed Student's *t*-test or analysis of variance (ANOVA) followed by Bonferroni-Dunn's post hoc test (Statview 4.12) was used for statistical comparison (Ludbrook, 1994). A P value less than 0.05 was considered as statistically significant.

Drugs

Acetylcholine chloride, endothelin-1, 11,12-epoxyeicosatrienoic acid, ETYA and N^{ω}-nitro-L-arginine were purchased from Sigma, St Louis, MO, U.S.A.; anandamide and 2-arachidonoylglycerol from RBI, Natick, U.S.A.; apamin from Alomone labs, Jerusalem, Israel; synthetic charybdotoxin from Latoxan, Rosans, France; indomethacin (Confortid) from Dumex, Copenhagen, Denmark; 17-octadecynoic acid from Tocris Cookson, Bristol, U.K.; palmitoylethanolamide from Biomol Res. Lab., Plymouth Meeting, U.S.A.; prostaglandin $F_{2\alpha}$ (Prostin) from Upjohn, Kalamazoo, MI, U.S.A. S-nitroso-N-acetylpenicillamine was a gift from Schwarz Pharma, Monheim, Germany.

RESULTS

Inhibitors of the Metabolism of Arachidonic Acid

17-Octadecynoic acid (5×10^{-5}M) was devoid of inhibitory action on the acetylcholine-induced relaxation in the presence of 3×10^{-4}M N^{ω}-nitro-L-arginine (Figure 5-1a); pEC_{50} and E_{max} were 6.31 ± 0.20 and $97 \pm 1\%$ in the presence, and 6.37 ± 0.11 and $96 \pm 2\%$

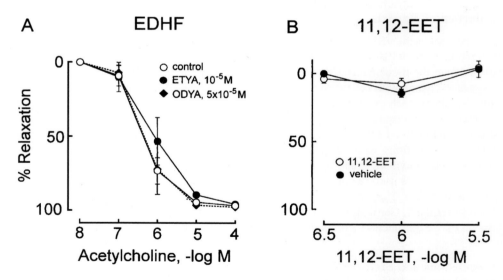

Figure 5-1. (a) Effects of 17-ODYA (5×10^{-5}M;●) or ETYA (10^{-5}M;◆) on the acetylcholine-induced relaxation mediated by EDHF in arteries contracted with prostaglandin $F_{2\alpha}$. Open circles (○) indicate controls. (b) Effects of 11,12-EET (●) or vehicle (0.3% ethanol, ○) on arteries contracted with prostaglandin $F_{2\alpha}$. All experiments (a, b) were performed in the presence of N^{ω}-nitro-L-arginine (3×10^{-4}M) and indomethacin (10^{-5}M). The level of contraction before addition of acetylcholine, 11,12-EET or vehicle was set to 100%. Data are presented as means $\pm$ s.e. mean (n = 4–6).

in the absence of 17-octadecynoic acid, respectively (n = 5). The arachidonic acid analogue ETYA (10^{-5}M) also had no effect on the acetylcholine-induced relaxation in the presence of N^{ω}-nitro-L-arginine (Figure 5-1a); pEC_{50} and E_{max} for acetylcholine were 6.16 ± 0.27 and $95 \pm 1\%$ in the presence, and 6.40 ± 0.14 and $96 \pm 3\%$ in the absence of ETYA, respectively (n = 4).

11,12-EET

Application of 11,12-EET ($3 \times 10^{-7} - 3 \times 10^{-6}$M) on arteries contracted with prostaglandin $F_{2\alpha}$ in the presence of N^{ω}-nitro-L-arginine induced a small relaxation ($E_{max} = 14 \pm 3\%$, n = 6), which did not differ significantly from the response to vehicle (0.3% ethanol; $E_{max} = 7 \pm 4\%$, n = 6), indicating that 11,12-EET had no relaxant effect of its own (Figure 5-1b).

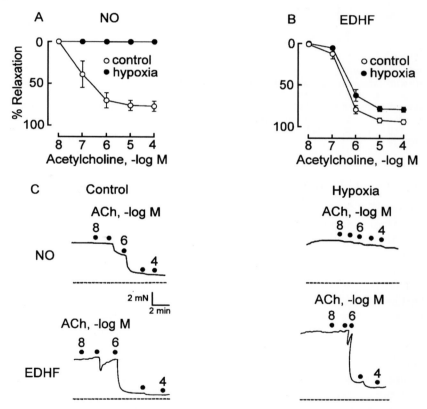

Figure 5-2. Effect of hypoxia on acetylcholine-induced relaxations in arteries contracted with endothelin-1 (a) in the presence of a combination of charybdotoxin and apamin (each 10^{-7}M), to study NO-mediated responses (n = 6) or (b) in the presence of N^{ω}-nitro-L-arginine (3×10^{-4}M) to reveal the action of EDHF (n = 10). Open symbols (O) denote oxygenized conditions and filled symbols (●) represent 30 min of hypoxia. Indomethacin (10^{-5}M) was present in all experiments. The level of contraction before addition of acetylcholine was set to 100%. Data are presented as means ± s.e. mean. (c) Traces from these experiments showing the effect of hypoxia on NO- and EDHF-mediated relaxations elicited by log molar concentrations of acetylcholine. Dashed lines indicate the basal tension level before contraction with endothelin-1.

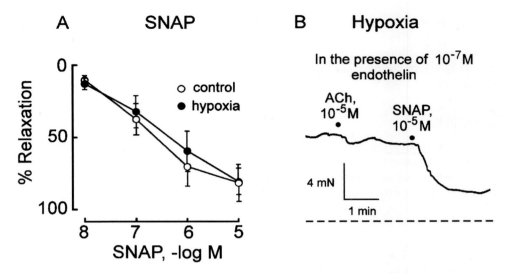

Figure 5-3. (a) Effect of hypoxia on relaxations induced by S-nitroso-N-acetylpenicillamine (SNAP) in arteries contracted with endothelin-1 in the presence of charybdotoxin (10^{-7}M), apamin (10^{-7}M) and indomethacin (10^{-5}M). Open symbols (O) denote oxygenized conditions and filled symbols (●) represent 30 min of hypoxia. The level of contraction before addition of SNAP was set to 100%. Data are presented as means ± s.e. mean (n = 7–8). (b) Trace after 30 min of hypoxia, showing the effects of acetylcholine and SNAP in the presence of charybdotoxin, apamin and indomethacin. Dashed line indicates the basal tension level before contraction with endothelin-1.

Hypoxia, NO- and EDHF-Mediated Responses

In the presence of the K channel inhibitors charybdotoxin (10^{-7}M) and apamin (10^{-7}M), exposure to hypoxia abolished the acetylcholine-induced relaxation (Figure 5-2). The NO donor SNAP also induced a concentration-dependent relaxation in the presence of charybdotoxin plus apamin; pEC_{50} and E_{max} for SNAP were 6.75 ± 0.19 and 92 ± 8%, respectively (n = 7). This relaxation was unchanged under hypoxic conditions (pEC_{50} = 6.51 ± 0.31; E_{max} = 80 ± 10%; n =7; Figure 5-2). Hypoxia marginally affected the acetylcholine-induced relaxation mediated by EDHF in the presence of N^{ω}-nitro-L-arginine; E_{max} was reduced by 16% ($P < 0.05$), whereas pEC_{50} was unchanged (Figure 5-2).

Cannabinoid Receptor Ligands

Anandamide induced consistent concentration-dependent relaxations in the presence of N^{ω}-nitro-L-arginine (Figure 5-4). SR141716 (3×10^{-6}M) depressed E_{max} ($P < 0.05$) but did not significantly affect pEC_{50} ; pEC_{50} and E_{max} for anandamide were 5.53 ± 0.02 and 43 ± 16% in the presence, and 5.69 ± 0.1 and 89 ± 5% in the absence of SR141716, respectively (n = 5–9). However, SR141716 did not affect the acetylcholine-induced relaxation mediated by EDHF in the presence of N^{ω}-nitro-L-arginine (Figure 5-4). While the combination of charybdotoxin and apamin abolished the acetylcholine-induced relaxation mediated by EDHF, the same treatment did not affect relaxations induced by anandamide (Figure 5-4). 2-arachidonoylglycerol (10^{-5}M) and palmitoylethanolamide

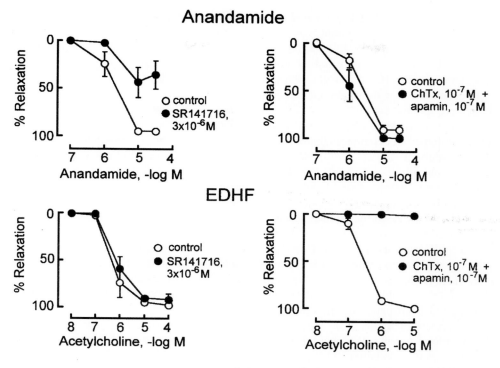

Figure 5-4.   Effects of SR141716 (3×10^{-6}M;●; **left**) or a combination of charybdotoxin and apamin (each 10^{-7}M;●; **right**) on relaxations induced by anandamide (**upper**) or acetylcholine (**lower**) in arteries contracted with prostaglandin $F_{2\alpha}$. Open circles (O) indicate controls. The level of contraction before addition of anandamide or acetylcholine was set to 100%. All experiments were performed in the presence of N^{ω}-nitro-L-arginine (3×10^{-4}M) and indomethacin (10^{-5}M). Data are presented as means $\pm$ s.e. mean (n = 5–9).

(10^{-5}M) both induced a small relaxation which did not differ from controls with vehicle (2-arachidonoylglycerol: E_{max} = 29 $\pm$ 14%, n = 5; palmitoylethanolamide: E_{max} = 3 $\pm$ 3%, n = 4; ethanol vehicle: E_{max} = 5 $\pm$ 2%, n = 5).

DISCUSSION

Involvement of Cytochrome P450 Monoxygenase

The four regioisomeric EETs (5,6-, 8,9-, 11,12- and 14,15-EET), which are metabolites of arachidonic acid derived from cytochrome P450 monoxygenases (Fitzpatrick and Murphy, 1989), have several properties in common with EDHF. They are formed in the endothelium and induce K channel-activation, hyperpolarization and relaxation of vascular smooth muscle, including that of cerebral arteries (Gebremedhin *et al.* 1992; Hu and Kim, 1993; Campbell *et al.* 1996; Rosolowsky and Campbell, 1996) However, in the present study, 17-octadecynoic acid and ETYA had no effect on EDHF-mediated relaxations, and 11,12-EET did not display any relaxing activity. 17-octadecynoic acid was also devoid of inhibitory activity on EDHF-mediated relaxations in guinea-pig coronary, and rat hepatic and mesenteric arteries (Zygmunt *et al.*, 1996; Van de Voorde and Vanheel, 1997; Yamanaka, Ishikawa and Goto, 1998). In addition, the hyperpolarization of vascular smooth muscle mediated by EDHF is unaffected by 17-octadecynoic acid in guinea-pig carotid and rat mesenteric arteries (Corriu *et al.*, 1996a; Fukao *et al.*, 1997b).

In concordance with the present study, EETs had no relaxant effect in rat hepatic and guinea-pig internal carotid arteries (Zygmunt *et al.*, 1996; Chataigneau *et al.*, 1998a). In rat mesenteric arteries, EETs elicited a small glibenclamide-sensitive hyperpolarization, while the hyperpolarization mediated by EDHF was unaffected by treatment with this K channel inhibitor (Fukao *et al.*, 1997b). By contrast, in porcine coronary arteries, EETs induce a concentration-dependent relaxation (Campbell *et al.*, 1996), and 17-octadecynoic acid inhibits EDHF-mediated hyperpolarization in the same artery (Popp *et al.*, 1996). Although this indicates that EETs may represent EDHF in the latter artery, this does not seem to be the case in the guinea-pig basilar artery and several other arteries. A possible explanation for the lack of effect of 17-octadecynoic acid could be that endothelial cells can store EETs as phospholipids (VanRollins *et al.*, 1993). Phospholipase activation may subsequently lead to release of EETs from the endothelial cell membrane. However, even if this mechanism is essential for the release of EETs, it still cannot explain why 11,12-EET itself was unable to mimic the action of EDHF in the guinea-pig basilar and internal carotid, and rat hepatic and mesenteric arteries.

The effects of hypoxia on endothelium-dependent relaxations have been studied in various blood vessels (Wadsworth, 1994). It has been shown that EDRF/NO-mediated relaxations and endothelial NO synthase are both inhibited during hypoxia (Furchgott and Zawadzki, 1980; De Mey and Vanhoutte, 1983; Johns, Linden and Peach, 1989; Myers, Muller and Tanner, 1991; Félétou, Girad and Canet, 1995; Rengasamy and Johns, 1996). It is, however, not known whether the relaxations mediated by EDHF are affected by oxygen deprivation. The oxygen sensitivities of microsomal cytochrome P450 monoxygenase and NO synthase, which requires oxygen for catalytic conversion of L-arginine to NO and citrulline, are similar (McGiff, 1991; Moncada and Higgs, 1993; Harder *et al.*, 1996). For cytochrome P450 monoxygenase, K_mO_2 values between 4.3×10^{-6}M and 9.3×10^{-6}M have been measured in isolated enzyme systems (Jones, 1986). The K_mO_2 value for NO synthase isolated from cultured bovine aortic endothelial cells has been estimated at 7.7×10^{-6}M (Rengasamy and Johns, 1996).

In the guinea-pig basilar artery, the endothelium-dependent relaxation induced by acetylcholine is mediated solely by NO in the combined presence of the K channel inhibitors charybdotoxin and apamin and may therefore be studied separately from the EDHF-component (Petersson, Zygmunt & Högestätt, 1997; Petersson *et al.*, 1998). The observation that the NO-mediated relaxation in the guinea-pig basilar artery was abolished during hypoxia indicates that the oxygen tension achieved in the present study was low enough to block NO synthesis. To ensure that the effect of hypoxia on NO-mediated relaxations was not due to inhibition of other mechanisms, such as guanylate cyclase, in the vascular smooth muscle cells, we also examined the relaxation induced by the NO donor SNAP under hypoxic conditions. However, the guanylate cyclase-cGMP pathway was still functional during hypoxia in the guinea-pig basilar artery, since the relaxant response to exogenous NO was not significantly different during oxygenation and hypoxia. This finding is in agreement with previous reports, suggesting that vascular guanylate cyclase is functionally intact at low oxygen tensions in the rabbit pulmonary artery and canine coronary microvessels (Johns, Linden and Peach, 1989; Myers, Muller and Tanner, 1991)

In the present study, a P_{O2} value of 6 mm Hg was measured in the organ chambers during experimental hypoxia. This corresponds to an O_2 concentration of $7.6 \times 10^{-6}M$. Since an oxygen gradient over the cell membrane and cytosol has been demonstrated, the P_{O2} value at the site of the microsomal enzymes might have been even lower (Jones, 1986; Rengasamy and Johns, 1996). The oxygen tension achieved in the present study would thus be expected to significantly affect microsomal cytochrome P450 monoxygenases. In a study on renal microvessels, the formation of EETs was decreased by more than 30% at a P_{O2} level of 20 mm Hg (Harder *et al.*, 1996). In the guinea-pig basilar artery, the EDHF-mediated relaxation was, however, almost intact during hypoxia. This strengthens the conclusion that EDHF is not an EET, produced by the cytochrome P450 monoxygenase system, and further suggests that formation of EDHF in this artery may not be dependent on (an) oxygen-sensitive enzyme(s).

Could EDHF Be an Endocannabinoid?

Another arachidonic acid-related lipid, the endocannabinoid anandamide, may be EDHF in the rat mesenteric arterial bed (Randall *et al.*, 1996). After the discovery of a receptor in the brain, responsible for the neuronal activity of cannabis (Devane *et al.*, 1988), two cannabinoid receptor types (CB1 and CB2) have been identified (Howlett, 1995). Anandamide and mRNA transcripts for CB1 receptors have been demonstrated in endothelial and vascular smooth muscle cells, respectively (Deutsch *et al.*, 1997; Sugiura *et al.*, 1998). Systemically applied anandamide induces hypotension by CB1 receptor-stimulation, as shown in anesthesized guinea-pigs (Calignano *et al.*, 1997). Anandamide also induces vasodilatation in isolated blood vessels (Randall and Kendall, 1998b), including cerebral arteries and arterioles (Ellis, Moore and Willoughby, 1995; Petersson *et al.*, 1997). In the perfused rat mesenteric arterial bed, anandamide induces a vasodilatation, which is antagonized by the CB1 receptor antagonist SR141716 (Randall *et al.*, 1996). In the same preparation, the acetylcholine-induced relaxation in the presence of blockers of both cyclo-oxygenase and NO synthase was inhibited by SR141716, suggesting that anandamide may act as EDHF in mesenteric arteries (Randall *et al.*, 1996).

In the present study, the anandamide-induced relaxation was inhibited by SR141716. Since indomethacin and N^{ω}-nitro-L-arginine were both present in these experiments, the relaxant effect of anandamide was not mediated by endothelium-derived NO or prostacyclin. Two main discrepancies were seen when the EDHF-mediated relaxation induced by acetylcholine was compared with the anandamide-induced response in the guinea-pig basilar artery. First, inhibition of CB1 receptors had no effect on the EDHF-mediated response, in contrast to responses induced by anandamide. Secondly, a combination of the K channel inhibitors charybdotoxin and apamin, which abolishes EDHF-mediated relaxations in this artery (Petersson, Zygmunt and Högestätt, 1997), had no effect on the relaxation induced by anandamide. These results do not support a role for anandamide or another CB1 receptor-related mechanism in EDHF-mediated relaxations in the guinea-pig basilar artery.

EDHF-mediated responses in rat isolated mesenteric and hepatic, porcine and bovine coronary arteries also do not involve CB1 receptor stimulation (Plane *et al.*, 1997; Zygmunt *et al.*, 1997b; Chataigneau *et al.*, 1998b; Pratt *et al.*, 1998). Furthermore, anandamide-induced hyperpolarization in rat hepatic and mesenteric artery is dependent on an intact endothelium (Zygmunt *et al.*, 1997b; Chataigneau *et al.*, 1998b), whereas removal of the endothelium does not affect the anandamide-induced relaxation in the same blood vessels (White and Hiley, 1997; Zygmunt *et al.*, 1997b).

2-Arachidonoylglycerol and palmitoylethanolamide have been described as possible endocannabinoids. 2-Arachidonoylglycerol, which was originally isolated from intestinal tissue (Mechoulam *et al.*, 1995), is present in vascular endothelium (Sugiura *et al.*, 1998) and in brain tissue, where it activates CB1 receptors with a potency similar to that of anandamide (Stella, Schweitzer and Piomelli, 1997). These two endocannabinoids are unlikely EDHF candidates in the guinea-pig basilar artery, since neither compound produced a significant relaxation in the present study. Furthermore, the lack of effect of palmitoylethanolamide, which has a higher affinity for CB2 than CB1 receptors (Facci *et al.*, 1995), suggests that CB2 receptors do not mediate relaxation in the guinea-pig basilar artery. This is in agreement with a previous report showing that palmitoylethanolamide does not produce any hyperpolarization in the rat mesenteric artery (Chataigneau *et al.*, 1998b).

Conclusion

The results suggest that in the guinea-pig basilar artery, EDHF is not produced by an oxygen-dependent enzymatic step such as metabolization of arachidonic acid by cytochrome P450 monoxygenases. Although EDHF and NO both contribute to endothelium-dependent relaxations in the presence of oxygen, EDHF alone mediates these relaxations during severe hypoxia. Although anandamide may be an endogenous vasodilator of cerebral arteries, EDHF is not an endocannabinoid acting at CB1 or CB2 receptors in the guinea-pig basilar artery.

6 Endothelium-Dependent Relaxation and Hyperpolarization in the Guinea-Pig Coronary Artery: Role of Epoxyeicosatrienoic Acid

D.M. Eckman[@], V. Mutafova[*], N. Hopkins[*], C. McBride[*] and K.D. Keef[*]

Department of Physiology & Cell Biology, University of Nevada, Reno, Reno, NV 89557, USA; @Department of Pharmacology, University of Vermont, Burlington, VT 05446, USA Address for correspondence: Dr. Kathleen D. Keef, Department of Physiology & Cell Biology, 352, Anderson Building, University of Nevada School of Medicine, Reno, Nevada 89557, USA. ph: (702) 784-4302, fx: (702) 784-6903 e-mail: kathy@physio.unr.edu Dr. Delrae M. Eckman, Department of Pharmacology, Given Building, B-326, School of Medicine, University of Vermont, Burlington, Vermont 05405, USA. ph: (802) 656-8037, fx: (802) 656-4523, e-mail: eckman@salus.med.uvm.edu

The present study investigates the role of NO and the cytochrome P450 pathway in the endothelium-dependent response to acetylcholine in the guinea-pig coronary artery. Acetylcholine and Cys-NO both increased cyclic GMP levels. This effect was abolished by combined treatment with L-NMMA plus hemoglobin, while the relaxation and hyperpolarization to acetylcholine remained. Scorpion venom abolished the relaxation and hyperpolarization to acetylcholine in the presence of L-NMMA plus hemolysate, but did not reduce relaxations to Cys-NO nor did it block cyclic GMP production. Acetylcholine (1×10^{-6} M) relaxation was unchanged by the guanylate cyclase inhibitor 1H-[1,2,4]Oxadiazolo[4,3-a]quinoxalin-1-one (10^{-5} M) while the relaxation to sodium nitroprusside (1×10^{-6} M) was abolished. Acetylcholine, arachidonic acid and 11,12-epoxyeicosatrienoic acid (11,12-EET) each relaxed and hyperpolarized vessels. The arachidonic acid-induced relaxation was enhanced significantly by eicosatetranynoic acid (3×10^{-5} M) or cinnamyl-3,4-dihydroxy-(cyanocinnamate (2×10^{-5} M) plus 5,8,11-eicostriynoic acid (2×10^{-5} M) suggesting that arachidonic acid metabolism can generate contractile products via the lipoxygenase pathway. Iberiotoxin (1×10^{-7} M) did not inhibit acetylcholine-induced relaxations and hyperpolarizations but significantly reduced responses to arachidonic acid and 11,12-EET. 4-Aminopyridine (5×10^{-3} M) significantly reduced relaxation and hyperpolarization to acetylcholine, but not arachidonic acid and 11,12-EET responses. Apamin (1×10^{-6} M) glibenclamide (1×10^{-5} M) and $BaCl_2$ (5×10^{-5} M) were without effect on responses to acetylcholine, arachidonic acid or 11,12-EET. These results suggest that a component of the acetylcholine-induced response is entirely independent of the NO/guanylate cyclase /cyclic GMP pathway. They further suggest that the coronary endothelium releases a factor upon application of arachidonic acid which hyperpolarizes the smooth muscle. The similarity in sensitivity of pharmacological agents between arachidonic acid and 11,12-EET suggest that the factor is an EET. However, the pharmacological disparity between responses to acetylcholine versus those to 11,12-EET do not support the hypothesis that EETs represent the predominant factor(s) released by acetylcholine from the endothelium leading to NO- and prostaglandin-independent hyperpolarizations and relaxations.

KEYWORDS: cyclic GMP, arachidonic acid, epoxyeicosatrienoic acid, membrane potential, potassium channel, cytochrome P450, coronary artery

INTRODUCTION

Acetylcholine produces relaxation and hyperpolarization in the guinea-pig coronary artery that is due to the release of multiple vasoactive factors from the endothelium. Two of these factors are nitric oxide (NO) and prostacyclin (see review by Thiemermann, 1991). However, endothelium-dependent relaxation and hyperpolarization can still be elicited following blockade of the formation of NO, prostaglandin and cyclic GMP (Chen, Suzuki, and Weston 1988; Parkington *et al.*, 1993; Eckman *et al.*, 1994) suggesting that a third factor or group of factors is also released from the endothelium. The NO- and prostaglandin-independent response has been attributed to a mediator that activates potassium channels in the cell membranes of vascular smooth muscle leading to its hyperpolarization and relaxation. This factor has been termed *endothelium-derived hyperpolarizing factor* (EDHF; Chen, Suzuki, and Weston 1988; Feletou and Vanhoutte, 1988; Cohen and Vanhoutte, 1997).

In some blood vessels EDHF could be a metabolite of arachidonic acid formed by cytochrome P450 monooxygenases (Hecker *et al.*, 1994; Bauersachs, Hecker and Busse 1994; Campbell *et al.*, 1996). Such metabolism results in the formation of epoxyeicosatrienoic acids (EETs, Fitzpatrick and Murphy, 1989). There are four regioisomers of EETs: 5,6-, 8,9-, 11,12- and 14,15-epoxyeicosatrienoic acids. Exogenous application of EETs produce dilatation of pial arteries (Ellis *et al.*, 1990; Gebremedhin *et al.*, 1992) and coronary arteries of the cow, pig and dog (Rosolowsky *et al.*, 1990; Rosolowsky and Campbell, 1993; Campbell *et al.*, 1996; Graier *et al.*, 1996) at concentrations ranging from 1×10^{-7} to 1×10^{-4} M. In addition to relaxation, EETs also increase K^+ channel activity in smooth muscle of the cat cerebral artery, rabbit portal vein, rat caudal artery, guinea pig aorta, porcine coronary artery and bovine coronary arteries (Gebremedhin *et al.*, 1992; Hu & Kim, 1993 and Campbell *et al.*, 1996). EETs also elicit hyperpolarization of vascular smooth muscle cells of intact arteries (bovine coronary, Campbell *et al.*, 1996). These data taken together suggest that EETs represent a likely candidate for EDHF.

However, other studies do not support this view. Indeed, some cytochrome P450 blockers do not inhibit EDHF-induced responses (Corriu *et al.*, 1996a; Zygmunt *et al.*, 1996; Fukao *et al.*, 1997a; Ohlmann *et al.*, 1997) and EETs do not relax some blood vessels (Zygmunt *et al.*, 1996). Furthermore, EET- and EDHF-induced responses are blocked by different potassium channel blockers in some tissues (Fukao *et al.*, 1997a). Many of the cytochrome P450 blockers also inhibit responses to potassium channel openers (Graier *et al.*, 1996; Edwards *et al.*, 1996; Fukao *et al.*, 1997; Zygmunt *et al.*, 1996; Vanheel and Van de Voorde, 1997) making interpretation of results with these drugs difficult.

The responses attributed to EDHF might in fact be due to residual released NO following incomplete block of NO synthase (Cohen *et al.*, 1997). Thus, in rabbit carotid arteries, the concentration-response relationship for the NO donor, Sin-1 is similar for both the relaxation and hyperpolarization (Cohen *et al.*, 1997). In contrast, in the guinea-pig coronary artery, the concentration of NO-donors needed to elicit hyperpolarization is higher than that needed to cause relaxation (Parkington *et al.*, 1995).

The present study was undertaken to determine whether EETs can account for the acetylcholine-induced relaxation and hyperpolarization which occurs in the guinea-pig coronary artery in the presence of inhibitors of the production of NO and prostaglandins. Additional experiments were designed to explore the possibility that residual activity of the NO/guanylate cyclase/cyclic GMP pathway can account for the response observed.

METHODS

Tissue Preparation

Male albino guinea pigs (350–550 g) were killed by CO_2 inhalation followed by exsanguination. The heart was removed immediately, placed in cold (4 to 6°C) oxygenated Krebs solution (in mM): NaCl 118.5, KCl 4.7, $MgCl_2$ 1.2, $NaHCO_3$ 23.8, KH_2PO_4 1.2, dextrose 11, $CaCl_2$ 2.5 and aerated with 95% 0_2/5% CO_2. Segments (1–1.5 cm) of both the left descending and circumflex coronary arteries were dissected out and cleaned of all adhering fat and myocardial tissue. Rings (3 mm long and 200–300 μm in diameter) were prepared for use in either contractile or electrophysiological experiments.

Organ Chamber Experiments

Tissues were suspended onto two triangular tungsten wires (89 μm diameter) and hung vertically in an isolated organ chamber (either 3 or 10 ml volume). The bottom triangle was mounted to a stable hook while the top triangle was attached to a Grass (FT03c) strain gauge. The organ chamber contained Krebs solution and was maintained at 37°C.

A resting force of 0.3 g was applied. In preliminary experiments this was found to stretch the rings to near the optimal length for tension development. The arteries were equilibrated for 1 to 2 hours with 4 minute exposures to the histamine H1-receptor agonist 2-(2-aminoethyl)pyridine (1×10^{-3} M, Durant, Ganellin and Parsons 1975) or 60 mM KCl.

For experiments which tested the effects of inhibitors of cytochrome P450 or K$^+$-channel inhibitors on acetylcholine, arachidonic acid, lemakalim or 11,12-EET relaxations, each tissue served as its own control. Two sequential concentration-relaxation curves were obtained for a particular vasodilator before beginning an experiment. If reproducible responses were not obtained the tissue was discarded. Preliminary experiments showed that tissues that had comparable concentration relaxation-relationship with two sequential applications of a given vasodilator exhibited a similar relationship on a third application. Following control responses the tissues were exposed to either a cytochrome P450- or K$^+$-channel blocker for 20 to 30 minutes. A third concentration-response curve was then obtained in the presence of blocker. Only one blocker was tested per tissue. In experiments investigating the role of arachidonic acid and EETs, Krebs solution containing 1×10^{-4} M L-NA and 1×10^{-5} M indomethacin was used throughout the experiment. Previous experiments have determined that indomethacin (1×10^{-5} M) is without effect on acetylcholine-induced relaxation and hyperpolarization (data not shown).

Electrophysiological Studies

Segments (5 mm long and 200 to 250 μm diameter) of left circumflex or left descending coronary artery were used. The rings were mounted onto two parallel wires and a resting force of 0.3 g was applied to mimic the conditions present in the organ chambers. All intracellular measurements were made through the adventitial surface with microelectrodes filled with 3M KCl (resistance between 70 and 100 MΩ). Successful impalements were judged on the basis of a rapid drop in potential upon entering the cell, a low noise level and minimal change in the electrode resistance and zero potential before and after impalement. Signals were viewed on a digital oscilloscope (Hitachi) and stored on tape with a Vetter PCM Recording Adapter attached to a video cassette recorder.

Cyclic GMP

Measurements of cyclic GMP were made under three different conditions including basal levels, during contraction with 2-(2-aminoethyl)pyradine and during relaxations with either acetylcholine or Cys-NO. The arterial segments were flash-frozen in liquid nitrogen while still attached to the stainless steel triangles. Cyclic GMP was assayed with commercially available reagents (Caymen Chemical Company, Ann Arbor, MI, U.S.A.). Samples were prepared for assay by homogenization in 6% trichloroacetic acid with glass Duall tissue grinders followed by extraction with water-saturated diethyl ether. Aqueous phases were then lyophylized to dryness and resuspended in 1M potassium phosphate buffer (1 M, pH 7.4) before addition to duplicated microtiter plate wells. The cyclic GMP-levels in samples and standards were detected following acetylation and competition between cyclic GMP and acetylcholinesterase-linked cyclic GMP tracer for specific antiserum binding sites. The antiserum complex, linked to acetylcholinesterase, was used to cleave Ellman's reagent (5,5'-dithio-bis-(2-nitrobenzoic acid), and absorbance was measured at 412 nm. The cyclic GMP content of samples was determined from a standard curve constructed from determination of known amounts of cyclic GMP added to the plate. Values of cyclic GMP are expressed as pmol cyclic GMP mg^{-1} protein (determined by method of Bradford, 1976). Duplicated variation in the cyclic GMP assay was less than 3%.

Drugs Used

Acetylcholine HCl, iberiotoxin, scorpion venom (*Leiurus quinquestriatus habraeus*), proadifen (SKF-525A), clotrimazole, 17-octadecynoic acid (17-ODYA), 7-ethoxyresorufin, N^{ω}-nitro-L-arginine (L-NA), U46619, indomethacin , eicosatetranynoic acid (ETYA), 4-aminopyridine, arachidonic acid (dissolved in bicarbonate buffer) were purchased from Sigma (St. Louis, MO). N^{ω}-Monomethyl-L-arginine, HOAc salt was purchased from Calbiochem. Cinnamyl-3,4-dihydroxy-∝-cyanocinnamate (dissolved in bicarbonate buffer) and 5,8,11-eicostriynoic acid (dissolved in ethanol) were purchased from Biomol (Plymouth Meeting, PA). 1H-[1,2,4]Oxadiazolo[4,3-a]quinoxalin-1-one (made in DMSO) was purchased from Tocris (Ballwin, MO). 2-(2-aminoethyl)pyridine was purchased from Aldrich (Milwaukee, WI). 11,12-epoxyeicosatrienoic acid (11,12-EET) was supplied in 100% ETOH at a concentration of 1×10^{-1} M (Cayman Chemical, Ann Arbor, MI). Lemakalim was a generous gift from Smith Kline Beecham. Hemolysate was made according to the method of Bowman and Gillespie (1982) and was stored at 2°C for up to 3 days. Three percent volume of hemolysate solution was added to the bath to the organ chambers to reach a final concentration of approximately 2×10^{-5} M. Unless otherwise specified all compounds were made in distilled water and stored on ice throughout the experiment.

Statistics

Statistical significance was determined by the two-tailed paired or unpaired *t*-test. The changes were considered significant at p < 0.05. The response to vasodilators was determined as the percent reduction of 2-(2-aminoethyl)pyradine-induced contraction. Thus 100% relaxation is equivalent to complete reversal of the 2-(2-aminoethyl)pyradine-induced contraction. Vasodilators rarely reduced tone below the pre-contracted level of 0.3 g. Data are expressed as mean ± s. e. mean, *n* values reflect the number of animals studied. Because we were unable to obtain complete concentration-response relationships in the presence of some cytochrome P450- or K^+-channel inhibitors, these responses were analyzed using a repeated measures ANOVA (SASS). This technique allowed us to compare the whole curve rather than a specific point such as the traditional EC_{50} calculation.

RESULTS

NO/guanylate Cyclase/cyclic GMP Pathway

N^{ω}-Monomethyl-L-arginine, hemolysate and scorpion venom on contractile responses

In the presence of combined N^{ω}-Monomethyl-L-arginine (3.5×10^{-4} M) and hemolysate (2×10^{-5} M) the concentration-relaxation curve to acetylcholine was significantly shifted to the right, but not abolished (control IC50 -7.24 ± 0.08, n = 14; NO-block IC50 -6.67 ± 0.15, n = 5). Scorpion venom (8.7 µg/ml), a non-selective K^+ channel blocker (Strong, 1990), significantly shifted the concentration-response relationship of acetylcholine to the right (control, -7.14 ± 0.11, n = 9; scorpion venom, $-6.39 \pm .22$, n = 5). The acetylcholine-induced relaxation was abolished by the combination of N^{ω}-Monomethyl-L-arginine, hemolysate plus scorpion venom (n = 4). The relaxation produced with the NO-donor CysNO (3.5×10^{-9} M to 1×10^{-5} M, n = 12) was abolished by hemolysate but unchanged by scorpion venom (n = 4) (Eckman *et al.*, 1994).

Guanylyl cyclase inhibition and the production of cyclic GMP

The guanylate cyclase inhibitor ODQ (1×10^{-5} M) did not significantly reduce the relaxation to acetylcholine (1×10^{-6} M) in the presence of L-NA (1×10^{-4} M) and indomethacin (1×10^{-5} M) while the relaxation elicited by sodium nitroprusside (1×10^{-6} M) was reduced to less than 2.5% of control (Figure 6-1).

N^{ω}-Monomethyl-L-arginine alone decreased acetylcholine induced accumulations of cyclic GMP to levels similar to that of 2-(2-aminoethyl)pyridine alone (n = 5). The acetylcholine-stimulated accumulations of cyclic GMP were reduced to less than 1% of the levels measured under control conditions when both N^{ω}-Monomethyl-L-arginine and hemolysate were added (n = 6). Scorpion venom did not reduce cyclic GMP production (Figure 6-1, Eckman *et al.*, 1994).

N^{ω}-Monomethyl-L-arginine, hemolysate and scorpion venom on acetylcholine-induced hyperpolarization

Under control conditions, acetylcholine (3.5×10^{-7} M) hyperpolarized tissues by 16 ± 1 mV (n = 8). The hyperpolarizations elicited with acetylcholine in the presence of either N^{ω}-Monomethyl-L-arginine (3.5×10^{-4} M) alone or in combination with hemolysate (2×10^{-5} M) were not significantly different from control (i.e., N^{ω}-Monomethyl-L-arginine, 16 ± 1 mV, n = 5; N^{ω}-Monomethyl-L-arginine plus hemolysate, 17 ± 3, n = 3). In contrast, scorpion

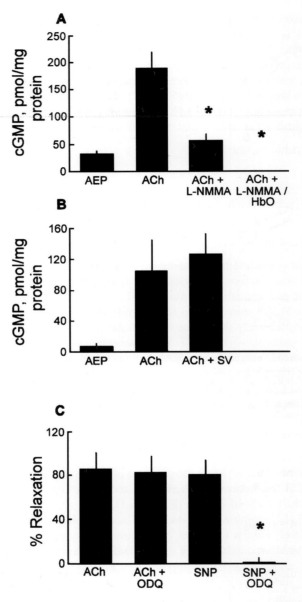

Figure 6-1. Role of the NO pathway in endothelium intact artery segments of the guinea-pig coronary (all experiments performed in the presence of the contractile agent, AEP). *A*: Effect of NO inhibitors on cyclic GMP accumulation. Shown are cyclic GMP levels with either 2-(2-aminoethyl)pyradine alone, acetylcholine alone, acetylcholine plus L-NMMA or acetylcholine plus combined L-NMMA and hemolysate. Asterisk indicates significant difference (P < 0.05) from acetylcholine response. *B*: Back of effect of crude scorpion venom on cyclic GMP accumulation. Shown are cyclic GMP levels with either 2-(2-aminoethyl)pyradine alone, acetylcholine alone or acetylcholine plus scorpion venom. *C*: Effect of the guanylate cyclase inhibitor 1H-[1,2,4]Oxadiazolo[4,3-a]quinoxalin-1-one (ODQ, 1×10^{-5} M) on acetylcholine (1×10^{-6} M) and sodium nitroprusside-(SNP, 1×10^{-6} M) induced relaxations. The ODQ experiments were performed in the presence of L-NA (1×10^{-4} M) and indomethacin (1×10^{-5} M). Asterisk indicates significant difference (P < 0.05) from control response.

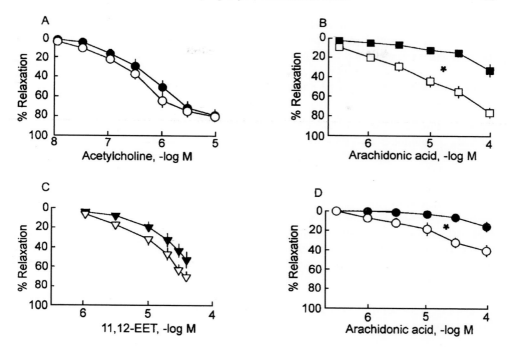

Figure 6-2. Comparison of acetylcholine (ACh; *A*), arachidonic acid (AA; *B*) and 11,12-epoxyeicosatrienoic acid (11,12-EET; *C*)-induced relaxations in the absence (filled symbol) and presence (open symbol) of ETYA $(1 \times 10^{-5}$ to 3×10^{-5} M) in the guinea-pig coronary artery. *D*: Effect of the combined lipoxygenase inhibitors cinnamyl-3,4-dihydroxy-∝-cyanocinnnamate (2×10^{-5}) and 5,8,11-eicostrienoic acid (2×10^{-5}) on arachidonic acid induced relaxations. L-NA $(1 \times 10^{-4}$ M) and indomethacin $(11 \times 10^{-5}$ M) present throughout. Shown are mean values ± s.e. mean. (n = 9–15). The asterisk indicates curves are statistically different from one another F > 0.05.

venom nearly abolished the acetylcholine-induced hyperpolarization (control 17.7 ± 1 mV, scorpion venom 1.1 ± 0.8, n = 4) (Eckman *et al.*, 1992; 1994).

Cytochrome P450 Pathway

Acetylcholine, arachidonic acid and 11,12-EET relaxations

The effects of cumulative additions of acetylcholine $(1 \times 10^{-8}$ M to 1×10^{-5} M), arachidonic acid $(1 \times 10^{-7}$ M to 1×10^{-4} M) and 11,12-EET $(1 \times 10^{-6}$ M to 4×10^{-5} M) were measured in guinea-pig coronary artery contracted with 2-(2-aminoethyl)pyridine (Figure 6-2). Both acetylcholine and 11,12-EET elicited concentration-dependent relaxations. The response to arachidonic acid was significantly smaller [the highest concentration of arachidonic acid tested $(1 \times 10^{-4}$ M) produced only a 30% relaxation]. ETYA $(3 \times 10^{-5}$ M) significantly increased the amplitude of relaxation induced with arachidonic acid and shifted the con-centration-response relationship to the left (Figure 6-2). ETYA did not significantly affect the response to acetylcholine or 11,12-EET (Figure 6-2). Because of the marked increase in arachidonic acid-induced response, ETYA was included in the remaining contractile experiments with arachidonic acid (Eckman *et al.*, 1998). The combination of cinnamyl-3,4-dihydroxy-∝cyanocinnamate plus 5,8,11-eicostriynoic acid also significantly potentiated the response to arachidonic acid (n = 6, Figure 6-2).

K+-channel inhibitors on acetylcholine, arachidonic acid and 11,12-EET induced relaxations

Concentration-response curves were determined for acetylcholine (1×10^{-8} M to 1×10^{-4} M), arachidonic acid (3×10^{-7} M to 3×10^{-4} M) and 11,12-EET (1×10^{-7} M to 4×10^{-5} M) in the presence and absence of iberiotoxin (1×10^{-7} M), 4-aminopyridine (5×10^{-3} M), apamin (1×10^{-6} M), glibenclamide (1×10^{-5} M) and $BaCl_2$ (5×10^{-5} M). Combined iberiotoxin plus 4-aminopyridine was also tested as well as a combination of all five blockers. Iberiotoxin had no significant effect on acetylcholine-induced relaxations whereas 4-AP significantly shifted its concentration-response curve to the right (Figure 6-3). Responses to arachidonic acid and 11,12-EET were shifted significantly to the right by iberiotoxin while 4-AP had no effect (Figure 6-3). Apamin, glibenclamide and $BaCl_2$ were without significant effect on the responses to the three vasodilators (data not shown, Eckman *et al.*, 1998).

K+-channel inhibitors on acetylcholine, arachidonic acid and 11,12-EET induced hyperpolarizations

Upon obtaining a stable microelectrode impalement, the tissues were exposed to acetylcholine (1×10^{-6} M), arachidonic acid (3×10^{-5} M) or 11,12-EET (3.3×10^{-5} M) for 2.5 to 3.0 minutes. Acetylcholine hyperpolarized cells by 26.4 ± 1.5 mV (n = 12) compared to 10.7 ± 1.2 mV (n = 10) for arachidonic acid and 15.6 ± 1.2 mV (n = 10) for 11,12-EET (Figure 6-4, Table 6-1). Iberiotoxin (1×10^{-9} M) did not significantly depolarize the membrane potential (-41.1 ± 1.6 mV vs -38.5 ± 1.5 mV respectively, n = 9, Table 6-1) but reduced the hyperpolarizations elicited by either 11,12-EET or arachidonic acid. The acetylcholine-induced hyperpolarization was not significantly affected by iberiotoxin. 4-aminopyridine (5×10^{-3} M) had no significant effect on resting membrane potential (-42.1 ± 0.5 versus -40.1 ± 1.3 mV, n = 16, Table 6-1), however it significantly reduced the acetylcholine-induced hyperpolarization while having no significant effect on the electrical response to arachidonic acid and 11,12-EET (Figure 6-4).

DISCUSSION

The endothelium-dependent hyperpolarization observed in some blood vessels in response to agonists such as acetylcholine and bradykinin, in the presence of inhibitors of NO synthase, may be due either to an incomplete blockade of the production of NO (Cohen *et al.*, 1997) or to a metabolite of AA produced by cytochrome P450 (Hecker *et al.*, 1994, Bauersachs, Hecker and Busse 1994, Campbell *et al.*, 1996). The present study re-addresses the role of NO/guanylate cyclase/cyclic GMP pathway in the EDHF induced response as well as investigates whether the cytochrome P450 pathway could account for the acetylcholine-induced relaxation and hyperpolarization which occurs in the presence of NO synthase and prostaglandin synthesis inhibitors. Our results suggest that acetylcho-

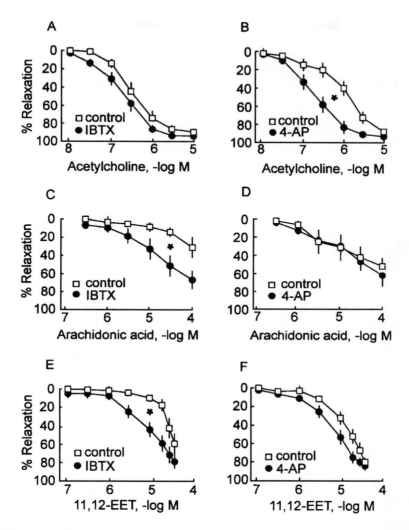

Figure 6-3. Effect of various K$^+$ channel blockers on concentration-relaxation curves to acetylcholine (*parts A and B*), arachidonic acid (*parts C and D*) and 11,12-epoxyeicosatrienoic acid (*parts E and F*) in the guinea-pig coronary artery. Responses to vasodilators were obtained in the absence (filled symbols) and presence (open symbols) of iberiotoxin (1 × 10^{-7} M; A,C,E) and 4-aminopyridine (4-AP, 5 × 10^{-3} M; B,D,F). L-NA (1 × 10^{-4} M) and indomethacin (1 × 10^{-5} M) were present throughout. Shown are mean values ± s.e. mean (n = 3–11). The asterisk indicates curves statistically different from one another F > 0.05.

line is capable of eliciting relaxation and hyperpolarization independent of the NO/guanylate cyclase/cyclic GMP pathway and that EET formation alone is insufficient to explain much of the action of acetylcholine in this vessel.

Acetylcholine elicited a relaxation and hyperpolarization in the presence of inhibition of NO synthase and cyclooxygenase. Although measurable levels of cyclic GMP are still present under these circumstances the increase in cyclic GMP is reduced from control

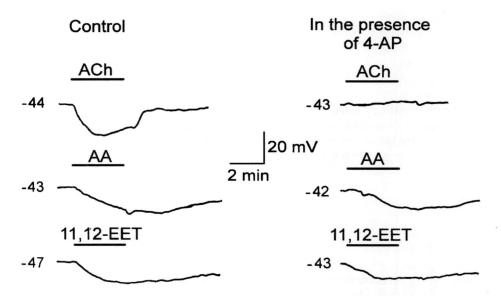

Figure 6-4. Comparison of acetylcholine (1×10^{-6} M), arachidonic acid (3×10^{-5} M) and 11,12-epoxyeicosatrienoic acid (11,12-EET, 3.3×10^{-5} M)-induced hyperpolarizations in the absence (control) or presence of 4-aminopyridine (4-AP, 5×10^{-3} M) in the guinea-pig coronary artery. Acetylcholine, arachidonic acid and 11, 12 EET each applied for three minutes. L-NA (1×10^{-4} M) and indomethacin (1×10^{-5} M) present throughout. 4-AP reduced the acetylcholine-induced hyperpolarization while leaving the responses to arachidonic acid and 11, 12 EET intact (see Table 6-1 for statistics).

Table 6-1 Effect of the K+-channel inhibitors on acetylcholine, arachidonic acid and 11,12-EET-induced hyperpolarizations in the guinea pig coronary artery

Experimental condition	Control RMP	Control Hyperpol.	Exp. RMP	Exp. Hyperpol.
4-AP (5×10^{-3} M)				
ACh (1×10^{-6} M, n = 7)	-42.1 ± 0.8	25.3 ± 2.1	-42.5 ± 1.9	$2.5 \pm 1.5^*$
AA (3×10^{-5} M, n = 4)	-42.8 ± 1.6	11.5 ± 2.7	-36.1 ± 2.6	11.8 ± 3.5
11,12-EET (3.3×10^{-5} M, n = 5)	-42.1 ± 1.0	16.3 ± 1.2	-40.0 ± 1.8	12.2 ± 2.1
IBTX (1×10^{-4} M)				
ACh (1×10^{-6} M, n = 3)	-43.0 ± 2.5	27.0 ± 2.3	-37.1 ± 0.5	31.3 ± 2.9
AA (3×10^{-5} M, n = 3)	-42.3 ± 1.4	12.0 ± 1.5	-38.6 ± 2.0	$6.3 \pm 1.9^*$
11,12-EET (3.3×10^{-5} M, n = 3)	-38.0 ± 4.0	12.3 ± 2.4	-37.0 ± 2.0	$3.6 \pm 1.9^*$

RMP	Resting membrane potential
Exp.	In the presence of K+ channel blocker
Hyperpol.	Hyperpolarization
4-AP	4-aminopyridine
IBTX	Iberiotoxin
ACh	Acetylcholine
AA	Arachidonic acid
11,12-EET	11,12-epoxyeicosatrienoic acid

Resting membrane potential and hyperolarization measurements are expressed as mean ± s.e.mean.
*significantly different from control at $P < 0.05$

suggesting that the production of NO is also decreased. In spite of the reduced production of cyclic GMP the hyperpolarization elicited with acetylcholine was unchanged from control. If the hyperpolarization were due to NO one might expect some kind of concentration-dependency, i.e., less hyperpolarization with less production of NO. Indeed, the acetylcholine-induced hyperpolarization was also unchanged when the cyclic GMP production was reduced to less than 1% of basal levels by combining N^{ω}-Monomethyl-L-arginine with the NO scavenger hemolysate. In the guinea-pig coronary artery large concentrations of NO-donors are required to elicit any hyperpolarization of the muscle at all (Eckman *et al.*, 1994; Parkington *et al.*, 1995). This makes it highly improbable that the amount of NO which is present with the combination of N^{ω}-Monomethyl-L-arginine and hemolysate could elicit any measurable hyperpolarization of the tissue. Similar conclusions were reached from the experiments with the guanylate cyclase inhibitor ODQ which abolished relaxations to the NO-donor sodium nitroprusside, but was without effect on acetylcholine-induced relaxation. The results with scorpion venom also provide evidence that a separate NO-independent pathway is involved in the actions of acetylcholine. Scorpion venom, which contains a number of components in addition to K^+ channel blockers, clearly can not be attributed to a specific block. However, scorpion venom abolished the acetylcholine-induced hyperpolarization, while having negligible effects on the NO/guanylate cyclase/cyclic GMP pathway. This conclusion was reached since relaxation with the NO-donor Cys-NO was unchanged in the presence of scorpion venom as was the acetylcholine-induced elevation in cyclic GMP levels. Finally, the effects of scorpion venom and NO-blockade were additive, i.e., each alone produced only a small shift in the concentration-relaxation relationship to acetylcholine while together the response was virtually abolished. All of these observations taken together strongly argue that a factor(s) separate from NO is released from the endothelium, which hyperpolarizes and relaxes the smooth muscle.

To investigate the possible contribution of products of the cytochrome P450 pathway to the EDHF-response, the actions of 11,12-EET were compared to those of arachidonic acid and acetylcholine. The response to 11,12-EET mimicked that of acetylcholine in several ways. Of particular relevance, 11,12-EET hyperpolarized and relaxed blood vessels. EETs have vasodilating properties (Ellis *et al.*, 1990; Pfister, Falck and Campbell 1991; Gebremedhin *et al.*, 1992; Rosolowsky & Campbell, 1993; Hecker *et al.*, 1994; Campbell *et al.*, 1996) and cause hyperpolarization (Campbell *et al.*, 1996; Fukao *et al.*, 1997a). Arachidonic acid, the putative precursor molecule for EET formation relaxed and hyperpolarized the guinea-pig coronary artery, which further supports the proposed role of EET as an EDHF. Although a number of studies have described the vasodilatory properties of arachidonic acid (Rosolowsky & Campbell, 1993; Lonigro *et al.*, 1994; Campbell *et al.*, 1996) fewer have measured its effect on cell membrane potential. Arachidonic acid (4×10^{-6} M) has previously been reported to hyperpolarize the GPCA in an indomethacin-sensitive manner (Parkington *et al.*, 1993). The present study found that a higher concentration of arachidonic acid (3×10^{-5} M) was required to hyperpolarize cells in the presence of indomethacin. Taken together these results suggest the addition of arachidonic acid may lead to formation of both cyclooxygenase as well as epoxygenase products, which can hyperpolarize vascular smooth muscle. The contractile and electrical characteristics of 11,12-EET and arachidonic acid in the GPCA are thus in keeping with the proposed role of EET as an EDHF. Likewise, the endothelium-dependence of arachidonic

D.M. Eckman et al.

acid-induced relaxations in the coronary artery and the endothelial independence of the response to EET (Rosolowsky *et al.*, 1990; Campbell *et al.*, 1996) favor that interpretation. Similar effects have been obtained in the coronary artery of monkey (n = 3, unpublished observation). Although 11,12-EET mimicked the electrical and contractile behavior of EDHF in the guinea-pig coronary artery the potency of 11,12-EET was less than that reported for either the bovine or canine coronary artery (Campbell *et al.*, 1996; Rosolowsky *et al.*, 1990) suggesting that EETs may play a greater role in canine and bovine coronary arteries than in the guinea-pig coronary artery.

An enhancement of arachidonic acid-induced relaxations occurred when ETYA was included in the bathing solution. ETYA blocks cyclooxygenase (IC_{50} 8×10^{-6} M), lipoxygenase (IC_{50} 2×10^{-7} to 1×10^{-5} M; Salari, Braquet and Borgeat, 1984; Bokoch and Reed, 1981; Tobias & Hamilton, 1979; Moncada, Flower and Vane, 1985) and at higher concentrations EET-production (IC_{50} 4×10^{-5} to 1×10^{-4} M; Capdevila *et al.*, 1988; Revtyak, Johnson and Campbell, 1988). The present results suggest that the predominant effect of ETYA was to block the lipoxygenase pathway. This pathway metabolizes arachidonic acid to at least three vasoconstrictor compounds; leukotrienes (Badr, Brenner and Ichikawa, 1987), 12-HETE (Ma *et al.*, 1991; Saito *et al.*, 1992) and 15-HETE (Van Diest, Verbeuren, and Herman, 1991). Simultaneous release of one or more of these contractile substances along with EETs would be predicted to limit the relaxing action of the latter. The results with the more selective lipoxygenase inhibitors cinnamyl-3,4-dihydroxy-α-cyanocinnamate and 5,8,11-eicostriynoic acid (Cho *et al.*, 1991, Salari, Braquet and Borgeat, 1984, Blomgren, Hammarstrom and Wasserman, 1987) also support the notion that the lipoxygenase pathway contributes vasoconstrictor products when arachidonic acid is added.

Both EETs (Hecker *et al.*, 1994, Hu & Kim, 1994) and EDHF (for review see Brayden, 1993) lead to activation of K^+ channels in vascular smooth muscle. K^+ channel blockers were found to have similar effects on arachidonic acid and 11,12-EET responses providing additional evidence for the proposed endothelial conversion of arachidonic acid to EETs (Rosolowsky *et al.*, 1990; Gebremedhin *et al.*, 1992; Campbell *et al.*, 1996). However, since the experiments were performed in the presence of indomethacin the observed effects may overestimate the activity of the epoxygenase pathway, since, indomethacin will shift arachidonic acid away from the cyclooxygenase toward the epoxygenase pathway.

In contrast to the similarities observed between arachidonic acid and 11, 12-EET, there were substantial differences in the potency of K^+ channel blockers on responses to acetylcholine. Iberiotoxin which blocks large conductance Ca^{2+} activated K^+ channels (BKCa) produced a significant reduction of the 11,12-EET and arachidonic acid-induced relaxation and hyperpolarization, in agreement with previous studies (Hu and Kim, 1993; Campbell *et al.*, 1996; Li, Zou and Campbell, 1997). In contrast, iberiotoxin was without effect on acetylcholine responses while 4-aminopyridine, which blocks delayed rectifier type K^+ channels (Kv), reduced relaxation and hyperpolarization while having no effect on responses to 11,12-EET and arachidonic acid. These data do not support the hypothesis that EETs represent the predominant compounds released from the endothelium to hyperpolarize the smooth muscle.

Little or no inhibition of EDHF-induced responses have been reported when low concentrations of 4-aminopyridine (i.e., 1×10^{-3} M or less) were used (e.g., Petersson *et al.*, 1997). In isolated rabbit coronary artery cells 4-aminopyridine (3×10^{-3} M) is required for complete inhibition of Kv currents (Ishikawa, Eckman and Keef, 1997).

Hence, 1×10^{-3} M 4-aminopyridine may not be sufficient to entirely block Kv currents in intact tissues. In the basilar artery of the guinea-pig the EDHF-induced relaxation is reduced by the Kv blocker ciclazindol and abolished when ciclazindol is combined with apamin (Petersson, Zygmunt and Hogestatt, 1997). Partial inhibition of the relaxation was also observed when 4-aminopyridine (1×10^{-3} M) was combined with apamin. The findings suggested that either two channels are involved in the actions of EDHF or alternatively that a single channel, structurally related to Kv and allosterically regulated by apamin is a target for EDHF (Petersson, Zygmunt and Hogestatt, 1997). A similar study in the rat hepatic artery suggested that the target K-channel for EDHF was neither Kv nor BKCa but rather was structurally related to both (Zygmunt *et al.*, 1997a). Thus, in both of these studies a "Kv-like" conductance may be activated by EDHF. It is possible that a similar conductance is activated by EDHF in the guinea-pig coronary artery.

In some blood vessels the NO- and prostaglandin-independent responses can be blocked by a combination of charybdotoxin plus apamin while addition of either blocker alone is ineffective (Corriu *et al.*, 1996a; Zygmunt and Hogestatt, 1996; Petersson, Zygmunt and Hogestatt, 1997; Zygmunt *et al.*, 1997a). Iberiotoxin plus apamin does not produce the same effect as charybdotoxin plus apamin (Petersson, Zygmunt and Hogestatt, 1997; Zygmunt *et al.*, 1997a). Since Kv channels are also inhibited by charybdotoxin (e.g., Chandy and Gutman, 1995) the actions of charybdotoxin may involve inhibition of Kv. Preliminary studies in the guinea-pig coronary artery also show that the actions of charybdotoxin and iberiotoxin differ (unpublished observation). Since iberiotoxin is the more selective blocker of BKCa all subsequent experiments in the present study were carried out with iberiotoxin. However, this observation again suggests that the EDHF-induced response in the guinea-pig coronary artery has similarities to the responses observed in the rat hepatic artery and guinea-pig basilar artery (Petersson, Zygmunt and Hogestatt, 1997; Zygmunt *et al.*, 1997a).

4-aminopyridine sensitive channels are present on both the smooth muscle (Volk and Shibata, 1993) and the endothelium (Chen and Cheung, 1992). Because of this complication it is not possible to definitively identify the site of action of the drug. This question can only be adequately addressed with donor-recipient style experiments in which 4-aminopyridine is applied exclusively to the recipient smooth muscle. However, in spite of the uncertainties regarding 4-aminopyridine, the marked difference in sensitivity of acetycholine versus 11,12-EET to iberiotoxin argues that the predominant hyperpolarizing factor released by acetylcholine is not an EET. Since iberiotoxin produced some inhibition of the ACh-induced relaxation when combined with 4-AP the cytochrome P450 pathway can not be entirely be excluded as a possible source of EDHFs, but the present results suggest that another factor(s) plays an important role.

7 Hyperpolarizing Factors and Membrane Potential in Cerebral Arterioles: A Role for Astrocytes and Cytochrome p450 Genes

David R. Harder, Richard J. Roman, Raymond C. Koehler[†], Nabil Alkayed[†], Debebe Gebremedhin, Diane Munzenmaier, Meetha Medhora and Jayashree Narayanan

The Cardiovascular Research Center and Department of Physiology, Medical College of Wisconsin, Milwaukee, Wisconsin, USA, and
[†]Department of Anesthesiology and Critical Care Medicine, Johns Hopkins University, School of Medicine, Baltimore Maryland, USA
Address for correspondence: David R. Harder, Ph.D., Cardiovascular Research Center, Medical College of Wisconsin, 8701 Watertown Plank Road, Milwaukee, WI 53226. Phone: (414) 456-5611, Fax: (414) 456-6515, dharder@mcw.edu

Neuronal activity relies almost exclusively on oxidative metabolism and is, therefore, critically dependent upon nutritive cerebral blood flow sufficient to maintain adequate substrate (oxygen) levels. Under a baseline level of metabolic activity there occurs redistribution of blood flow to regions of increased neural activity, i.e., upon movement of a limb blood flow increases to those neurons which control movement of that limb. This process is referred to as functional hyperemia. This short communication will describe novel mechanisms through which functional hyperemia might occur based on regulation of membrane potential by metabolites of arachidonic acid. We have cloned and sequenced a cDNA in rat cortex and astrocytes which is homologous to cytochrome P450 (P450) 2C11 previously identified in liver. The P450 2C11 encodes for an enzyme which catalyzes epoxygenation of arachidonic acid across the unsaturated carbon double bonds forming 5,6-, 8-9-, 11-12-, or 14-15 epoxyeicosatrienoic acids (EETs). These EETs are potent dilators of cerebral arteries, and act to hyperpolarize vascular muscle by increasing outward K^+ current. Pharmacological inhibition of P450 epoxygenase activity results in a 30% reduction in cerebral blood flow. Glutamate induces release of EETs from cultured astrocytes. Infusion of glutamate into the parietal cortex of experimental animals induces a dilatation which is inhibited by P450 antagonists. Similarly, a large portion of the *in vivo* vasodilator response to neuronal activation in anesthetized rats is blocked by inhibitors of P450 epoxygenase activity. These findings form the underpinnings of a hypothesis that release of glutamate during neuronal activity stimulates release of P450 derived EETs from astrocytes situated between neurons and arterioles which dilate these microvessels , thereby, increasing nutritive blood flow. In addition, at normal physiological pressures cerebral arterial muscle is depolarized as part of the mechanism through which cerebral blood flow is autoregulated. Recent findings suggest that nitric oxide (NO) hyperpolarizes cerebral arterial muscle both directly, and by binding to the heme portion and inhibiting P450 enzymes which catalyze formation of metabolites which are partly responsible for autoregulation of cerebral blood flow.

KEYWORDS: astrocytes, astrocytic P450, membrane potential, epoxides, autoregulation, functional hyperemia

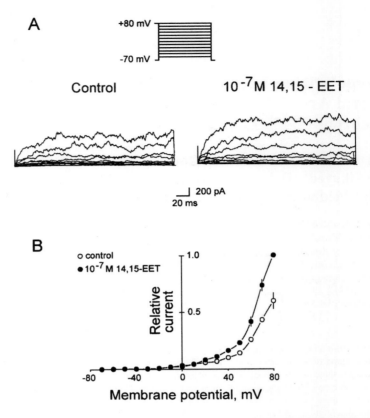

Figure 7-1. A) Whole-cell K$^+$ current recordings in single smooth muscle cells enzymatically dispersed from rat cerebral microvessels. Pulse protocol and original tracings of whole-cell K$^+$ currents under control conditions and after application of 100 nmol/L 14,15-EET to the bath solution. **B)** Summary of the effect of 14,15-EET (100×10^{-6}M) on control peak K$^+$ currents elicited over a range of membrane potentials. Relative current at each voltage is calculated as a percentage of the peak current at +80 mV.

FORMATION OF DILATOR EPOXIDES BY ASTROCYTES RELEASE OF EPOXIDES BY GLUTAMATE

A P450 2C11 cDNA has been cloned and sequenced from rat astrocytes which encodes for an enzyme catalyzing epoxygenation of arachidonic acid. The epoxygenation occurs across the double bonds of arachidonic acid leading to formation of 5,6-, 8,9-, 11,12-, and 14,15 epoxyeicosatrienoic acid (EETs). All of the EETs are dilators of cerebral arteries, however, those with the greatest potency are 11,12-, and 14,15-EETs. Both 11,12-, and 14,15-EETs dilate cerebral arteries, in part, by increasing the activity of the large conductance, Ca^{2+}-activated, K$^+$ channel (K$_{Ca}$) leading to membrane hyperpolarization (Alkayed *et al.*, 1996b, 1997; Harder, Campbell and Roman, 1995a). All four regioisomers of EETs are formed in homogenates of whole brain (Alkayed *et al.*, 1996b, 1997, 1996a). In cultured rat astrocytes the predominant P450 derived epoxides are 8, 9-, 11,12-, and 14,15-EET (Harder, Campbell and Roman, 1995a; Alkayed *et al.*, 1996a). 14,15-EET potently increase the activity of K$_{Ca}$ when studied in the whole-cell mode (Figure 7-1). When astrocytes are incubated with ^{14}C-arachidonic acid and stimulated with exogenous glutamate, there is a rapid (shortest time tested was 2 min) and significant release of EETs into the bathing media (Alkayed *et al.*, 1997).

Astrocytes are anatomically located between neurons and the microvasculature. There is close juxtaposition of astrocytic foot processes to arterioles and capillaries of the brain. Glutamate is the most extensively distributed excitatory amino acid neurotransmitter in the cerebral cortex (Heistad and Kontos, 1983; Kuschinsy, 1987; Hansson and Rönnbäck, 1995). While glutamate has no direct action on cerebral vascular smooth muscle, it induces dilatation when applied intracranially. This dilatation is blocked by pharmacological inhibitors of P450 epoxygenase (Alkayed *et al.*, 1996a, 1997; Harder, Campbell and Roman, 1995a), and by subdural infusion of antisense oligonucleotides against P450 2C11 (Harder, Campbell and Roman, 1995a). These data form the basis for the hypothesis that astrocytes may act as an intermediary cell type which monitors neuronal activity and shunts blood flow to match that activity. A recent review summarizes this hypothesis (Harder, Campbell and Roman, 1995a).

ARTERIOLAR MEMBRANE POTENTIAL IS THE FINAL CONTROLLER REGULATING BLOOD FLOW TO METABOLICALLY ACTIVE NEURONS

One of the hallmark characteristics of the cerebral circulation is the ability to autoregulate blood flow despite wide fluctuations in arterial pressure. The mechanism of this autoregulatory behavior is at the level of the vascular muscle cell. Studies over the past decade have extended studies initiated over 90 years ago demonstrating that isolated arteries contract actively, and reduce their diameter in response to elevation of arterial pressure (Bayliss, 1902; Harder, 1984). This pressure-induced activation is accompanied by depolarization of the plasma membrane together with activation of phospholipase C and protein kinase C (Laher *et al.*, 1989; Narayanan *et al.*, 1994; Osol, Laher and Kelly, 1993). A number of communications have identified a family of P450 cDNAs in arterial muscle which code for enzymes catalyzing ω-hydroxylation of arachidonic acid and formation of 20-hydroxyeicosatetraenoic acid (20-HETE) (Gebremedhin *et al.*, 1998; Harder, Campbell and Roman, 1995a; Lange *et al.*, 1997). These ω-hydroxylase enzymes are encoded by the 4A family of P450 cDNAs, of which P450 4A1, 4A2 and 4A3 have been identified in arterial muscle (13,15,16). 20-HETE is a potent activator of small cerebral arteries (ED_{10} around 10^{-11}M). The action of 20-HETE includes membrane depolarization via inhibition of K_{Ca} and activation of L-type Ca^{2+} channels (13,15). One of the intracellular mechanisms of action of 20-HETE is activation of protein kinase C (Figure 7-2) (Lange *et al.*, 1997). The protein kinase C isoforms activated by 20-HETE inhibit K_{Ca} channels by phosphorylation of the α subunit (Lange *et al.*, 1997). Inhibition of the K_{Ca} channel is necessary for the maintenance of arterial muscle depolarization in the face in high intracellular Ca^{2+}. In that the conductance of this channel is around 200pS, only a relatively small number of these channels need to be activated to bring the membrane potential of vascular smooth muscle cells to the K^+ equilibration potential. Indeed, the more K_{Ca} channel activity is reduced, the greater the myogenic response to pressure (Brayden and Nelson, 1992). However, some K^+ channel activity must still occur during pressure-induced activation to serve as a "brake", not allowing the membrane

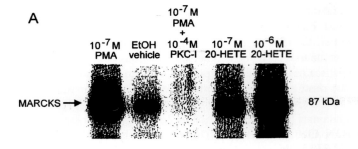

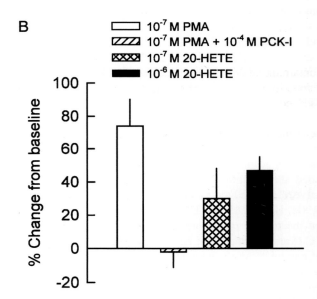

Figure 7-2. 20-HETE increases 87 kDa MARCKS phosphorylation in intact cat cerebral VSMC. A) Representative autoradiogram depicting an increase in 87 kDa MARCKS phosphorylation following administration of phorbol ester (100×10^{-9}M) and 20- HETE (100×10^{-9}M and 1×10^{-6}M) or vehicle (EtOH). The phorbol ester induced increase in MARCKS phosphorylation was abolished by the pseudosubstrate protein kinase C inhibitor (Myr-ΨPKC-I$_{(19-27)}$; 100×10^{-6}M). **B)** Bar graphs showing summary of data from six such experiments. Data are presented as percent change in MARCKS phosphorylation from vehicle treated levels. Asterisks denote significant difference from control ($P < 0.05$). Vertical bars denote SEM.

potential to become so small as to induce a level of activation which would close the blood vessel down.

From the above it is apparent that the membrane potential is a common factor in determining the active state of cerebral arteries and cerebral blood flow, and that P450 enzymes are the progenitor of a number of endogenous metabolites which control membrane potential of the cell. Cerebral arterial muscle depends upon the influx of extracellular

Ca^{2+} for the regulation of contractile elements. The entry of Ca^{2+} in cerebral vascular muscle largely rests on movements through voltage sensitive L-type Ca^{2+} channels (Pearce and Harder, 1996). Given that the cerebral circulation is surrounded by a variety of electrically excitable cell types, it is not difficult to understand that, by being sensitive to small changes in membrane potential, the cerebral vasculature can be responsive to a variety of enviromental conditions which act by modulating conductances.

REGULATION OF MEMBRANE POTENTIAL THROUGH NITRIC OXIDE: INHIBITION OF THE DEPOLARIZING ACTION OF 20-HETE

The mechanism of action of NO is generally thought to involve activation of soluble guanylate cyclase and formation of cyclic GMP (Cohen and Vanhoutte, 1995; Loscalzo and Welch, 1995). While controversial, there are a number of reports that NO acts to increase K$^+$ channel activity and hyperpolarize arterial muscle via a cyclic GMP in dependent mechanism (Cohen and Vanhoutte, 1995). Another action of NO is to bind to and inhibit heme containing enzymes (Sewer, Koop and Morgan, 1996, 1997). Most of this work has been done in hepatic xenobiotic P450 systems. NOS1 (bNOS) and NOSlll (eNOS) are expressed constitutively in the brain. bNOS immunoreactivity is found in neurons, astrocytes and autonomic nerve fibers surrounding cerebral arteries (Loesch and Burnstock, 1996; Schmidt *et al.*, 1992). eNOS protein is expressed in the endothelium of

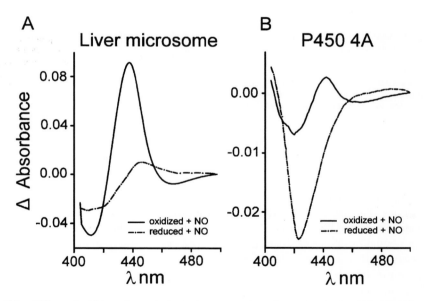

Figure 7-3. Effects of addition of an NO donor that generated an NO concentration of 0.25×10^{-6} M in the cuvette on the visible light spectrum of oxidized and reduced microsomes prepared from the liver of a rat and a recombinant P4504A2 protein expressed in insect cells using a baculo virus system. Samples were reduced with sodium dithionite. The results indicate that liver microsomes (a rich source of P450 proteins) and the recombinant P4504A2 protein binds to NO and increase absorbance at 440 nm.

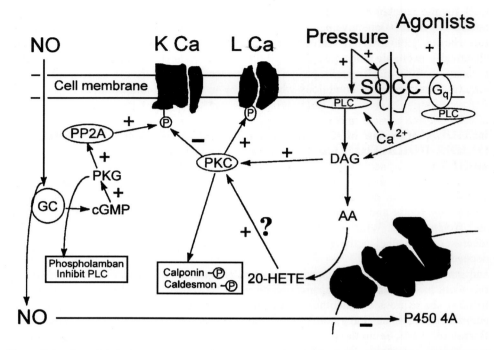

Figure 7-4. From left to right, pressure and/or stretch activates phospholipases (the present studies focused on phospholipase C) to generate diacylglycerol (DAG) and arachidonic acid. Arachidonic acid can be metabolized by membrane-bound P4504A protein catalyzing formation of 20-HETE. 20-HETE then activates protein kinase C phosphorylation of ion channels, and in the case of large conductance, calcium activated potassium channel (K_{Ca}) reducing its activity, while enhancing Ca^{2+} entry through L-type Ca^{2+} channels. These ionic mechanisms depolarize and activate arteriolar smooth muscle which is maintained as long as the pressure stimulus is applied. Activation of endothelial or parenchymal nitric oxide synthase generates NO which inhibits this pressure-induced vasoactive process by 1) increasing cyclic GMP and 2) inhibiting P4504A enzyme, reducing endogenous 20-HETE formation which enhances K_{Ca} and reduces inward Ca^{2+} current, thereby dilating arterioles. There appears to be a balance of ion channel phosphorylation between cyclic GMP-mediated PP2A activity and protein kinase C activation; NOS and P4504A activity are critical to the regulation of this process. (Definitions: SOCC-stretch operated cationic channels; PLC-phospholipase C; PKC-protein kinase C; PKG-protein kinase G; PP2A-phosphoprotein phosphatase 2A; GC-guanylate cyclase; K_{Ca}-large conductance Ca^{2+}-activated K^+ channel; Lca-L type Ca^{2+} channel diacylglycerol).

the cerebral microcirculation (Loesch and Burnstock, 1996) and in pyramidal cells of the hippocampus (Dinerman *et al.*, 1994). Inhibition of NOS lowers resting cerebral blood flow (Iadecola *et al.*, 1994; Iadecola and Zhang, 1994). Thus, the tonic release of NO plays a role in regulation of cerebral vascular tone. The conclusion that an elevation of cyclic GMP levels mediates the vasodilator effects of NO is based on pharmacological studies with inhibitors of guanylate cyclase (Iadecola *et al.*, 1994; Iadecola and Zhang, 1994). However, an increasing number of studies show that a significant portion of the dilator action of NO is cyclic GMP independent (Wong, Roman and Fleisch, 1995; Wang *et al.*, 1998).

Preliminary data suggest that NO donors selectively activate K_{Ca} in vascular smooth muscle from small cerebral arteries. This NO-dependent activation of K_{Ca} was not sensitive to ODQ, and roughly 50% of the dilatation of pressurized cerebral arteries to NO was ODQ sensitive (Alonso-Galacia *et al.*, 1997). Furthermore, NO donors dose-dependently block the formation of 20-HETE in pressurized arterial preparations (Alonso-Galacia *et al.*, 1997). NO reversibly binds to recombinant P450 4A ω-hydroxylase (the enzyme catalyzing formation of 20-HETE from AA) resulting in a spectral phase shift (Figure 7-3). These data demonstrate that NO can bind to, and inhibit P450 ω-hydroxylase and the formation of 20-HETE. Given that 20-HETE inhibits the K_{Ca} channel and depolarizes cerebral arterial smooth muscle under normal conditions, inhibition of 20-HETE by NO would increase the open state of K_{Ca} channels and hyperpolarize arterial smooth muscle cells (Figure 7-4). At present it is unknown whether or not all of the hyperpolarization observed in the presence of NO is due to inhibition of P450 4A enzyme activity. However, such a mechanism is compelling, and can account for the cyclic GMP independent dilation to NO.

CONCLUSION

Hyperpolarizing factors coming from the vascular endothelium are important determinants of vascular tone. As discussed in great detail in this monograph, there does not appear to be a single class of endothelium-derived hyperpolarizing agent. However, the data summarized here and those by Campbell and colleagues provide biochemical and functional data that one of the primary classes of hyperpolarizing factors coming from the endothelium are metabolites of P450 epoxygenases. The present findings suggest that in the brain, this same class of P450 epoxides is present in astrocytes and released by glutamate. As in the endothelium, the P450 metabolites coming from astrocytes function to regulate cerebral blood flow by hyperpolarizing cerebral arterial smooth muscle. These observations support the hypothesis that astrocytes are an intermediary cell type, situated between neurons and the microcirculation to distribute hyperemic blood flow to metabolically active neurons. In addition, in the brain, NO may function as a hyperpolarizing metabolite to inhibit P450 enzymes important in mediating autoregulation of cerebral blood flow.

8 Intracellular Action of Epoxyeicosatrienoic Acids in Endothelial Cells

Wolfgang F. Graier*, Bernhard G. Hoebel, Sascha Frank and Gert M. Kostner

Department of Medical Biochemistry, Karl-Franzens University of Graz, Graz, Austria
*Address for correspondence: Prof. Wolfgang F. Graier, Department of Medical Biochemistry, Karl-Franzens University of Graz, Harrachgasse 21/III, A-8010 Graz. Tel.: A43-316-380-7560, Fax: A43-316-380-9615, e-mail: wolfgang.graier@kfunigraz.ac.at

Besides their contribution to endothelium-dependent hyperpolarization, the understanding of the physiological function of endothelium-generated epoxyeicosatrienoic acids is limited. In the present work it is intended to verify intracellular properties of these cytochrome P450 epoxygenase-derived metabolites of arachidonic acid in porcine endothelial cells. Evidence is provided that epoxyeicosatrienoic acids contribute to the regulation of intracellular Ca^{2+} concentration by affecting the driving force for Ca^{2+} influx through the capacitative Ca^{2+} entry in endothelial cells, which is due to hyperpolarization of the cell membrane. Enzyme induction by dexamethasone/clofibrate increase cytochrome P450 epoxygenase activity, Ca^{2+}/Mn^{2+} entry and membrane hyperpolarization in response to thapsigargin, a powerful stimulus for cytochrome P450 epoxygenase. Epoxyeicosatrienoic acids stimulated endothelial tyrosine kinase. Arachidonic acid, mimicked the effect of the epoxyeicosatrienoic acids on tyrosine kinase activity at high concentration. Since the effect of arachidonic acid but not that of the epoxyeicosatrienoic acids on tyrosine kinase was prevented by thiopentone sodium (an inhibitor of cytochrome P450 epoxygenase), arachidonic acid probably has to be converted to epoxyeicosatrienoic acids in order to stimulate endothelial tyrosine kinase. These data strongly indicate that epoxyeicosatrienoic acids not only serve as putative endothelium-derived hyperpolarizing factor(s) but also have important intracellular functions in endothelial cells.

Key Words: Ca^{2+} homeostasis, cytochrome P450 epoxygenase hyperpolarization, tyrosine kinase

INTRODUCTION

Endothelial cells produce various vasoactive compounds, such as nitric oxide, prostacyclin and the endothelins (Furchgott and Zawadzki, 1980; Busse, Trogisch and Bassenge, 1985; Graier, Sturek and Kukovetz, 1994). Besides these powerful paracrines causing either relaxation or contraction, metabolites of arachidonic acid derived from the cytochrome P 450 epoxygenase pathway are potent activators of K^+ channels in smooth muscle cells (Gebremedhin *et al.*, 1992; Hu and Kim, 1993). This activation may involve G proteins (Graier *et al.*, 1996; Li and Campbell, 1997). These compounds are produced in endothelial cells by various isoforms of cytochrome P450 epoxygenases [e.g. CYP2J2 in human lung; (Zeldin *et al.*, 1996)] and represent the 5,6-, 8,9-, 11,12- and 14,15- isoforms of epoxyeicosatrienoic acid. Whereas attention has focused on their putative role as endothelium-derived hyperpolarizing factor (Campbell *et al.*, 1996; Graier *et al.*, 1996; Hecker *et al.*,

Wolfgang F. Graier et al.

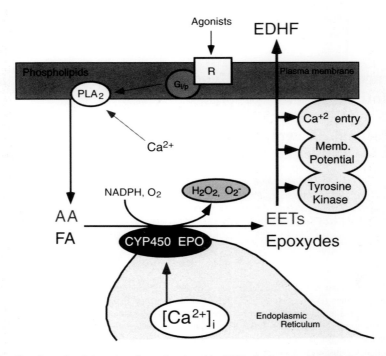

Figure 8-1. Putative role of the epoxyeicosatrienoic acids (EETs) in the vascular wall. Agonists initiate the formation of inositol 1,4,5-trisphosphate (IP₃) by phospholipase C (PLC). The depletion of the Ca²⁺ stores results in activation of cytochrome P450 epoxygenase. This enzyme converts arachidonic acid (AA) liberated by Ca²⁺/G protein-stimulated phospholipase A₂ (PLA₂) to epoxyeicosatrienoic acids, in the presence of NADPH and oxygen. Besides their putative action as endothelium-derived hyperpolarizing factors, these products serve as powerful intracellular signalling molecules for endothelial Ca²⁺ homeostasis, cell membrane potential and tyrosine kinase-mediated cascades.

1994), little is known about the action of epoxyeicosatrienoic acids within the endothelial cells. Modulation by epoxyeicosatrienoic acids of Ca^{2+} signaling has been repeated in cells such as rat hepatocytes (Karara *et al.*, 1991), epithelial cells (Madhun *et al.*, 1991) and vascular endothelial cells (Graier, Simecek and Sturek, 1995a; Hoebel, Kostner and Graier, 1997).

The present chapter summarizes data on the intracellular action of the epoxyeicosatrienoic acids in endothelial cells, which demonstrate that, in addition to their function as putative endothelium-derived hyperpolarizing factors, these acids may act as intracellular signal molecules (Figure 8-1).

METHODS

Cell Isolation and Culture

Endothelial cells were isolated from porcine aortae by enzymatic digestion (200 U/ml collagenase [type II] and trypsin inhibitor (soybean type I; 1 mg/ml) in Dulbecco's minimal essential medium containing 2 and 1% of commercially available stock solutions of amino acids and vitamins) as previously described (Graier, Simecek and Sturek, 1995a). Cells were grown in Opti-minimal essential medium containing 2% fetal calf serum.

Ca²⁺ Measurement

Intracellular free Ca^{2+} concentration ($[Ca^{2+}]_i$) was determined in porcine aortic endothelial cells using the fura-2 technique (Graier *et al.*, 1998; Graier, Simecek and Sturek, 1995a). Cells were incubated in serum-free culture medium containing 2×10^{-3} M fura-2/am for 45 to 30 min at 37C in the dark. Prior to the experiment, they were centrifuged and resuspended in nominally Ca^{2+} free (i.e. app. 10 μmol/L free extracellular Ca^{2+}) Hepes-buffered solution containing in 10^{-3} M: 145 NaCl, 5 KCl, 1 $MgCl_2$, 10 Hepes-acid, pH 7.4. Intracellular Ca^{2+} was monitored every 0.25s as the ratio of 340 or 360 and 380 nm excitation at 510 nm emission. Stimulation of capacitative Ca^{2+} entry was also studied by Mn^{2+} quench experiments (Paltauf-Dobruzynska and Graier, 1997).

Data Acquisition

In view of the reported errors of the $[Ca^{2+}]_i$ calibration in this system (Graier, Simcek and Sturek, 1995a) and the general uncertainties of the calibration techniques (Morgan, 1993), intracellular free Ca^{2+} is expressed as a ratio of 340/380 emission.

Measurement of Cytochrome P450 Epoxygenase Activity

Activity of microsomal cytochrome P450 epoxygenase activity was measured by fluoro-metrically monitoring of the dealkylation of 1-ethoxypyrene-3,6,8-tris-(dimethyl-sulfonamide) (EPSA) to 1-hydroxy-pyrene-3,6,8-tris-(dimethyl-sulfonamide) (HPSA; Graier *et al.*, 1996; Hoebel, Kostner and Graier, 1997). Cultured endothelial cells were harvested by trypsin digestion (Graier, Simecek and Sturek., 1995a) and were were permeabilized with 1 mg/ml saponin, followed by the addition of a NADPH-regenerating system [i.e. 25 I.U./ml isocitric dehydrogenase, 8×10^{-3} M D,L-isocitric acid, 1×10^{-3} M NADP⁺; (Capdevila *et al.*, 1983)]. Enzyme activity was stimulated by depletion of intracellular Ca^{2+} stores with the compound indicated. The resultant increase in fluorescence by the con-version of EPSA to HPSA was monitored at excitation 495 nm and 550 nm emission. Activity of cytochrome P450 epoxygenase was calculated using the standard calibration curve and was expressed as 10^{-9} M $\times$ min⁻¹ $\times$ (10^6 cells)⁻¹.

Tyrosine Kinase Activity

Tyrosine kinase activity was measured in cell lysates using a photometric assay (Hoebel and Graier, 1998; Hoebel, Kostner and Graier, 1997). Cell lysates were obtained by sonication in chilled buffer containing in 10^{-3} M: 20 Tris, 150 NaCl, 1 EDTA, 1 EGTA, 0.2 phenylmethylsulfonyl fluoride, 0.2 Na_3VO_4, 5 mercaptoethanol, 1 μg/ml pepstatin-A and 0.5 μg/ml leupeptin, pH adjusted at 7.4. Phosphorylation was determined by a horse-radish peroxidase-labeled phosphotyrosine-specific antibody and monitored at 450 nm.

Materials

All tissue culture media, antibiotics and cell culture dishes were obtained from Life Technologies (Vienna, Austria) and fetal calf serum was from PAA, (Linz, Austria). 1-Ethoxypyrene-3,6,8-tris-(dimethyl-sulfonamide), 1-hydroxy-pyrene-3,6,8-tris-(dimethyl-sulfonamide) and fura-2/acetoxymethyl ester were from Lambda Fluorescence Technology (Graz, Austria). Bis-(1,3-dibutylbarbituric acid) pentamethine oxonol (DiBac$_4$(5)) was from Molecular Probes Inc. (Leiden, The Netherlands). Epoxy-eicosatrienoic acids were obtained from Cascade Ltd. (Reading, United Kingdom). Cyclopiazonic acid was from Aldrich (Vienna, Austria). The protein tyrosine kinase assay kit was purchased from Calbiochem-Novabiochem International (Vienna, Austria). All other chemicals used were obtained from Sigma (Vienna, Austria).

Statistics

Analysis of variance was performed and statistical significance of differences were estimated by Scheffes F test. Differences were considered to be statistically significant when P was less than 0.05.

RESULTS

In single endothelial cells, the cytochrome P450 epoxygenase product, 5,6-epoxyeicosatrienoic acid (156×10^{-9} M) evoked activation of a Ca^{2+} entry, indistinguishable from that activated by bradykinin (Graier, Simecek and Sturek, 1995a). Inhibition of endothelial cytochrome P450 epoxygenase by thiopentone sodium (3×10^{-4}) diminished store-operated Ca^{2+} entry by about 40% (Hoebel, Kostner and Graier, 1997). Incubation with arachidonic acid (10^{-5} M) for 10 min to achieve loading of endothelial cells with the substrate of cytochrome P450 epoxygenase, significantly increased ATP-induced Ca^{2+} (data not shown) and Mn^{2+} entry (Figure 8-2).

Exposure of endothelial cells with a mixture of dexamethasone and clofibrate increased the expression of various cytochrome P450 enzymes including CYP2E1 and CYP3A (Hoebel, Steyrer and Graier, 1998). The combination of 3×10^{-6} M dexamethasone plus 5×10^{-5} M clofibrate for 72 h, augmented significantly the effect of thapsigargin (10^{-6} M) on microsomal cytochrome P450 epoxygenase, Ca^{2+} entry, Mn^{2+} influx and cell membrane hyperpolarization (Figure 8-3).

Thiopentone sodium [an inhibitor for endothelial cytochrome P450 epoxygenase (Hoebel, Kostner and Graier, 1997; Lischke, Busse and Hecker, 1995c) without additional inhibitory properties on tyrosine kinase, endothelial Ca^{2+} and K_{Ca} channels (Hoebel, Kostner and Graier, 1997)] diminished the capacitive Ca^{2+} entry in endothelial cells evoked by either bradykinin, ATP or thapsigargin (Hoebel, Kostner and Graier, 1997). Inhibition of endothelial cytochrome P450 epoxygenase by thiopentone sodium (3×10^{-4} M) reduced ATP-induced membrane hyperpolarization by 68% (Figure 8-4). In contrast to the ATP-mediated membrane hyperpolarization, the hyperpolarization induced 8,9-epoxyeicosatrienoic acid (0.75×10^{-6} M) remained unchanged in the presence of thiopentone sodium (Figure 8-4).

In addition, 8,9-epoxyeicosatrienoic acid (and 11,12-epoxyeicosatrienoic acids: (Hoebel and Graier, 1998)) stimulated endothelial tyrosine kinase activity in a concentration-

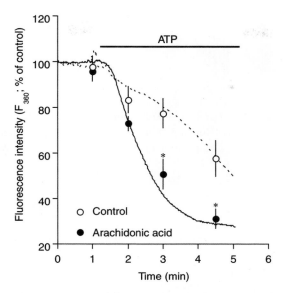

Figure 8-2. Loading the endothelial cells with the precursor of epoxyeicosatrienoic acids, arachidonic acid increased Mn^{2+} entry in response to ATP. Freshly isolated porcine artery endothelial cells were loaded for 10 min with arachidonic acid (10^{-5} M). After removal of the arachidonic acid, the cells were stimulated in the presence of 2×10^{-4} M Mn^{2+} with ATP (10^{-4} M). Each point represents the mean ± SEM (n = 3–5). The asterisks indicates a statistically significant (P < 0.05) difference from data obtained in cells not loaded with arachidonic acid.

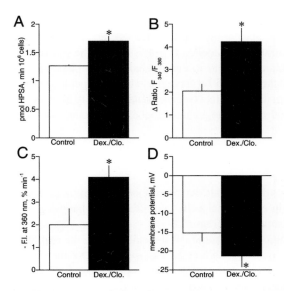

Figure 8-3. Effect of incubation with dexamethasone plus clofibrate on thapsigargin-induced cytochrome P450 epoxygenase activity (A), Ca^{2+} entry (B), Mn^{2+} quench (C) and membrane hyperpolarization (D) in endothelial cells. Cells were treated for 72 h with the solvent (DMSO 0.2%; Control; n = 6–12) or 3×10^{-6} M dexamethasone plus 5×10^{-5} M clofibrate (Dex./Clo., n = 4–9). Each column represents the effect of thapsigargin (10^{-6} M). Data shown as means ± SEM. The asterisks indicates a statistically significant (P < 0.05) difference from data obtained in cells incubated with DMSO.

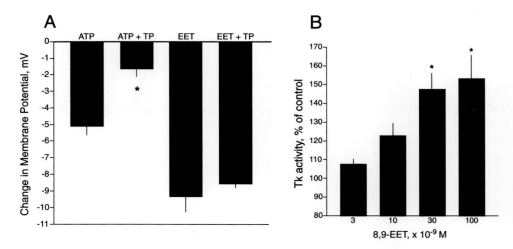

Figure 8-4. Effect of epoxyeicosatrienoic acid on endothelial membrane potential (A) and tyrosine kinase (B). **A:** Changes in membrane potential to stimulation with 10^{-6} M ATP and 0.75×10^{-6} M 8,9-epoxyeicosatrienoic acid (EETs; mixed in 1 mg/ml bovine serum albumin) in the absence or presence of 3×10^{-4} M thiopentone sodium (TP). Changes in membrane potential were monitored with $DiBAC_4(5)$ (Hoebel, Kostner and Graier, 1997). **B:** Effect of various concentrations of 8,9-epoxyeicosatrienoic acids (8,9-EET; mixed in 1 mg/ml bovine serum albumin) on tyrosine kinase activity (Hoebel and Graier, 1998). Each column represents the mean ± SEM (n = 3). The asterisks indicates a statistically significant (P < 0.05) difference from data obtained in cells in the absence of thiopentone sodium (A) or 8,9-epoxyeicosatrienoic acid (B).

dependent manner (Figure 8-4). This effect was prevented by erbstatin (10^{-5} M) (data not shown). Arachidonic acid mimicked the effect of epoxyeicosatrienoic acids in about 20-fold higher concentration (Hoebel and Graier, 1998). The stimulatory effect of arachidonic acid on tyrosine kinase activity in endothelial homogenates was prevented by inhibition of endothelial cytochrome P450 epoxygenase with thiopentone sodium (3×10^{-4} M), while thiopentone sodium did not affect tyrosine kinase activity activated by epoxyeicosatrienoic acids in endothelial cell homogenate (Hoebel and Graier, 1998).

DISCUSSION

The present findings demonstrate that epoxyeicosatrienoic acids act as powerful intracellular signal molecules in endothelial cells and affect Ca^{2+}-handling, cell membrane potential and tyrosine kinase activity (Figure 8-1).

The involvement of cytochrome P450 metabolites in the regulation of capacitative Ca^{2+} entry pathways in non-excitable cells was suggested by the observations that inhibitors of the cytochrome P450 pathway prevent capacitative Ca^{2+} entry in a variety of cells (Alonso *et al.*, 1991; Alonso-Torre *et al.*, 1993; Alvarez, Montero and Garcia-Sancho, 1991; 1992b; Montero, Alvarez and Garcia-Sancho, 1992; Montero, Garcia-Sancho and Alvarez, 1993). In those studies, the inhibitors used were far from selective due to their very high potency to inhibit K^+ channels (Alvarez, Montero and Garcia-Sancho, 1992b) or tyrosine kinase activity (Sargeant, Farndale and Sage, 1994). Although thiopentone sodium is a somewhat more selective inhibitor of endothelial cytochrome P450 epoxygenase (Hoebel, Kostner and Graier, 1997; Lischke, Busse and Hecker, 1995c), additional experimental approaches are needed to elucidate the role of the epoxyeicosatrienoic acids in endothelial signal transduction.

Endothelial cells express CYP2B1, CYP2E1 and CYP3A but not CYP1A1 and CYP4A in (Hoebel, Steyrer and Graier, 1998). These findings are consistent with those obtained in the rabbit aorta (Irizar and Ioannides, 1995). Among the observed cytochrome P450 enzymes, CYP2E1 and CYP3A can be induced by dexamethasone plus clofibrate (Hoebel, Steyrer and Graier, 1998). In agreement with these earlier observations, in the present study incubation with dexamethasone plus clofibrate increased the thapsigargin-induced Ca^{2+}/Mn^{2+} entry and membrane hyperpolarization. Since these changes were associated with increased cytochrome P450 epoxygenase activity in cells treated with dexamethasone plus clofibrate, these findings support the hypothesis that epoxyeicosatrienoic acids are involved in endothelial Ca^{2+} signaling. This conclusion is strengthened by the finding that loading of endothelial cells with the precursor for synthesis of epoxyeicosatrienoic acids, arachidonic acid enhanced autacoid-induced capacitative Ca^{2+} entry, as indicated by an enhanced Mn^{2+} influx in ATP-stimulated cells.

In addition to their action on endothelial Ca^{2+} signaling pathways, epoxyeicosatrienoic acids evoke hyperpolarization of endothelial cells (Hoebel, Kostner and Graier, 1997; Hoebel, Steyrer and Graier, 1998) by opening endothelial Ca^{2+}-activated K^+ channels (Baron, Frieden and Beny, 1997). While the ATP-mediated membrane hyperpolarization was sensitive to thiopentone sodium, those to 8,9- (this study) and 11,12-epoxyeicosatrienoic acid (Hoebel, Kostner and Graier, 1997; Hoebel, Steyrer and Graier, 1998) remain unchanged in the presence of the inhibitor of cytochrome P450 epoxygenase. These data suggest the involvement of epoxyeicosatrienoic acids in the cell membrane hyperpolarization in response to autacoids, such as ATP and bradykinin.

Tyrosine kinase is involved in endothelial Ca^{2+} signaling (Fleming, Fisselthaler and Busse, 1995; 1996) and shear stress-mediated activation (Corson *et al.*, 1996). The findings that 8,9- (this study) and 11,12-epoxyeicosatrienoic acid (Hoebel, Steyrer and Graier, 1998; Hoebel and Graier, 1998) increase tyrosine kinase activity in endothelial homogenates suggests a novel type of physiological action of the epoxyeicosatrienoic acids in vascular cells. Since the (similar) effect of arachidonic acid on tyrosine kinase activity was prevented by inhibition of endothelial cytochrome P450 epoxygenase, arachidonic acid may have to be converted to the epoxyeicosatrienoic acids first to activate the endothelial enzyme. Activation of the serine/threonine extracellular-regulated protein kinases (ERK1/2) by arachidonic acid also depends on the conversion by lipoxygenase and/or cytochrome P450 epoxygenase(s) in renal epithelial cells (Cui and Douglas, 1997).

In conclusion, the vascular function of metabolites of arachidonic acid derived from cytochrome P450 epoxygenase exceeds that of being one of the endothelium-derived hyperpolarizing factors (Figure 8-1). Indeed, epoxyeicosatrienoic acids are powerful intracellular mediators for two major events of the endothelial signal-transduction cascade: Ca^{2+} signaling and activation of tyrosine kinase.

9 Epoxyeicosatrienoic Acids Potentiate Ca^{2+} Signaling in Both Endothelial and Vascular Smooth Muscle Cells

*Jean-Vivien Mombouli, §Darryl Zeldin, #Timothy Scott-Burden, Sigrid Holzmann, Gert M. Kostner and Wolfgang F. Graier

*Baylor College of Medicine and #Texas Heart Institute, Houston, Texas, USA;
§NIH/NIEHS Research Triangle Park, North Carolina, USA;
Karl Franzens University of Graz, Graz, Austria.
Address for correspondence: Jean-Vivien Mombouli, Ph. D., Center For Experimental Therapeutics, Baylor College of Medicine, One Baylor Plaza, room 802 E, Houston, TX 77030, U.S.A. Tel: (1-713) 798 5187. Fax: (1-713) 799 2469. E-mail: ombouli@bcm.tmc.edu
Darryl Zeldin, MD., Laboratory of Pulmonary Pathobiology, NIH/NIEHS, 111 T.W. Alexander Dr. Bldg 101, D236, Research Triangle Park, NC 27709, USA.
Timothy Scott-Burden, Ph. D., Texas Heart Institute, University of Texas at Houston, Houston, TX77030, USA.
Sigrid Holzmann, Ph. D., Department of Pharmacologie and Toxicologie, Karl Franzens University of Graz, Graz A8010 Austria.
Gert M. Korstner, Ph. D., Department of Medical Biochemistry, Karl Franzens University of Graz, Graz A8010 Austria.
Wolfgang F. Graier, Ph. D., Department of Medical Biochemistry, Karl Franzens University of Graz, Graz A8010 Austria.

Experiments were designed to assess the impact on Ca^{2+} signaling of sustained increases in the concentration epoxyeicosatrienoic acids (EETs) in cultured endothelial or vascular smooth muscle cells. Cultured human umbilical vein-derived endothelial cells (EA.hy926) and bovine aortic vascular smooth muscle cells were processed for conventional spectrofluorometric measurement of cytosolic Ca^{2+}, using Fura 2/AM. A bovine aortic vascular smooth muscle cell line expressing the CYP2J2 epoxygenase was developed to determine the impact of chronic enhancement of epoxygenase activity. In EA.hy926 endothelial cells, the histamine-induced mobilization of Ca^{2+} from intracellular stores, as well as the stimulation of Ca^{2+} influx were inhibited by the cytochrome P450 inhibitor thiopental. However, incubation of the cells with 11,12-EET amplified the mobilization of Ca^{2+} from both intracellular and extracellular pools. In this case, thiopental no longer affects significantly the mobilization of Ca^{2+} induced by histamine in EA.hy926 cells. A similar potentiation of Ca^{2+} signaling was obtained in bovine aortic vascular smooth muscle cells stimulated with thapsigargin. In vascular smooth muscle cells overexpressing the epoxygenase CYP2J2, an increased spontaneous Ca^{2+} influx was obtained, as well as an amplification of endothelin 1-induced capacitative Ca^{2+} entry. These results show that sustained increases in EET levels result in an amplification of the mobilization of cytosolic Ca^{2+} in cultured endothelial and vascular smooth muscle cells.

KEYWORDS: Cytochrome P450, capacitative Ca^{2+} entry, arachidonic acid, thapsigargin, endothelin.

INTRODUCTION

Epoxyeicosatrienoic acids (EETs) are generated in blood vessels by certain cytochrome P450 monooxygenases called epoxygenases which use arachidonic acid as a substrate. These epoxygenases are expressed mostly in endothelial cells (Rosolowski and Campbell, 1996). However, cultured rat aortic smooth muscle microsomes can process arachidonic acid to form EETs (Hasunuma *et al.*, 1991). Actually, epoxygenases are expressed constitutively in smooth muscle cells of certain vascular beds (Zeldin *et al.*, 1996). EET metabolism may be altered during cardiovascular diseases. Thus, EET levels increase in pregancy-induced hypertension (Catella *et al.*, 1991). Also, oxidized low-density lipoporoteins, which are recognized as a risk factor for coronary artery disease, enhance the synthesis of EETs by cultured endothelial cells *in vitro* (Pritchard, Wong and Sternerman, 1990). Moreover, the atherogenic molecule formed from nitric oxide and superoxide anions, peroxynitrite reacts with arachidonic acid to form EETs (Balazy *et al.*, 1994). Therefore, it is possible that vascular cells are exposed *in vivo* to elevated concentrations of EETs for prolonged periods, the consequences of which are not known.

The demonstration that EETs activate K^+ channels (Gebremedhin *et al.*, 1992; Hu and Kim, 1993; Zou *et al.*, 1996; Campbell *et al.*, 1996), and induce relaxation of vascular smooth muscle (Hecker *et al.*, 1994; Campbell *et al.*, 1996) suggests that they could represent endothelium-derived hyperpolarizing factor (EDHF) (Campbell *et al.*, 1996). Hence, an increased production of EETs in rabbit aortas from hypercholesterolemic animals is believed to compensate for the loss of vasodilator efficacy of endothelium-derived nitric oxide (Pfister *et al.*, 1996). However, EETs are also involved in Ca^{2+} signaling in endothelial cells (Graier *et al.*, 1995). For instance, EETs activate K^+ channels in endothelial cells to elicit hyperpolarization (Baron, Frieden, Beny, 1997). This ensuing hyperpolarization augments the driving force for Ca^{2+} entry in endothelial cells (Adams *et al.*, 1990). 5,6-EET induces a Ca^{2+} influx that is sensitive to the same channel blockers as the Ca^{2+} entry evoked by agonists, such as bradykinin (Graier *et al.*, 1995).

While endothelium-derived EETs may act as EDHF, the intracrine role of EETs produced by vascular smooth muscle cells themselves is not clear. In cultured vascular smooth muscle cells, EETs induce the mobilization of cytosolic Ca^{2+} (Thibonnier *et al.*, 1993). The inhibitory action of certain cytochrome P450 inhibitors in native blood vessels suggests that EETs are involved in hormone-induced vasoconstriction (Vazquez, Rios and Escalante, 1993). Moreover, EETs induce the proliferation of cultured vascular smooth muscle cells (Sheu *et al.*, 1995). These findings are not consistent with the vasodilator role attributed to EETs.

The present study was designed to assess how increased EET concentrations in the environement of endothelial and vascular smooth muscle cells impact on Ca^{2+} signaling in these cells. Aside from the involvement of EETs as EDHF, this impact on Ca^{2+} signaling may have consequences on the ability of the blood vessels to respond to endothelium-dependent vasodilators or to vasoconstrictor hormones.

METHODS

A human umbilical vein-derived endothelial cell line EA. hy926 (generously provided by Dr. Edgell, University of North Carolina, Chapell Hill, NC, USA) was subcultured in the laboratory. Freshly isolated bovine aortas were collected from a nearby slaughterhouse. After cleaning the vessels, primary cultures of bovine aortic smooth muscle cells were obtained using the explant method, after isolation of the medial layer of the blood vessels. The vascular smooth cells were utilized for the Ca^{2+} signaling experiments after the first passage. For some experiments, bovine aortic smooth muscle cells were stably transfected using plasmid vector constructs $pREP_9$ and pcDNA (Invitrogen, San Diego, CA). As a consequence of their maintenance in the host cell as episomes, these vectors minimize insertional mutagenesis commonly associated with vectors that integrate into chromosomes (Scott-Burden *et al.*, 1996 and 1997). Full-length cDNA (*cyp2j2*) encoding for the active human epoxygenases CYP2J2 was cloned into the $pREP_9$ plasmid vector, which contains coding sequences for ampicilin and neomycin resistance for selection of transformates. Transfection of cultured vascular smooth muscle cells with $pREP_9$ plasmid vectors (containing or not the *cyp2j2* cDNA sequence) was performed by square-wave electroporation. Cells (2×10^{-6}) were suspended in 200 μl of 0.27 M sucrose containing 2.5 mM Tris-HCl buffer, pH 7.4, 10 mM $MgCl_2$, and 0.25 mM EDTA and incubated on ice with 2 to 8 μg DNA for 15 minutes before electroporation at 400 V for 20 μs. This process was repeated 20 times for each incubate, and samples were returned to ice for a further 10 minutes.

Determination of changes in intracellular Ca^{2+} levels was assessed spectrofluorimetrically using fura-2 as a probe. Cell suspensions were obtained after moderate trypsinisation of subconfluent endothelial or vascular smooth muscle cells. The cell suspensions were incubated at 37°C in Dulbecco's minimum essential medium containing 2.5×10^{-6} M Fura-2/AM to load the fluorescent probe. During this fura-2 loading procedure, the cell suspension was divided in two parts to study the effects of prior incubation with EETs in cultured EA.hy296 and bovine aortic smooth muscle cells. One part was treated with solvent (ethanol 0.3% vol/vol) and the other was treated with 5×10^{-6} M 11,12-EET for 30 min. After the incubations, the cells were centrifuged, washed and resuspended in 2 ml of Ca^{2+}-free physiological salt solution (composition in mM): 135 NaCl, 1 $MgCl_2$, 5 KCl, 10 Hepes, 10 D-Glucose, adjusted to pH 7.4 with NaOH. The cell suspension was transferred to a cuvette, and stirred continuously. Experiments were carried out following a 4 min equilibration period at room temperature. In some tests, the cell supensions were first challenged 20 seconds into recording of data with a stimulant to release Ca^{2+} from intracellular stores. In all cases, 2.5 mM $CaCl_2$ was added to the suspension to achieve standard Ca^{2+} concentration in the extracellular milieu.

Drugs

±11,12-epoxyeicosatrienoic acid, histamine hydrochloride, thapsigargin (Sigma), thiopental (Tyrol Pharma), fura-2/AM (Lambda Fluorescence Technology).

Statistics

Due to the uncertainties inherent to measurement of intracellular Ca^{2+} with fluorescent probes, the relative cytosolic concentration of Ca^{2+} is determined as the ratio of fluorescence of fura-2 due to excitation at 340 nm relative to that due to excitation at 380 nm (F_{340}/F_{380}). Indeed, the dissociation constant for the fura-2/Ca^{2+} complex in the cytosol is difficult to determine accurately. In addition, the intensity of fura-2 fluorescence in the

Jean-Vivien Mombouli et al.

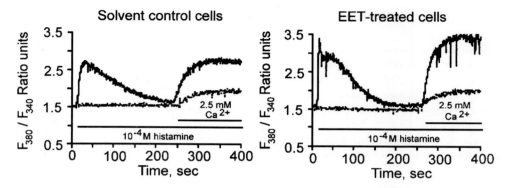

Figure 9-1. **Effects of thiopental on histamine-induced Ca²⁺ signaling in EA.hy926 cells.** Recordings from Fura 2-loaded EA.hy926 endothelial cell suspensions that were either stimulated (continuous line, upper traces) or not (broken line, lower traces) with histamine. Prior to the experiments, the cells were equilibrated for 4 min (controls, broken line, full circles) or treated with 3×10^{-4} M thiopental (continuous line, open circles). Horizontal bars indicate application of 10^{-4} M histamine and 2.5 mM Ca^{+2}, respectively. Circles represent means ± SEM of fluorescence ratio values determined at indicated times. Asterisks represent *P* values less than 0.05, as determined by Student's *t* test for unpaired observations ($n = 8$–9).

cytoplasm may not be identical to that determined in cell-free calibration solutions. Evaluation of the statistical significance between treatments was performed either by Student's t test for unpaired observations or by analysis of variance when more than two treatment groups were considered. P values less than 0.05 were considered to represent statistically significant differences.

RESULTS

Endothelial Cells

In EA. hy926 endothelial cells incubated in Ca²⁺-free physiologcal salt solution, the mobilization of Ca²⁺ evoked by histamine (10^{-4} M) was biphasic (Figure 9-1). Under these experimental conditions, the addition of 2.5 mM CaCl₂ to the extracellular bath elicited a large Ca²⁺ entry (Figure 9-1). This influx of Ca²⁺ was larger in the presence of histamine as compared to the Ca²⁺ influx that is obtained following addition of extracellular Ca²⁺ in non-stimulated endothelial cells (data not shown). The cytochrome P450 inhibitor thiopental (3×10^{-4} M) reduced significantly the mobilization of Ca²⁺ obtained in Ca²⁺-free physiological salt solution. Also, the Ca²⁺ influx following addition of Ca²⁺ to the extracellular milieu that was obtained in the presence of histamine was depressed significantly in EA.hy926 cells treated with thiopental (Figure 9-1). In the absence of histamine, the Ca²⁺ influx following addition of 2.5 mM CaCl₂ was not affected significantly by thiopental (data not shown).

In EA. hy926 endothelial cells that had been incubated with 5×10^{-6} M 11,12-EET, the mobilization of Ca²⁺ evoked by histamine had similar kinetic characteristics (data not shown). However, the magnitude of the mobilization of Ca²⁺ obtained both in absence and after introduction of Ca²⁺ in the extracellular medium was enhanced significantly (Table 9-1). By contrast, treatment of the EA. hy926 cells with solvent alone selectively

Table 9-1 Effects of EETs on Ca^{2+} signaling induced by histamine

Treatment	Basal Ca^{2+}	Ca^{2+} peak I	Ca^{2+} peak II	Ca^{2+} entry	n
None	1.59 ± 0.02	1.42 ± 0.05	1.19 ± 0.05	1.88 ± 0.09	9
Solvent	1.62 ± 0.04	0.89 ± 0.11*	1.15 ± 0.06	1.65 ± 0.17	6
11,12-EET	1.60 ± 0.02	1.67 ± 0.04*	1.40 ± 0.05*§	2.04 ± 0.08§	6

Prior to the experiments, the cells were incubated or not with either 5×10^{-6} M 11,12-EET or its solvent ethanol (0.3% vol/vol). After determination of fluorescence ratio values for basal cytosolic Ca^{2+} concentration, the cells were exposed to 10^{-4} M histamine. Ca^{2+} peaks I and II correspond to the mobilization of Ca^{2+} evoked by histamine in Ca^{2+}-free physiological salt solution, as depicted in Figure 9-1. Results are presented in arbitrary ratio units of fluorescence measured at wave lengths 340/384 nm. Ca^{2+} peak I and peak II values represent Δfluorescence ratio above basal of the rapid and secondary release of Ca^{2+} from intracellular stores. Ca^{2+} entry represents increases in fluorescence ratios above Ca^{2+} levels achieved prior to introduction of Ca^{2+} to the extracellular milieu. Statistical significance of differences in Ca^{2+} signals between naive and solvent-controls (*), naive and EET-treated (*), and solvent-control versus EET-treated (§) were determined by analysis of variance ($P < 0.05$). n represents the number of experiments.

attenuated the initial rapid rise in intracellular Ca^{2+} without affecting the secondary peak of the mobilization of Ca^{2+} induced by histamine in Ca^{2+}-free physiological salt solution (Table 9-1). The solvent did not significantly depress Ca^{2+} entry as compared to naive EA.hy926 cells. Thiopental did not significantly inhibit Ca^{2+} signaling (both intracellular release and Ca^{2+} entry) in cells that had been previously incubated with 11,12-EET (n = 4, data not shown).

Vascular Smooth Muscle Cells

The Ca^{2+} entry induced by thapsigargin was greater in bovine aortic smooth muscle cells treated with 5×10^{-6} M EET than in solvent-control cells (Figure 9-2A). The Ca^{2+} entry obtained in the absence of stimulation was the same in both EET-treated and solvent-control cells (Figure 9-2B). In bovine aortic vascular smooth muscle cells that were

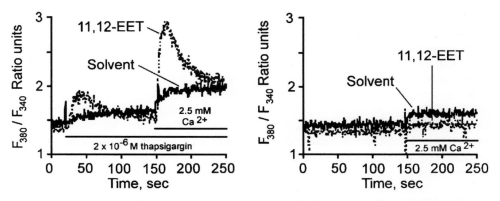

Figure 9-2. **Effects of incubation with authentic EET on Ca^{2+} signaling in cultured bovine aortic smooth muscle cells.** Recordings from Fura 2-loaded bovine aortic vascular smooth muscle cell suspensions that were either stimulated (left panel) or not (right panel) with thapsigargin. Prior to the experiments, the cells were treated for 4 min with either solvent (0.3% vol/vol ethanol, continuous lines) or 5×10^{-6} M 11,12-EET (broken lines). Horizontal bars indicate application of 2×10^{-6} M thapsigargin (panel A) and 2.5 mM Ca^{+2} (panels A and B), respectively.

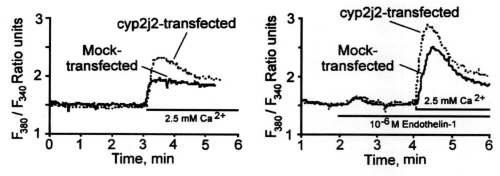

Figure 9-3.   Effects of epoxygenase-induction on Ca^{2+} signaling in cultured bovine aortic smooth muscle cells. Recordings from Fura 2-loaded bovine aortic vascular smooth muscle cell suspensions that were transfected with either *cyp2j2* gene (broken lines, upper traces) or the vector alone (mock-transfected, continuous lines). Ca^{2+} entry was examined in the absence (panel A) or in the presence (panel B) of 10^{-9} M endothelin-1. Horizontal bars indicate application of 2.5 mM Ca^{+2} (panels A and B) and 10^{-9} M endothelin-1 (panel A) respectively.

transfected with the epoxygenase *cyp2j2* gene, Ca^{2+} entry in the absence of exogenous stimulants was greater than in cells that were only transfected with the vector devoid of the *cyp2j2* gene (Figure 9-3A). Stimulation of the cells with a threshold concentration of endothelin-1 induced a mobilization of Ca^{2+} from intracellular stores which had a similar magnitude in both *cyp2j2* -and mock-transfected control cells (Figure 9-3B). However, Ca^{2+} entry was stimulated in the presence of endothelin-1. This stimulation was biphasic and greater in *cyp2j2* -transfected bovine aortic vascular smooth cells (Figure 9-3B).

DISCUSSION

The present findings show that incubation of cultured endothelial and vascular smooth muscle cells with EETs augments the mobilization of Ca^{2+} from intracellular stores and enhances the stimulation of Ca^{2+} influx induced by agonists. As checked in the endothelial cells, treatment of the endothelial cells with EETs overcomes the inhibitory action of the cytochrome P450 inhibitor thiopental on Ca^{2+} signaling. Induction of an epoxygenase in vascular smooth muscle cells increases the mobilization of Ca^{2+} evoked by the vasoconstrictor peptide endothelin-1. Thus, this study confirms that EETs facilitate Ca^{2+} signaling in cultured endothelial and vascular smooth muscle cells.

11,12-EET induced a significant potentiation of Ca^{2+} signaling in the human umbilical vein-derived EA.hy926 endothelial cells. The cytochrome P450 inhibitor thiopental inhibited Ca^{2+} mobilization in naive EA.hy926 cells. This suggests that cytochrome P450 products may modulate Ca^{2+} signaling in response to histamine in these cells. Incubation of the cells with 11,12-EET was able to overcome the inhibition by thiopental. This is consistent with the interpretation that the target for the facilitatory action of 11,12-EET is beyond that for thiopental, which is presumably the epoxygenase.

A priming effect of 11,12-EET could also be demonstrated on Ca^{2+} signaling induced by the Ca^{2+} ATPase-inhibitor thapsigargin, in first passage cultured vascular smooth muscle cells. In vascular smooth cells transfected with a gene encoding for the epoxygenase

CYP2J2 (Zeldin *et al.*, 1996), non-stimulated Ca^{2+} entry was changed both in magnitude and kinetics. Secondly, endothelin-1 induced a mobilization of intracellular Ca^{2+} of equal magnitude in both transfected and control cells; however, the peptide-induced Ca^{2+} entry was enhanced. These results suggest that molecular induction of epoxygenase in vascular smooth muscle cell potentiates capacitative Ca^{2+} entry in response to agonists, and may induce a basal active state of this Ca^{2+} entry mechanism. Parallel experiments showed that induction of epoxygenase overexpression in endothelial cells, also results in a similar amplification of Ca^{2+} signaling (see chapter by Graier and collaborators in this monograph).

However, since vascular smooth muscle cells undergo significant phenotypic changes when cultured, it is not clear what the functional significance of these findings may be. For instance, receptor-dependent Ca^{2+} influx in vascular smooth muscle involves both voltage-dependent and -independent channels. In our cultured smooth muscle cells the voltage-dependent channels were not functional. Moreover, the vascular wall contains a heterogenous population of smooth muscle cells in respect to ion channel expression. Thus, it is possible that depending of the vascular beds, or the physiopathological status of the vascular wall, intracrine or paracrine EETs would act to inhibit the mobilization of cytosolic Ca^{2+}, in some, or conversely to amplify Ca^{2+} signaling in other instances. In view of the central role of Ca^{2+} in the vascular smooth muscle contractile or throphic functions, resolution of the molecular determinants of these antagonistic actions of EETs may be relevant to adaptative mechanisms in the blood vessel wall.

ACKNOWLEDGEMENTS

The authors are indebted to Dr. Edgell from the University of North Carolina, Chapell Hill, NC, USA, from providing the human umbilical vein-derived EA. hy926 endothelial cell line. JVM was supported in part by an Austrian Science Funds's Lise Meitner Visiting Scientist Fellowship (No M00442).

10 Influence of Various Cytochrome P450 Inhibitors on Hyperpolarizations and Relaxations in the Main Mesenteric Artery of the Rat

Bert Vanheel and Johan Van de Voorde

Department of Physiology and Physiopathology, University of Gent, De Pintelaan 185, B-9000 Gent, Belgium
Address for correspondence: Prof. Dr. B. Vanheel, Dept. of Physiology and Physiopathology, University of Gent, U.Z.-Blok B, De Pintelaan 185, B-9000 Gent, Belgium. Phone: (+32) 9 240 3341, Fax: (+32) 9 240 3059,
Email: Bert.Vanheel@rug.ac.be

In the main mesenteric artery isolated from rats, measurements of cell membrane potential and isometric tension were made to assess the influence of four structurally and mechanistically different inhibitors of cytochrome P450 (proadifen, miconazole, 17-octadecynoic acid and 1-aminobenzotriazole) on hyperpolarizations and relaxations elicited by endothelium-derived hyperpolarizing factor (EDHF), liberated by acetylcholine. Responses to the endothelium-independent potassium channel opener levcromakalim (applied either as such or as the racemic cromakalim) were also studied. Proadifen completely and reversibly inhibited EDHF-induced hyperpolarizations and relaxations. However, the substance also blocked electrical and mechanical responses to levcromakalim. Miconazole also inhibited both EDHF-induced and levcromakalim-induced hyperpolarizations and relaxations. By contrast, neither 17-octadecynoic acid nor 1-aminobenzotriazole affected EDHF-induced and levcromakalim-induced hyperpolarizations. EDHF- and cromakalim-induced relaxations were not influenced by 17-octadecynoic acid and were potentiated by 1-aminobenzotriazole. These results suggest that proadifen and miconazole interfere with the activation of ATP-regulated potassium channels in mesenteric arterial cells, which makes these substances less suitable to study the involvement of endothelial cytochrome P450 metabolism in the synthesis of EDHF in intact blood vessels. Moreover, the experiments with 17-octadecynoic acid and 1-aminobenzotriazole are not consistent with EDHF being a metabolite of arachidonic acid formed by cytochrome P450-dependent monooxygenases in the main mesenteric artery of the rat.

KEYWORDS: vascular smooth muscle, potassium channels, EDHF, cytochrome P450, proadifen, miconazole

INTRODUCTION

The tone of vascular smooth muscle is modulated by factors liberated from the endothelium (Furchgott and Vanhoutte, 1989). Endothelium-dependent relaxations of vascular smooth muscle are elicited not only by activation of soluble guanylate cyclase induced by nitric oxide (NO), but also by a hyperpolarization of the vascular smooth muscle cells (Cohen and Vanhoutte, 1995). The contribution of endothelium-dependent hyperpolarization seems to increase with smaller diameter of the vessel and might be of major importance, therefore, in the regulation of vascular resistance (Cohen and Vanhoutte, 1995; Garland *et al.*, 1995).

The chemical nature of EDHF is still debated. Studies in which the N^{ω}-nitro-L-arginine-resistant component of vasodilatation was measured in bovine, porcine and rat coronary arteries suggest that the substance liberated by the endothelium in response to bradykinin might be a metabolite of arachidonic acid derived from cytochrome P450-dependent monooxygenase activity (Bauersachs, Hecker and Busse, 1994; Hecker *et al.*, 1994). Epoxyeicosatrienoic acids (EETs) indeed have a potent vasorelaxing action, hyperpolarize bovine coronary smooth muscle cells (Campbell *et al.*, 1996) and increase the activity of Ca^{2+}-dependent K^+ channels in smooth muscle cells from several arteries (Gebremedhin *et al.*, 1992; Hu and Kim, 1993, Campbell *et al.*, 1996). However, studies on other blood vessels do not support the hypothesis that EDHF is a cytochrome P450-dependent monooxygenase metabolite of arachidonic acid (Corriu *et al.*, 1996a; Zygmunt *et al.*, 1996).

The aim of the present study was to investigate, using inhibitors of cytochrome P450, the potential role of cytochrome P450-dependent metabolism in the response to the EDHF liberated by acetylcholine in the isolated main mesenteric artery of the rat. Since some inhibitors of cytochrome P450 are non-selective and might influence other systems (including vascular smooth muscle potassium channels), four structurally and mechanistically different substances were used [proadifen, miconazole, 17-octadecynoic acid (17-ODYA) and 1-aminobenzotriazole]. In addition, the influence of these various inhibitors was assessed on the endothelium-independent hyperpolarizations and relaxations of smooth muscle elicited by the selective opener of ATP-regulated potassium channels, levcromakalim.

MATERIAL AND METHODS

Preparation

The mesentery from four to six weeks old female Wistar rats was excised and placed in cold, oxygenated Krebs-Ringer solution. The superior main mesenteric artery was dissected free and cut into rings (4–6 mm long). Electrophysiological parameters and tension were not measured simultaneously.

Electrophysiological Measurements

To measure membrane potential, an arterial segment was slit along its longitudinal axis and the strip was pinned down, luminal side upwards, to the bottom of a small recording chamber. The experimental chamber was perfused continuously (three bathvolumes/min) with warmed (35°C) and oxygenated Krebs-Ringer solution. The preparations were allowed to equilibrate for at least 60 min before starting the impalements with microelectrodes. Transmembrane potentials were measured using conventional microelectrodes, pulled with a vertical pipette puller (David Kopf, model 750, Tujunga, CA, USA) from 1 mm O.D. filamented borosilicate glass tubings (Hilgenberg, Germany). Microelectrodes were filled with 1 M KCl and their electrical resistances ranged from 40 to 80 MΩ. The measured potential was continuously followed on oscilloscope and traced with a pen recorder. Only continuous recordings of membrane potential changes produced by 10 min applications of either acetylcholine or levcromakalim were included. Moreover, the hyperpolarizations obtained under control conditions and after experimental interventions were usually compared during the same cell impalement. The various inhibitors were superfused for at least ten minutes before challenging the preparation with the hyperpolarizing vasodilators.

Tension Measurements

Arterial rings were suspended in an automated small vessel myograph (model 500 A, J.P. Trading, Aarhus, Denmark) between two stainless steel wires (40 μm diameter). One wire was fixed to a force-displacement transducer. The preparations were allowed to equilibrate for 30 minutes in the Krebs-Ringer bicarbonate solution bubbled with 95% O_2 and 5% CO_2 at 37°C. In order to obtain optimal conditions for active force development, the arteries were then set to an internal circumference corresponding to 90% of the value at a passive transmural pressure of 100 mm Hg (Mulvany and Halpern, 1977). After this normalization, the preparations were allowed to equilibrate again for at least half an hour. The arteries were then challenged three times with 10^{-5} M norepinephrine in high K^+ solution. When reproducible contractions were obtained, the preparations were exposed to norepinephrine (10^{-6} or 10^{-5} M) and relaxation curves to cumulative concentrations of acetylcholine, cromakalim or sodium nitroprusside were constructed. Dose-response curves were obtained in the continued presence of N^{ω}-nitro-L-arginine, and/or indomethacin, and/ or inhibitors of cytochrome P450. The incubation of the preparations with these agents lasted at least 10 minutes, except when using indomethacin or miconazole (30 minutes exposure).

Drugs and Solutions

All experiments were performed using a Krebs-Ringer bicarbonate solution of the following composition (in mmol/L): NaCl, 135; KCl, 5; NaHCO$_3$, 20; glucose, 10; CaCl$_2$, 2.5; MgSO$_4$.7H$_2$O, 1.3; KH$_2$PO$_4$, 1.2; EDTA, 0.026. High K^+ solution was prepared by equimolar replacement of 120 mM NaCl with KCl. Proadifen (SKF525a), miconazole nitrate, 1-aminobenzotriazole, 17-octadecynoic acid (17-ODYA), N^{ω}-nitro-L-arginine, indomethacin, norepinephrine bitartrate, acetylcholine chloride and cromakalim (racemate of + and − isomer) were all obtained from Sigma Chemical Co (St. Louis, MO, USA). Levcromakalim was kindly provided by Beecham Pharmaceuticals, Essex, UK.

All concentrations are expressed as final molar concentrations in the experimental chamber. Except for 1-aminobenzotriazole, all substances were added from appropriate stock solutions to the equilibrated Krebs-Ringer solution. Stock solutions were made in water, except for acetylcholine chloride (dissolved in phtalate-buffer pH 4.0), clotrimazole and miconazole (dissolved in dimethylsulfoxide), indomethacin and 17-ODYA (dissolved in pure ethanol) and cromakalim or levcromakalim (dissolved in 70% ethanol).

Statistics

Results are expressed as means ± SEM. Statistical significance was evaluated using Student's t-test for paired or unpaired observations, as appropriate. P-values smaller than 0.05 indicated significant differences; n indicates the number of preparations, obtained from different animals.

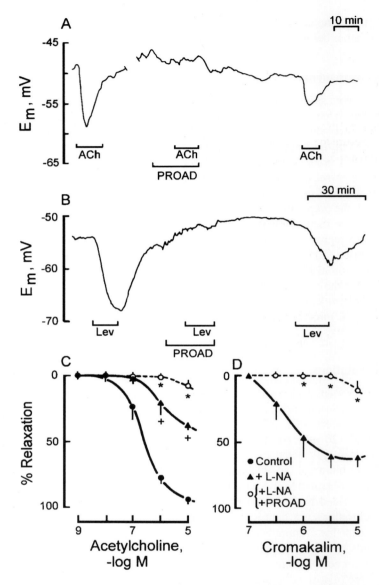

Figure 10-1. Influence of proadifen on endothelium-dependent and -independent responses. A,B: Original recordings of membrane potential (E_m) responses to acetylcholine (ACh, 10^{-6} M) (A) and to levcromakalim (Lev, 3×10^{-7} M) (B) in resting mesenteric artery cells before, during, and after application of proadifen (PROAD, 3×10^{-5} M). During the break in the record in A, the preparation was superfused with control solution. C,D: Relaxation in norepinephrine-contracted arteries induced by cumulative addition of increasing molar (M) concentrations of acetylcholine (C) or cromakalim (D) under control conditions, in the presence of N^{ω}-nitro-L-arginine (L-NA, 10^{-4} M) and in the simultaneous presence of N^{ω}-nitro-L-arginine and proadifen (3×10^{-5} M). The data are means ± SEM of five to eight experiments and are expressed as percent lowering of norepinephrine-induced tone. The crosses indicate that the difference between N^{ω}-nitro-L-arginine treatment and control groups, the asterisks between proadifen + N^{ω}-nitro-L-arginine treatment and N^{ω}-nitro-L-arginine treatment groups, is statistically significant (P < 0.05). Modified from Vanheel and Van de Voorde, J. Physiol. 1997, and Van de Voorde and Vanheel, J. Cardiovasc. Pharmacol. 1997, with permission.

RESULTS

Proadifen

The resting membrane potential (E_m) was stable and averaged -50.2 ± 0.5 mV (n = 49). Exposing the preparations to acetylcholine (10^{-6} M) in the superfusate produced a transient hyperpolarization to -68.3 ± 1.0 mV (n = 28). This peak hyperpolarization was not affected significantly by previous exposure of the preparation to indomethacin (5×10^{-5} M, not shown). Exposure to proadifen (3×10^{-5} M) virtually abolished the EDHF-mediated hyperpolarization (n = 5, Figure 10-1); the effect of proadifen was partly reversible (Figure 10-1).

Levcromakalim hyperpolarized the cell membrane in a concentration-dependent manner. In order to induce a hyperpolarization similar to that obtained with 10^{-6} M acetylcholine, 3×10^{-7} M levcromakalim was applied usually, producing a mean change in E_m of -15.6 ± 0.9 mV (n = 21) under control conditions. This hyperpolarization was endothelium-independent and was completely prevented by previous exposure to glibenclamide (10^{-5} M, not shown). In the presence of proadifen the levcromakalim-induced hyperpolarization was completely inhibited (n = 3). This influence was only partly reversible within the time limits of the experiments (Figure 10-1).

During contractions to norepinephrine, the cumulative addition of increasing concentrations of acetylcholine caused concentration-dependent relaxations which were partly inhibited by N^{ω}-nitro-L-arginine. The N^{ω}-nitro-L-arginine-resistant relaxation in response to acetylcholine was significantly inhibited by proadifen (Figure 10-1). The previous exposure to proadifen also prevented the relaxation to increasing concentrations of cromakalim (Figure 10-1). The drug had no influence on the relaxation induced by sodium nitroprusside (not shown).

Miconazole

Miconazole (10^{-4} M) depolarized the resting membrane by about 5.7 ± 0.6 mV (Figure 10-2). After exposing the preparation for at least 10 min to this inhibitor, the peak hyperpolarization in response to acetylcholine was decreased significantly to 8.9 ± 3.3 mV (n = 7). This inhibitory effect was partly reversible (Figure 10-2). Miconazole also inhibited the hyperpolarization in response to levcromakalim (3×10^{-7} M, n = 4). However, this inhibition was irreversible within the time limits of the experiments (Figure 10-2). Similar findings were obtained with another imidazole antimycotic, clotrimazole (3×10^{-5} M, not shown).

In the presence of N^{ω}-nitro-L-arginine, miconazole (3×10^{-5} M) increased basal tone (1.62 ± 0.9 mN), but norepinephrine-induced contractions decreased (from 7.9 ± 1.1 to 4.0 ± 0.9 mN). Under these conditions, the N^{ω}-nitro-L-arginine-resistant relaxation in response to increasing concentrations of acetylcholine was nearly abolished (Figure 10-2). Similarly, exposure to miconazole prevented cromakalim-induced relaxation (Figure 10-2).

17-Octadecynoic Acid

After previous exposure to 17-ODYA (5×10^{-6} M), which had no significant effect on the resting membrane potential, the acetylcholine-induced hyperpolarization was not

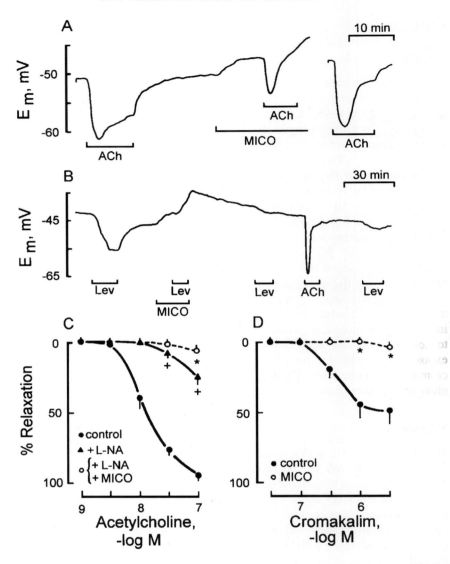

Figure 10-2. Influence of miconazole on endothelium-dependent and -independent responses. A,B: Original recordings of the membrane potential (E_m) response to acetylcholine (ACh, 10^{-6} M) (A) and to levcromakalim (Lev, 3×10^{-7} M) (B) in resting mesenteric artery cells in the absence and the presence of miconazole (MICO, 10^{-4} M). During the break in the trace in A, the preparation was superfused with control solution. Note, in B, intact peak hyperpolarization to acetylcholine after washout of miconazole, while responses to levcromakalim were still inhibited. C,D: Relaxation in norepinephrine-contracted arteries by cumulative addition of increasing molar (M) concentrations of (C) acetylcholine in control conditions, in the presence of N^{ω}-nitro-L-arginine (L-NA, 10^{-4} M) and in the simultaneous presence of N^{ω}-nitro-L-arginine and miconazole (MICO, 3×10^{-5} M) and (D) cromakalim in control conditions and in the presence of miconazole. The data are means ± SEM of five to six experiments and are expressed as percent lowering of norepinephrine-induced tone. The crosses indicate that the difference between N^{ω}-nitro-L-arginine treatment and control groups, the asterisks between miconazole + N^{ω}-nitro-L-arginine treatment and N^{ω}-nitro-L-arginine treatment groups (C) or between miconazole treatment and control groups (D), is statistically significant (P < 0.05). Modified from Vanheel and Van de Voorde, J. Physiol. 1997, and Van de Voorde and Vanheel, J. Cardiovasc. Pharmacol. 1997, with permission.

significantly influenced (n = 5, Figure 10-3). Similarly, 17-ODYA did not significantly affect the levcromakalim-induced hyperpolarization (n = 3, Figure 10-3).

In the contractile studies, 17-ODYA slightly increased basal tone (0.4 ± 0.1 mN) but had no influence on norepinephrine-induced contractions (from 11.9 ± 1.8 mN in the absence to 13.4 ± 1.7 mN in the presence of 17-ODYA). The N^{ω}-nitro-L-arginine-resistant relaxation to acetylcholine was not influenced by 17-ODYA, both in the absence (Figure 10-3) and in the presence of indomethacin (not shown, n = 6). 17-ODYA did not affect concentration-relaxation curves to cromakalim (Figure 10-3).

1-Aminobenzotriazole

In resting preparations, 1-aminobenzotriazole (2×10^{-3} M) increased the resting membrane potential on average by –2.6 ± 0.3 mV (Figure 10-4). The drug had no significant effect on the acetylcholine (n = 4, Figure 10-4) or on the levcromakalim-induced (n = 2, Figure 10-4) hyperpolarizations.

1-Aminobenzotriazole (10^{-3} M) decreased basal tone (by 0.3 ± 0.1 mN). The contraction to norepinephrine decreased from 14.9 ± 0.9 mN in the absence to 9.9 ± 0.9 mN in the presence of 1-aminobenzotriazole. The drug significantly augmented the N^{ω}-nitro-L-arginine-resistant relaxations to increasing concentrations of acetylcholine (Figure 10-4). A similar influence was observed when indomethacin was present throughout (not shown, n = 6). The cromakalim-induced relaxations also were potentiated by 1-aminobenzotriazole (Figure 10-4).

DISCUSSION

Soon after endothelium-dependent relaxations were described (Furchgott and Zawadzki, 1980), the involvement of cytochrome P450-metabolites of arachidonic acid was suggested (Singer and Peach, 1983b; Singer, Saye and Peach, 1984). When it became obvious that, in addition to NO, a hyperpolarizing factor partly underlies endothelium-mediated relaxations (Chen, Suzuki and Weston, 1988; Félétou and Vanhoutte, 1988; Taylor and Weston, 1988), this possibility has received renewed interest. Several inhibitors of cytochrome P450 were reported to inhibit endothelium-mediated, NO-independent relaxations in isolated blood vessels such as bovine and porcine coronary arteries, as well as the NO-independent component of the vasodilatation of the coronary and renal circulation of the rat (Fulton, McGiff and Quilley, 1992; Hecker *et al.*, 1994; Bauersachs, Hecker and Busse, 1994; Fulton *et al.*, 1995; Campbell *et al.*, 1996). Endothelial cells synthetize, by cytochrome P450-dependent monooxygenases, hydroxyeicosatetraenoic acids and epoxyeicosatrienoic acids (Moore, Spector and Hart, 1988; Rosolowski *et al.*, 1990b). Of these, 5,6-, 8,9-, 11,12- and 14,15-epoxyeicosatrienoic acids relax certain blood vessels (Pinto, Abraham and Mullane, 1987; Hecker *et al.*, 1994) and activate Ca^{2+}-dependent potassium channels in isolated smooth muscle cells from several species (Gebremedhin *et al.*, 1992; Hu and Kim, 1993; Campbell *et al.*, 1996). Epoxyeicosatrienoic acids (EETs), therefore, are strong candidates as EDHF.

However, other data do not support the hypothesis that EDHF might be a metabolite of arachidonic acid through the cytochrome P450 pathway. In the guinea-pig carotid artery,

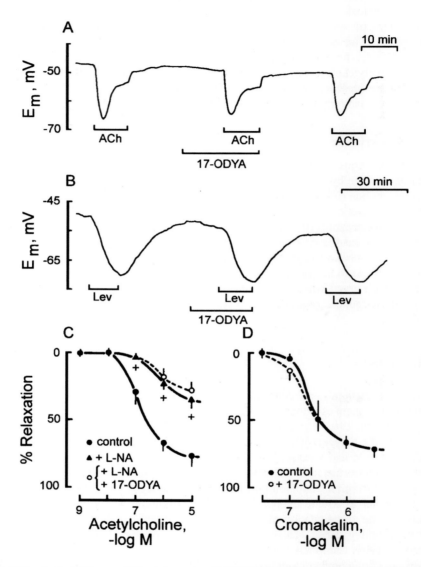

Figure 10-3. Lack of influence of 17-octadecynoic acid (17-ODYA) on endothelium-dependent and -independent responses. A,B: Original membrane potential (E_m) tracings in resting mesenteric artery cells in which the influence of exposure to 17-ODYA (5×10^{-6} M) on the hyperpolarization induced by acetylcholine (ACh, 10^{-6} M) (A) and levcromakalim (Lev, 3×10^{-7} M) (B) was tested. C,D: Relaxation in norepinephrine-contracted arteries by cumulative addition of increasing molar (M) concentrations of (C) acetylcholine under control conditions, in the presence of N^{ω}-nitro-L-arginine (L-NA, 10^{-4} M) and in the simultaneous presence of N^{ω}-nitro-L-arginine and 17-ODYA (5×10^{-6} M) and (D) cromakalim in control conditions and in the presence of 17-ODYA. The data are means ± SEM of eight experiments and are expressed as percent lowering of norepinephrine-induced tone. The crosses indicate that the difference between N^{ω}-nitro-L-arginine treatment and control groups is statistically significant (P < 0.05). Modified from Vanheel and Van de Voorde, J. Physiol. 1997, and Van de Voorde and Vanheel, J. Cardiovasc. Pharmacol. 1997, with permission.

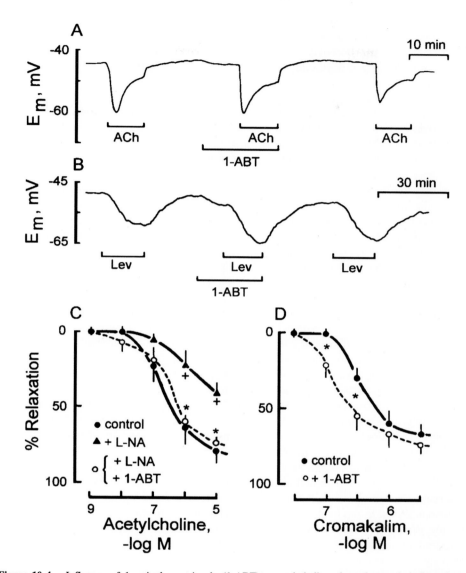

Figure 10-4. Influence of 1-aminobenzotriazole (1-ABT) on endothelium-dependent and -independent responses. A,B: Original membrane potential (E_m) tracings in resting mesenteric artery cells showing the influence of acetylcholine (ACh, 10^{-6} M) (A) and levcromakalim (Lev, 3×10^{-7} M) (B) in the absence and the presence of 1-aminobenzotriazole (2×10^{-3} M). Relaxation in norepinephrine-contracted arteries by cumulative addition of increasing molar (M) concentrations of (C) acetylcholine under control conditions, in the presence of N^{ω}-nitro-L-arginine (L-NA, 10^{-4} M), and in the simultaneous presence of N^{ω}-nitro-L-arginine and 1-aminobenzotriazole (1-ABT, 10^{-3} M) and (D) cromakalim in control conditions and in the presence of 1-aminobenzotriazole. The data are means ± SEM of six to seven experiments and are expressed as percent lowering of norepinephrine-induced tone. The crosses indicate that the difference between N^{ω}-nitro-L-arginine treatment and control groups, the asterisks between 1-aminobenzotriazole + N^{ω}-nitro-L-arginine treatment and N^{ω}-nitro-L-arginine treatment groups (C) or between 1-aminobenzotriazole treatment and control groups (D), is statistically significant (P < 0.05). Modified from Vanheel and Van de Voorde, J. Physiol. 1997, and Van de Voorde and Vanheel, J. Cardiovasc. Pharmacol. 1997, with permission.

the N$^\omega$-nitro-L-arginine-resistant membrane hyperpolarization elicited by acetylcholine was not affected by various inhibitors of the cytochrome P450-dependent monooxygenases (Corriu *et al.*, 1996a). Moreover, all EET-regioisomers failed to relax rat hepatic arteries denuded of endothelium (Zygmunt *et al.*, 1996). This discrepancy prompted the present study.

In the main mesenteric artery of the rat, the acetylcholine-induced peak hyperpolarization is glibenclamide-, indomethacin- and N$^\omega$-nitro-L-arginine-insensitive (Fujii *et al.*, 1992; 1993; Fukao *et al.*, 1995; Vanheel and Van de Voorde, 1997) and is therefore largely due to the action of EDHF. Similarly, the N$^\omega$-nitro-L-arginine-resistant relaxation is not affected by indomethacin and absent in high potassium solution (unpublished observations) and is, therefore, most probably exclusively mediated by the EDHF.

In a first series of experiments, the prototypical inhibitor of the cytochrome P450 systems, the alkylamine proadifen (SKF 525a), was used. Alkylamines are converted by cytochrome P450 to reactive intermediates that sequester the cytochrome in a catalytically inactive complexed state (Murray and Reidy, 1990). Proadifen fully blocked the acetylcholine-induced hyperpolarization and the N$^\omega$-nitro-L-arginine-resistant relaxation. These findings are consistent with the hypothesis that the endothelium-dependent hyperpolarizing factor is a cytochrome P450-derived metabolite (Hecker *et al.*, 1994; Fulton *et al.*, 1995). However, from these experiments an inhibitory influence of proadifen on the liberation of this factor and/or on the target for EDHF, the K$^+$ channels in the smooth muscle cell membrane, could not be excluded. Proadifen, e.g., possesses calmodulin antagonizing activity (Volpi *et al.*, 1981), and calmodulin antagonists inhibit endothelium-dependent hyperpolarizations (Nagao, Iliano and Vanhoutte, 1992b). In the present study, proadifen inhibited also the endothelium-independent hyperpolarization and relaxation produced by openers of ATP-regulated K$^+$ channels (levcromakalim and cromakalim). Similar observations have been made in the rat portal vein (Zygmunt *et al.*, 1996). This indicates that proadifen interferes not only with endothelial, but also with smooth muscle cells. Proadifen, however, did not block endothelium-independent relaxations to sodium nitroprusside. This excludes the possibility that it interferes with soluble guanylate cyclase (Bennet *et al.*, 1992) or with a more aspecific mechanism in the relaxation process.

Similar results were obtained in the experiments with the imidazole antimycotic miconazole. This molecule interferes with cytochrome P450 by direct formation of a nitrogenous ligand to the heme iron of P450 (Murray and Reidy, 1990). Exposure to this inhibitor diminished acetylcholine-induced and abolished levcromakalim-induced hyperpolarizations. Also the relaxations to both acetylcholine and cromakalim were inhibited by this compound.

These data suggest that proadifen and miconazole, in addition to blocking cytochrome P450, also interfere with hyperpolarizations and the resulting relaxations elicited by activation of ATP-regulated K$^+$-channels in the vascular smooth muscle cells. This interpretation is in line with several reports on interference of these substances with a variety of K$^+$-channel types. Thus, proadifen inhibits ATP-regulated K$^+$ channels in Xenopus oocytes (Sakuta and Yoneda, 1994, IC$_{50}$ 4×10^{-6} M) as well as the currents carried by the delayed rectifier and ATP-regulated K$^+$ channels in freshly isolated smooth muscle cells of the rat portal vein (Edwards *et al.*, 1996). Likewise, imidazole antimycotic agents have been reported to inhibit Ca^{2+}-dependent K$^+$ channels in human red cells (Alvarez, Montero and Garcia-Sancho, 1992b) and in rat carotid body cells (Hatton and Peers, 1996) as well as voltage-gated K^+ currents in these cells (Hatton and Peers, 1996) and in smooth muscle cells of the rat pulmonary artery (Yuan *et al.*, 1995). Thus, interpretation of the observations with proadifen and miconazole is difficult since they may block the actions of EDHF on the smooth muscle cells as well as the synthesis of the products of cytochrome P450-dependent monooxygenase.

17-ODYA was used as a third inhibitor of cytochrome P450. This substance is an effective inhibitor of those isoforms of cytochrome P450 monooxygenases which utilize long-chain fatty acids such as arachidonic acid as substrates. It inhibits both the ω-hydroxylase and epoxygenase pathways of the metabolism of arachidonic acid (Zou *et al.*, 1994). In the present experiments, neither endothelium-dependent nor endothelium-independent hyperpolarizations and relaxations were affected by the inhibitor, at a concentration which almost abolishes the formation of EETs in renal tissue (Zou *et al.*, 1994). This finding is in line with observations in the rat hepatic artery (Zygmunt *et al.*, 1996), but not in rat isolated perfused heart, where a slightly lower concentration of 17-ODYA inhibited bradykinin-induced vasodilator responses (Fulton *et al.*, 1995). The present findings are not in favor of the proposal that the endothelium-dependent hyperpolarizing factor liberated by acetylcholine is a cytochrome P450-derived metabolite from arachidonic acid in the main mesenteric artery of the rat.

In a last series of experiments 1-aminobenzotriazole was used. This substance causes an autocatalytic destruction of cytochrome P450 by forming benzyne, which alkylates the prosthetic heme group at the active center of cytochrome P450 (Murray and Reidy, 1990). In a concentration of 1 to 2×10^{-3} M, 1-aminobenzotriazole can be considered as an effective cytochrome P450-blocker. Indeed, in the isolated perfused rabbit lung, 10^{-3} M 1-aminobenzotriazole causes a 43% loss of cytochrome P450 activity within two minutes (Mathews, Dostal and Bend, 1985). In the presence of 2×10^{-3} M of 1-aminobenzotriazole, the acetylcholine- and levcromakalim-induced hyperpolarizations of the rat mesenteric artery were not inhibited. Moreover, the relaxations to both the endothelium-dependent and the endothelium-independent agents were not only not inhibited but potentiated. This potentiation might eventually have resulted from the reduction of the norepinephrine-induced contraction in the presence of 1-aminobenzotriazole. Indeed, N^ω-nitro-L-arginine-resistant relaxation is more prominent when the level of contraction is lower (Hatake, Wakabayashi and Hishida, 1995). Whatever the mechanism involved in this potentiation, however, these results combined with those of the 17-ODYA experiments strongly argue against EDHF being a cytochrome P450-dependent metabolite in the preparation studied.

The present results are in full agreement with the findings of Fukao *et al.* (1997a), who reported inhibition by proadifen, miconazole and clotrimazole of both acetylcholine-induced and pinacidil-induced hyperpolarization, but the lack of effect on these membrane potential changes of 17-ODYA and several other cytochrome P450 inhibitors in the same preparation. They are not consistent, however, with the increase and the decrease of the magnitude of the acetylcholine-induced endothelium-dependent hyperpolarization observed in the same preparation after induction and depletion of cytochrome P450, respectively (Chen and Cheung, 1996). In the same artery, however, induction of cytochrome P450 with phenobarbitone did not affect acetylcholine-induced endothelium-dependent hyperpolarizations (Fukao *et al.*, 1997a). The present findings do not permit to speculate further on the reasons for these discrepancies.

ACKNOWLEDGEMENTS

This work was supported by the Flemish Fund for Scientific Research (FWO-Vlaanderen). B.V. is a senior research associate of the FWO.

11 Role of Phospholipase A₂ in EDHF-Mediated Relaxation of the Porcine Coronary Artery

Neal L. Weintraub[1], Alan H. Stephenson[2], Randy S. Sprague[3] and Andrew J. Lonigro[4]

[1] Department of Internal Medicine, University of Iowa College of Medicine, 200 Hawkins Drive, Iowa City, IA 52242. Phone: 319-353-7807; Fax: 319-353-6343, E-mail: neal-weintraub@uiowa.edu

[2] Departments of Pharmacological & Physiological Science, Saint Louis University School of Medicine, 1402 S. Grand Blvd., St. Louis, Missouri, 63104. Phone: 314-577-8549; Fax: 314-577-8554; E-mail: stephens@sluvca.slu.edu;

[3] Department of Internal Medicine, Saint Louis University School of Medicine, 1402 S. Grand Blvd., St. Louis, Missouri, 63104. Phone: 314-577-8549; Fax: 314-577-8554; E-mail: spraguer@wpogate.slu.edu

[4] Departments of Pharmacological & Physiological Science and Internal Medicine, Saint Louis University School of Medicine, 1402 S. Grand Blvd., St. Louis, Missouri, 63104. Phone: 314-577-8549; Fax: 314-577-8554; E-mail: lonigro@sluvca.slu.edu.

Address for correspondence: Neal L. Weintraub, M.D., Department of Internal Medicine, The University of Iowa College of Medicine, 200 Hawkins Drive, E329 GH, Iowa City, IA 52246. Tel: (319) 353-7807; Fax: (319) 353-6343; E-mail: neal-weintraub@uiowa.edu

In porcine coronary artery, bradykinin-induced relaxation, in the presence of inhibitors of nitric oxide synthase and cyclooxygenase, is mediated by EDHF, which was proposed to be a non-cyclooxygenase metabolite of arachidonic acid. Unesterified arachidonic acid can be generated through activation of cytosolic phospholipase A₂ and/or by phospholipase C-mediated release of diacylglycerol, which is metabolized by lipases to yield arachidonic acid. To investigate the role of phospholipase A₂ and diacylglycerol lipase in bradykinin-induced EDHF production, porcine coronary rings treated with nitric oxide synthase and cyclooxygenase inhibitors were contracted with U46619. Relaxations to bradykinin or arachidonic acid were obtained before and after exposure to arachidonyl trifluoromethyl ketone or RHC 80267, inhibitors of phospholipase A₂ and diacylglycerol lipase, respectively. Arachidonyl trifluoromethyl ketone, but not RHC 80267, inhibited bradykinin-induced, EDHF-mediated relaxation. Under similar conditions, arachidonyl trifluoromethyl ketone did not inhibit the arachidonic acid-induced relaxation, suggesting that the compound did not block signaling pathways distal to arachidonic acid release. In porcine coronary arteries, arachidonyl trifluoromethyl ketone also blocked bradykinin-induced production of prostacyclin (an index of arachidonic acid release). When incubated with [³H]arachidonic acid, porcine coronary arteries formed products which co-migrated with prostaglandins and hydroxyeicosatetraenoic acids; treatment with nitric oxide synthase and cyclooxygenase inhibitors abolished the formation of prostaglandins, but not of hydroxyeicosatetraenoic acids. These findings suggest that the production of EDHF in the porcine coronary artery is dependent upon phospholipase A₂-mediated arachidonic acid release. Whether the arachidonic acid is metabolized to form EDHF, or whether it acts as a second messenger to stimulate EDHF production, remains to be determined.

KEYWORDS: Endothelium-derived hyperpolarizing factor, EDHF, bradykinin, phospholipase A₂, porcine coronary artery, arachidonic acid.

Neal L. Weintraub et al.

INTRODUCTION

The endothelium is an important regulator of the tone of vascular smooth muscle (Furchgott and Zawadzki, 1980). In addition to synthesizing nitric oxide and prostacyclin, the endothelium produces an unidentified factor, termed endothelium-derived hyperpolarizing factor (EDHF) (Beny and Brunet 1988a; Cowan and Cohen 1991; Nagao and Vanhoutte, 1992a). This factor relaxes vascular smooth muscle by activating potassium (K$^+$) channels, and its action can be prevented by depolarization with KCl or by the administration of K$^+$ channel blockers (Nagao and Vanhoutte, 1992a).

The observations that metabolites of arachidonic acid can activate K$^+$ channels, and, consequently, relax vascular smooth muscle (Jackson *et al.*, 1993; Campbell *et al.*, 1996), suggest that EDHF could be a metabolite(s) of arachidonic acid. Indeed, investigations in several blood vessels, including porcine coronary artery, suggest that relaxations attributable to EDHF are mediated by non-cyclooxygenase derivatives of arachidonic acid (Weintraub *et al.*, 1994; Hecker *et al.*, 1994). In particular, the cytochrome P-450-derived epoxyeicosatrienoic acids have been proposed to account for the activity of EDHF in the isolated, perfused rat heart and in bovine coronary arteries (Fulton *et al.*, 1995; Campbell *et al.*, 1996).

In the porcine coronary artery, both bradykinin-induced arachidonic acid release and EDHF-mediated relaxation are dependent upon activation of phospholipase C (Weintraub *et al.*, 1995). Phospholipase C can effect arachidonic acid-release by activating a cytosolic phospholipase A$_2$, which acts on membrane phospholipids to liberate arachidonic acid, and/or by providing diacylglycerol, which can be metabolized by lipases to yield unesterified arachidonic acid (Fulton, McGiff, and Quilley 1996).

In the present study, inhibitors of cytosolic phospholipase A$_2$ (arachidonyl trifluoromethyl ketone, a trifluromethyl ketone analogue of arachidonic acid) and diacylglycerol lipase (RHC 80267) were used to investigate the mechanisms of arachidonic acid release in bradykinin-induced, EDHF-mediated relaxation of porcine coronary artery. Also, to examine the profile of arachidonic acid metabolites produced by porcine coronary arteries, incubations were performed with [^{3}H]arachidonic acid in the absence and presence of inhibitors of nitric oxide synthase and cyclooxygenase.

METHODS

Measurement of Reactivity

Coronary arteries were dissected from pig hearts within 10 minutes after removal at a local slaughterhouse. The arteries were placed into ice-cold modified Krebs-Ringer bicarbonate solution ([composition in mmol/l]: NaCl, 118.3; KCl, 4.7; CaCl$_2$, 2.5; MgSO$_4$, 1.2; KH$_2$PO$_4$, 1.2; NaHCO$_3$, 25.0; Na-EDTA, 0.026; and glucose, 11.1), aerated with 95%O$_2$/5%CO$_2$, and transported to the laboratory. The arteries were stored at 4°C for a maximum of three days and cut into rings (3–5 mm in width) just prior to use. The rings were mounted onto stainless steel triangles, which were attached to isometric force transducers coupled to a polygraph for continuous recording of ring tension. Each ring was suspended in a water-jacketed (37°C) organ bath containing 10 or 20 ml Krebs solution which was aerated with

$95\%O_2/5\%CO_2$. Basal ring tension was gradually adjusted to 10 grams, a value which yields a maximal contraction to KCl (60 mmol/l) (Weintraub *et al.*, 1994). Each ring was contracted with KCl (60 mmol/l) until a stable degree of tension was achieved. After the final contraction to KCl, indomethacin (10^{-6} M) and N^ω-nitro-L-arginine methyl ester (10^{-4} M) were introduced into the organ chamber for 45 minutes prior to contracting with U46619 and were, thereafter, included for the duration of each experiment. Each ring was then contracted with U46619 (a thromboxane mimetic; $6 \times 10^{-9} - 2 \times 10^{-8}$ M) to achieve 40–80% of the maximal contraction obtained with KCl (60 mmol/l). When contraction was stable, the ring was relaxed in a cumulative-dose fashion with bradykinin ($3 \times 10^{-10} - 1 \times 10^{-7}$ M). The organ chamber was rinsed, and the ring was exposed to arachidonyl trifluoromethyl ketone ($3 \times 10^{-5} - 5 \times 10^{-5}$ M), RHC 80267 (5×10^{-5} M), or arachidonic acid (3×10^{-5} M) for 30 minutes. U46619 was re-applied to achieve a similar level of tension as the previous contraction, and the relaxation to bradykinin was repeated. In some experiments, following completion of concentration-response curves to bradykinin in the presence of arachidonyl trifluoromethyl ketone, the organ chambers were rinsed, arachidonyl trifluoromethyl ketone was added with or without RHC 80267, and the concentration-response curves were repeated. In other experiments, the effects of treatment with arachidonyl trifluoromethyl ketone on relaxations to arachidonic acid ($3 \times 10^{-7} - 1 \times 10^{-4}$ M) were determined.

Relaxations were expressed as the percent decrease from the U46619-induced tension.

Prostacyclin Production

In separate experiments, porcine coronary artery rings were equilibrated in 20 ml organ chambers and treated with N^ω-nitro-L-arginine methyl ester (10^{-4} M), but not indomethacin, for the duration of the experiments. The rings were contracted with U46619 and relaxed with bradykinin (10^{-7} M) to document adequate endothelial function; rings which relaxed less than 50% were discarded. The organ chambers were rinsed, and either arachidonyl trifluoromethyl ketone (5×10^{-5} M) or vehicle was introduced. After 30 minutes, the Krebs solution was removed and replaced with fresh solution containing arachidonyl trifluoromethyl ketone or vehicle. The rings were then contracted with U46619 for 20 minutes (which resulted in a stable level of tension in all experiments), at which time a 500 μl aliquot of Krebs solution was withdrawn from the bath for estimation of basal prostacyclin production, measured as 6-keto-PGF$_{1\alpha}$, the non-enzymatic hydrolysis product. The rings were then exposed serially to 10^{-8} M and 10^{-7} M bradykinin for 10 minutes, and 500 μL aliquots of Krebs solution were removed at the end of each dose-response for the calculation of bradykinin-induced production of prostacyclin.

After withdrawal from the organ chamber, the aliquots of Krebs solution were placed in 2 ml vials containing 50 μl of a solution of indomethacin (100 mg/ml) and EDTA (0.5 mg/ml) in 0.1 M Na_2CO_3 (pH = 7.0). The vials were vortexed and stored at $-20°C$ for analysis. Concentrations of 6-keto-PGF$_{1\alpha}$ were determined by means of an enzyme-linked immunoassay (Weintraub *et al.*, 1995) and normalized to ring wet weights. 6-keto-PGF$_{1\alpha}$ production in response to 10^{-8} M and 10^{-7} M bradykinin was calculated by subtracting the basal 6-keto-PGF$_{1\alpha}$ values from those values obtained following treatment with 10^{-8} M and 10^{-7} M bradykinin, respectively.

[³H]Arachidonic Acid

Porcine coronary arteries were removed, placed into sterile phosphate buffered saline containing 2 ml of a solution of penicillin (5000 U/ml) and streptomycin (5 mg/ml) per 100 ml phosphate buffered saline, and transported to the laboratory. In a laminar flow hood, the arteries were trimmed of fat and connective tissue and cut into strips, which were placed into 12-well plates containing 1 ml medium M-199 supplemented with 20% fetal calf serum per well. After 12 hours, N^{ω}-nitro-L-arginine methyl ester (10^{-4} M) and indomethacin (10^{-5} M), or vehicle, were added, and the incubation was continued for 2 more hours. The medium was removed and replaced with fresh medium supplemented with 1% fetal calf serum, to which 10^{-6} M [³H]arachidonic acid was added. N^{ω}-nitro-L-arginine methyl ester and indomethacin, or vehicle, were re-added, and the incubation was continued for 6 hours. The medium was then removed, acidified to pH 3–4 with H_2PO_4, and medium-associated lipids were extracted twice with 4 vol of ethyl acetate saturated with water. After evaporating the solvent under N_2, the lipid residue was dissolved in acetonitrile for separation by reverse-phase high-performance liquid chromatography (Fang *et al.*, 1996).

Chemicals

Bradykinin (acetate salt), arachidonic acid (sodium salt), U46619, indomethacin, and N^{ω}-nitro-L-arginine methyl ester were purchased from Sigma Chemical Co, arachidonyl trifluoromethyl ketone was from Cayman Chemical Co, and RHC 80267 was from Biomol. [³H]arachidonic acid was obtained from American Radiolabeled Chemicals Inc. Indomethacin, U46619, and arachidonyl trifluoromethyl ketone were dissolved in ethanol, and RHC 80267 was dissolved in dimethyl sulfoxide (DMSO). All other compounds were dissolved in distilled water. Final bath concentrations of ethanol and DMSO did not exceed 0.1%.

Statistical Analyses

All data are expressed as the mean ± SEM. The half-maximal effective concentrations (EC_{50}) of vasodilator compounds (expressed as -log[M]) were calculated for each concentration-response curve, and the EC_{50} values before and after treatment with inhibitor compounds were analyzed by Student's t test for paired observations. 6-keto-$PGF_{1\alpha}$ values among vehicle- and arachidonyl trifluoromethyl ketone-treated groups were analyzed by repeated measures analysis of variance followed by a Newman-Keul's post hoc analysis. P values of 0.05 or less were considered to be statistically significant.

RESULTS

Effects of Inhibitors of Arachidonic Acid Release on Bradykinin-Induced Relaxation

In porcine coronary artery rings contracted with U46619, in the presence of indomethacin and N^{ω}-nitro-L-arginine methyl ester, the application of bradykinin resulted in concentration-dependent relaxation (Figure 11-1). Repetitive responses to bradykinin, in the absence

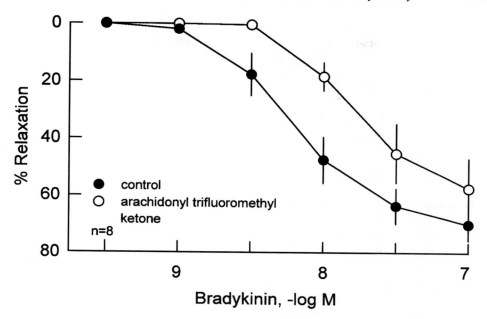

Figure 11-1. Effects of arachidonyl trifluoromethyl ketone (ATK) on bradykinin-induced, EDHF-mediated relaxation of porcine coronary artery rings. Rings were contracted with U46619 following treatment with indomethacin (10^{-5} M) and N^{ω}-nitro-L-arginine methyl ester (10^{-4} M), and bradykinin-induced relaxation was determined before (control) and after (ATK) incubation with arachidonyl trifluoromethyl ketone (3×10^{-5} M) for 30 minutes. Data shown as mean ± SEM (n = 8).

of inhibitor compounds, were unchanged (data not shown). Following the administration of the cytosolic phospholipase A_2 inhibitor, arachidonyl trifluoromethyl ketone (3×10^{-5} M), U46619-induced contraction was not diminished (data not shown). However, bradykinin-induced relaxation was inhibited by arachidonyl trifluormethyl ketone [EC_{50} = 8.2 ± 0.1 (vehicle) versus 7.7 ± 0.1 (arachidonyl trifluormethyl ketone), P < 0.05] (Figure 11-1). Incubation with higher concentrations of arachidonyl trifluoromethyl ketone (up to 5×10^{-5} M) did not result in additional inhibition of bradykinin-induced relaxation (data not shown).

To investigate the selectivity of the inhibitory effects of arachidonyl trifluoromethyl ketone, the compound was tested against arachidonic acid-induced relaxation of porcine coronary artery rings contracted with U46619. In the presence of indomethacin and N^{ω}-nitro-L-arginine methyl ester, arachidonic acid-induced relaxation was not inhibited by 3×10^{-5} M arachidonyl trifluoromethyl ketone (Figure 11-2).

The effects of unmodified arachidonic acid on bradykinin-induced relaxation were determined. Unlike arachidonyl trifluoromethyl ketone, arachidonic acid (3×10^{-5} M) failed to inhibit bradykinin-induced relaxation (Figure 11-2).

The diacylglycerol lipase inhibitor RHC 80267 (at concentrations up to 5×10^{-5} M) failed to inhibit bradykinin-induced relaxation of U46619-contracted porcine coronary artery rings in the presence of indomethacin and N^{ω}-nitro-L-arginine methyl ester (Figure 11-3). In some experiments, RHC 80267 (5×10^{-5} M) was administered to rings which

Neal L. Weintraub et al.

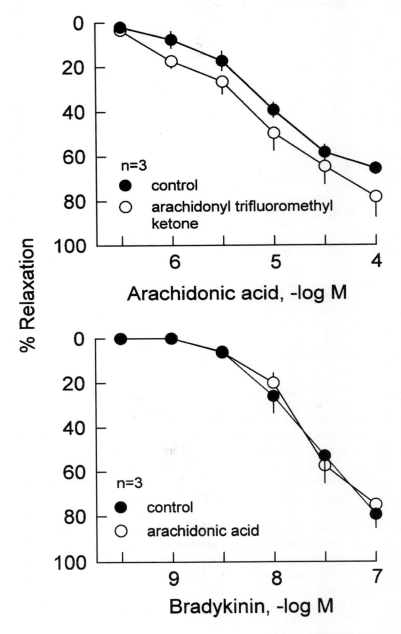

Figure 11-2. Effects of 3×10^{-5} M arachidonyl trifluoromethyl ketone on arachidonic acid-induced relaxation (upper panel) or 3×10^{-5} M unmodified arachidonic acid on bradykinin-induced relaxation (lower panel) of porcine coronary artery rings. All experiments were conducted in the presence of indomethacin (10^{-5} M) and N^{ω}-nitro-L-arginine methyl ester (10^{-4} M). Data shown as mean ± SEM (n = 3).

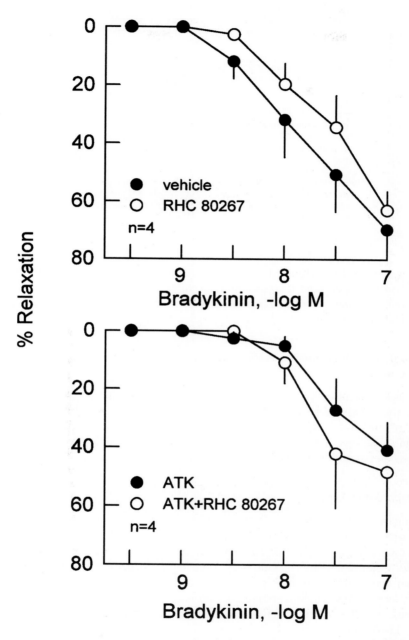

Figure 11-3. Effects of 5×10^{-5} M RHC 80267 (upper panel) and the combination of RHC 80267 and 3×10^{-5} M arachidonyl trifluoromethyl ketone (lower panel) on bradykinin-induced, EDHF-mediated relaxation of porcine coronary artery rings. Rings were contracted with U46619 following treatment with indomethacin (10^{-5} M) and N^{ω}-nitro-L-arginine methyl ester (10^{-4} M), and bradykinin-induced relaxation was determined before (control) and after treatment with the inhibitor compounds for 30 minutes. Data shown as mean $\pm$ SEM ($n = 4$).

Neal L. Weintraub et al.

Table 11.1   Effects of arachidonyl trifluoromethyl ketone on 6-keto-$PGF_{1\alpha}$ production by porcine coronary artery rings

	6-keto-$PGF_{1\alpha}$ (ng/gram)	
[Bradykinin]	*Vehicle*	*ATK*
10^{-8} M	19.2 ± 7	9.3 ± 3
10^{-7} M	68.1 ± 16	35.0 ± 6*

Porcine coronary artery rings with endothelium were treated with arachidonyl trifluoromethyl ketone (ATK; 5×10^{-5} M) or vehicle, and contracted with U46619 for 20 minutes, and then exposed serially to 10^{-8} M and 10^{-7} M bradykinin for 10 minutes. Before and after exposure to bradykinin, 500 μl aliquots of Krebs solution were removed for analysis of 6-keto-$PGF_{1\alpha}$. The bradykinin-induced values were normalized to ring weight and are expressed as the mean ± SEM (n = 6). The asterisk denotes a statistically significant difference versus vehicle (P < 0.05).

had been exposed to arachidonyl trifluoromethyl ketone (3×10^{-5} M). Under these conditions, bradykinin-induced relaxation did not differ among rings treated with both RHC 80267 and arachidonyl trifluoromethyl ketone as compared with those treated with arachidonyl trifluoromethyl ketone alone (Figure 11-3).

Arachidonyl Trifluoromethyl Ketone and Bradykinin-Induced Prostacyclin Production

Following treatment with arachidonyl trifluormethyl ketone, the amount of 6-keto-$PGF_{1\alpha}$ formed after exposure to 10^{-7} M bradykinin was inhibited by 50% as compared with vehicle (Table 11-1). The amount of 6-keto-$PGF_{1\alpha}$ formed after exposure to 10^{-8} M bradykinin also tended to be reduced by arachidonyl trifluoromethyl ketone, although the value was not statistically significant as compared with vehicle.

[³H]Arachidonic Acid

Following a 12 hour incubation in tissue culture media, porcine coronary arteries were treated with vehicle or the combination of indomethacin (10^{-5} M) and N^{ω}-nitro-L-arginine methyl ester (10^{-4} M) for 2 hours, after which they were incubated with [³H]arachidonic acid. Six hours later, the medium was removed, and medium-associated lipids were extracted and analyzed by high-performance liquid chromatography. In vehicle-treated arteries, large peaks which co-migrated with arachidonic acid and prostaglandins were observed, as well as small peaks which co-migrated with hydroxyeicosatetraenoic acids (Figure 11-4). No products which co-migrated with epoxyeicosatrienoic acids or their major metabolites, the dihydroxyeicosatrienoic acids, were detected. In the presence of indomethacin and N^{ω}-nitro-L-arginine methyl ester, prostaglandin production was abolished, while the peak which co-migrated with hydroxyeicosatetraenoic acids was unaffected (Figure 11-4). Similar results were obtained after washing the arteries, incubating with the calcium ionophore A23187 (2×10^{-6} M) for 30 minutes, and analyzing the radioactive

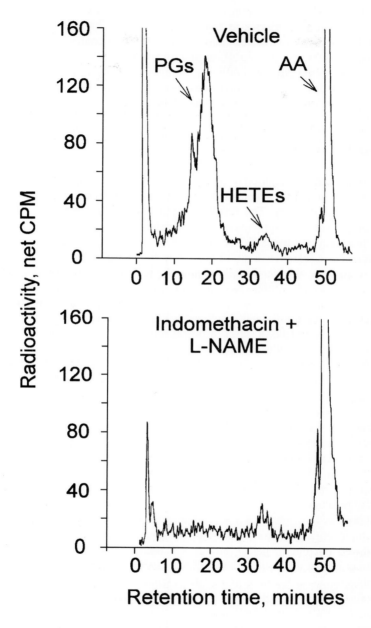

Figure 11-4.   Chromatogram showing production of arachidonic acid metabolites by porcine coronary arteries. Porcine coronary arteries were treated with vehicle (upper panel) or N^{ω}-nitro-L-arginine methyl ester (10^{-4} M) and indomethacin (10^{-5} M) (lower panel) for 2 hours in medium M-199, followed by incubation with 10^{-6} M [^{3}H]arachidonic acid. After 6 hours, the medium was removed and analyzed by reverse-phase high-performance liquid chromatography. The major radiolabeled peaks co-migrated with prostaglandins (PGs), hydroxy-eicosatetraenoic acids (HETEs), and arachidonic acid (AA). Similar results were obtained in duplicate experiments performed under identical conditions.

products released into the medium (data not shown). To exclude the possibility that maintaining the coronary arteries in tissue culture conditions abolished their ability to produce EDHF, bradykinin-induced relaxations of U46619-contracted rings, in the presence of indomethacin and N^ω-nitro-L-arginine methyl ester, were determined. In rings maintained under tissue culture conditions, bradykinin-induced, EDHF-mediated relaxations were enhanced as compared with those observed in freshly-harvested rings, suggesting that EDHF production was preserved (data not shown).

DISCUSSION

In the present study, arachidonyl trifluoromethyl ketone, an inhibitor of cytosolic phospholipase A_2, selectively blocked bradykinin-induced, EDHF-mediated relaxation of the porcine coronary artery. The results suggest that arachidonic acid release through activation of cytosolic phospholipase A_2 participates in the signal transduction pathway leading to bradykinin-induced EDHF production.

Until recently, a selective inhibitor of cytosolic phospholipase A_2, which has been suggested to mediate the hormonally-regulated arachidonic acid release in cell signal transduction (Lin *et al.*, 1992), has not been available. Traditional phospholipase A_2 inhibitors, such as quinacrine, are inhibitory against secretory types of phospholipase A_2, which differ both structurally and functionally from cytosolic phospholipases A_2 (Mayer and Marshall, 1993). Indeed, a series of potent inhibitors of secretory phospholipases A_2 did not inhibit thrombin-induced arachidonic acid release from platelets, suggesting that the inhibitors were not active against cytosolic forms of the enzyme (Bartoli *et al.*, 1994). In contrast, concentrations of arachidonyl trifluoromethyl ketone which inhibit cytosolic phospholipases A_2 did not inhibit secretory forms of the enzyme (Street *et al.*, 1993), suggesting that the compound can be used to selectively investigate the role of cytosolic phospholipases A_2 in cell signal transduction.

In porcine coronary artery rings pretreated with indomethacin and N^ω-nitro-L-arginine methyl ester, the bradykinin-induced relaxation, which is mediated by EDHF (Weintraub *et al.*, 1994), was inhibited by arachidonyl trifluoromethyl ketone. That the compound blocked phospholipase A_2 is suggested by its inhibitory effects on bradykinin-induced production of prostacyclin. Thus, one interpretation of these observations is that arachidonyl trifluoromethyl ketone inhibited a cytosolic phospholipase A_2 which is required for bradykinin-induced EDHF production. This interpretation is supported by the observation that unmodified arachidonic acid, which is structurally identical to arachidonyl trifluoromethyl ketone except that it lacks the trifluoromethyl ketone moiety required to inhibit cytosolic phospholipase A_2 (Street *et al.*, 1993), failed to block bradykinin-induced, EDHF-mediated relaxation. Alternatively, it is possible that arachidonyl trifluoromethyl ketone inhibited bradykinin-induced EDHF production through effects unrelated to phospholipase A_2 inhibition, such as through inhibition of phospholipase C (which is required for bradykinin-induced EDHF production in porcine coronary artery) (Weintraub *et al.*, 1995). However, the observation that arachidonyl trifluoromethyl ketone did not inhibit U46619-induced contraction, which is mediated through phospholipase C activation (Dorn and Becker, 1993), suggests that this explanation is unlikely. Finally, arachidonyl trifluoromethyl ketone might have inhibited bradykinin-induced, EDHF-mediated relaxation by non-specifically interfering with relaxation of smooth muscle (i.e., by inhibiting K^+ channels). Since arachidonyl trifluoromethyl ketone did not inhibit arachidonic acid-induced relaxations, which, like EDHF, is mediated by a K^+-sensitive mechanism (Weintraub *et al.*, 1994), this explanation also seems improbable. The absence of an inhibitory effect of arachidonyl trifluoromethyl ketone on arachidonic acid-induced relaxation also suggests that the compound did not block signal transduction pathways distal to arachidonic acid release. Therefore, arachidonyl trifluoromethyl ketone most likely attenuated bradykinin-induced, EDHF-mediated relaxation by inhibiting a cytosolic phospholipase A_2, which participates in the signal transduction pathway leading to EDHF production.

In the isolated, perfused rat heart, bradykinin-induced vasodilatation, in the presence of inhibitors of nitric oxide synthase and cyclooxygenase, was blocked by arachidonyl trifluoromethyl ketone (Fulton, McGiff, and Quilley 1996). Likewise, in the isolated, perfused rat mesenteric bed pretreated with inhibitors of nitric oxide synthase and cyclooxygenase, vasodilator responses to acetylcholine and histamine were inhibited by arachidonyl trifluoromethyl ketone (Adeagbo and Henzel, 1998). In both of these studies, arachidonyl trifluoromethyl ketone also blocked release of 6-keto-$PGF_{1\alpha}$ from the respective vascular beds. The potency of arachidonyl trifluoromethyl ketone to inhibit EDHF-mediated vasorelaxation and 6-keto-$PGF_{1\alpha}$ production in the present study was far less than that reported in the aforementioned studies. These differences could relate to differences in species, vascular beds, or blood vessel size. Nevertheless, our these findings, when taken together with the previous reports, support the general view that phospholipase A_2-mediated arachidonic acid release is a fundamental step in EDHF signal transduction.

In contradistinction to arachidonyl trifluoromethyl ketone, RHC 80267, a diacylglycerol lipase inhibitor (Chuang and Severson, 1990), at concentrations of up to 5×10^{-5} M, failed to inhibit bradykinin-induced, EDHF-mediated relaxation. Similar findings were reported with this compound in bradykinin-induced, EDHF-mediated dilation of the isolated, perfused rat heart (Fulton, McGiff, and Quilley 1996). Although an assay of diacylglycerol lipase activity was not employed to confirm that this concentration of RHC 80267 is inhibitory, the compound was reported to inhibit diacylglycerol lipase activity in cardiac myocytes with an IC_{50} of 3.5×10^{-6} M (Chuang and Severson, 1990). Thus, the observation that 5×10^{-5} M RHC 80267 did not inhibit bradykinin-induced, EDHF-mediated relaxation suggests that diacylglycerol activity is not required for the production of EDHF by bradykinin in porcine coronary artery. Moreover, since the administration of RHC 80267 to rings treated with arachidonyl trifluoromethyl ketone did not result in additional inhibition of bradykinin-induced, EDHF-mediated relaxation (beyond that observed with arachidonyl trifluoromethyl ketone alone), a role for diacylglycerol lipase in EDHF production under circumstances in which cytosolic phospholipase A_2 activity is inhibited also appears unlikely.

To begin to address whether EDHF in porcine coronary artery is a non-cyclooxygenase metabolite of arachidonic acid, porcine coronary arteries were incubated with [^{3}H]arachidonic acid, and the products released into the medium were analyzed by high-performance liquid chromatography. Large peaks which co-migrated with prostaglandins and small peaks which co-migrated with hydroxyeicosatetraenoic acids were detected. In rings pretreated with indomethacin and N^ω-nitro-L-arginine methyl ester, only the hydroxyeicosatetraenoic acid peaks were observed. Hydroxyeicosatetraenoic acids can be produced through lipoxygenase- or cytochrome P-450-mediated metabolism of arachidonic acid. Lipoxygenase metabolism results in the initial formation of unstable hydroperoxyeicosatetraenoic acid intermediates, which are rapidly reduced to their corresponding hydroxyeicosatetraenoic acids by peroxidases (Feinmark *et al.*, 1992). Lipoxygenase-derived hydroxyeicosatetraenoic acids activate potassium channels in neural and adrenal chromaffin cells and in cardiomyocytes (Kurachi *et al.*, 1989; Feinmark *et al.*, 1992; Twitchell, Pena, and Rane 1997). The specific hydroxyeicosatetraenoic acids produced by porcine coronary arteries, the pathways of their production, and the contribution of these pathways to EDHF-mediated relaxation remain to be determined.

Epoxyeicosatrienoic acids, or their diol metabolites, were not detected under any incubation conditions, either in medium- or cell-associated lipids (data not shown). Nevertheless, it is possible that these metabolites are produced from endogenous sources of arachidonic acid, thereby leading to formation of non-radiolabeled products which would not be detectable by the methods employed in the present study. However, the mechanism of epoxyeicosatrienoic acid-induced relaxation of porcine coronary arteries differs from bradykinin-induced, EDHF-mediated relaxation, suggesting that EDHF in porcine coronary artery is not an epoxyeicosatrienoic acid (Weintraub *et al.*, 1997). The arachidonic acid metabolite anandamide may be an EDHF in the isolated, perfused rat mesentary (Randall *et al.*, 1996). However, when administered to porcine coronary artery rings, anandamide produced a weak contraction (data not shown), suggesting that production of anandamide is also unlikely to account for EDHF-mediated vasorelaxation in porcine coronary artery. Thus, while the results of this study suggest that EDHF production in porcine coronary artery requires phospholipase A_2-mediated arachidonic acid release, whether the arachidonic acid is metabolized to form EDHF, or whether it acts as a second messenger to stimulate the production of a chemically-unrelated hyperpolarizing factor, remains to be determined.

ACKNOWLEDGEMENTS

The authors gratefully acknowledge Jo Schreiweis and Terry Kaduce for assistance in conducting the experiments, and Dr. Arthur Spector for helpful discussions and review of the manuscript. The authors also thank Ruzicka's Meat Processing in Solon, IA, for supplying the porcine coronary arteries.

12 K+ is an Endothelium-Derived Hyperpolarizing Factor: An Alternative Working Hypothesis

G. Edwards, M.J. Gardener and A.H. Weston

School of Biological Sciences, G38 Stopford Building, University of Manchester, Oxford Rd, Manchester M13 9PT, UK.
Address for correspondence: G. Edwards, Tel. (++44) 161 275 5488;
Fax: (++44) 161 275 5600; email: gedwards@man.ac.uk
M.J. Gardener, Tel. (++44) 161 275 5488; Fax: (++44) 161 275 5600;
email: matthew.gardener@man.ac.uk
A.H. Weston, Tel. (++44) 161 275 5490; Fax: (++44) 161 275 5600;
email: aweston@man.ac.uk

In the rat hepatic artery, exposure to acetylcholine produced an endothelium-dependent hyperpolarization of the smooth muscle which was inhibited by a mixture of charybdotoxin plus apamin whereas the combination of iberiotoxin plus apamin was ineffective. Increasing the extracellular concentration of K^+ by 5×10^{-3} M to 9.6×10^{-3} M also hyperpolarized the smooth muscle cells of hepatic artery, but this effect was endothelium-independent and was not inhibited by either charybdotoxin plus apamin or iberiotoxin plus apamin. The addition of either ouabain or Ba^{2+} to the physiological solution produced some inhibition of the hyperpolarizations to acetylcholine and K^+ but the combination of ouabain plus Ba^{2+} abolished the acetylcholine- and K^+-induced increases in membrane potential. These results suggest that the endothelium-derived hyperpolarization produced by acetylcholine is generated by the K^+ effluxing via charybdotoxin- and apamin-sensitive potassium channels in the endothelium following their activation by an acetylcholine-induced increase in intracellular Ca^{2+}. The resulting accumulation of K^+ between the endothelial and vascular smooth muscle layers stimulates the ouabain-sensitive Na^+/K^+-ATPase and increases the outward movement of K^+ through Ba^{2+}-sensitive inwardly-rectifying potassium channels in the smooth muscle cells. Thus, endothelium-derived hyperpolarizing factor is K^+ derived from the vascular endothelium.

KEYWORDS: Na^+/K^+-ATPase, K^+-channels, K^+, endothelium-derived hyperpolarizing factor, inward rectifier

INTRODUCTION

The relaxing and hyperpolarizing properties of endothelium-derived hyperpolarizing factor (EDHF) have been well-characterised (Garland *et al.*, 1995; Mombouli & Vanhoutte, 1997; Edwards & Weston, 1998) but neither the identity of EDHF nor the mechanism by which it hyperpolarizes vascular smooth muscle is known. Early observations (Félétou and Vanhoutte, 1988) suggested that EDHF might stimulate the smooth muscle Na^+/K^+-ATPase, although other laboratories were unable to substantiate these findings (Chen, Hashitani & Suzuki, 1989; Zygmunt & Högestätt, 1996b). However, an unequivocal observation is that in all preparations in which there is evidence for EDHF release, its effects are abolished by apamin (an inhibitor of the small-conductance Ca^{2+}-sensitive K^+-channel, SK_{Ca}), charybdotoxin (a non-selective inhibitor of the large-conductance Ca^{2+}-sensitive

K⁺-channel, BK_{Ca}), or a combination of apamin plus charybdotoxin (Waldron & Garland, 1994b; Zygmunt & Högestätt, 1996b; Corriu *et al*., 1996b; Zygmunt *et al*., 1997a; see Edwards & Weston, 1998). The finding that iberiotoxin, the selective inhibitor of BK_{Ca}, is unable to substitute for charybdotoxin (at least in the rat hepatic artery in which apamin + charybdotoxin is required for full inhibition of the response to EDHF; Zygmunt & Högestätt, 1996b; Zygmunt *et al*., 1997a), suggests that the K⁺-channel involved is not BK_{Ca}.

In most of the evaluations of EDHF-induced opening of K⁺ channels, it has been assumed that these are located on the vascular smooth muscle. This is true for the charybdotoxin- and apamin-sensitive conductances, inhibition of which abolishes EDHF-induced relaxations (see Zygmunt & Högestätt, 1996b; Zygmunt *et al*., 1997a). Furthermore, the loss of ^{86}Rb on exposure of vessels to acetylcholine was interpreted as an indicator of K⁺ loss from the vascular smooth muscle (Chen, Suzuki & Weston, 1988). However, the efflux of ^{86}Rb from cultured endothelial cells on exposure to bradykinin (an agent which also releases EDHF; Bény, 1990a) was clearly described by Gordon and Martin (1983) and K⁺ channels sensitive to charybdotoxin and to apamin, respectively, have been reported on the endothelial cells of rat aorta (Marchenko & Sage, 1996). Since small increases in extracellular K⁺ are known to hyperpolarize vascular smooth muscle cells (Knot, Zimmermann & Nelson, 1996; Quayle, Dart & Standen, 1996), the present microelectrode study was undertaken to determine whether the hyperpolarizations attributed to EDHF could in fact be generated by K⁺ effluxing from the vascular endothelium.

METHODS

All results were obtained from hepatic arteries removed from male Sprague-Dawley rats weighing 150–200g. The whole liver was excised and the hepatic artery was cleaned of major components of connective tissue under a dissecting microscope. A segment of artery was removed and carefully fixed without stretching onto the Sylgard base of a thermostatically-controlled bath using fine dissecting pins inserted through the remaining connective tissue. In some experiments, the endothelium was removed by cannulating and then perfusing the artery for 20s with distilled water before its removal from the liver.

The microelectrode organ chamber (volume 3 ml) was superfused at 3 ml/min with Krebs solution (37°C) comprising (10^{-3} M) NaCl 119, KCl 3.4, $MgCl_2$ 1.2, $CaCl_2$ 1.5, KH_2PO_4 1.2, $NaHCO_3$ 15, glucose 11 and containing 10^{-5} M indomethacin and 3×10^{-4} M N^{ω}-nitro-L-arginine. Electrodes, which were filled with 3 M KCl and had resistances of 35–60 MΩ, were flexibly mounted to the head stage of a high-impedance amplifier.

The general experimental design was to obtain long-term recordings from vascular smooth muscle cells which acted as their own controls. This often involved exposure to acetylcholine and then K⁺ followed by repeat exposures to both these agents in the presence of first one and then a combination of modifying drugs. For this reason, no attempt was made to achieve equilibrium conditions with acetylcholine and K⁺. Instead, a small bolus of stock solution, calculated to achieve instantaneously the desired concentration, was pipetted into the bath. In contrast, modifying drugs such as Ba^{2+}, ouabain, apamin, charybdotoxin and iberiotoxin were added to the thermostatically-controlled reservoirs which fed the recording chamber. A period of 5 min was then allowed before the effects of acetylcholine and K⁺ were re-examined under the modified conditions.

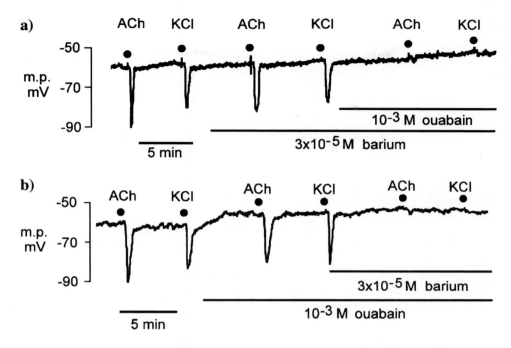

Figure 12-1. Comparison of the hyperpolarizations induced by 10^{-5} M acetylcholine (ACh) with those produced by elevation of extracellular K^+ by 5×10^{-3} M (KCl) in endothelium-intact segments of rat hepatic artery. In the presence of indomethacin (10^{-5} M) and N^{ω}-nitro-L-arginine (3×10^{-4} M), initial exposure to barium (a) or ouabain (b) had a small inhibitory effect on the hyperpolarizations produced by ACh and K^+. However, the changes in membrane potential were abolished in the presence of both inhibitors (both panels). Each trace is part of a continuous microelectrode recording from two smooth muscle cells isolated from two different arteries.

Membrane potential data were stored on magnetic tape and the digitised for subsequent analysis using a MacLab A-D converter.

RESULTS

Ba^{2+} and Ouabain

In segments of artery with an intact endothelium, the membrane potential was increased in the presence of 10^{-5} M acetylcholine when the extracellular K^+ concentration was increased by 5×10^{-3} M to 9.6×10^{-3} M (Figure 12-1). The addition of Ba^{2+} (3×10^{-5} M) to the physiological salt solution usually resulted in a small depolarization although there was no effect in some individual vessels. Under these conditions, acetylcholine- and K^+-induced hyperpolarizations were reduced and the resistant component was inhibited by subsequent exposure to ouabain (10^{-3} M) in the continuing presence of Ba^{2+} (Figure 12-1). Similarly, exposure to ouabain depolarized the vessels and the increase in membrane potential produced by acetylcholine and K^+ was reduced but not abolished (Figure 12-1). On addition of Ba^{2+}, however, acetylcholine- and K^+-induced hyperpolarizations were essentially abolished (Figure 12-1).

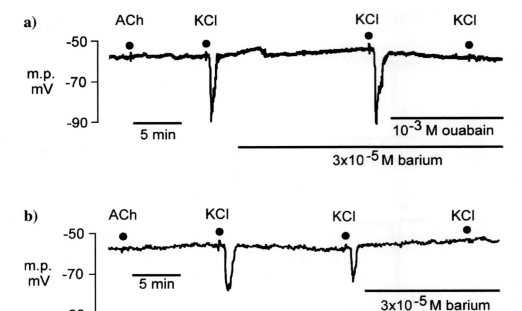

Figure 12-2. Effects of barium and ouabain on hyperpolarization induced by 10^{-5} M acetylcholine (ACh) or by elevation of extracellular K^+ by 5×10^{-3} M (KCl) in rat hepatic arteries without endothelium. In the presence of indomethacin (10^{-5} M) and N^ω-nitro-L-arginine (3×10^{-4} M), the response to KCl was similar to that in the intact artery (see Figure 12-1), although acetylcholine was without effect. Neither barium (a) nor ouabain (b) had a large inhibitory effect on the K^+-induced hyperpolarizations which were abolished in the presence of both inhibitors (a,b).

In preparations without endothelium (Figure 12-2), both the resting membrane potential and the magnitude of the K^+-induced hyperpolarization were similar to those of intact vessels although there was no essentially no hyperpolarization in response to acetylcholine (0.4 ± 0.4 mV, n = 7). Under these conditions, the combination of Ba^{2+} and ouabain abolished the response to elevation of extracellular K^+ whereas only small inhibitory effects were produced by either Ba^{2+} or ouabain alone (Figure 12-2).

Effects Of K^+ Channel Toxins

On addition of charybdotoxin plus apamin (each 10^{-7} M) to the physiological salt solution, the responses to acetylcholine were essentially abolished whereas those to K^+ were unaffected or even increased (Figure 12-3). In contrast, the combined presence of iberiotoxin plus apamin (each 10^{-7} M) did not modify the hyperpolarizing effects of either acetylcholine or K^+ (Figure 12-3).

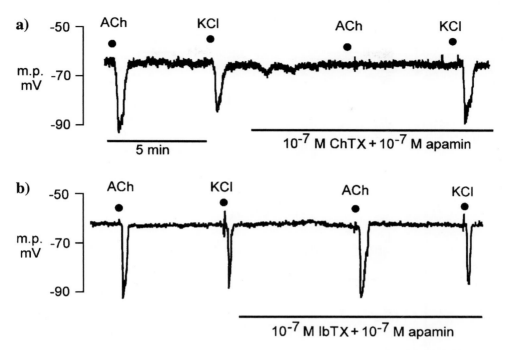

Figure 12-3. Effect of K^+ channel toxins on the hyperpolarizations induced by 10^{-5} M acetylcholine (ACh) or by elevation of extracellular K^+ by 5×10^{-3} M (KCl) in endothelium-intact segments of rat hepatic artery. a) On exposure to 10^{-7} M apamin + 10^{-7} M charybdotoxin, responses to acetylcholine were abolished but the hyperpolarization induced by application of KCl was unaffected. b) On exposure to 10^{-7} M apamin + 10^{-7} M iberiotoxin, responses to both acetylcholine and to KCl were unaffected.

DISCUSSION

The New Hypothesis

The results of the present study strongly indicate that K^+ liberated from the vascular endothelium is identical to EDHF, at least in the rat hepatic artery (see Figure 12-4). Thus, a 5×10^{-3} M increase in extracellular K^+ ($[K^+]_o$) produced a membrane hyperpolarization with characteristics very similar to those of EDHF (liberated by acetylcholine). The K^+-induced response, like that to EDHF, was inhibited by both Ba^{2+} and ouabain and abolished by a combination of these agents. This suggests that the hyperpolarizing actions of K^+ are generated by activation of both Na^+/K^+-ATPase and inwardly-rectifying K^+ channels on the vascular smooth muscle. A critical finding was that neither Ba^{2+} nor ouabain alone was a very effective inhibitor in contrast to the combination of these agents. A possible explanation for this is that if only one 'arm' of this activation pathway is inhibited, the other is able almost fully to compensate (Figure 12-4).

For the novel hypothesis to be tenable, there must be evidence a) that K^+ is liberated from the endothelium following exposure to agonists which generate an 'EDHF' response and b) that exogenous K^+ can generate a membrane hyperpolarization with characteristics identical to those of the 'EDHF'-induced change in membrane potential.

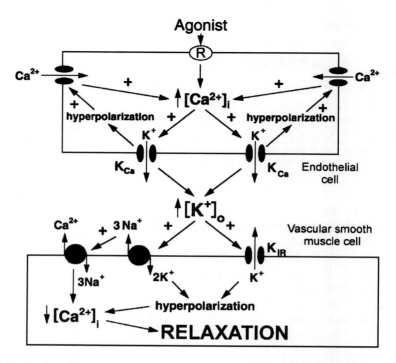

Figure 12-4. EDHF is identical to K+ derived from the endothelium in the rat hepatic artery. Hypothetical pathway linking agonist-induced effects on the endothelial cells (e.c.) of hepatic artery to hyperpolarization and relaxation of the underlying vascular smooth muscle cells (v.s.m.). It is proposed that the agonist-induced elevation of intracellular calcium in endothelium stimulates the opening of two types of calcium-sensitive K+-channel (K_{Ca}), one of which is inhibited by apamin and the other by charybdotoxin. K+ efflux through these channels hyperpolarizes endothelial cells which stimulates calcium influx through voltage-insensitive calcium channels and further increases the intracellular calcium concentration. The increase in extracellular K+ (putatively identical to EDHF) is thought to increase outward current through barium-sensitive inwardly-rectifying K+-channels (K_{IR}) and to stimulate ouabain-sensitive Na+/K+-ATPase on the adjacent smooth muscle cells. The effect of each of these actions is to hyperpolarize the smooth muscle, opposing calcium influx through voltage-sensitive K+-channels and thus producing a relaxation. The reduction in intracellular Na+ which results from an increase in the activity of the Na+/K+-ATPase may stimulate the Na+/Ca2+ exchanger, leading to a further reduction in intracellular calcium.

Liberation Of K+ From The Endothelium

Early ion flux studies showed that there was a detectable loss of the K+ marker ^{86}Rb from blood vessels exposed to acetylcholine (Chen, Suzuki & Weston, 1988; Taylor *et al.*, 1988). These data, together with the observation that a combination of charybdotoxin and apamin fully inhibits the EDHF-induced relaxations in a variety of blood vessels, suggest that the K+ channels involved are located on the vascular smooth muscle (see Edwards *et al.*, 1998a). One possible explanation involved the EDHF-induced opening of a K+ channel comprising a heteromultimeric assembly of α-subunits, the combination of which conferred the combined toxin-sensitivity profile. Alternatively, two different K+ channel subtypes could be involved, one charybdotoxin-sensitive and the other inhibited by apamin (see Zygmunt *et al.*, 1997a). Efflux studies showed that several agents, including bradykinin, produced a concentration-dependent loss of ^{86}Rb from cultured endothelial cells (Gordon & Martin, 1983), an indication that the endothelium could be the source of this tracer ion in the experiments (Chen, Suzuki & Weston, 1988; Taylor *et al.*, 1988) involving whole blood vessels. Moreover, K^+ channels sensitive to apamin and to charybdotoxin have been identified on the endothelial cells of rat aorta (Marchenko & Sage, 1996).

Taking these previous observations together with the findings of the present study, there is a clear indication that the inhibition of EDHF by charybdotoxin and apamin (Waldron & Garland, 1994b; Corriu *et al.*, 1996b; Zygmunt & Högestätt, 1996b) is likely to result from an action of the toxins on K^+ channels in the endothelial cells and not on those in vascular smooth muscle. Thus, EDHF-induced hyperpolarizations were inhibited by charybdotoxin plus apamin, a combination which had no effect on K^+-induced increases in membrane potential in spite of the fact that the hyperpolarizations induced by both EDHF and by K^+ were inhibited by ouabain plus Ba^{2+}. Furthermore, the EDHF-induced hyperpolarizations were insensitive to iberiotoxin, an agent which has no inhibitory effect on charybdotoxin-sensitive K^+-currents in endothelial cells (Edwards *et al.*, 1998b).

If the K^+-channels which are inhibited by charybdotoxin and apamin are indeed situated on endothelial cells, these agents might depolarize the endothelium and reduce the release of mediators such as nitric oxide. However, in the absence of a nitric oxide synthase inhibitor, the relaxing effects of acetylcholine are not modified, a finding which could indicate that the toxin-sensitive channels must be located on the smooth muscle (Zygmunt & Högestätt, 1996b). This interpretation assumes that the source of the calcium which activates nitric oxide synthase is the same as that which stimulates the release of EDHF and that these two processes have a similar calcium sensitivity. If, however, nitric oxide synthase is activated by a global increase in the endothelial intracellular calcium concentration, whereas channel activity is more sensitive to a local concentration increase in the vicinity of channels through which calcium is entering the cells [as suggested by Robitaille *et al.* (1993) and Daut (1997) for other calcium-sensitive K^+-channels], relaxations to EDRF may not be modified by apamin and charybdotoxin.

K^+-induced Hyperpolarizations

Vasodilatation induced by small increases in $[K^+]_o$ is a well-described phenomenon and two mechanistic explanations have been proposed. In some vessels, stimulation of Na^+/K^+-ATPase seems to be involved, since the effects of K^+ can be inhibited by ouabain (Hendrickx & Casteels, 1974; Webb & Bohr, 1978). In others, a Ba^{2+}-sensitive inwardly-rectifying K^+ channel, which increases its conductance when $[K^+]_o$ is increased, mediates the hyperpolarization (Edwards & Hirst, 1988; Knot, Zimmermann & Nelson, 1996; Quayle, Dart & Standen, 1996). Additionally, both mechanisms may play a role (see McCarron & Halpern, 1990).

In the present study, a 5×10^{-3} M increase in $[K^+]_o$ produced a marked hyperpolarization of vascular smooth muscle cells in hepatic arteries with an intact endothelium and this could only be completely inhibited by a combination of Ba^{2+} and ouabain. This suggests a stimulation of both inwardly-rectifying K^+ channels and a Na^+/K^+-ATPase but these data do not necessarily indicate that these are located on the smooth muscle. Presumably, elevation of $[K^+]_o$ could stimulate a Na^+/K^+-ATPase and the opening of inwardly-rectifying K^+ channels on the endothelial cells. This should enhance K^+ efflux, hyperpolarize the endothelial cells and thus enhance calcium influx which could be the trigger for the release of the 'real' EDHF. However, in de-endothelialized hepatic arteries, a 5×10^{-3} M increase in $[K^+]_o$ still produced hyperpolarization of the smooth muscle, the magnitude of which was the same as that seen when the endothelium was present. This finding clearly substantiates the view that EDHF is identical with K^+ liberated from the vascular endothelium, at least in the hepatic artery of the rat.

Importantly, this view offers an explanation for the universal finding that the effects of EDHF are not seen if extracellular K^+ is increased to $25 - 30 \times 10^{-3}$ M (Zygmunt, Waldeck & Högestätt, 1994). Under these conditions, the smooth muscle Na^+/K^+-ATPase will be maximally stimulated (Hendrickx & Casteels, 1974). Furthermore, since outward currents through the Ba^{2+}-sensitive inward rectifier are small (relative, for example, to those associated with the opening of K_{ATP}), these are unable to clamp the cell at the new E_K (see Quayle, Dart & Standen, 1996) and prevent agonist-induced depolarization associated with the activation of Ca^{2+}-activated chloride currents (Large & Wang, 1996).

CONCLUSION

The inescapable conclusion from the present study is that EDHF is identical with the K^+ liberated from the vascular endothelium of rat hepatic artery when this layer of cells is stimulated by agonists such as acetylcholine (Figure 12-4). K^+ can be released from endothelial cells (Gordon & Martin, 1983), K^+-induced vasodilatation and hyperpolarization have been detected in many vessels (present study; Knot, Zimmermann & Nelson, 1996; Quayle, Dart & Standen, 1996) and the pharmacology of these changes (inhibition by Ba^{2+} and /or ouabain) is identical to that of 'EDHF' (present study). Further studies using other blood vessels are in progress to determine whether the findings of the present study can be applied to the vasculature in general.

ACKNOWLEDGEMENTS

Gillian Edwards was supported by the British Heart Foundation; Matthew Gardener was funded by the MRC.

13 Hyperpolarization and Relaxation of Vascular Smooth Muscle to Endothelium-derived Nitric Oxide

Richard A. Cohen[1], Robert M. Weisbrod[1], Mark C. Griswold[1], Frances Plane[2], Christopher J. Garland[2], Tadeusz Malinski[3], Padmini Komalavilas[4] and Thomas M. Lincoln[4]

[1] Department of Medicine, Boston University, Boston, Massachusetts 02218 USA,
[2] Department Pharmacology, University of Bristol, Bristol, B58 1TD United Kingdom,
[3] Department of Chemistry, Institute of Biotechnology, Oakland University, Rochester, Michigan 48309 USA,
[4] Department of Pathology, University of Alabama, Birmingham, Alabama 35294 USA
Address correspondence to: Richard A. Cohen, MD, Vascular Biology Unit, R408, Boston University School of Medicine, 80 E. Concord Street, Boston, MA 02118. Tel/fax: 617-638-7115/7113; e-mail: racohen@med-med1.bu.edu

The existence of an EDHF distinct from nitric oxide is based on pharmacological studies showing that its actions are not interfered with by inhibitors of nitric oxide synthase, that hyperpolarization is not a prominent response observed with nitric oxide donors, and that EDHF does not depend on cyclic GMP, the primary second messenger system responsible for the effects of nitric oxide in smooth muscle. In the rabbit carotid artery, measurements of nitric oxide release show that acetylcholine-induced nitric oxide release is only inhibited by 50 to 70% by high concentrations of inhibitors of nitric oxide synthase. Hyperpolarization and relaxation of the smooth muscle correlates with the nitric oxide released. Furthermore, authentic exogenous nitric oxide both hyperpolarizes and relaxes the contracted and depolarized artery. Studies with a guanylate cyclase inhibitor show that unlike with a nitric oxide donor, the hyperpolarization and relaxation caused by acetylcholine or authentic nitric oxide does not depend strictly on stimulation of the production of cyclic GMP. Therefore, at least in this artery, direct measurements of nitric oxide, membrane hyperpolarizing actions of nitric oxide, and cyclic GMP-independent actions of nitric oxide all support the identity of nitric oxide as the principal EDHF.

KEYWORDS: Nitric oxide, Cyclic GMP

INTRODUCTION

The postulated existence of an endothelium-derived hyperpolarizing factor which is distinct from nitric oxide is based on pharmacological evidence in multiple vascular beds and isolated blood vessels. This evidence differentiates the characteristics of EDHF from nitric oxide on the basis of; (a) the failure of inhibitors of nitric oxide synthase to the prevent the actions of EDHF; (b) the failure of exogenous nitric oxide or nitric oxide donors to mimic the actions of EDHF; and (c) the apparent lack of involvement of cyclic GMP, the principle intracellular mediator of nitric oxide action. However, in the rabbit carotid artery, in which all the same principal pharmacological discrepancies between EDHF and nitric oxide have been observed, released nitric oxide does account for the actions of EDHF (Cohen and Vanhoutte, 1995). These studies explain the apparent discrepancies between

Richard A. Cohen et al.

Table 13-1   Criteria for identification of an endothelium-derived mediator as an EDHF

1. An EDHF should be capable of hyperpolarizing smooth muscle.
2. EDHF should be released in high enough concentrations by native endothelial cells to hyperpolarize underlying smooth muscle.
3. Inhibitors of the synthesis of an EDHF should decrease endothelium-dependent hyperpolarization.
4. EDHF should be identified chemically.
5. EDHF should have an identifiable molecular mechanism for causing endothelium-dependent hyperpolarization.

the expected properties of nitric oxide and those which appear to be accounted for by a distinct EDHF, by demonstrating unexpected results with inhibitors of nitric oxide synthase and guanylate cyclase, and by suggesting a wider than previously recognized spectrum of action of nitric oxide on vascular smooth muscle cells. They indicate that nitric oxide possesses all the characteristics of an EDHF (Table 13-1), and in the rabbit carotid artery is the principal EDHF.

INHIBITORS OF NITRIC OXIDE SYNTHASE DO NOT FULLY BLOCK NITRIC OXIDE RELEASE

Many studies have implicated a non-nitric oxide mediator as EDHF by showing that inhibitors of nitric oxide synthase do not fully block endothelium-dependent relaxation of vascular smooth muscle (Cohen and Vanhoutte, 1995). Conclusions of these *in vivo* and *in vitro* studies are based on the assumption that the inhibitors fully block the release of nitric oxide, thus implying the existence of other factors mediating the response. The following studies were performed to test the effects on the release of nitric oxide of two nitric oxide synthase inhibitors, L-N$^\omega$-nitroarginine methyl ester and N$^\omega$-nitro-L-arginine, applied in concentrations at or above those commonly used.

In the isolated rabbit carotid artery contractile agonists such as phenylephrine cause coordinated contraction and depolarization of the smooth muscle (Cohen *et al.*, 1997). Acetylcholine ($10^{-8} - 10^{-5}$M) causes a concentration-dependent relaxation which is accompanied by hyperpolarization (Figure 13-1). The relaxation and hyperpolarization are highly correlated and are accompanied by the concentration-dependent release of nitric oxide (Cohen *et al.*, 1997). High concentrations of L-N$^\omega$-nitroarginine methyl ester and N$^\omega$-nitro-L-arginine either alone, or in combination, inhibit, but do not block the release of nitric oxide (Figure 13-1). Either of two independent techniques to measure nitric oxide show that approximately 30 to 50% of the nitric oxide released under control conditions persists in the presence of the inhibitors (Cohen *et al.*, 1997). This raises the possibility that nitric oxide released in the presence of the inhibitors of nitric oxide synthase mediates the residual physiological response.

The relaxation and hyperpolarization of the carotid artery to acetylcholine is reduced, but not blocked by the inhibitors of nitric oxide synthase (Figure 13-1). L-N$^\omega$-nitroarginine methyl ester (30 μM) alone significantly reduces the responses, and they are further reduced by the combination of L-N$^\omega$-nitroarginine methyl ester and N$^\omega$-nitro-L-arginine (300 μM). The reduction in the physiological responses correlates well with the reduction

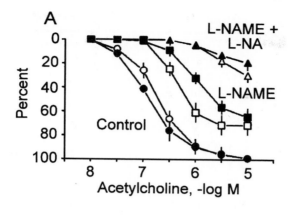

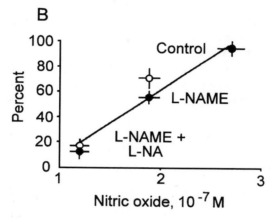

Figure 13-1. Effect of L-N$^\omega$-nitroarginine methyl ester and N$^\omega$-nitro-L-arginine on acetylcholine-induced hyperpolarization, relaxation, and nitric oxide release. Rings of rabbit carotid artery were mounted for isometric tension measurement and membrane potential was recorded with a microelectrode inserted from the adventitial aspect. Rings were untreated or pretreated with L-N$^\omega$-nitroarginine methyl ester (L-NAME) or N$^\omega$-nitro-L-arginine (L-NNA), prior to contracting and depolarizing with phenylephrine and administering acetylcholine. (a) Acetylcholine-evoked relaxations (open symbols) and repolarizations (closed symbols) under control conditions (circles), after treatment with L-N$^\omega$-nitroarginine methyl ester (3×10^{-5} M, squares), or after treatment with L-N$^\omega$-nitroarginine methyl ester (3×10^{-5} M) together with N$^\omega$-nitro-L-arginine (3×10^{-4} M, triangles). (b) Close correlation between the maximal relaxation (filled circles) or repolarization (open circles) and the release of nitric oxide from rabbit carotid arteries caused by acetylcholine (3×10^{-6} M) measured with a porphyrinic microsensor under control conditions or after treatment with the same concentrations of L-N$^\omega$-nitroarginine methyl ester, or L-N$^\omega$-nitroarginine methyl ester combined with N$^\omega$-nitro-L-arginine. Reprinted with permission from (Cohen *et al.*, 1997).

in the release of nitric oxide as measured by an electrochemical electrode (Figure 13-1) or as measured by the accumulation in the media of the nitric oxide oxidation product, nitrite (Cohen *et al.*, 1997). The lower amounts of nitric oxide released in the presence of the inhibitors have been confirmed to be physiologically active by showing that 3-morpholino-sydnonimine (SIN-1), a nitric oxide donor, caused relaxation and hyperpolarization by releasing nitric oxide in the same concentration range as that released from

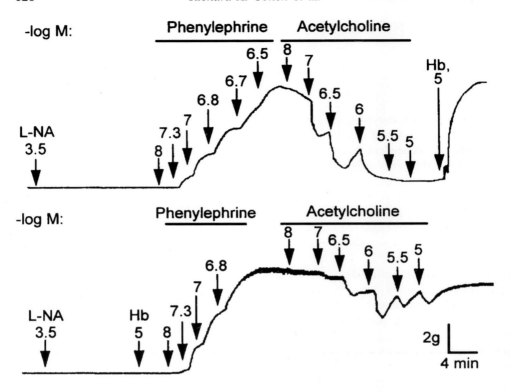

Figure 13-2. Effect of hemoglobin on response to acetylcholine of the rabbit carotid artery in the presence of a nitric oxide synthase inhibitor. Recording of isometric tension of two rabbit carotid artery rings pretreated with N^{ω}-nitro-L-arginine (3×10^{-4} M). The rings were contracted with phenylephrine (PE) in increasing concentrations, indicated as the negative logarithm, and then acetylcholine was administered. In the upper ring the response to acetylcholine was reversed when hemoglobin (10^{-5} M) was added. In the lower tracing, hemoglobin was added prior to phenylephrine. In the presence of hemoglobin, the response to acetylcholine which persists in the presence of the inhibitor of nitric oxide synthase is greatly reduced, but not eliminated by hemoglobin. Prior to its use, hemoglobin was reduced with sodium dithionite, and the dithionite was removed by dialysis.

the endothelium (Cohen *et al.*, 1997). Further studies showed that the response to acetylcholine which persists in the presence of the inhibitors of nitric oxide synthase is further significantly decreased by the scavenger of nitric oxide, hemoglobin, indicating that the persistent release of nitric oxide mediates the response (Figure 13-2).

Although these studies do not exclude other potential substances that could act as an EDHF in the rabbit carotid artery, they do show that endogenously released nitric oxide accounts primarily for the relaxation and hyperpolarization of the smooth muscle. In addition, they show that the persistence of physiological responses in the presence of even high concentrations of inhibitors of nitric oxide synthase should not be considered as evidence of a factor other than nitric oxide, unless the persistent release of nitric oxide is excluded by its measurement.

CYCLIC GMP IS NOT ESSENTIAL TO THE ACTIONS OF NITRIC OXIDE

Previous studies showed the association between relaxation caused by nitric oxide and stimulation of guanylate cyclase and increases in smooth muscle levels of cyclic GMP. Thus, endothelium-dependent vasodilators increased cyclic GMP levels in smooth muscle, and the relaxation caused by nitric oxide was mimicked by cell-permeable analogs of cyclic GMP. An inhibitor of guanylate cyclase, methylene blue, inhibited the relaxation to endothelium-dependent vasodilators and nitric oxide donors suggesting the essential role of cyclic GMP in the action of nitric oxide. The similarities between the actions of nitric oxide and endothelium-dependent vasodilators were important in identifying nitric oxide as the principle endothelium-derived vasodilator (Ignarro *et al.*, 1988a; Ignarro *et al.*, 1988b; Furchgott, 1988; Martin *et al.*, 1985). However, in the rabbit carotid artery, stimulation of cyclic GMP was prevented in arteries treated with inhibitors of nitric oxide synthase, whereas endothelium-dependent relaxations persisted (Cowan *et al.*, 1993). Prior to the above mentioned measurements of nitric oxide release, the apparent block the accumulation of cyclic GMP after treatment with L-N$^\omega$-nitroarginine methyl ester in the rabbit carotid artery was considered to be evidence that the release of nitric oxide was blocked by the inhibitor, and that another endothelium-derived mediator accounted for the relaxation. Having shown that the release of nitric oxide persists in the presence of L-N$^\omega$-nitroarginine methyl ester, it is not clear why the reduced levels of nitric oxide fail to stimulate cyclic GMP levels. This may be explained by a requirement for a relatively high threshold level of nitric oxide to stimulate guanylate cyclase.

Further studies have utilized the guanylate cyclase inhibitor, 1H-(1,2,4)oxadiazolo (4,3-a)quinoxalin-1-one (ODQ). This agent has been shown to inhibit guanylyl cyclase without inhibiting nitric oxide synthase or scavenging nitric oxide (Olson *et al.*, 1997). In the rabbit carotid artery, ODQ, at concentrations above 10^{-6} M, completely blocks the increases in cyclic GMP caused by either acetylcholine or nitric oxide (Plane *et al.*, 1998 and unpublished studies). In carotid arteries treated with ODQ (10^{-5} M), acetylcholine-induced relaxations and hyperpolarizations are partially inhibited, but not blocked (Figure 13-3) (Plane *et al.*, 1998). Similar to acetylcholine, exogenous nitric oxide relaxes and hyperpolarizes carotid artery rings, and ODQ also only partially inhibits the response (Figure 13-3). These observations indicate that stimulation of cyclic GMP is not essential for the relaxation and hyperpolarization caused by nitric oxide, whether released by acetylcholine or administered exogenously. In contrast to the responses to acetylcholine or authentic nitric oxide, the relaxations and hyperpolarizations caused by the nitric oxide donor, SIN-1, are completely blocked by ODQ (Figure 13-3). This indicates that while this nitric oxide donor relaxes and hyperpolarizes the carotid artery, it does not do so by the same mechanism as authentic nitric oxide, and apparently relies entirely on cyclic GMP for its action. The apparent discrepancy between SIN-1 and authentic nitric oxide may be explained by differences in the rate at which nitric oxide is delivered to the artery. Nitric oxide, when released by acetylcholine, or when administered exogenously, reaches a peak concentration within seconds, whereas that released by SIN-1 requires up to four minutes to reach a stable level (Cohen *et al.*, 1997). This difference draws attention to the fact that nitric oxide donors may not mimic all the properties of endogenously released nitric oxide, and that authentic nitric oxide is a more suitable analog of the endogenously released

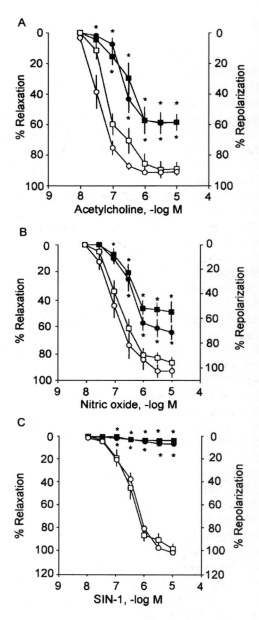

Figure 13-3.   Effect of 1H-(1,2,4)oxadiazolo(4,3-a)quinoxalin-1-one (ODQ) on the response of rabbit carotid artery to acetylcholine, nitric oxide, and SIN-1. Artery rings were pretreated for 10 minutes with ODQ (10^{-5} M) and contracted and depolarized with phenylephrine. Acetylcholine (ACh), nitric oxide (NO) and SIN-1 were administered in increasing concentrations (indicated as the logarithm). All three nitric oxide-dependent agonists caused complete relaxation (circles) and repolarization (squares). Although ODQ significantly reduced relaxation and repolarization to acetylcholine and nitric oxide, significant responses persisted. In contrast, the response to SIN-1 was completely blocked by ODQ. Under these conditions, ODQ completely prevented the increase in cyclic GMP and protein kinase G activation stimulated by acetylcholine, nitric oxide, or SIN-1 (Plane *et al.*, 1998 and unpublished observations). Reprinted by permission (Plane *et al.*, 1998).

vasodilator. The failure of nitric oxide donors to completely mimic the actions of authentic nitric oxide may also help to explain why cyclic GMP became viewed previously as essential to the action of nitric oxide.

Cyclic GMP normally acts primarily through activation of protein kinase G and subsequent phosphorylation of a number of protein targets including ion channels and transporters which mediate response of the smooth muscle (Gupta *et al.*, 1994; Bolotina *et al.*, 1994; Cornwell *et al.*, 1991; Raeymaekers *et al.*, 1988; Furukaws *et al.*, 1988; Furukawa *et al.*, 1991; Quignard *et al.*, 1997; Blatter and Wier, 1994; Koga *et al.*, 1994; Komalavilas and Lincoln, 1994; Hirata *et al.*, 1990; Rapoport, 1986). Protein kinase G activity assayed by back phosphorylation of the inositol trisphosphate receptor was increased by nitric oxide in carotid arteries. In arteries treated with ODQ in which nitric oxide-induced relaxation persisted, no changes were observed in protein kinase G activity, confirming that cyclic GMP stimulation was completely blocked by ODQ (unpublished observations).

Further studies have demonstrated cyclic GMP-independent actions of authentic nitric oxide in primary culture of rabbit aortic smooth muscle cells, in which free intracellular calcium concentration has been measured with fura-2 fluorometry. In cells in which angiotensin II increases calcium, nitric oxide causes concentration-dependent transient decreases in calcium similar in time-course to nitric oxide-induced relaxations of intact arteries. The maximal decrease in calcium level is reached in 3–5 seconds. Cyclic GMP levels peak in 10 seconds and protein kinase G activation is significantly increased by 20–30 seconds. The much more rapid decrease in intracellular calcium levels suggests a response which occurs independently of cyclic GMP and protein kinase G.

ODQ (10^{-5} M) blocks the rise in cyclic GMP and completely prevents the activation of protein kinase G in cultured rabbit aortic smooth muscle cells, but only partially inhibits the response to nitric oxide. The decrease in intracellular free calcium caused by 10^{-7} M nitric oxide is of significantly shorter duration, but the magnitude of the initial decrease in calcium is unchanged. Cyclic GMP analogs also decrease intracellular calcium levels, and the response to these analogs is blocked by protein kinase G inhibitors. These studies indicate that although authentic nitric oxide increases cyclic GMP, and cyclic GMP can decrease intracellular calcium, the decrease in calcium can be only partially explained by effects mediated by cyclic GMP. Thus, these studies demonstrate cyclic GMP-independent effects of nitric oxide on smooth muscle tone, membrane potential, and intracellular calcium of vascular smooth muscle. These cyclic GMP-independent actions of nitric oxide represent those of an EDHF.

CONCLUSIONS

These studies indicate that endothelium-dependent relaxation and hyperpolarization of the smooth muscle in the rabbit carotid artery most closely correlate with the measured release of vasoactive concentrations of nitric oxide from the endothelium. This indicates that nitric oxide, rather than another factor, is the primary mediator of endothelium-dependent relaxation and hyperpolarization in this artery. In addition, nitric oxide, when released from the endothelium or when applied exogenously, can mediate its effects by mechanisms additional to guanylate cyclase stimulation, increases in cyclic GMP, and protein kinase G activation. These cyclic GMP-independent actions of nitric oxide include its ability to decrease intracellular calcium, and to hyperpolarize and relax vascular smooth muscle fulfilling the definition of an EDHF. Important in these studies has been the use of authentic nitric oxide, whose actions most closely mimic the action of nitric oxide released from the endothelium. Given that nitric oxide mediates endothelium-dependent hyperpolarization and can fulfill all the criteria for EDHF in the rabbit carotid artery (Table 13-1), the mechanism by which nitric oxide causes hyperpolarization would be important to consider in future studies of EDHF. At least one mechanism, the activation of calcium-dependent potassium channels, which can occur both indirectly via cyclic GMP and independently of the cyclic nucleotide, has been proposed (Bolotina *et al.*, 1994).

ACKNOWLEDGEMENTS

The studies were supported by grants to R.A.C., NIH HL 31607 and HL 55993; to C.J.G., Wellcome Trust; to T.M., NIH HL55397; and to T.M.L., NIH HL 34646 and HL 54326.

14 Endothelium-Dependent Responses in Endothelial Nitric Oxide Synthase Knockout Mice: Contribution of NO, EDHF and Products of Cyclooxygenase

Thierry Chataigneau[1], Michel Félétou[1], Paul L. Huang[2], Mark C. Fishman[2], Jacques Duhault[1] and Paul M. Vanhoutte[3]

[1] Département de Diabétologie, Institut de Recherches Servier, 92150 Suresnes, France;
[2] Cardiovascular Research Center and Department of Medicine, Massachusetts General Hospital and Harvard Medical School, Charlestown, Massachusetts 02129, USA;
[3] Institut de Recherche International Servier, 92410 Courbevoie, France
Address for correspondence: Michel Félétou: Dpt Diabétologie, Institut de Recherche Servier, 92150 Suresnes, France.

Experiments were designed to determine whether or not alternative pathway(s) for endothelium-dependent responses can be demonstrated in blood vessels from endothelial nitric oxide synthase knockout mice [eNOS (–/–) mice]. Isometric tension was recorded in isolated rings of aorta, carotid, mesenteric and coronary arteries taken from eNOS(–/–) mice and the corresponding wild-type strain [C57BL6, eNOS(+/+) mice]. The membrane potential of smooth muscle cells was measured in coronary arteries by means of intracellular microelectrodes. In the isolated aorta, carotid and coronary arteries from the wild-type strain, acetylcholine induced endothelium-dependent relaxations which were abolished by N^ω-L-nitro-arginine. In mesenteric arteries, the cholinergic relaxation was inhibited partially by N^ω-L-nitro-arginine and blocked fully by the combination of N^ω-L-nitro-arginine plus indomethacin. The isolated aorta, carotid and coronary arteries from the eNOS(–/–) mice did not relax in response to acetylcholine. However, acetylcholine produced an indomethacin-sensitive relaxation in the mesenteric arteries of the genetically transformed animals. Sodium nitroprusside induced relaxations in all the arteries studied from both strains. The resting membrane potential of smooth muscle cells from isolated coronary arteries was less negative in eNOS(–/–) mice when compared to eNOS(+/+) mice. In both strains, acetylcholine, bradykinin and substance P did not induce endothelium-dependent hyperpolarizations whereas cromakalim consistently produced a significant hyperpolarization. In most blood vessels, SIN-1 did not affect the resting membrane potential; however, in some cells from both strains, the nitrovasodilator produced a small hyperpolarization. Thus, in the murine blood vessels studied, endothelium-dependent relaxations involve either NO or the combination of NO and a product of cyclooxygenase, but endothelium-derived hyperpolarizing factor (EDHF) does not seem to contribute. In the eNOS(–/–) mice, NO-dependent responses and EDHF-like responses were not observed suggesting that the latter response is not upregulated by eNOS gene disruption. The only back-up mechanism which is observed in some blood vessels from the eNOS(–/–) mice is dependent on cyclooxygenase.

KEYWORDS: Cyclooxygenase, EDHF, endothelium, eNOS knockout mice, NO, smooth muscle.

INTRODUCTION

Chronic inhibition of nitric oxide synthase induces an increase in vascular resistance which is associated with an increase in blood pressure in certain (rat, pig) but not all (rabbit, dog) species (Oliveira *et al.*, 1992; Cayatte *et al.*, 1994; Ito *et al.*, 1995; Puybasset *et al.*,

1995). However, inhibitors of nitric oxide synthase are not selective and the three isoforms of nitric oxide synthase may be either partially or totally inhibited by the available agents (Huang & Fishman, 1996 ; Steudel *et al.*, 1997). Therefore, it is difficult to evaluate the exact participation of the type III isoform (eNOS) to the physiological alterations observed during chronic administration of nitric oxide synthase inhibitors. By contrast, disruption of eNOS gene in mice produces a specific suppression of endothelial NO production (Huang *et al.*, 1995; Faraci *et al.*, 1998; Gödecke *et al.*, 1998) and provides an experimental model to investigate the precise cardiovascular role of endothelial NO and its interactions with other endothelial mediators such as prostacyclin and endothelium-derived hyperpolarizing factor (EDHF).

NO may regulate the production and/or the effects of EDHF (Olmos *et al.*, 1995; Mombouli *et al.*, 1996; Popp *et al.*, 1996a; Kristof *et al.*, 1997). Alternatively, EDHF might provide a compensatory mechanism of relaxation when the production of NO is impaired (Kilpatrick & Cocks, 1994; Corriu *et al.*, 1998). The present experiments were designed to investigate these two possibilities by studying endothelium-dependent relaxations and hyperpolarizations in various isolated arteries of eNOS gene knock-out mice [(eNOS(–/–)] and their corresponding control wild-type strain [(eNOS(+/+)].

MATERIAL AND METHODS

Organ Chambers

C57BL6 mice [wild-type; eNOS(+/+)] and mice lacking the gene for endothelial NO synthase [eNOS(–/–)] were killed by CO_2 inhalation. The aorta, carotid, mesenteric (first order) and left anterior coronary arteries were removed and dissected free of adherent connective tissues. Segments of the blood vessels were suspended in organ chambers (aorta) or microvessel myographs (carotid, coronary and mesenteric arteries; Mulvany & Halpern, 1977) filled with modified Krebs-Ringer bicarbonate solution (composition in mM: NaCl 118.3, KCl 4.7, $CaCl_2$ 2.5, $MgSO_4$ 1.2, KH_2PO_4 1.2, $NaHCO_3$ 25, calcium-disodium EDTA 0.026 and glucose 11.1, 37°C, aerated with a 95% O_2/5% CO_2 gas mixture, pH 7.4) for recording of isometric tension. After a 45 min equilibration period, isolated rings of aorta and carotid arteries from both strains were studied under control conditions or treated either with N^ω-L-nitroarginine (10^{-4} M), indomethacin (5×10^{-6} M), or the combination of both inhibitors and then contracted with norepinephrine in order to obtain a level of tension corresponding to 50 to 80 % of the reference contraction to KCl (60 mM). Mesenteric and coronary arteries were contracted with norepinephrine (3×10^{-6} M) and U 46619 (10^{-7} M), respectively. In these two blood vessels, administration of N^ω-L-nitroarginine and/or indomethacin was performed at the equilibrium state of the contraction. Then, endothelium-dependent relaxations induced by acetylcholine were studied. At the end of the experiment, maximal relaxation was obtained with sodium nitroprusside (10^{-3} M).

Electrophysiological Experiments

Segments of the left anterior descending coronary artery were pinned to the bottom of an organ chamber (0.5 ml in volume) superfused at constant flow (2.5 ml/min) with modified

Krebs-Ringer bicarbonate solution. In preliminary experiments, the presence of the endothelial lining was verified histologically. The cell membrane potential was recorded using intracellular microelectrodes. Impalements of the smooth muscle cells were performed from the adventitial side of the vessels with glass microelectrodes (tip resistance of 30 to 90 MΩ) filled with KCl (3M) (Corriu *et al.*, 1996a).

Statistics

Data are shown as mean $\pm$ s.e. mean; n indicates the number of animals from which arteries were taken or the number of cells in which membrane potential was recorded. Statistical analysis was performed using Student's *t*-test for paired or unpaired observations. Differences were considered to be statistically significant when P was less than 0.05.

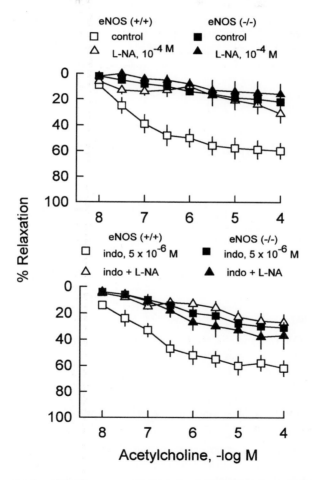

Figure 14-1. Concentration-relaxation curves to acetylcholine in aorta with endothelium from eNOS(+/+) and eNOS(–/–) mice, in the presence or absence of N$^\omega$-Nitro-L-arginine (L-NA, 10^{-4} M, upper panel) and in the presence of indomethacin (5×10^{-6} M) with and without N$^\omega$-L-nitroarginine (lower panel). Data are shown as means $\pm$ s.e. mean (modified from Chataigneau *et al.*, 1999).

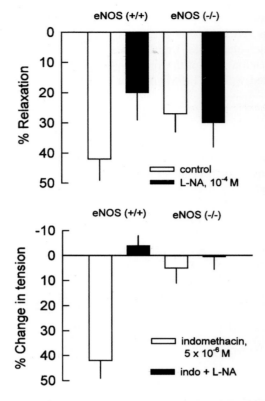

Figure 14-2. Acetylcholine (10^{-5} M)-induced relaxations in mesenteric arteries with endothelium from eNOS(+/+) and eNOS(–/–) mice, in the presence or absence of N^{ω}-Nitro-L-arginine (L-NA, 10^{-4} M, upper panel) and in the presence of indomethacin (5×10^{-6} M) with and without N^{ω}-L-nitroarginine (lower panel). Data are shown as means ± s.e. mean (modified from Chataigneau *et al.*, 1999).

RESULTS

Vascular Reactivity

In arteries from eNOS(+/+) and eNOS(–/–) mice, the contractions induced by KCl (60 mM) did not differ significantly (data not shown).

In isolated aortas and carotid arteries (not shown) contracted with norepinephrine, endothelium-dependent relaxations induced by acetylcholine were obtained only in vessels from the wild-type strain (Figure 14-1). These relaxations were unaltered in the presence of indomethacin but abolished by N^{ω}-L-nitroarginine (10^{-4} M) or the combination of N^{ω}-L-nitroarginine (10^{-4} M) plus indomethacin (5×10^{-6} M).

In mesenteric arteries, the contraction to norepinephrine was significantly larger in eNOS(–/–) than in eNOS(+/+) mice (67 ± 7 mg, n = 31 and 48 ± 4 mg, n = 33, respectively). The application of N^{ω}-L-nitroarginine or N^{ω}-L-nitroarginine plus indomethacin induced an additional increase in tension in mesenteric arteries from eNOS(+/+) but not from eNOS(–/–) mice. In both strains, acetylcholine (10^{-5} M) induced a relaxation of the mesenteric arteries (Figure 14-2). In eNOS(+/+) mice, this relaxation was not affected

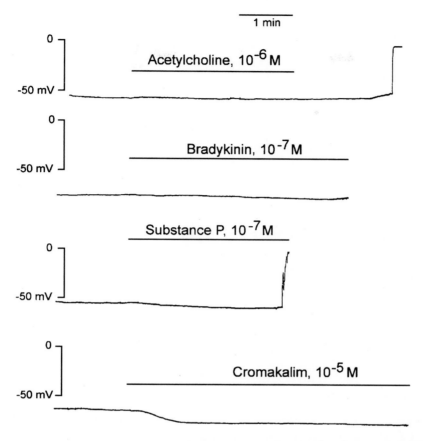

Figure 14-3. Effects of acetylcholine (ACh, 10^{-6} M), bradykinin (10^{-7} M), substance P (10^{-7} M) and cromakalim (10^{-5} M) on the resting membrane potential of smooth muscle cells in isolated coronary arteries with endothelium from eNOS(+/+) mice. Acetylcholine, substance P and bradykinin did not alter the membrane potential whereas cromakalim produced a consistent hyperpolarization (data from Chataigneau *et al.*, 1999).

significantly by N^{ω}-L-nitroarginine or indomethacin but was abolished by the combination of the two inhibitors. In eNOS(–/–) mice, the acetylcholine-induced relaxation was not affected by N^{ω}-L-nitroarginine but was significantly inhibited either by indomethacin or the combination of both inhibitors.

In coronary arteries, the contractile response produced by U 46619 (10^{-7} M) was significantly larger in eNOS(–/–) than in eNOS(+/+) mice (45 ± 7 mg, n = 28 and 27 ± 3 mg, n = 25, respectively). The addition of N^{ω}-L-nitroarginine or N^{ω}-L-nitroarginine plus indomethacin induced a further increase in tension in coronary arteries from eNOS(+/+) but not eNOS(–/–) mice. Acetylcholine (10^{-5} M) induced a relaxation in the coronary artery from eNOS(+/+) mice (not shown). This relaxation was inhibited significantly by N^{ω}-L-nitroarginine, unaffected by indomethacin and abolished by the combination of both inhibitors. In coronary arteries from eNOS(–/–), acetylcholine induced no or minimal relaxations. In the few cases where relaxation occured in response to acetylcholine, the response was inhibited by the addition of indomethacin.

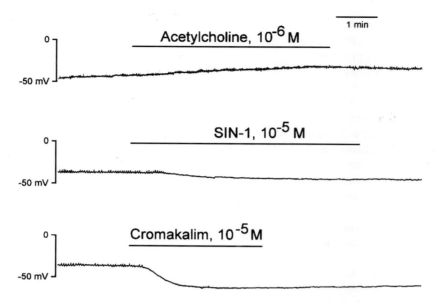

Figure 14-4. Repolarizing effect of SIN–1 (10^{-5} M) in a coronary smooth muscle cell from a eNOS(–/–) mouse. Acetylcholine (ACh, 10^{-6} M) induced a significant depolarization (9 mV) whereas SIN–1 (10^{-5} M) significantly repolarized (–7 mV) the cell. After washout, cromakalim (10^{-5} M) induced a consistent hyperpolarization (–26 mV; data from Chataigneau *et al.*, 1999).

Electrophysiological Experiments

The resting membrane potential of eNOS(+/+) coronary arterial smooth muscle cells was –64.8 ± 1.8 mV (n = 20), under control conditions. This value was not altered significantly by treatments with N^{ω}-L-nitroarginine or/and indomethacin. In eNOS(–/–) mice, the control resting membrane potential of the coronary artery smooth muscle cells was significantly less negative when compared to eNOS(+/+) mice (–58.4 ± 1.9 mV, n = 17).

In coronary smooth muscle cells from eNOS(+/+) or eNOS(–/–) mice, acetylcholine (10^{-6} and 10^{-5} M), bradykinin (10^{-7} M) and substance P (10^{-7} M) did not induce significant hyperpolarizations, in the absence or presence of N^{ω}-L-nitroarginine and/or indomethacin (Figure 14-3). SIN-1 (10^{-5} M) hyperpolarized 4 smooth muscle cells out of 24 but induced no changes in the other cells (Figure 14-4). By contrast, cromakalim (10^{-5} M) induced a significant hyperpolarization in both strains under all experimental conditions tested (Figure 14-4).

DISCUSSION

The present results show that, in isolated murine aorta, carotid, mesenteric and coronary arteries, endothelium-dependent relaxations elicited by acetylcholine, which are resistant to inhibition of both nitric oxide synthase and cyclooxygenase and thus can be attributed to EDHF, do not occur. Furthermore, in eNOS(–/–) mice, the release of EDHF does not constitute an alternative mechanism of endothelium-dependent relaxations. Instead, in both eNOS(–/–) and eNOS(+/+) strains, a cyclooxygenase-dependent component of acetylcholine-induced relaxation can be observed in blood vessels such as the mesenteric artery and possibly the coronary artery.

The disruption of the eNOS gene did not alter the intrinsic contractile properties of the smooth muscle cells, since isolated rings of aorta, carotid, mesenteric and coronary arteries from eNOS(–/–) and eNOS(+/+) mice developed a similar level of tension in response to KCl (60 mM). Furthermore, the occurrence of relaxation and hyperpolarization evoked by sodium nitroprusside and cromakalim, respectively, indicate that the vascular smooth muscle responses are not affected by the eNOS gene disruption.

Under control conditions, the coronary smooth muscle cells from eNOS(–/–) mice are depolarized when compared to those from eNOS(+/+) mice. This cannot be attributed to the disappearance of endothelial NO *per se*. Indeed, in coronary arteries of eNOS(+/+) mice, as well as in various blood vessels from different species, inhibitors of NO synthase and/or cyclooxygenase do not affect the resting membrane potential, excluding the participation of the basal production of NO and/or products of cyclooxygenase in the control of that potential (Parkington *et al.*, 1993, 1995; Zygmunt *et al.*, 1994c; Corriu *et al.*, 1996b). eNOS (–/–) mice have systemic arterial and moderate pulmonary hypertension (Huang *et al.*, 1995; Huang & Fishman, 1996; Steudel *et al.*, 1997). This in agreement with previous reports indicating that depolarization of the arterial smooth muscle cells accompanies various types of experimental hypertension (Tomobe *et al.*, 1991; Fujii *et al.*, 1992; Morel *et al.*, 1996; Vanheel *et al.*, 1996).

In mesenteric and coronary arteries from eNOS(+/+) mice, the addition of N^{ω}-L-nitroarginine at the equilibrium state of the contractile response resulted in an additional increase in tension revealing a basal endothelial release of NO. In the corresponding arteries from eNOS(–/–) mice, the addition of N^{ω}-L-nitroarginine did not produce contraction, confirming the absence of endothelial NO production in the mutant animals. The endothelium-dependent relaxations induced by acetylcholine in the aorta and carotid arteries from eNOS(+/+) mice were abolished by inhibition of NO synthase indicating that, in these blood vessels, NO is the sole endogeneous endothelium-derived vasodilator. This conclusion is strengthened by the disappearance of these relaxations in the corresponding eNOS(–/–) mice vessels, in agreement with previous studies (Huang *et al.*, 1995; Faraci *et al.*, 1998).

The endothelium-dependent relaxation induced by acetylcholine in the mesenteric arteries from eNOS(+/+) mice was blocked partially by the nitric oxide synthase inhibitor, as previously shown in the murine perfused mesenteric bed (Berthiaume *et al.*, 1997). However, by contrast with the aorta and carotid arteries, total inhibition of acetylcholine-induced relaxation was only achieved with the combination of N^{ω}-L-nitroarginine plus indomethacin. These results demonstrate that in response to acetylcholine, besides NO, a product of cyclooxygenase (presumably prostacyclin) contributes to the endothelium-dependent relaxation of the murine mesenteric artery. The contribution of the product of cyclooxygenase is unmasked only after inhibition of nitric oxide synthase, and the effects of the two endothelial mediators are not additive. Similar observations have been reported previously as regards the relative contribution of NO and prostacyclin to the endothelium-dependent relaxations and hyperpolarizations in rat and guinea-pig arteries (Corriu *et al.*, 1996b; Zygmunt *et al.*, 1998). In the mesenteric arteries from eNOS(–/–) mice, acetylcholine induced relaxations which were insensitive to N^{ω}-L-nitroarginine but abolished by indomethacin, confirming the absence of NO production and the contribution of a product of cyclooxygenase in this murine blood vessel. An upregulation of endothelial cyclo-oxygenases (COX) has been shown in dogs (COX-1) and rats (COX-2) chronically treated with inhibitors of NO synthase (Puybasset *et al.*, 1996; Beverelli *et al.*, 1997; Henrion *et al.*, 1997). However, the present experimental protocols were not designed to determine whether or not the the cyclooxygenase pathway is upregulated in eNOS(-/-) mice.

In murine coronary arteries from eNOS(+/+) mice, the acetylcholine-induced relaxation is reduced by inhibition of nitric oxide synthase, confirming previous findings (Ku *et al.*, 1996), and is abolished by the combined inhibition of nitric oxide synthase and cyclooxygenase. Thus both NO and a product of cyclooxygenase account for the endothelium-dependent relaxation to acetylcholine. In eNOS(-/-) mice, an indomethacin-sensitive relaxation could be observed in some coronary arteries, suggesting the contribution of cyclooxygenase, as seen in the mesenteric arteries. This is in agreement with the dem-onstration that in the isolated perfused murine heart, the vasodilatation produced by acetylcholine involves both NO and a product of cyclooxygenase in eNOS(+/+) mice but only the latter in eNOS(-/-) mice (Gödecke *et al.*, 1998).

The nitrovasodilator, sodium nitroprusside, induced consistent relaxations in all the vessels studied from both murine strains, in agreement with previous studies, demonstrat-ing that the sensitivity of vascular smooth muscle to NO is conserved in the eNOS(-/-) mouse (Huang *et al.*, 1995; Faraci *et al.*, 1998).

Acetylcholine, bradykinin and substance P did not hyperpolarize mice coronary arterial smooth muscle cells. This finding, taken in conjunction with the absence of relaxations resistant to the inhibitors of nitric oxide synthase and cyclooxygenase, indicates that the release of EDHF plays no role in the endothelium-dependent relaxations of the various murine blood vessels examined. NO can suppress or regulate the expression of EDHF (Olmos *et al* 1995; Mombouli *et al.*, 1996; Popp *et al.*, 1996a; Kristof *et al.*, 1997); however, studies performed in eNOS(+/+) mice, in the presence of an inhibitor of NO synthase, and in eNOS (-/-) mice did not unmask an EDHF-like response. Therefore, EDHF, which may be responsible of a backup mechanism in case of dysfunction of the endothelial NO pathway (Kilpatrick et Cocks, 1994; Corriu *et al.*, 1998; McCulloch *et al.*, 1997), is absent in both eNOS(+/+) and eNOS(-/-) mice. This lack of endothelium-dependent hyperpolarization is not likely to be due to a deficiency in vascular smooth muscle function as cromakalim hyperpolarized the coronary artery smooth muscle cells of both murine strains. However, it remains to be determined whether the absence of EDHF-mediated responses lies in the absence of synthesis or release of the mediator by the endothelial cells or in the absence of the target for EDHF such as a specific population of potassium channel on the vascular smooth muscle cells.

15 Inhibition by Tetrabutylammonium of Cyclic Guanosine Monophosphate-Independent Vasorelaxation Induced by Interaction of Thiols with Nitric Oxide Stores

Bernard Muller, Andrei L. Kleschyov, Sophie Malblanc and Jean-Claude Stoclet

Université Louis Pasteur de Strasbourg, Faculté de Pharmacie, Pharmacologie et Physico-Chimie des Interactions Cellulaires et Moléculaires, UMR CNRS (ex ERS 653, ex URA 491), 67401 Illkirch-cedex, France
Address for correspondence: Dr Bernard Muller, Faculté de Pharmacie, Pharmacologie et Physico-Chimie des Interactions Cellulaires et Moléculaires, UMR CNRS (ex ERS 653, ex URA 491), BP 24, 67401 ILLKIRCH cedex, France. Phone: 33-3-88-67-69-67, Fax: 33-3-88-66-46-33, email: bmuller@pharma.u-strasbg.fr

The existence of nitric oxide (NO) stores and the potential contribution of dinitrosyl-iron complexes were investigated in lipopolysaccharide-treated vascular tissue. The low molecular weight thiol N-acetylcysteine evoked a relaxation in rat aortic rings incubated with lipopolysaccharide and L-arginine. This effect was dependent on NO production during the incubation period with lipopolysaccharide and was reproduced in aortic rings exposed to exogenous dinitrosyl-iron complexes and extensively washed-out. The relaxing effect of N-acetylcysteine in rings incubated with lipopolysaccharide and L-arginine was not associated with an elevation of aortic cyclic GMP content, but was significantly attenuated in the presence of tetrabutylammonium or by the use of elevated concentration of KCl as contracting agent. By contrast, the relaxation produced by exogenous low molecular weight dinitrosyl-iron complexes in rat aortic rings was not affected by tetrabutylammonium or by the use of high KCl and was associated with a large increase in cyclic GMP content. Protein-bound dinitrosyl-iron complexes were detected by electron paramagnetic resonance spectroscopy in aorta incubated with lipopolysaccharide and L-arginine. The metal chelator diethyldithiocarbamate both destroyed dinitrosyl-iron complexes and abolished the relaxing effect of N-acetylcysteine. In conclusion, low molecular weight thiols such as N-acetylcysteine may interact with NO stores and promote vascular relaxation. Protein-bound dinitrosyl-iron complexes are likely candidate for N-acetylcysteine-sensitive NO stores, even though the contribution of other forms of NO derivatives cannot be excluded. The mechanism of the relaxation evoked by N-acetylcysteine in lipopolysaccharide-treated vessels is cyclic GMP-independent and may involve the activation of tetrabutylammonium-sensitive potassium channels.

KEYWORDS: nitric oxide stores, dinitrosyl iron complexes, cyclic GMP, potassium channels, N-acetylcysteine, vascular relaxation

INTRODUCTION

The overproduction of nitric oxide (NO) and the subsequent accumulation of cyclic guanosine monophosphate (cyclic GMP) are involved in the vascular failure observed during endotoxemia. This results from the expression of an inducible form of NO synthase (iNOS) in vascular wall (see Stoclet *et al.*, 1993; 1998). In the rat thoracic aorta exposed to lipopolysaccharide (*in vitro* or *in vivo*), the adventitial layer can be the major source of NO and NO may diffuse from the adventitia, leading to activation of soluble guanylate-cyclase and accumulation of cyclic GMP in the medial layer (Kleschyov *et al.*, 1998a).

In situations where large quantities of NO are produced over prolonged periods, NO may also react with other targets than soluble guanylate-cyclase. For instance, the formation of protein-bound dinitrosyl non-haeme iron complexes occurs in different cell types or tissues during the activation of the iNOS pathway (see Henry *et al.*, 1993; Vanin and Kleschyov, 1998). However, this has never been documented in vascular tissue exposed to lipopolysaccharide. This might be of particular importance, since these complexes may represent stores of vasoactive, releasable NO. Indeed, low molecular weight thiols such as N-acetylcysteine interact with NO stores such as protein-bound dinitrosyl-iron complexes and generate (through ligand exchange mechanisms) low molecular weight dinitrosyl-iron complexes which can activate soluble guanylate-cyclase and promote vascular relaxation (Mülsch *et al.*, 1991; Mülsch, 1994).

In the present study, it was firstly assessed whether or not NO stores exist in vascular tissue treated with lipopolysaccharide. For this purpose, the vasoactive properties of N-acetylcysteine were characterized in lipopolysaccharide-treated rat aorta and the formation of protein-bound dinitrosyl-iron complexes was evaluated by electron paramagnetic resonance spectroscopy. Furthermore, the contribution of dinitrosyl-iron complexes was also investigated by exposure of rat aorta to exogenous low molecular weight dinitrosyl-iron complexes.

METHODS

Aortic Preparations

The thoracic aorta (from male Wistar rats, 10–12 weeks old) was removed, cleaned of fat and connective tissues. Whole thoracic aortas (about 30 mm long) or aortic rings (2 to 3 mm long) were used for contraction experiments, cyclic GMP determination or electron paramagnetic resonance studies. Some vascular preparations were incubated for 18 hours (at 37°C in an incubator gassed with 5% CO_2/95% air) in Minimal Essential Medium in the absence or in the presence of lipopolysaccharide (10 μg/ml), L-arginine (10^{-3} M) or N^{ω}-nitro-L-arginine methyl ester (10^{-3} M).

Contractile Studies

At the end of the incubation period, aortic rings were suspended (2 g of passive tension) in an organ chamber filled with Krebs solution (kept at 37°C and bubbled with 95% O_2/5% CO_2). Tension was measured with an isometric force transducer. The effect of N-acetylcysteine was studied in rings contracted either with norepinephrine (3×10^{-6} M) or KCl (50 mM, in which NaCl was substituted by an equimolar concentration of KCl) plus N^{ω}-nitro-L-arginine methyl ester (3×10^{-3} M). In some experiments, the effect of N-acetylcysteine was studied after addition of tetrabutylammonium (3×10^{-3} M, a non selective blocker of potassium channels) or the metal chelator diethyldithiocarbamate (6×10^{-6} to 6×10^{-4} M).

The effect of exogenous dinitrosyl-iron complexes (with thiosulfate as ligand) was studied in aortic rings without endothelium (not incubated with lipopolysaccharide) which were contracted either with norepinephrine (10^{-7} M, in the absence or in the presence of 3×10^{-3} M tetrabutylammonium) or KCl (50 mM). In another set of experiments, aortic rings were exposed for 30 min to dinitrosyl-iron complexes (10^{-6} M) and then extensively washed out (during 90 min). The rings were then contracted with norepinephrine (10^{-7} M, in the absence or in the presence of 3×10^{-3} M tetrabutylammonium) or by KCl (50 mM). N-acetylcysteine was added subsequently.

Measurement of Cyclic GMP

Aortic rings were placed for 30 min in Krebs solution (at 37°C and gassed with a mixture of 95% O_2/5% CO_2) containing isobutylmethylxanthine (10^{-4} M) in the absence or presence of N^ω-nitro-L-arginine methyl ester (3×10^{-3} M for 15 min), N^ω-nitro-L-arginine methyl ester plus N-acetylcysteine (10^{-2} M for 5 min) or dinitrosyl-iron complexes (10^{-7} M for 5 min). After the incubation period, rings were transferred into a 1 ml ice-cold HCl solution (0.1 N). Following homogenisation and centrifugation, the cyclic GMP content of the supernatant was determined by radioimmunoassay (Cailla, Vannier and Delaage, 1976). The DNA content of the pellet was measured (Brunk, Jones and James, 1979).

Electron Paramagnetic Resonance Spectroscopy

Recordings were made on an x-band spectrometer Bruker 300E either at 293K or 77K. Parameters were following: microwave power 10 mW (for 77K recordings) or 102 mW (for 293K recordings), modulation amplitude 6 Gauss and modulation frequency 100 kHz. The aortas treated with lipopolysaccharide were opened longitudinally and placed with care in a flat vial filled with Krebs solution. Addition of N-acetylcysteine (final concentration 2×10^{-2} M) was made using a microcatheter inserted into the vial. Some lipopolysaccharide-treated rings were exposed to N^ω-nitro-L-arginine methyl ester (3×10^{-3} M for 15 min) and diethyldithiocarbamate (10^{-3} M for 1 min) and then studied at 77K.

Chemicals

L-arginine, diethydithiocarbamate, N^ω-nitro-L-arginine methyl ester, norepinephrine, tetrabutylammonium (from Sigma Chemical Co, Saint Quentin-Fallavier, France), [^{125}I]-cyclic GMP and antibodies against cyclic GMP (supplied from Dr B. Lutz-Bucher, CNRS URA 1446, Strasbourg), dinitrosyl-iron complexes (synthetized by treatment of 2×10^{-3} M $FeSO_4$ $7H_2O$ and 4×10^{-2} M $Na_2S_2O_3$ solution for 5 min with NO in oxygen free conditions followed by 1 min evaporation to remove unbound NO), lipopolysaccharide (from *Escherichia coli*, Difco, Detroit, USA), N-acetylcysteine (Fluimucil® for parenteral use from Zambon Laboratory, Antibes, France).

Expression of Results and Statistical Analysis

Results are expressed as mean ± s.e. mean of n experiments. Relaxing effects are expressed in percentage of contraction. Concentration that produced 50% relaxation (EC_{50} values)

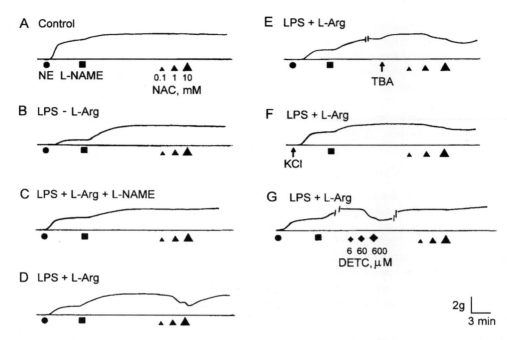

Figure 15-1. Representative traces showing the effect of N-acetylcysteine (NAC) on rat aortic rings incubated for 18 h in the absence [A] or in the presence of lipopolysaccharide (LPS) [B], lipopolysaccharide + L-arginine (L-arg) + N^{ω}-nitro-L-arginine methyl ester (L-NAME) [C], or lipopolysaccharide + L-arginine [D-G]. The rings were precontracted with 3×10^{-6} M norepinephrine (NE) [A-E and G] or 50 mM KCl [F] plus 3×10^{-3} M N^{ω}-nitro-L-arginine methyl ester. The effect of N-acetylcysteine was studied in the presence of 3×10^{-3} M tetrabutylammonium (TBA) [F], or after addition of diethyldithiocarbamate (DETC) [G].

were determined by log-logit regression. The cyclic GMP content is expressed as pmol per μg DNA. Statistical comparisons were performed by the analysis of variance. P values less than 0.05 were considered to be statistically significant.

RESULTS

Effect of N-Acetylcysteine in Rat Aorta Treated with Lipopolysaccharide

Incubation of aortic preparations for 18 h with lipopolysaccharide produced a decrease in the amplitude of the contractile response to norepinephrine (Figure 15-1), an elevation in basal cyclic GMP content (Figure 15-2) and the appearance of a distinct anisotropic electron paramagnetic resonance signal with $g\perp = 2.04$ and $g\| = 2.015$, attributed to dinitrosyl-iron complexes with sulfur groups of proteins (Figure 15-3A). The addition of N^{ω}-nitro-L-arginine methyl ester (3×10^{-3} M) produced a large contractile effect in lipopolysaccharide-treated rings (Figure 15-1) and significantly decreased their cyclic GMP content (Figure 15-2).

The effect of N-acetylcysteine was studied on rings contracted with norepinephrine (3×10^{-6} M) and N^{ω}-nitro-L-arginine methyl ester (3×10^{-3} M). N-acetylcysteine (10^{-4}

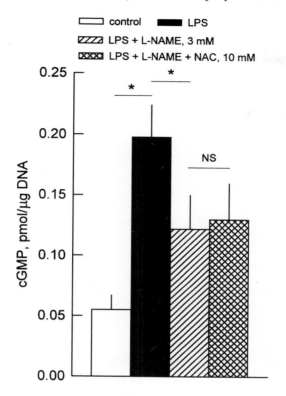

Figure 15-2. Histograms showing the cyclic GMP content of rat aortic rings incubated in the absence ($\square$) or in the presence of LPS + L-arginine ($\blacksquare$) and then exposed to L-NAME (3×10^{-3} M, $\boxtimes$) or L-NAME (3×10^{-3} M) + N-acetylcysteine (NAC, 10^{-2} M, $\boxtimes$). NS: not significant, * P < 0.05.

to 10^{-2} M) did not significantly modify the tension of control rings, of rings incubated for 18 h with lipopolysaccharide in the absence of L-arginine or in the presence of L-arginine + N$^\omega$-nitro-L-arginine methyl ester (Figure 15-1). However, in rings incubated during 18 h with lipopolysaccharide and L-arginine, N-acetylcysteine evoked a concentration-dependent relaxation which was rapid in onset and transient (Figure 15-1D) and still was observed following removal of the endothelium (after the incubation period with lipopolysaccharide (Muller, Kleschyov and Stoclet, 1996)). The relaxation evoked by N-acetylcysteine was not associated with an elevation in cyclic GMP content (Figure 15-2) but was attenuated significantly in the presence of tetrabutylammonium (3×10^{-3} M) or by the use of an elevated concentration of external KCl (50 mM, instead of norepinephrine) as constricting agent (Figure 15-1). The relaxation produced by N-acetylcysteine in lipopolysaccharide-treated rings was not affected by the selective potassium channels blockers glibenclamide (10^{-5} M), charybdotoxin (10^{-7} M), apamin (5×10^{-7} M) or 4-aminopyridine (3×10^{-3} M), or by inhibitors of cyclooxygenases (indomethacin, 10^{-5} M) or cytochromes P450 monooxygenases (17-octadecynoic acid, 10^{-5} M) (Muller *et al.*, 1998). In rings incubated with lipopolysaccharide + L-arginine and contracted with

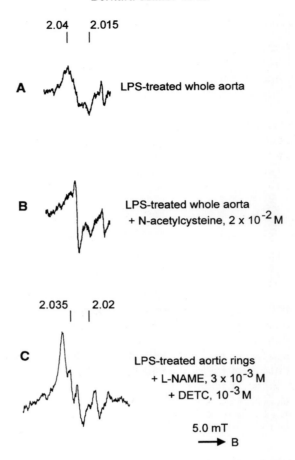

Figure 15-3. Electron paramagnetic resonance spectra of whole thoracic aorta (A, B) or aortic rings (C) incubated with lipopolysaccharide (LPS) + L-arginine. Spectra were recorded at room temperature before (A) or after (B) the addition of N-acetylcysteine (2×10^{-2} M) or (C) at 77K after the addition of N^{ω}-nitro-L-arginine methyl ester (3×10^{-3} M) and diethyldithiocarbamate (10^{-3} M).

norepinephrine and N^{ω}-nitro-L-arginine methyl ester, the metal chelator diethyldithio-carbamate (6×10^{-6} to 6×10^{-4} M) also produced a transient relaxation, with the tension returning back to the initial values within 10 minutes (Figure 15-1). Subsequent addition of N-acetylcysteine failed to produce a relaxing effect (Figure 15-1).

Addition of N-acetylcysteine (2×10^{-2} M) to aortas incubated with lipopolysaccharide and L-arginine led to the appearance of a narrow electron paramagnetic resonance singlet with g = 2.03, which is characteristic of low molecular weight dinitrosyl-iron complexes, in addition to the anisotropic signal (Figure 15-3B). In preparations incubated with lipo-polysaccharide + L-arginine and exposed to N^{ω}-nitro-L-arginine methyl ester (3×10^{-3} M), the metal chelator diethyldithiocarbamate (10^{-3} M) induced the transformation of dinitrosyl-iron complexes into mononitrosyl-iron complexes with diethyldithiocarbamate (Figure 15-3C).

Effect of Exogenous Low Molecular Weight Dinitrosyl-Iron Complexes

In rat aortic rings without endothelium, exogenous low molecular weight dinitrosyl-iron complexes (10^{-10} to 10^{-6} M) evoked a concentration-dependent relaxation which was not modified in the presence of tetrabutylammonium (3×10^{-3} M) nor by the use of 50 mM KCl as contracting agent (EC_{50} values of 1.0 ± 0.2, 1.2 ± 0.6 and $1.0 \pm 0.3 \times 10^{-7}$ M, respectively, Figure 15-4). The relaxation evoked by dinitrosyl-iron complexes was associated with a large (11 fold at 10^{-7} M) increase in cyclic GMP content (Figure 15-4).

Some rings (without endothelium) were exposed to dinitrosyl-iron complexes (10^{-6} M for 30 min) and then washed-out extensively. This resulted in the formation of protein-bound dinitrosyl-iron complexes in the aorta (not shown) which was due to the transition of iron-$(NO)_2$ groups from low molecular weight ligands to cysteine-residues of proteins (Mülsch *et al.*, 1991). In rings exposed to dinitrosyl-iron complexes, N-acetylcysteine also produced a relaxation which was attenuated by tetrabutylammonium (3×10^{-3} M) or by the use of 50 mM KCl as contracting agent (Figure 15-4).

DISCUSSION

This study shows that N-acetylcysteine evoked a relaxation in rat aortic rings incubated with lipopolysaccharide and L-arginine. However, N-acetylcysteine had no effect in rings incubated with lipopolysaccharide in the absence of the substrate of the enzyme, L-arginine, or in the presence of L-arginine and an inhibitor of NOS activity, N^{ω}-nitro-L-arginine methyl ester. Removal of the endothelium (after the incubation period with lipopolysaccharide) did not modify the relaxation induced by N-acetylcysteine (Muller, Kleschyov and Stoclet, 1996). Taken in conjunction, these observations suggest that the relaxing effect of N-acetylcysteine is related to an interaction with NO stores formed in the media and/or the adventitia via lipopolysaccharide-induced NOS activity.

An electron paramagnetic resonance signal which is characteristic of protein-bound dinitrosyl-iron complexes was detected in aortas incubated with lipopolysaccharide and L-arginine. These complexes were mainly localized in the media (Kleschyov, Muller and Stoclet, 1998b). No characteristic signal was found when the aortas were incubated without lipopolysaccharide or L-arginine, or in the presence of N^{ω}-nitro-L-arginine methyl ester or dexamethasone (Kleschyov, Muller and Stoclet, 1998b). These data indicate that protein-bound dinitrosyl-iron complexes were generated via lipopolysaccharide-induced, iNOS activity. The association between the appearance of protein-bound dinitrosyl-iron complexes and a relaxing effect of N-acetylcysteine suggests that dinitrosyl-iron complexes may represent N-acetylcysteine-sensitive NO stores in vessels treated with lipopolysaccharide. Accordingly, exposure of rings to exogenous dinitrosyl-iron complexes (which resulted in the formation of protein-bound dinitrosyl-iron complexes) reproduced the N-acetylcysteine-induced relaxation. Furthermore, the metal chelator diethyldithiocarbamate both destroyed dinitrosyl-iron complexes and prevented the relaxing effect of N-acetylcysteine in aorta treated with lipopolysaccharide.

Low molecular dinitrosyl-iron complexes are potent activators of soluble guanylate-cyclase (Mülsch *et al.*, 1991) and display relaxing (Vedernikov *et al.*, 1992; Vanin, Stukan and Manukhina, 1996) and hypotensive properties (Kleschyov, Mordvintcev and Vanin,

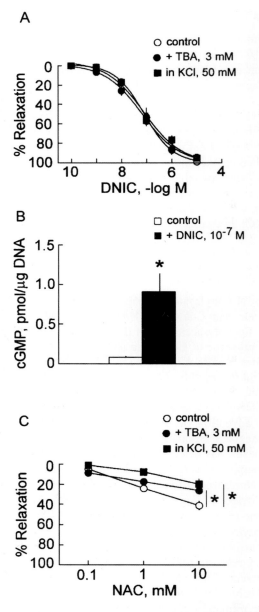

Figure 15-4. (A) Concentration-response curves of dinitrosyl-iron complexes on rat aortic rings contracted with KCl (50 mM) or with norepinephrine (10^{-7} M) in the absence or in the presence of tetrabutylammonium (TBA, 3×10^{-3} M). (B) Histograms showing the cyclic GMP content of rat aortic rings exposed or not to dinitrosyl-iron complexes (DNIC, 10^{-7} M for 5 min). The asterisk indicates a statistically significant difference (P < 0.01) between groups. (C) Concentration-response curves of N-acetylcysteine on rat aortic rings exposed to dinitrosyl-iron complexes (10^{-6} M for 30 min) and extensively washed-out. The rings were precontracted with KCl (50 mM) or with norepinephrine (10^{-7} M) and the effect of N-acetylcysteine was studied in the absence or in the presence of tetrabutylammonium (TBA, 3×10^{-3} M). The asterisks indicate a statistically significant difference (P < 0.05) between groups.

1985). Therefore, the generation of low molecular weight dinitrosyl-iron complexes from protein-bound species may account for the relaxing effect of N-acetylcysteine via the activation of soluble guanylate-cyclase. Accordingly, the addition of N-acetylcysteine to aortas incubated with lipopolysaccharide and L-arginine led to the appearance of a narrow singlet by electron paramagnetic resonance which is characteristic of low molecular weight dinitrosyl-iron complexes. However, such evidence for the partial conversion of protein-bound dinitrosyl-iron complexes into low molecular weight species was obtained with a relatively large concentration of N-acetylcysteine (> 10^{-2} M), questioning the contribution of low molecular weight dinitrosyl-iron complexes to the relaxation induced by N-acetylcysteine.

To further investigate this possibility, the mechanism of the relaxation evoked by N-acetylcysteine in lipopolysaccharide-treated aorta was compared with that produced by exogenous low molecular dinitrosyl-iron complexes. Two major differences were observed: (a) the relaxing effect of low molecular weight dinitrosyl-iron complexes was associated with a large increase in aortic cyclic GMP content, whereas the relaxation evoked by N-acetylcysteine in lipopolysaccharide-treated rings was not; and (b) tetrabutylammonium (a non selective potassium channels blocker) or an elevated concentration of external KCl (which decreases the driving force for potassium ions, thereby decreasing potassium efflux and subsequent hyperpolarisation) attenuated the relaxation evoked by N-acetylcysteine, whereas such treatments did not affect the relaxing effect of exogenous dinitrosyl-iron complexes. These observations strongly suggest that in the rat aorta, the activation of the cyclic GMP pathway is not associated with an activation of potassium channels. They also do not support the hypothesis that in lipopolysaccharide-treated rat aorta, the generation of low molecular weight dinitrosyl-iron complexes from protein-bound species and the subsequent activation of guanylate-cyclase play a major role in the relaxation produced by N-acetylcysteine. They rather suggest that the relaxation induced by N-acetylcysteine involves the opening of tetrabutylammonium-sensitive potassium channels. A mechanism which can be proposed is that N-acetylcysteine may in some way facilitate the transfer of NO^+ from protein dinitrosyl-iron complexes to critical thiol groups of tetrabutylammonium-sensitive potassium channels. Interestingly, NO may elicit hyperpolarisation and relaxation in vascular smooth muscle by a cyclic GMP-independent mechanism involving a direct activation of calcium-activated potassium channels (Bolotina *et al.*, 1994). The lack of effect of various selective potassium channel blockers on N-acetylcysteine-induced relaxation (Muller *et al.*, 1998) might indicate the ineffectiveness of the selective blockade of a single type of potassium channel to achieve a significant inhibition of the responses in intact blood vessels. Indeed the use of non specific potassium channels blockers or of a combination of different selective blockers is sometimes necessary for such inhibition in various vascular preparations (Corriu *et al.*, 1996b; Zygmunt and Högestätt, 1996; Ohlmann *et al.*, 1997).

16 Is L-Citrulline an EDHF in the Rabbit Aorta?

E. Ruiz and T. Tejerina

*Department of Pharmacology, School of Medicine, Complutense University,
28040 Madrid, Spain.
Address for correspondence: Emilio Ruiz, Department of Pharmacology,
School of Medicine, Complutense University 28040 Madrid, Spain. Tel: 34-91-3941476;
Fax: -34-91-3941463; E-mail address:Teje@ucmax.sim.ucm.es*

Several endothelium-dependent relaxing factors such as nitric oxide, prostacyclin and Endothelium-Derived Hyperpolarizing Factor (EDHF) contribute to the control of vascular tone. In previous work, L-citrulline, the byproduct of NO synthesis, relaxed aortic rings contracted with norepinephrine, and this relaxation was blocked by apamin (a K_{ca}-channel blocker) and by HS-142-1 (an inhibitor of particulate guanylate cyclase). The present experiments were designed to test the physiological importance of L citrulline in rabbit vascular smooth muscle-relaxation. Both apamin and HS-142-1 decreased the endothelium-dependent relaxation induced by acetylcholine in rabbit aortic rings contracted with norepinephrine. Incubation with oxyhemoglobin also decreased this relaxation, which was enhanced by HS-142-1. No differences were observed between the incubation with N^{ω}-nitro-L-arginine methyl ester or N^{ω}-nitro-L-arginine methyl ester plus HS-142-1 as regards endothelium-dependent relaxations. This suggests that, although endothelium-dependent relaxations in the rabbit thoracic aorta is mainly due to NO, L-citrulline may play a role by acting on the atrial natriuretic peptide receptor, leading to the hyperpolarization of vascular smooth muscle.

KEYWORDS: endothelium, rabbit aorta, L-citrulline, EDHF, ANP-receptor.

INTRODUCTION

Since the seminal observation of Furchgott and Zawadzki (1980) many studies have emphasized the importance of the endothelium in the control of vascular tone. Thus, several endothelium-dependent relaxing factors such as nitric oxide, prostacyclin and endothelium-derived hyperpolarizing factor (EDHF) have been implicated in this control. The chemical structure of EDHF has not been identified, but several candidates have been proposed including epoxyeicosatrienoic acid (Campbell *et al.*, 1996).

On the other hand, L-citrulline is the byproduct of the reaction catalyzed by NO synthase, which converts L-arginine into nitric oxide plus L-citrulline. Previous work demonstrated that L-citrulline relaxes the rabbit aorta by a mechanism which involves the activation of the HS-142-1-sensible atrial natriuretic peptide receptor. In the vascular smooth muscle of the rabbit aorta this leads to hyperpolarization (Ruiz and Tejerina, 1998).

Vascular smooth muscle is one of the major target tissues of natriuretic peptides. These comprise a family of three distinct peptides, atrial natriuretic peptide, brain natriuretic

143

peptide and C-type natriuretic peptide and are involved in body fluid homeostasis and blood pressure control. The ANP-A and ANP-B receptors involve the particulate guanylate cyclase and elicit an increase in intracellular cyclic GMP that mediates relaxation of vascular smooth muscle, whereas the ANP-C receptor (also called "clearance receptor") probably is involved in the clearance of the natriuretic peptides (Chinkers et al., 1992). The present study investigated the possible relationship between L-arginine-nitric oxide pathway and natriuretic peptides in the relaxation of the rabbit aorta, caused by L-citrulline.

MATERIALS AND METHODS

General Procedure

Male New Zealand White rabbits weighing 2.5 to 3.0 kg were obtained from Biocentre S.A. (Barcelona, Spain). The animals were anesthetized with ethyl ether and killed by exsanguination from the common carotid.

All protocols concerning animals were approved by the Complutense University of Madrid (EEC official registration 28079-15ABC).

Isolated Aortic Preparations

The thoracic aorta was rapidly removed and placed in physiological salt solution of the following composition (mM): NaCl 121, KCl 5.8, NaHCO$_3$ 14.9, MgCl$_2$ 1.22, glucose 11 and CaCl$_2$ 1.25. Adherent fat and surrounding tissue were cleaned off and the arteries were cut into rings (approximately 2 to 3 mm wide). The rings were suspended between two stainless steel hooks in organ chambers containing 10 ml of physiological salt solution. The solution was kept at 36 ± 0.5°C and gassed continuously with a 95% O$_2$–5% CO$_2$ gas mixture. The aortic rings were streched to 2 g tension. Each preparation was allowed to equilibrate for 90 to 120 min. Contractile responses were measured isometrically by means of force-displacement transducers (Grass FT 03) and were recorded on a Grass polygraph (Tejerina et al., 1988). The isometric force was also digitalized by a MacLab A/D converter (Chart v3.2, A.D Instruments Pty. Ltd., Castle Hill, Australia) and stored and displayed on a Mackintosh computer (Ruiz and Tejerina, 1998).

Experimental Procedure

After the equilibration period, a contraction to norepinephrine (10^{-6} M) was obtained, and when a steady response was reached, a cumulative concentration-response curve to acetylcholine (10^{-8}–10^{-5} M) was made. After washing out the arteries, the rings were incubated for 30 min in the presence of the following agents and the procedure was repeated: apamin (3 × 10^{-6} M) (a K$_{ca}$-channel blocker), HS-142-1 (10^{-5} M) (a inhibitor of particulate guanylate cyclase), oxyhemoglobin (3 × 10^{-5} M) (in order to scavenge NO) or oxyhemoglobin plus HS-142-1, or N$^\omega$-nitro-L-arginine methyl ester (10^{-4} M) (in order to block NO and L-citrulline production) and L-NAME plus HS 142-1.

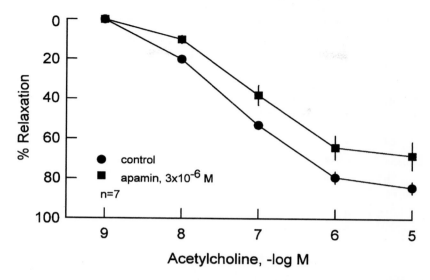

Figure 16-1. Effect of apamin (3×10^{-6} M) on the concentration-response curve to acetylcholine (10^{-8}–10^{-5} M) in isolated aortic rings of the rabbit contracted with norepinephrine (10^{-6} M). Each point represents the mean ± s.e. mean of seven experiments. The contraction induced by norepinephrine (10^{-6} M) in control and in the presence of apamin were 5.0 ± 0.3 g and 5.0 ± 0.4 g respectively. *$P < 0.05$.

Drugs

The following drugs were used: acetylcholine chloride (Sigma), apamin (Sigma), L-citrulline (Sigma), HS-142-1 was supplied by Kyowa Hakko Kogyo Co., Ltd., N^{ω}-nitro-L-arginine methyl ester (Sigma), norepinephrine bitartrate (Sigma). Oxyhemoglobin was a generous gift from Dr. Carlos Sanchez Ferrer. Stock solutions were prepared by dissolving the compound in distilled water. Ascorbic acid was added to the norepinephrine solution in order to avoid oxidation. Working solutions were made in physiological salt solution.

Statistical Analysis

All values used in analyses represent mean ± s.e. mean of six to seven rabbits in each experimental group. Comparisons between the different groups were performed by two way ANOVA test and differences were considered significant when p was less than 0.05.

RESULTS

In aortic rings contracted with norepinephrine (10^{-6} M), acetylcholine (10^{-8}–10^{-5} M) caused relaxation in a concentration-dependent manner. The maximal relaxation induced by acetylcholine averaged $84.1 \pm 2.0\%$. Incubation in the presence of apamin (3×10^{-6} M) shifted the concentration-response curve to acetylcholine upward and to the right [$F(1,36) = 4,59$], reaching $70.2 \pm 6.0\%$ of the maximal relaxation (Figure 16-1).

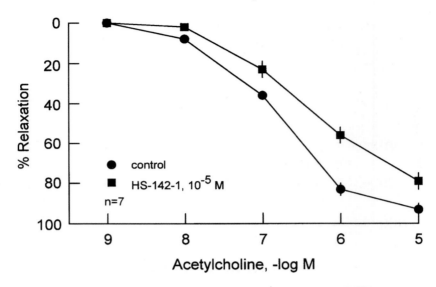

Figure 16-2. Effect of HS-142-1 (10^{-5} M) on the concentration-response curve to acetylcholine (10^{-8}–10^{-5} M) in isolated aortic rings of the rabbit contracted with norepinephrine (10^{-6} M). Each point represents the mean ± s.e. mean of seven experiments. The contraction induced by norepinephrine 10^{-6} M in control and in the presence of HS-142-1 were 5.0 ± 0.3 g and 4.7 ± 0.5 g respectively. ***P < 0.001.

HS-142-1 (10^{-5} M) decreased the maximal relaxation induced by acetylcholine (10^{-5} M) from 93.1 ± 2.0% to 80.1 ± 2.2% and shifted the concentration-response curve to acetyl-choline upward and to the right [$F(1, 8) = 48,8$] (Figure 16-2).

The maximal relaxation induced by acetylcholine (10^{-5} M) decreased in the presence of 3×10^{-5} M oxyhemoglobin (from 84.1 ± 2.0% in control to 59.0 ± 7.0%). In the presence of oxyhemoglobin plus HS-142-1 (10^{-5} M) the relaxation decreased to 38.0 ± 9.0% ($F(1,10) = 10,2$) (Figure 16-3).

The maximal relaxation induced by acetylcholine decreased from 84.2 ± 3.1% to 47.0 ± 3.0% in arteries with 10^{-4} M N^{ω}-nitro-L-arginine methyl ester. No differences were found between N-nitro L-arginine methyl ester and N^{ω}-nitro-L-arginine methyl ester plus HS-142-1 (10^{-5} M) treated arteries (Figure 16-4).

DISCUSSION

The present study demonstrates that both apamin and HS-142-1 decrease the endothelium-dependent relaxation induced by acetylcholine in the rabbit-isolated aorta artery; the degree of inhibition is comparable for both compounds. In a previous study, the addition of exogenous L-citrulline caused relaxantoin in rabbit aortas contracted to norepinephrine, but not to high-K$^+$, and this relaxation was blocked by apamin and by HS-142-1 (Ruiz and Tejerina, 1998). The present experiments were designed to determine whether or not endogenous L-citrulline has a role in the endothelium-dependent relaxation in the rabbit aorta.

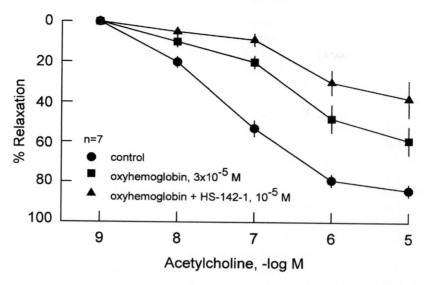

Figure 16-3. Effect of oxyhemoglobin (3×10^{-5} M) or oxyhemoglobin plus HS-142-1 (10^{-5} M) on the concentration-response curve to acetylcholine (10^{-8}–10^{-5} M) in isolated aortic rings of the rabbit contracted with norepinephrine (10^{-6} M). Each point represents the mean ± s.e. mean of seven experiments. The contractions induced by norepinephrine (10^{-6} M) in control and in the presence of oxyhemoglobin or oxyhemoglobin plus HS-142-1 were 5.0 ± 0.3 g; 5.2 ± 0.7 g and 4.8 ± 0.6 g respectively. *P < 0.05.

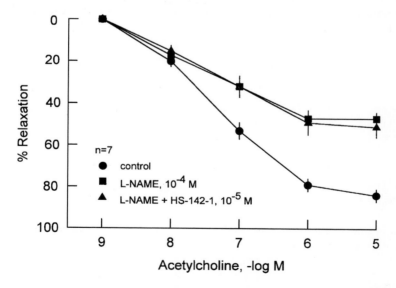

Figure 16-4. Effect of N^{w}-nitro-L-arginine methyl ester (L-NAME) (10^{-4} M) or N^{ω}-nitro-L arginine methyl ester (L-NAME) (10^{-4} M) plus HS-142-1 (10^{-5} M) on the concentration-response curve to acetylcholine (10^{-8}–10^{-5} M) in isolated aortic rings of the rabbit contracted with norepinephrine (10^{-6} M). Each point represents the mean ± s.e. mean of seven experiments. The contraction induced by norepinephrine (10^{-6} M) in control and in the presence of L-NAME or L-NAME plus HS-142-1 were 5.0 ± 0.3 g; 5.4 ± 0.4 g and 5.1 ± 0.7 g respectively.

The inhibition of endothelium-dependent relaxations by oxyhemoglobin was enhanced by HS-142-1. There are at least two possible explanations for these data presented. First, the L-citrulline released by the endothelium, together with nitric oxide, plays a role in endothelium-dependent relaxations. Second, some type of natriuretic peptide (more than likely the C-type natriuretic peptide, Chen and Burnett, 1997) is released by the endothelium stimulated by acetylcholine, and the observed inhibitory effect of HS-142-1 and apamin is due to the C-type natriuretic peptide and not L-citrulline. C-type natriuretic peptide has been localized in human, bovine and canine vascular endothelial cells (Heublein *et al.*, 1992; Stingo *et al.*, 1992; Suga *et al.*, 1992).

The inhibition exerted by N^{ω}-nitro-L-arginine methyl ester on endothelium dependent relaxations was not enhanced by HS-142-1. This suggests that the factor responsible for the enhancement by HS-142-1 is a product of NO synthase (nitric oxide or L-citrulline). The only natriuretic peptide receptor coupled to particulate guanylate cyclase identified so far in the rabbit is the ANP-A receptor (Vandlen, Arcur and Napier, 1985; Vandlen *et al.*, 1986; Gunning *et al.*, 1988; Gunning *et al.*, 1989). The present data which support the hypothesis is that nitric oxide and L-citrulline are released together in an equimolecular way from endothelial cells after stimulation by acetylcholine (Moncada and Palmer, 1990) and thus, that L-citrulline could acts on vascular smooth muscle cells (presumably through activation of ANP-A receptors).

L-citrulline hyperpolarizes vascular smooth muscle, since the relaxation it induces is abolished by high-K^+ (Ruiz and Tejerina, 1998) and decreased by apamin, (a low conductance K_{ca}-channel blocker), which also decreased the relaxation induced by acetylcholine. EDHF induces hyperpolarization and relaxation of vascular smooth muscle cells (Vanhoutte and Cohen, 1996). One of the EDHFs may be an activator of low-conductance K_{ca}-channel (Chataigneau *et al.*, 1998). The present data would be explained if L-citrulline acted as an EDHF.

17 Endocannabinoids: Endothelium-Derived Vasodilators

Michael D. Randall*, David Harris, Iain T. Darker, Paul J. Millns and David A. Kendall

School of Biomedical Sciences, University of Nottingham Medical School, Queen's Medical Centre, Nottingham NG7 2UH
Address for correspondence: Michael D. Randall. Telephone: +44 115 9709484; Fax: + 44 115 9709259, E-mail: michael.randall@nottingham.ac.uk

Endogenous cannabinoids (endocannabinoids), which were first identified in the central nervous system, exert cardiovascular actions. The prototypic endocannabinoid, derived from arachidonic acid, anandamide, is a dilator in the resistance vasculature but not in conduit vessels. The relaxation it causes is either endothelium-independent or -dependent, depending on the vascular bed. An endocannabinoid may mediate the nitric oxide- and prostanoid-independent component of endothelium-dependent relaxations, as these responses are sensitive to cannabinoid receptor antagonists and show similarities to anandamide-induced relaxations. This hypothesis has generated much controversy with substantial conflicts emerging in the literature. Despite this controversy, endothelial cells have been shown to produce endocannabinoids, and cannabinoid receptors in the vasculature have been identified.

KEYWORDS: cannabinoids, anandamide, endothelium-derived hyperpolarizing factor (EDHF), endothelium.

ENDOCANNABINOIDS

In 1992 the first endocannabinoid, anandamide (N-arachidonoylethanolamine), which is the ethanolamide of arachidonic acid, was isolated from the porcine brain (Devane *et al.*, 1992). It was shown both to occupy cannabinoid receptors and to mimic the functional effects of Δ^9-tetrahydrocannabinol. Anandamide is the prototype of a family of N-acylethanolamines and other polyunsaturated N-acylethanolamines (Hanus *et al.*, 1993), which have similar effects via activation of G protein-linked cannabinoid (CB_1 or CB_2) receptors. To date, CB_1 receptors have been found predominantly in the brain and peripheral nervous system, and the CB_2 receptor appears to be exclusive to immune tissues.

CARDIOVASCULAR EFFECTS OF ENDOCANNABINOIDS

Actions *in vivo*

In urethane-anaesthetized rats, anandamide causes bradycardia (with secondary hypotension) and a transient pressor effect which is followed by a longer lasting depressor effect

149

(Varga *et al.*, 1995; Lake *et al.*, 1997). This depressor effect is believed to be mediated by CB_1 receptor-dependent inhibition of sympathetic tone via a presynaptic mechanism, as the effect was independently attenuated by cervical spinal transection, α-adrenoceptor blockade and cannabinoid receptor blockade. The pressor component of the response to anandamide was not, however, sensitive to cannabinoid receptor blockade, perhaps reflecting a non-CB_1 receptor-mediated response (Lake *et al.*, 1997). By contrast, in the conscious rat anandamide caused bradycardia, and a transient depressor response, followed by a longer pressor phase, and only at high doses was there delayed hypotension (Stein *et al.*, 1996).

Vascular Actions of Endocannabinoids

Anandamide was first shown to be a dilator in the rat cerebral vasculature, but these effects were sensitive to indomethacin, suggesting that cannabinoids may cause relaxation through the stimulation of the metabolism of arachidonic acid (Ellis, Moore, Willoughby, 1995). Dependence on prostanoids was also found for the relaxing effects of Δ^9-tetrahydro-cannabinol.

In the rat isolated mesenteric and coronary vasculatures, anandamide is also a vasodilator, and endocannabinoids may be endothelium-derived agents (Randall *et al.*, 1996; Randall & Kendall, 1997). In mesenteric blood vessels (Randall *et al.*, 1996; Randall, McCulloch and Kendall, 1997; Plane *et al.*, 1997; White & Hiley, 1997), and the coronary vasculature (Randall & Kendall, 1997), anandamide induces relaxation in the presence of blockers of both nitric oxide synthase and cyclooxygenase, and also in the absence of the endothelium (Randall *et al.*, 1996; White & Hiley, 1997), and so acts independently of endothelial autacoids. Earlier studies have reported that responses to anandamide were abolished by high extracellular potassium, and proposed that endocannabinoids may be hyperpolarizing agents (Randall *et al.*, 1996).

The relaxant effects of anandamide also show tissue selectivity, as it does not relax rabbit or rat carotid arteries (Holland *et al.*, 1997) or the rat aorta (Darker & Kendall, unpublished observations). Indeed, it may be that the actions of endocannabinoids are localised to the resistance vasculature.

Endocannabinoids and EDHF

The observations that anandamide was an endothelium-independent vasodilator, which appeared to act via a hyperpolarizing mechanism, led to the hypothesis that an endocannabinoid could account for EDHF activity. In this respect, an examination has been made of the effects of cannabinoid receptor antagonists against EDHF-mediated relaxation in the mesenteric and coronary vasculatures of the rat. In the mesenteric vascular bed, EDHF-mediated responses were found to be sensitive to the CB_1-receptor antagonists, SR141716A and LY320135 but were not affected by the CB_1 ligand, AM630, or the CB_2 receptor antagonist SR144528 (Figure 17-1; Randall *et al.*, 1996; Harris, Kendall and Randall, 1998a). The finding that SR141716A opposes EDHF-mediated responses has been confirmed by others (White & Hiley, 1997; Hewitt, Plane and Garland, 1997; Rowe, Garland and Plane, 1998; Sedhev, Garland and Plane, 1998) but this has not been the case in all studies (Plane *et al.*, 1997; Zygmunt *et al.*, 1997b; Holland *et al.*, 1997; Chaigneau

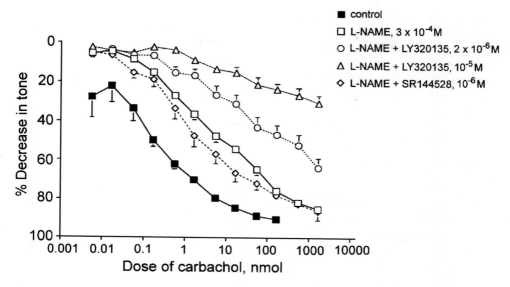

Figure 17-1. The relaxing effects of carbachol in the rat isolated perfused mesenteric arterial bed under control conditions (with 1×10^{-5} M indomethacin) and in the presence of 3×10^{-4} M N$^\omega$-nitro-L-arginine methyl ester. Additional dose-response curves are shown for EDHF-mediated relaxations (relaxations to carbachol in the presence of 3×10^{-4} M N$^\omega$-nitro-L-arginine methyl ester and 1×10^{-5} M indomethacin) following addition of the CB$_1$ receptor antagonist, LY320135, and the CB$_2$ receptor antagonist, SR144528. EDHF-mediated relaxations to carbachol (ED$_{50}$ = $(3.3 \pm 0.6) \times 10^{-9}$ mol; R$_{max}$ = $87.0 \pm 2.5\%$) were significantly (P < 0.05, analysis of variance) opposed by the selective cannabinoid CB$_1$ antagonist, LY320135 (2×10^{-6} M, ED$_{50}$ = $(1.0 \pm 0.3) \times 10^{-8}$ mol and R$_{max}$ = $66.9 \pm 6.2\%$; 1×10^{-5} M, ED$_{50}$ = $(1.6 \pm 0.4) \times 10^{-8}$ mol and R$_{max}$ = $34.0 \pm 4.3\%$). However, these responses were unaffected by the CB$_2$ selective antagonist, SR144528 (1×10^{-6} M). Data shown as mean ± s.e.m.

et al., 1998b; Pratt *et al.*, 1998). Chataigneau *et al.* (1998b) reported that SR141716A opposed EDHF-mediated hyperpolarization in some preparations, but ascribed this effect to the antagonist interacting with potassium channels rather than CB receptors. Pratt *et al.* (1998) provided evidence that SR141716A interferes with arachidonic acid metabolism, and this could, potentially, explain the effects against EDHF, which they proposed to be a non-cannabimimetic metabolite of arachidonic acid. Nevertheless, earlier results point to the involvement of CB$_1$-like receptors in EDHF-mediated vasorelaxation (Randall *et al.*, 1996).

Do Endothelial Cells Produce Endocannabinoids?

This question is fundamental for endocannabinoids to be regarded as endothelium-derived autacoids. Cultured rat renal endothelial cells contain anandamide, together with synthase and amidase activities (Deutsch *et al.*, 1997). This study confirmed that anandamide is a vasodilator of renal afferent arterioles but acts via the release of endothelium-derived nitric oxide. More recently cultured human umbilical vein endothelial cells were shown to release the endocannabinoid, 2-arachidonylglycerol, on stimulation with a calcium ionophore (Sugiura *et al.*, 1998). However, anandamide was not produced in bovine coronary endothelial cells, and furthermore, these cells metabolized exogenous anandamide, possibility via a cytochrome P450 monooxygenase to vasoactive metabolites, which accounted for the relaxing effects of anandamide (Pratt *et al.*, 1998). This is, however, unlikely to be the case in the rat mesentery, since methanandamide, a metabolically stable analogue of anandamide, is equipotent with anandamide (Randall, unpublished observations) and exogenous arachidonic acid is without effect (Randall *et al.*, 1996). These conflicts raise the possibility that the actions of anandamide, and other endocannabinoids, may be dependent on the vasculature in question.

Mechanism(s) of Action of Endocannabinoids

The original findings suggested that anandamide might act via a hyperpolarizing mechanism. This proposal was confirmed by the demonstration that anandamide causes hyperpolarization or repolarization of vascular smooth muscle, but in both cases this effect was insensitive to cannabinoid CB_1 receptor blockade (Plane *et al.*, 1997, Chataigneau *et al.*, 1998b). The hyperpolarization was found to be endothelium-dependent (Chataigneau *et al.*, 1998b), an observation also made in the rat hepatic artery (Zygmunt *et al.*, 1997b). Indeed, the latter study provided evidence that anandamide acted via inhibition of calcium mobilization in vascular smooth muscle cells, without direct effects on potassium conductance.

In rat mesenteric vessels anandamide-induced relaxation is sensitive to non-specific potassium channel blockers, including cytochrome P450 inhibitors (Randall, McCulloch and Kendall, 1997). In isolated mesenteric arterial segments the relaxation to anandamide was blocked by selective inhibitors of large conductance calcium-activated K^+-channels (charybdotoxin and iberiotoxin), while EDHF responses, and not those to anandamide, were sensitive to apamin, thereby indicating important differences in their pharmacology (Plane *et al.*, 1997). Furthermore, in similar mesenteric vessels, the anandamide–induced relaxation was insensitive to the combination of charybdotoxin and apamin, which blocked EDHF-mediated relaxation (White & Hiley, 1997). Similar observations have also been made in the rat hepatic artery (Zygmunt *et al.*, 1997b). In the guinea-pig carotid artery, the anandamide-induced hyperpolarization was insensitive to charybdotoxin plus apamin, but was blocked by the ATP-sensitive potassium channel inhibitor, glibenclamide. By contrast, glibenclamide does not affect anandamide-induced relaxation in the rat mesentery (Randall *et al.*, 1997; White & Hiley, 1997). Neither charybdotoxin nor apamin alone affected either EDHF-mediated or anandamide-induced responses in the isolated mesentery, but when used in combination, charybdotoxin and apamin abolished both responses (Randall & Kendall, 1998). There are clearly many discrepancies in the literature to date, with only our findings pointing to EDHF and endocannabinoids acting at a common site.

Vascular Cannabinoid Receptors

It is clearly essential to demonstrate the presence of cannabinoid receptors in vascular tissue to support the findings of functional studies. However, the very lipophilic nature of the cannabinoid receptor ligands available has made it very difficult to use either radioligand binding or autoradiographic techniques to identify CB receptors in the rat vasculature (Kendall & Millns, unpublished).

Two other approaches have therefore been used, the reverse transcriptase polymerase chain reaction (RT-PCR) and immunohistochemistry in an attempt to locate CB_1 receptor transcripts and protein, respectively.

Reverse transcriptase — polymerase chain reaction

Cerebella, primary, secondary and tertiary branches of mesenteric resistance arterioles, cerebral micro-vessels, mesenteric fat excluding arterioles, hearts, kidneys and peritoneal fat were removed from male Hooded Lister or Wistar rats. A human uterus sample (myometrium, endometrium and sclera) was taken after a caesarean birth. Human mesentery was taken from *post-mortem* tissue. For each sample, 70–300 mg were processed to yield 1–5 μg of total RNA (purified with the RNeasy Midi Kit, QIAGEN UK Ltd), which was reverse transcribed (using the "Super Script™ Preamplification System for First Strand cDNA Synthesis", Gibco BRL, Life Science Technologies). The cDNA was amplified by polymerase chain reaction (PCR) using two runs of 35 cycles (1 cycle = 1 min, 94°C; 1 min, 55°C; 1 min, 72°C), each run using a different primer set. The first run utilised primers based on the rat CB_1 transcript (Genbank accession number: x55812 (Matsuda *et al.*, 1990)), designed to amplify a 170 base pair (bp) sequence. Forward primer1:5'TTC ACG GTT CTG GAG AAC CT3', reverse primer1:5'TTC CAC CGT AAA GAC AGC C3'. The second run utilised primers designed to amplify a 140 bp sequence, nested inside the first set to solve the problem of ragged ends on the initial PCR products. Forward primer2:5' CTG GTG CTG TGT GTC ATC CT 3', reverse primer2:5' CGG TGG AAT ACA TGG AAG TCA 3'. PCR products were analysed by electrophoresis on a 4% agarose gel stained with ethidium bromide against a 100–1000 bp standard ladder. PCR products were visualised under ultra violet light.

The first PCR run yielded 170 bp products from rat cerebral microvessels, kidney and cerebellum and human uterus. The second PCR run yielded 140 bp products from rat mesenteric resistance arterioles, cerebellum, cerebral microvessels, kidney, heart and peritoneal fat pads and human mesentery and uterus. Rat mesenteric fat excluding resistance arterioles did not yield a CB_1-related PCR products.

Thus, a CB_1 like transcript is present in rat and human resistance vessels suggesting local CB_1 receptor expression.

Immunohistochemistry

The distribution of CB_1 receptors in the brain has been mapped using specific antibodies (Dove Pettit *et al.*, 1998; Tsou *et al.*, 1998). A domain-specific polyclonal antibody to amino acids 83–98 of the CB_1 receptor was utilised to define the expression of the receptor in some rat blood vessels at the cellular level.

Frozen sections of rat mesenteric vascular bed and thoracic aorta were cut, slide mounted and fixed with formaldehyde and acetone. The primary CB_1 receptor antibody (rabbit anti-human) was incubated on the slide overnight at 4°C and visualisation of the bound CB_1 antibody was achieved using a Vectastain Elite ABC kit.

Antibody staining associated with both the endothelium and smooth muscle was observed in mesenteric vessels (50–200 μm) but not in the mesenteric fat or thoracic aorta. While anandamide relaxes the rat mesenteric vasculature (Randall *et al.*, 1996), it was without effect (11.6 ± 10.2% relaxation, n = 5) on the tone of rat thoracic aorta rings

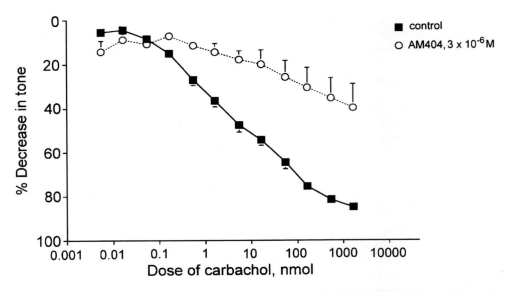

Figure 17-2.   EDHF-mediated relaxations to carbachol in the rat isolated perfused mesenteric arterial bed, in the presence of 3×10^{-4} M N^{ω}-nitro-L-arginine methyl ester and 1×10^{-5} M indomethacin and in the additional presence of the cannabinoid reuptake inhibitor, AM404. Addition of AM404 significantly ($P < 0.01$, analysis of variance) increased the ED_{50} value from $(3.3 \pm 0.6) \times 10^{-9}$ mol to $(1.0 \pm 0.2) \times 10^{-8}$ mol and the maximum relaxation was reduced significantly ($P < 0.001$) from $87.0 \pm 2.5\%$ to $43.6 \pm 9.3\%$. Data shown as mean $\pm$ s.e.m.

constricted with methoxamine (50 μM) in the presence of indomethacin (10 μM). There is, therefore, objective evidence of the presence of cannabinoid CB_1 receptors in the resistance vasculature, but not in conduit vessels, which supports their proposed role in the control of the arterial blood pressure.

Cannabinoid Reuptake Mechanisms and EDHF

Both anandamide-induced and EDHF-mediated relaxations were potentiated by inhibition of amidase, a route for anandamide metabolism (White & Hiley, 1997). Similarly, the anandamide-induced hypotension in guinea-pigs, was potentiated by inhibition of anandamide reuptake with AM404 (Beltramo *et al.*, 1997, Calignano *et al.*, 1997). This suggests that the cannabinoid reuptake transporter terminates the cardiovascular actions of cannabinoids *in vivo*. The actions of AM404 against EDHF-mediated relaxations to carbachol was examined in the rat isolated mesentery (Harris, Kendall and Randall, 1998b). In these experiments, the EDHF-mediated responses were selectively opposed by 3 μM AM404 (Figure 17-2). On the basis of this interesting, and unexpected finding, it is proposed that the cannabinoid transporter may be involved in the actions of EDHF. In this respect, there are two possibilities. Firstly, released EDHF is rapidly taken up by the cannabinoid transporter and recycled, such that in the presence of AM404, EDHF becomes depleted as this cycle is interrupted. Alternatively the cannabinoid transporter may be essential for EDHF release from the endothelium.

Endocannabinoids and Pathophysiology

A role for endocannabinoids in hypotension associated with haemorrhagic shock has been advanced (Wagner *et al.*, 1997). In this respect, in a rat model of haemorrhagic shock, the accompanying hypotension was reversed by the cannabinoid receptor antagonist, SR141716A, whilst activated macrophages were found to release anandamide. The role of the endothelium in this response was not established, but it is conceivable that the activated macrophages could also stimulate the release of endocannabinoids from the endothelium, contributing towards the hypotension.

CONCLUDING REMARKS

There are already substantial conflicts in the literature, with regard to the vascular actions of endocannabinoids, with evidence both for and against an endocannabinoid being an EDHF. Nevertheless, cannabinoid receptors have been identified within the vasculature and others have produced evidence that endocannabinoids are indeed produced and released by the endothelium. It is therefore reasonable to conclude that endocannabinoids are a novel class of endothelium-derived vasoactive substances.

ACKNOWLEDGEMENTS

MDR thanks the British Heart Foundation and the University of Nottingham for financial support. DH holds an MRC Studentship and ITD is supported by Pharmagene PLC.

18 SR141716A Does Not Inhibit Non-Nitric Oxide, Non-Prostanoid Mediated Relaxations to Ethanol and Cyclopiazonic Acid in the Isolated Porcine Pulmonary Artery

Rebecca N. Lawrence, William R. Dunn and Vince G. Wilson

School of Biomedical Science, University of Nottingham Medical School, Queens Medical Centre, Nottingham, NG7 2UH, UK. Tel: +44 (0) 115 970 9480; Fax: +44 (0) 115 970 9259; E mail: MQXRL@MQN1.phpharm.nottingham.ac.uk

Ethanol and cyclopiazonic acid, a Ca^{2+}-ATPase inhibitor, produce endothelium-dependent relaxations of the porcine isolated pulmonary artery that involve nitric oxide and prostanoids. However, following inhibition of both nitric oxide synthase and cyclooxygenase, using N^{ω}-nitro-L-arginine methyl ester and flurbiprofen, respectively, a small relaxation was still detectable. One possible candidate for mediating this residual response is endothelium-derived hyperpolarizing factor (EDHF). The nature of this response was therefore examined following (a) removal of the endothelium and (b) depolarization of the membrane using 3×10^{-2} M KCl. Following reports that in the rat mesenteric artery, EDHF appears to be a cannabinoid-related substance, possibly anandamide, the flurbiprofen-insensitive, N^{ω}-nitro-L-arginine methyl ester-resistant relaxations to ethanol and cyclopiazonic acid in the porcine isolated pulmonary artery were investigated to see whether a similar substance mediates these responses.

In this study, it has been demonstrated that the non-nitric oxide, non-prostanoid component of the relaxations to ethanol and cyclopiazonic acid are (a) endothelium-dependent, and (b) can be abolished by raising extracellular K^+ levels using 3×10^{-2} M KCl. This is consistent with the release of an endothelium-derived hyperpolarizing factor (EDHF). It has also been demonstrated that the cannabinoid receptor-antagonist, SR141716A does not attenuate the remaining relaxation to either ethanol or cyclopiazonic acid. These observations therefore suggest that endocannabinoids are not directly involved in the non-nitric oxide, non-prostanoid relaxation responses of the porcine isolated pulmonary artery to ethanol and cyclopiazonic acid.

KEYWORDS: Endothelium-derived hyperpolarizing factor, cyclopiazonic acid, ethanol, cannabinoids, pulmonary arteries, pig

INTRODUCTION

A number of factors can be released from the endothelium to cause relaxation, including various prostanoids, nitric oxide and a non-prostanoid, non-nitric oxide hyperpolarizing factor, otherwise known as endothelium-derived hyperpolarizing factor(s) (EDHF), (Vanhoutte *et al.*, 1996). However, as yet, the chemical identity of EDHF is unknown. Current knowledge suggests that there may be several EDHFs in different vascular beds and/or in different species. One suggestion for the identity of EDHF is a cytochrome P_{450} metabolite of arachidonic acid, such as epoxyeicosatrienoic acids (Hecker *et al.*, 1994; Campbell *et al.*, 1996). However, these products are not responsible for non-nitric oxide, non-prostanoid relaxations in some vascular preparations such as the rat hepatic artery (Zygmunt *et al.*, 1996) and human small omental arteries (Ohlmann *et al.*, 1997). The suggestion has been made that EDHF may be an endocannabinoid such as anandamide (Randall *et al.*, 1996), based on the observation that non-nitric oxide and non-prostanoid responses to carbachol and A23187 (a calcium ionophore), in the rat perfused mesentery, were attenuated by the presence of the cannabinoid-receptor antagonist, SR141716A.

It has been previously demonstrated that ethanol and the Ca^{2+}-ATPase inhibitor, cyclopiazonic acid cause endothelium-dependent relaxations in the porcine isolated pulmonary artery, which were partially reduced by the presence of the nitric oxide synthase inhibitor N^{ω}-nitro-L-arginine methyl ester and the cyclooxygenase inhibitor, flurbiprofen (Lawrence, Dunn and Wilson, 1997; Lawrence *et al.*, 1998). One possibility for these remaining relaxations is the release of an EDHF. In order to investigate this possibility, responses to ethanol and cyclopiazonic acid have been examined in the presence of flurbiprofen and N^{ω}-nitro-L-arginine methyl ester, following removal of the endothelium, after depolarization of the vascular smooth muscle, and in the presence of SR141716A.

METHODS

Lungs were removed from pigs recently killed in the abattoir (within 30 min) and transported back to the laboratory in ice-cold, modified Krebs-Henseleit solution. The main pulmonary artery was located at the point where it enters the lung, and was carefully followed down to the end of the lobe, cutting away any branches. The arteries were placed in 20 ml of Krebs-Henseleit solution containing 2% ficoll, which had previously been gassed for 5–10 min with 95% O_2: 5% CO_2, and stored overnight at 4°C; ficoll was used to produce a hyperosmolar solution to prevent osmotic swelling of the cells (Lot, Starke and Wilson, 1993).

The following day, vessels (approximately 2–4 mm in diameter) were cleaned of excess connective tissue and cut into 5–6 mm segments. The endothelium of some segments was removed by gently rubbing the lumen of the vessel with a roughened metal rod. Two stainless steel wires (0.2 mm thick) were then placed in the artery, one being linked to a glass support while the other was connected by cotton to a Grass force-displacement transducer (Model FT03) connected to a Grass Polygraph. The segments were then placed in an isolated organ chamber containing 5 ml Krebs-Henseleit solution maintained at 37°C and gassed with 95% O_2: 5% CO_2.

Experimental Protocol

An initial resting tension of 4 g wt was applied to each segment 30 min after equilibration, which levelled off to 1.5 to 2 g wt after a further 30 min. Preparations were then stimulated with 6×10^{-2} M KCl until reproducible responses were obtained.

Experiment 1

Responses to increasing concentrations of either ethanol or cyclopiazonic acid were examined in the absence or presence of the cyclooxygenase inhibitor, flurbiprofen (10^{-6} M) and the nitric oxide synthase inhibitor, N^{ω}-nitro-L-arginine methyl ester (10^{-4} M). N^{ω}-nitro-L-arginine methyl ester, which produced a contraction on its own, was added first, before the addition of flurbiprofen. Subsequently, the thromboxane-mimetic, U46619 was added at a concentration which produced a final level of tone equivalent to approximately 60% of the response to 6×10^{-2} M KCl. In control experiments, tissues were constricted to a similar level of tone using U46619 alone (10^{-9} to 3×10^{-8} M). The dilators were then added after a minimum of 40 min.

Experiment 2

Responses to a single concentration of ethanol (1.7% v/v) or cyclopiazonic acid (10^{-5} M), in the presence of 10^{-6} M flurbiprofen and 10^{-4} M N^{ω}-nitro-L-arginine methyl ester, were compared in endothelium-intact and endothelium-denuded segments. All segments were first exposed to 10^{-7} M acetylcholine in the presence of U46619, to assess endothelial function. Tissues were then washed, before reconstricting using N^{ω}-nitro-L-arginine methyl ester, flurbiprofen and U46619. Once a stable response was achieved, the respective concentrations of ethanol and cyclopiazonic acid were added.

Experiment 3

Responses to 1.7% v/v ethanol and 10^{-5} M cyclopiazonic acid (in the presence of 10^{-6} M flurbiprofen and 10^{-4} M N^{ω}-nitro-L-arginine methyl ester) were then examined in the absence and presence of Krebs-Henseleit solution containing 3×10^{-2} M KCl (prepared by equimolar exchange of K^+ ions for Na^+ ions). In these experiments, high KCl was added prior to the addition of N^{ω}-nitro-L-arginine methyl ester and flurbiprofen. Tissues were then left to equilibrate for at least 40 min, after which preparations were reconstricted with U46619, to produce a level of tone which was approximately 60% of the response to 6×10^{-2} M KCl. Once a stable response was achieved, a single concentration of either ethanol or cyclopiazonic acid was added.

Experiment 4

Responses to 1.7% v/v ethanol and 10^{-5} M cyclopiazonic acid (in the presence of 10^{-6} M flurbiprofen and 10^{-4} M N^{ω}-nitro-L-arginine methyl ester) were examined in the absence and presence of the CB_1-receptor antagonist, SR141716A (N-(piperidin-1-yl)-5-(4-chlorophenyl)-1-(2,4-dichlorophenyl)-4-methyl-1*H*-pyrazole-3-carboxamide HCl; Rinaldi-Carmona *et al.*, 1994), (10^{-6} or 10^{-5} M). In these experiments, N^{ω}-nitro-L-arginine methyl ester was added first, followed by flurbiprofen and then SR141716A. Tissues were then left to equilibrate for at least 40 min, after which preparations were reconstricted with U46619, to produce a level of tone which was approximately 60% of the response to 6×10^{-2} M KCl. Once a stable response was achieved, a single concentration of either ethanol or cyclopiazonic acid was added.

Finally, to assess the ability of the tissue to respond to cannabinoid-related agents, anandamide was added in increasing concentrations to rubbed segments of the porcine pulmonary artery.

Data Analysis

Unless otherwise stated, the effect of the relaxing drugs have been expressed in terms of the percentage of vasoconstrictor-induced tone and are shown as the mean ± s.e. mean. Differences between mean values have been examined using unpaired Student's t-tests. Differences were considered to be statistically significant when P was less than 0.05.

Drugs and Solutions

The following chemicals were used: anandamide (arachidonoylethanolamide; synthesised from arachidonoyl chloride and ethanolamine (Devane *et al.*, 1992) and dissolved in an inert oil/water emulsion by Dr EA Boyd, Nottingham University), cyclopiazonic acid (Sigma Chemical Co), ethanol and ficoll (BDH Laboratory Supplies), flurbiprofen and N^{ω}-nitro-L-arginine methyl ester (Sigma Chemical Co), potassium chloride (BDH Laboratory Supplies), SR141716A (N-(piperidin-1-yl)-5-(4-chlorophenyl)-1-(2,4-dichlorophenyl)-4-methyl-1*H*-pyrazole-3-carboxamide HCl; a kind gift from Dr MD Randall, Nottingham University), U46619 (9,11-dideoxy-9α,11α-methanoepoxy Prostaglandin F$_{2\alpha}$ methyl acetate; Sigma Chemical Co). Stock solutions of U46619 (10^{-2} M), flurbiprofen (10^{-3} M) and SR141716A (10^{-2} M) were prepared in absolute ethanol while cyclopiazonic acid (10^{-2} M) was dissolved in dimethylsulphoxide. Further dilutions of all drugs were made in distilled water. The composition of the Krebs-Henseleit solution in mM was: NaCl 119; KCl 4.7; MgSO$_4$.7H$_2$O 1.17; NaHCO$_3$ 24; CaCl$_2$ 1.25; KH$_2$PO$_4$ 1.17; Glucose 5.5. The potassium Krebs-Henseleit solution was made by equimolar substitution of potassium ions for sodium ions.

RESULTS

KCl (6×10^{-2} M) produced contractions which were not significantly different in segments without endothelium (3.04 ± 0.50 g wt, n = 8) and with endothelium (2.25 ± 0.44 g wt, n = 8) of the porcine isolated pulmonary artery. There was also no significant difference between the level of U46619-induced tone in rings with and without endothelium ($53.1 \pm 5.6\%$ and $54.0 \pm 14.8\%$ (n = 8), respectively, of the response to 6×10^{-2} M KCl).

Flurbiprofen and N^{ω}-nitro-L-arginine methyl ester

Ethanol (0.1% to 1.2% v/v) and cyclopiazonic acid (3×10^{-7} to 10^{-5} M) elicited concentration-dependent relaxations during contractions to U46619 (Figure 18-1). These relaxations to ethanol and cyclopiazonic acid were attenuated, but not abolished, by the presence of 10^{-6} M flurbiprofen and 10^{-4} M N^{ω}-nitro-L-arginine methyl ester (Figure 18-1; n = 7).

Removal of the Endothelium

Relaxations to 10^{-7} M acetylcholine ($54.6 \pm 6.5\%$, n = 14) were abolished in segments which had been rubbed to remove the endothelium, indicating that the endothelium had been adequately removed (Figure 18-2). In segments with endothelium, a single concentration of both ethanol (1.7% v/v) and cyclopiazonic acid (10^{-5} M) produced relaxation

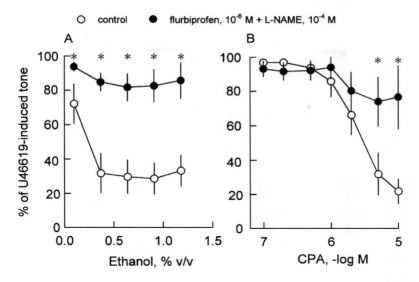

Figure 18-1. Comparison of responses to increasing concentrations of **A** ethanol and **B** cyclopiazonic acid (CPA) in the absence and presence of 10^{-6} M flurbiprofen and 10^{-4} M N^{ω}-nitro-L-arginine methyl ester in segments of porcine isolated pulmonary artery. Where necessary, tissues were also contracted with the thromboxane-mimetic, U46619 to produce approximately 60% of the response to 6×10^{-2} M KCl. Responses have been expressed as a percentage of starting tone and are shown as the mean ± s.e. mean of 7 observations. The asterisks denote a statistically significant difference versus the control ($P < 0.05$).

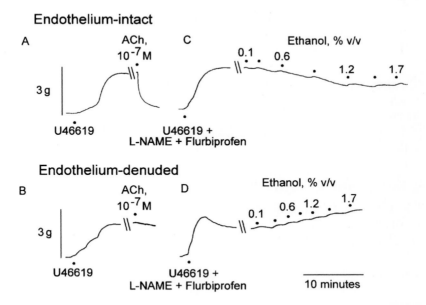

Figure 18-2. Responses to 10^{-7} M acetylcholine (ACh) on rings **A** with and **B** without endothelium of porcine isolated pulmonary artery, contracted with the thromboxane-mimetic, U46619. Responses to increasing concentrations of ethanol (0.1 to 1.7% v/v), (in the presence of 10^{-6} M flurbiprofen and 10^{-4} M N^{ω}-nitro-L-arginine methyl ester) in rings **C** with and **D** without endothelium contracted with the thromboxane-mimetic, U46619 to produce approximately 60% of the response to 6×10^{-2} M KCl.

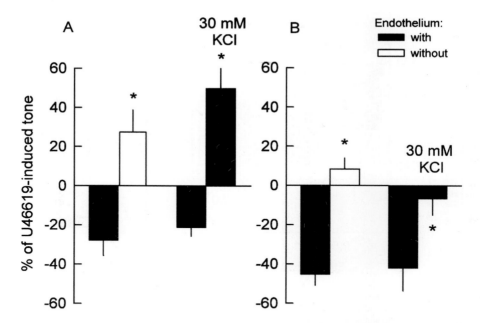

Figure 18-3. Responses to **A** 1.7% v/v ethanol and **B** 10^{-5} M cyclopiazonic acid in rings of porcine isolated pulmonary artery (in the presence of 10^{-6} M flurbiprofen and 10^{-4} M N^{ω}-nitro-L-arginine methyl ester) with and without the endothelium and in the absence and presence of Krebs-Henseleit solution containing 3×10^{-2} M KCl. Where necessary, these segments were contracted with the thromboxane-mimetic, U46619 to produce approximately 60% of the response to 6×10^{-2} M KCl. Responses have been expressed as a percentage of initial tone and are shown as the mean ± s.e. mean of 6 to 7 observations. The asterisks denote a statistically significant difference versus the control (P < 0.05).

responses, in the presence of 10^{-6} M flurbiprofen and 10^{-4} M N^{ω}-nitro-L-arginine methyl ester, which were abolished by removal of the endothelium (ethanol: 27.9 ± 7.7% relaxation with and 27.3 ± 11.5% contraction without endothelium (n = 7); cyclopiazonic acid: 45.3 ± 5.5% relaxation with and 8.5 ± 5.4% contraction without endothelium (n = 6); Figures 18-2 and 18-3).

Raising Extracellular K+ Levels

Raising extracellular K+ levels by using Krebs-Henseleit solution containing 3×10^{-2} M KCl abolished the relaxations to both ethanol and cyclopiazonic acid, converting the response to ethanol into a contraction (ethanol: without KCl 21.2 ± 4.3% relaxation, with KCl 49.9 ± 10.3% contraction (n = 6); cyclopiazonic acid: without KCl 42.0 ± 11.5% relaxation, with KCl 6.7 ± 8.2% relaxation (n = 6); Figure 18-3).

SR141716A

SR141716A (10^{-6} M) neither enhanced nor attenuated responses to cyclopiazonic acid in the presence of flurbiprofen and N^{ω}-nitro-L-arginine methyl ester (data not shown).

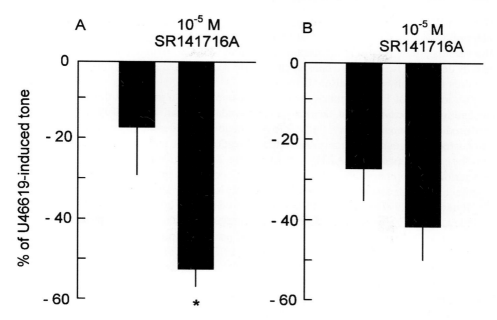

Figure 18-4. Responses of segments of porcine isolated pulmonary artery to **A** 1.7% v/v ethanol and **B** 10^{-5} M cyclopiazonic acid (in the presence of 10^{-6} M flurbiprofen, 10^{-4} M N^{ω}-nitro-L-arginine methyl ester) in the absence and presence of 10^{-5} M SR141716A. Where necessary, segments were also contracted with the thromboxane-mimetic, U46619 to produce approximately 60% of the response to 6×10^{-2} M KCl. Responses have been expressed as a percentage of initial tone and are shown as the mean ± s.e. mean of 6 to 8 observations. The asterisk denotes a statistically significant difference versus the control (P < 0.05).

The higher concentration of SR141716A (10^{-5} M) elicited a small relaxation in U46619-contracted tissues (18.6 ± 7.4%, n = 8). SR141716A (10^{-5} M), did not inhibit the relaxation to either ethanol or cyclopiazonic acid in the presence of 10^{-6} M flurbiprofen and 10^{-4} M N^{ω}-nitro-L-arginine methyl ester, and actually enhanced significantly the relaxation response to ethanol (ethanol: without SR141716A 17.4 ± 11.7% relaxation, with SR141716A 52.5 ± 4.3% relaxation (n = 8); cyclopiazonic acid: without SR141716A 27.3 ± 8.0% relaxation, with SR141716A 41.6 ± 8.3% relaxation (n = 6); Figure 18-4).

Anandamide (10^{-9} to 10^{-5} M) had no effect on U46619-induced tone (n = 6), while there was a very variable relaxation seen at 3×10^{-5} M anandamide (48.0 ± 17.0%, n = 6).

DISCUSSION

In the present study, it has been shown that the non-nitric oxide, non-prostanoid-induced relaxations are endothelium-dependent and abolished by depolarization of the vascular smooth muscle following elevation of the extracellular K^+ concentration. Qualitatively similar results, have been demonstrated in bovine coronary and rat hepatic arteries, where incubation with K^+ abolished relaxations to methacholine and acetylcholine, respectively

Rebecca N. Lawrence et al.

(Campbell *et al.*, 1996; Zygmunt *et al.*, 1996). These studies also demonstrated, using electrophysiological techniques, that these relaxations were due to a hyperpolarization of the vascular smooth muscle membrane, consistent with the involvement of an EDHF. The results described here are also consistent with the release of an EDHF, although this clearly requires electrophysiological confirmation.

The identity of EDHF remains unclear, but one suggestion is that this factor may be an endogenous cannabinoid, such as anandamide (Randall *et al.*, 1996). Thus in the perfused rat mesenteric artery, the selective cannabinoid-receptor antagonist, SR141716A attenuated non-prostanoid, non-nitric oxide relaxations to the muscarinic agonist, carbachol and the calcium ionophore, A23187. Furthermore, anandamide has been shown to produce relaxations of preparations of rat mesenteric, coronary and hepatic arteries without endothelium (Randall *et al.*, 1996; Randall & Kendall, 1997; White & Hiley, 1997; Zygmunt *et al.*, 1997b), that are also attenuated by SR141716A (Randall *et al.*, 1996; White & Hiley, 1997; Zygmunt *et al.*, 1997b). The intravenous administration of anandamide to anaesthetised guinea-pigs and rats lowers blood pressure, and this effect is also attenuated by SR141716A (Calignano *et al.*, 1997; Lake *et al.*, 1997a and b).

In the present study, the possibility that a cannabinoid-related substance mediates the flurbiprofen-insensitive, N^{ω}-nitro-L-arginine methyl ester-resistant relaxations to ethanol and cyclopiazonic acid was investigated. In the porcine isolated pulmonary artery, neither 10^{-5} nor 10^{-6} M SR141716A altered the residual response to cyclopiazonic acid, but the higher concentration of the cannabinoid receptor antagonist enhanced similar responses to ethanol. Although there is no obvious explanation for the latter observation, neither response is mediated by an endogenous cannabinoid. The concentration of SR141716A employed in this study is similar to that used in the rat mesenteric bed (Randall *et al.*, 1996), being 100 to 1000-fold greater than the affinity for CB_1 receptors. A pA_2 of 8.2 for SR141716A against anandamide-induced inhibition of electrically-evoked contraction has been obtained in the mouse vas deferens (Rinaldi-Carmona *et al.*, 1994). Thus, there is no pharmacological evidence for an endogenous cannabinoid contributing directly to non-nitric oxide, non-prostanoid-induced relaxation of the porcine isolated pulmonary artery.

An indirect role for endogenous cannabinoids in the above phenomenon remains a possibility since anandamide may be released, but subsequently degraded into a bioactive metabolite, such as epoxyeicosatrienoic acids, which can then itself produce a hyper-polarization of the vascular smooth muscle and cause relaxation of the vessel (Chataigneau *et al.*, 1998b). This has been further supported by evidence that radiolabelled anandamide can be broken down into arachidonic acid and then to prostaglandins and epoxyeicosatrienoic acids in the presence of bovine coronary arteries (Pratt *et al.*, 1998). The latter study also demonstrated that SR141716A attenuates the release of arachidonic acid when tissues are stimulated with A23187, but has no effect on relaxations to anandamide or 5,6-epoxyeicosatrienoic acid, raising questions about the specificity of action of the cannabinoid receptor antagonist.

The endothelium releases a variety of substances with the potential to cause hyperpolarization of the underlying smooth muscle. The involvement of these substances may vary between species and vascular beds. It would appear, for example, that epoxyeicosatrienoic acids may be mainly responsible for EDHFs released from the coro-nary and renal circulation (Fulton, McGiff and Quilley, 1992; Bauersachs, Hecker and

Busse, 1994; Hecker *et al.*, 1994; Campbell *et al.*, 1996). Meanwhile, in the porcine coronary artery, endothelium-dependent hyperpolarizations were unaffected by SR141716A, whilst anandamide itself did not hyperpolarize the coronary smooth muscle cells from the pig (Chataigneau *et al.*, 1998b). These results are similar to those obtained in the porcine pulmonary artery, where very high concentrations of anandamide failed to elicit a response, suggesting a lack of involvement of endocannabinoids in this preparation. The differences in the identity of the EDHF released by the rat mesenteric artery may be accounted for, in part, by the fact that some studies have used arteries of the second to fourth order (Plane *et al.*, 1997; White & Hiley, 1997; Chataigneau *et al.*, 1998b), whilst others have used the whole perfused mesenteric bed, therefore incorporating arteries and small veins (Randall *et al.*, 1996). In venules of the rat small intestine, EDRF is released on stimulation with acetylcholine, and this EDRF can then diffuse across to produce a dilatation of associated arterioles (Falcone & Bohlen, 1990).

In conclusion, ethanol and cyclopiazonic acid can stimulate the release of an endothelium-derived non-prostanoid, non-nitric oxide relaxing factor in the porcine isolated pulmonary artery, which shares characteristics with EDHF. There is no evidence that the non-prostanoid, non-nitric oxide factor in this preparation is a cannabinoid-related substance.

ACKNOWLEDGEMENTS

The technical support of Nigel Blaylock is gratefully acknowledged. We also thank Dr EA Boyd for the synthesis of the anandamide and Drs MD Randall and DA Kendall for their helpful criticism and the kind gift of SR141716A.

19 EDHF Does Not Activate Cannabinoid CB$_1$ Receptor in Rat Mesenteric, Guinea-Pig Carotid and Porcine Coronary Arteries

Thierry Chataigneau[1], Catherine Thollon[2], Nicole Villeneuve[2], Jean-Paul Vilaine[2], Jacques Duhault[1], Michel Félétou[1] and Paul M. Vanhoutte[3]

[1] *Département de Diabétologie,*
[2] *Département de Pathologies Cardiaques et Vasculaires, Institut de Recherches Servier, 11 rue des Moulineaux, 92150 Suresnes and*
[3] *Institut de Recherches Internationales Servier, 6 place des Pléiades, 92410 Courbevoie, France*
Address for correspondence: Michel Félétou: Dpt Diabétologie, Institut de Recherche Servier, 92150 Suresnes, France.

Experiments were designed to determine whether or not the endothelium-dependent hyperpolarizations of vascular smooth muscle cells (observed in the presence of inhibitors of nitric oxide synthase and cyclooxygenase) can be attributed to the production of an endogenous cannabinoid. Membrane potential was recorded in vascular smooth muscle cells of the rat mesenteric, guinea-pig carotid and porcine coronary arteries using intracellular microelectrodes. In the rat mesenteric artery, the cannabinoid receptor antagonist, SR 141716, did not modify the resting membrane potential of smooth muscle cells or the endothelium-dependent hyperpolarization induced by acetylcholine. Anandamide induced an endothelium-dependent hyperpolarization of the smooth muscle cells which was not reproducible in the same tissue whereas acetylcholine was still able to hyperpolarize the preparation in the presence of the cannabinoid. The hyperpolarization induced by anandamide was not significantly influenced by SR 141716. HU-210, a synthetic CB$_1$ receptor agonist, and palmitoylethanolamide, a CB$_2$ receptor agonist, did not influence the membrane potential of the vascular smooth muscle cells. In the rat mesenteric artery, the endothelium-dependent hyperpolarization induced by acetylcholine was not altered by glibenclamide. However, the combination of charybdotoxin plus apamin abolished the acetylcholine-induced hyperpolarization and under these conditions, acetylcholine evoked a depolarization. The hyperpolarization induced by anandamide was significantly inhibited by glibenclamide but not significantly affected by the combination of charybdotoxin plus apamin. In the guinea-pig carotid artery, anandamide induced a non-reproducible hyperpolarization with a slow onset which was not influenced by SR 141716. In the porcine coronary artery, anandamide did not hyperpolarize or relax the smooth muscle cells. The endothelium-dependent hyperpolarizations and relaxations induced by bradykinin were not influenced by SR 141716. These results indicate that the endothelium-dependent hyperpolarizations observed in rat mesenteric, guinea-pig carotid and porcine coronary arteries cannot be attributed to the activation of cannabinoid CB$_1$ receptors.

KEYWORDS: Anandamide, cannabinoid receptors, EDHF, potassium channels.

INTRODUCTION

Randall *et al.* (1996) have reported that, in the rat perfused mesenteric arterial bed, anandamide, an endogenous ligand for the cannabinoid CB$_1$ receptor derived from arachidonic acid (Devane *et al.*, 1992; Di Marzo *et al.*, 1994), induces dilatation. In the same vascular bed, the dilatation to carbachol, observed in the presence of inhibitors of nitric oxide synthase and cyclooxygenase, is inhibited by a specific CB_1 receptor antagonist (Randall *et al.*, 1996). These observations suggest that an endogeneous cannabinoid, possibly anandamide, is EDHF. The present experiments were designed to determine whether or not the endothelium-dependent hyperpolarizations, observed in the rat mesenteric, guinea-pig carotid and porcine coronary arteries and attributed to EDHF, can be related to the production of an endogenous cannabinoid.

MATERIAL AND METHODS

Electrophysiological Experiments

Male Sprague Dawley rats, Hartley guinea-pigs and Large-White pigs of either sex were anaesthetized. Branches (second to fourth order, 400 to 250 μm) of the rat mesenteric arteries, guinea-pig internal carotid arteries and porcine coronary arteries were dissected free of connective tissues and pinned to the bottom of an organ chamber perfused with modified Krebs-Ringer bicarbonate solution (37°C, aerated with a 95% O_2/5% CO_2 gas mixture, pH 7.4), of the following composition (in mM): NaCl 118.3, KCl 4.7, $CaCl_2$ 2.5, $MgSO_4$ 1.2, KH_2PO_4 1.2, $NaHCO_3$ 25, calcium-disodium EDTA 0.026 and glucose 11.1. Transmembrane potential was recorded using glass capillary microelectrodes (tip resistance of 30 to 90 MΩ) filled with KCl (3 M). The incubation times were at least 30 min with the various potassium channel inhibitors studied and 60 min with the cannabinoid receptor antagonist, SR 141716. Acetylcholine was infused for no longer than five minutes to avoid desensitization of the preparation. With the exception of the experiments designed to study the reproducibility of the effects of anandamide, each preparation was exposed only once to one concentration of a single cannabinoid agonist. All experiments were performed in the presence of N^{ω}-L-nitroarginine (10^{-4} M) and indomethacin (5×10^{-6} M) to inhibit nitric oxide synthase and cyclooxygenase, respectively. Care was taken to preserve the integrity of the endothelium but in some experiments the endothelium was destroyed by a rapid infusion of saponin (1mg/ml) in the lumen of the blood vessel (Corriu *et al.*, 1996b).

Organ Chambers

Rings of porcine coronary arteries were suspended in organ chambers filled with Krebs-Ringer bicarbonate solution for recording of isometric tension. The tissues were connected to a force transducer and isometric changes in tension were recorded on a polygraph.

Statistics

Data are shown as mean ± s.e. mean ; n indicates the number of cells in which membrane potential was recorded. Statistical analysis was performed using Student's *t*-test for paired or unpaired observations.

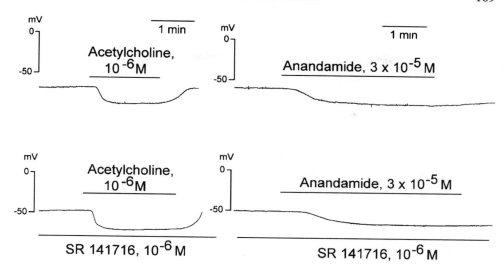

Figure 19-1. Effect of SR 141716 (10^{-6} M) on the endothelium-dependent hyperpolarizations induced by acetylcholine (10^{-6} M, left panel) and anandamide (3×10^{-5} M, right panel), in rat isolated mesenteric arteries, in the presence of N^{ω}-L-nitroarginine (10^{-4} M) and indomethacin (5×10^{-6} M); Data from Chataigneau *et al.* (1998b).

RESULTS

Rat Mesenteric Artery

SR 141716 (10^{-6} M) did not modify the resting membrane potential of the smooth muscle cells and the endothelium-dependent hyperpolarization induced by acetylcholine (10^{-6} M) (-17.3 ± 1.8 mV, n = 4 and -17.8 ± 2.6 mV, n = 4, in control and in the presence of SR 141716, respectively, Figure 19-1). Anandamide (3×10^{-5} M) induced endothelium-dependent hyperpolarizations (-12.6 ± 1.4 mV, n = 13 and -2.0 ± 3.0 mV, n = 6, in the presence and the absence of endothelium, respectively, Figure 19-2). HU-210 (3×10^{-5} M) and palmitoylethanolamide (3×10^{-5} M) did not significantly modify the membrane potential of the rat mesenteric arteries with endothelium. The hyperpolarization elicited by anandamide could not be reproduced in the same preparation, even after a washout period of 30 min, whereas acetylcholine could still induce an endothelium-dependent hyperpolarization. The hyperpolarization elicited by anandamide (3×10^{-5} M) was not influenced significantly by the presence of SR 141716 (10^{-6} M) (-17.3 ± 2.3 mV, n = 4, Figure 19-1).

The endothelium-dependent hyperpolarization produced by acetylcholine (10^{-6} M) was not affected by glibenclamide (10^{-6} M) but was abolished in the presence of charybdotoxin (10^{-7} M) plus apamin (5×10^{-7} M). In the presence of the latter two drugs, acetylcholine induced a depolarization. The hyperpolarization induced by anandamide (3×10^{-5} M) was significantly inhibited by glibenclamide (10^{-6} M) but was not influenced by the combination of charybdotoxin plus apamin (Figure 19-3).

With endothelium

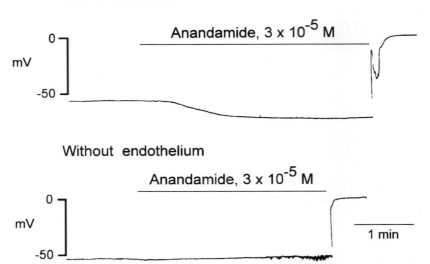

Figure 19-2. Endothelium-dependence of the hyperpolarization induced by anandamide (3×10^{-5} M) in rat isolated mesenteric arteries with or without endothelium, in the presence of N^{ω}-L-nitroarginine (10^{-4} M) and indomethacin (5×10^{-6} M); Data from Chataigneau *et al.* (1998b).

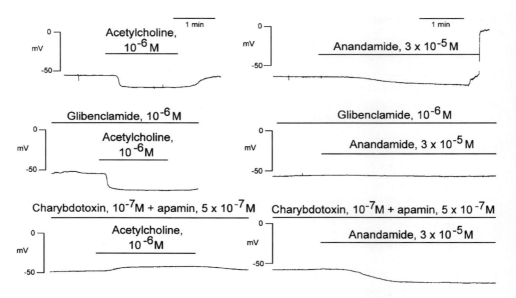

Figure 19-3. Effects of glibenclamide (10^{-6} M) and the combination of charybdotoxin (10^{-7} M) plus apamin (5×10^{-7} M) on the endothelium-dependent hyperpolarizations elicited by acetylcholine (ACh, 10^{-6} M, left panel) and anandamide (3×10^{-5} M, right panel), in rat isolated mesenteric arteries, in the presence of N^{ω}-L-nitroarginine (10^{-4} M) and indomethacin (5×10^{-6} M); Data from Chataigneau *et al.* (1998b).

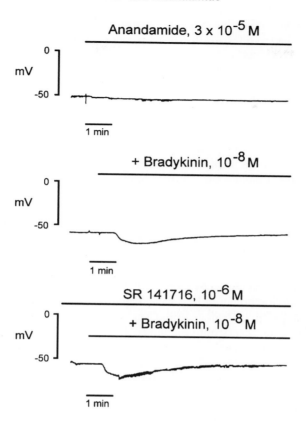

Figure 19-4. Effects of anandamide (3×10^{-5} M) and bradykinin (10^{-8} M), in porcine isolated coronary arteries with endothelium, in the presence of N^{ω}-L-nitroarginine (3×10^{-5} M) and indomethacin (10^{-5} M). Anandamide (upper panel) had no effect on the membrane potential whereas bradykinin (middle panel) elicited an endothelium-dependent hyperpolarization which was insensitive to the cannabinoid CB1 receptor antagonist, SR 141716 (10^{-6} M, lower panel); Data from Chataigneau *et al.* (1998b).

Guinea-pig Carotid Artery

Anandamide (3×10^{-5} M) induced a small non-reproducible hyperpolarization (-5.6 ± 1.3 mV, n = 10) characterized by a slower time course than the response to acetylcholine (time to reach the maximal amplitude of the hyperpolarization: 228 ± 40 s, n = 8 and 76 ± 8 s, n = 15, for anandamide and acetylcholine, respectively). The anandamide-induced hyperpolarization was not influenced by SR 141716 (10^{-6} M; -5.3 ± 1.5 mV, n = 3).

Porcine Coronary Artery

Anandamide (3×10^{-5} M) did not hyperpolarize the smooth muscle cells (Figure 19-4). By contrast, bradykinin (10^{-8} M) induced an endothelium-dependent hyperpolarization which was not altered by SR 141716 (10^{-6} M) (-12.4 ± 2.1 mV, n = 6 and -12.2 ± 1.9 mV, n = 6, in the absence and the presence of SR 141716, respectively).

In isolated segments of porcine coronary arteries contracted with U-46619, anandamide $(10^{-7}$ to 3×10^{-5} M) had no relaxant effect on preparations either with or without endothelium. Bradykinin $(10^{-10}$ to 10^{-6} M) induced endothelium-dependent relaxations which were not affected significantly by SR 141716.

DISCUSSION

The present experiments suggest that EDHF and anandamide do not share the same mechanism of action in the three blood vessels investigated.

Anandamide did not modify the cell membrane potential of porcine coronary arterial smooth muscle but hyperpolarized the smooth muscle cells of the rat mesenteric and guinea-pig carotid arteries. In these two blood vessels, these hyperpolarizations were not inhibited by the CB_1 receptor antagonist, SR 141716, at an effective and specific concentration (Rinaldi-Carmona et al., 1994). Furthermore, in the rat mesenteric artery, HU-210, a synthetic CB_1 receptor agonist, and palmitoylethanolamide, a preferential CB_2 receptor agonist, were ineffective. These results indicate that the responses produced by anandamide are not linked to CB_1 receptor activation.

The hyperpolarization produced by anandamide did not mimic that to acetylcholine in terms of amplitude or time-course and did not share a similar sensitivity to potassium channel blockers. The acetylcholine-induced hyperpolarization was not altered by glibenclamide but abolished by the combination of charybdotoxin plus apamin confirming previous studies indicating that ATP-sensitive potassium channels are not involved in EDHF-mediated response in the studied blood vessels (Corriu et al., 1996b; Fukao et al., 1997b; McCulloch et al., 1997; Petersson et al., 1997a). Indeed, it rather appears that EDHF activates a particular class of potassium channel(s) sensitive to the combination of the two toxins (Garland and Plane, 1996; Corriu et al., 1996b; Zygmunt and Högestätt, 1996b; Petersson et al., 1997a; Chataigneau et al., 1998a). By contrast, the hyperpolarization induced by anandamide was inhibited by glibenclamide but not significantly affected by the combination of charybdotoxin plus apamin. This indicates that anandamide and EDHF do not share a similar mechanism of action. In the rat mesenteric and guinea-pig carotid arteries, the hyperpolarizations to anandamide were not reproducible in the same preparation by contrast with those evoked by acetylcholine. Furthermore, while tissues were unable to respond to a second application of anandamide, acetylcholine still evoked hyperpolarization, confirming that anandamide is not the endothelial factor released by acetylcholine. In the rat mesenteric artery, as in the rat hepatic artery (Zygmunt et al., 1997b), the hyperpolarization induced by anandamide was endothelium-dependent, confirming that anandamide cannot be EDHF. The experiments were performed in the presence of inhibitors of nitric oxide synthase and cyclooxygenase, ruling out the participation of either nitric oxide or prostacyclin to the endothelium-dependent hyperpolarization produced by anandamide. Furthermore, the hyperpolarization produced by anandamide in the rat mesenteric artery cannot be attributed to the release of EDHF since it is blocked by glibenclamide. Anandamide is metabolized by amidohydrolase, into arachidonic acid and ethanolamine (Desarnaud et al., 1995). Other bioactive metabolites are formed either directly or after degradation into arachidonic acid, by various enzymes including cyclooxygenase, lipoxygenase and cytochrome P_{450} mono-oxygenase (Bornheim et al., 1993,

1995; Hampson *et al.*, 1995). In bovine coronary arteries, anandamide induces endothelium-dependent relaxation as a result of its conversion to vasodilatory eicosanoids such as prostacyclin or epoxyeicosatrienoic acids (Pratt *et al.*, 1998). Arachidonic acid does not alter the membrane potential of the smooth muscle cells of rat mesenteric arteries but 11,12-epoxyeicosatrienoic acid induces endothelium-independent hyperpolarization by activation of a glibenclamide-sensitive potassium conductance (Fukao *et al.*, 1997b). Therefore, the present results suggest that anandamide is degraded, by the endothelial cells, into an epoxyeicosatrienoic acid producing hyperpolarization of the vascular smooth muscle cells through the activation of ATP-sensitive potassium channels.

In the isolated porcine coronary artery, anandamide, even at a high concentration, did not relax or hyperpolarize the smooth muscle cells. Furthermore, the hyperpolarization and the relaxation provoked by bradykinin, which are attributed to EDHF, were not affected by SR 141716.

The present study is in agreement with others demonstrating that anandamide does not mimick the response to EDHF in a variety of isolated blood vessels (Plane *et al.*, 1997; Zygmunt *et al.*, 1997b; White and Hilley, 1997). The discrepancy between those studies and the results obtained in perfused organs (Randall *et al.*, 1996; Randall and Kendall, 1997, 1998a and b) cannot be explained by the present experiments.

ACKNOWLEDGEMENTS

The authors wish to thank Dr M. Rinaldi-Carmona (Sanofi, Montpellier, France) for the generous gift of SR 141716 and also Drs T. Bennett, D.A. Kendall, M.D. Randall (Faculty of Medicine, Nottingham, UK) and Dr J-P. Iliou (I. d R. Servier, Suresnes, France) for the gift of anandamide and constructive discussions.

20 Endocannabinoids as Endothelium-Derived Hyperpolarizing Factors: The End of the Road?

C.J. Garland

Department of Pharmacology, University of Bristol, University Walk, Bristol BS8 1TD, UK. Ph: 44(117)928–8666; Fax: 44(117)929–3194; e-mail: chris.garland@bris.ac.uk

Similarities between the action of the endocannabinoid, N-arachidonylethanolamide (anandamide), and endothelium-derived hyperpolarizing factor (EDHF) in the perfused mesenteric bed of the rat, led to the suggestion that these agents may be identical. For EDHF to be identified as anandamide, it must be able to hyperpolarize vascular smooth muscle. In isolated mesenteric resistance arteries of the rat, anandamide could indeed evoke both hyperpolarization and relaxation of smooth muscle during contraction with phenylephrine. However, the CB_1 antagonist, SR141716A did not alter these responses, nor the similar, endothelium-dependent responses to EDHF. Furthermore, the effect of anandamide was not mimicked by the selective CB_1 receptor agonists, HU210 and WIN 55,212–2 and had a different sensitivity profile to potassium channel blockers when compared with EDHF. In both the mesenteric and the hepatic arteries, the hyperpolarization to anandamide was found to be endothelium-dependent. Anandamide also had a direct relaxing action, which reflected a disruption of intracellular calcium handling, independent of any cannabinoid receptor activation. There is, therefore, no clear evidence, particularly from electrophysiological experiments, to indicate that endocannabinoids represent an EDHF. One explanation for the endothelium-dependent hyperpolarizing action of anandamide is that it reflects conversion to arachidonic acid (and ethanolamine) and then eicosanoids which can hyperpolarize and relax vascular smooth muscle cells directly.

INTRODUCTION

It has been suggested that smooth muscle relaxation evoked by endothelium-derived hyperpolarizing factor (EDHF) may be explained by the action of an endocannabinoid, the arachidonic acid derivative, N-arachidonylethanolamide or anandamide (Randall *et al.*, 1996). The suggestion was based on apparent similarities between the dilatation induced by either anandamide or EDHF in the perfused mesenteric bed of the rat. The relaxation to anandamide was apparently endothelium-independent and the EDHF responses were evoked with either carbachol or the ionophore A23187 in the presence of an inhibitor of NO synthase. In addition, carbachol also enhanced the release of a labelled metabolite of arachidonic acid from the mesenteric bed with an identical Rf value to anandamide (Randall *et al.*, 1996).

In order definitively to identify EDHF as anadanimde, a number of criteria need to be satisfied. First of course, anandamide must mimic the hyperpolarization of smooth muscle

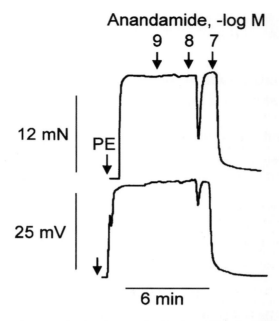

Figure 20–1. Representative trace showing that anandamide $(1 \times 10^{-9}M - 1 \times 10^{-7}M)$ reversed the depolarization and contraction induced with phenylephrine in the rat isolated mesenteric artery. Reproduced with the permission of Stockton Press (Plane *et al.*, 1997)

which characterises the action of EDHF. If hyperpolarization does in fact occur, then it must be mediated by the same membrane mechanisms as EDHF. Next, as the proposal of anandamide as EDHF was based mainly on the inhibitory action of the cannabinoid anatagonist, SR 141716A, it is reasonable to expect that this agent will have a similar action against hyperpolarizations and relaxations due to EDHF. Finally, and perhaps most crucial to the proposal, anandamide must act directly on vascular smooth muscle cells and evoke both hyperpolarization and relaxation. In addition, if anandamide is indeed EDHF then all of the above criteria must be satisfied in a range of different vascular preparations where the effects of "EDHF" have been described previously.

Does Anandamide Evoke Hyperpolarization, and in the Same Way as EDHF?

If anandamide is a hyperpolarizing factor it must cause hyperpolarization of vascular smooth muscle. In rat isolated mesenteric resistance arteries, $1 \times 10^{-9}M - 3 \times 10^{-7}M$ anandamide indeed caused hyperpolarization and relaxation during contractions to phenylephrine (Figure 20–1). Anandamide reversed the depolarization $(28.0 \pm 4.3$ mV, $n = 4)$ and contraction $(15.4 \pm 3.4$ mN, $n = 16)$ evoked by phenylephrine by 89% and 93%, respectively (Plane *et al.*, 1997). However, the effects of anandamide were not mimicked by the selective CB_1 receptor agonists, HU210 and WIN 55,212–2 (both $0.01 - 3 \times 10^{-6}M$), indicating that they were independent of cannabinoid receptor activation. An extensive study characterised the action of anandamide in a number of arteries,

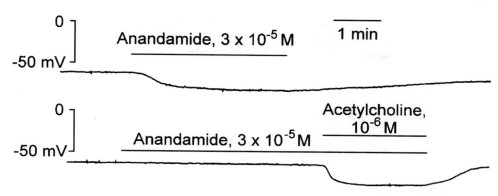

Figure 20–2. Original intracellular record of smooth muscle membrane potential in the rat isolated mesenteric artery with endothelium and in the presence of 100 μM L-nitroarginine and 5 μM indomethacin. Continuous recording from the same cell, showing hyperpolarization to exogenous anandamide. The subsequent application of the same concentration of anandamide, after a wahout period of 30 minutes, failed to evoke hyperpolarization, whereas acetylcholine was able to evoke endothelium-dependent hyperpolarization at this time. Reproduced with the permission of Stockton Press (Chataigneau *et al.*, 1998b).

and reported a number of major discrepancies with the effects of EDHF (Chataignaeu *et al.*, 1998b). Probably the most striking of these was the observation, in the rat mesenteric artery, that the hyperpolarization evoked with anandamide was not reproducible. Prolonged periods of washout are necessary in order to obtain reproducible hyperpolarization and relaxation with anandamide (Plane *et al.*, 1997). However, hyperpolarization to EDHF occurs at a time when hyperpolarization to exogenous anandamide is completely abolished (Figure 20–2) (Chataigneau *et al.*, 1998b).

Although it seems that anandamide can evoke hyperpolarization of smooth muscle, the mechanism underlying this effect is different from the action of EDHF. In a wide range of arteries, EDHF responses are inhibited by the combined application of the toxins apamin and charybdotoxin, with iberiotoxin unable to substitute for charybdotoxin (Waldron & Garland, 1994; Zygmunt & Hogestatt, 1996; Corriu *et al.*, 1996b; Zygmunt *et al.*, 1997a). However, this combination of toxins failed to inhibit either the relaxation or the hyperpolarization evoked with anandamide (Plane *et al.*, 1997; Chataigneau *et al.*, 1998b; Zygmunt *et al.*, 1997b).

As well as causing hyperpolarization by an action on different ion channels, the ability of anandamide to stimulate hyperpolarization in vascular smooth muscle cells is not a universal phenomenon. In cells from both the guinea-pig carotid and the porcine coronary arteries, anandamide in concentrations up to 30 μM failed to cause any measurable hyperpolarization (Chataigneau *et al.*, 1998b).

Does the Cannabinoid Antagonist SR 141 716A Inhibit the Action of Anandamide and EDHF?

In the rat mesenteric artery, SR 141716A in concentrations between 10^{-6}M and 5×10^{-6} M had no measurable effect on relaxation or hyperpolarization to either anandamide or EDHF (Plane *et al.*, 1997; Chataigneau *et al.*, 1998b). This contrasts with the inhibitory effect of this antagonist reported originally against relaxation to these agents in the *same* artery (Randall *et al.*, 1996; White and Hiley, 1997b). The reason for this discrepancy between these sets of observations is not clear. It certainly does not reflect any variation in the concentration of SR 141716A used, as in both cases this was in the range $1 - 5 \times 10^{-6}$ M. Nor is it likely that the discrepancy reflected a variation in the extent to which the antagonist dissolved (SR 141617A is fairly insoluble), as this ligand did inhibit the weak relaxation induced by the CB_1 receptor agonist HU210. By contrast, SR 141716A causes a 1.6–2.8 fold rightward shift in relaxation to either anandamide or carbachol (EDHF-release) in arteries of the rat contracted with methoxamine (White and Hiley, 1997). It is not clear what the precise explanation for this discrepancy is.

What is clear, however, is that the action of SR 141716A is not simply restricted to the CB receptor. At 10^{-6}M, SR 141716A blocked relaxation to levcromakalim in the mesenteric artery (White and Hiley, 1997), while in the guinea-pig carotid artery, SR 141716A evoked a direct hyperpolarization of the smooth muscle cells. The hyper-polarization was similar in magnitude to the effect of EDHF released by acetylcholine (Chataigneau *et al.*, 1998b). In the latter case, the ability of SR 141716A to *apparently* block the hyperpolarization to EDHF, may be explained by a common membrane mechanism for hyperpolarization. In this artery, anandamide itself did not alter the membrane potential of vascular smooth muscle.

Are the Actions of Anandamide Restricted to Cannabinoid Receptors on Vascular Smooth Muscle Cells?

If anandamide is EDHF, then its primary action must by definition be on the smooth muscle of the arterial wall. It is clear that this is not the case. In fact, the hyperpolarization to anandamide in the hepatic and the mesenteric arteries is endothelium-*dependent* (Zygmunt *et al.*, 1997b; Chataigneau *et al.*, 1998b). In the hepatic artery, hyperpolarization to exogenous anandamide was blocked by removal of the endothelium. However, anandamide still evoked relaxation of the smooth muscle. Whole cell patch-clamp measurements revealed that anandamide abolished the iberiotoxin-sensitive potassium current stimulated by caffeine, and spontaneous transient outward currents (reflecting the release of calcium from intracellular stores) (Figure 20–3). As anandamide did not alter the current generated by NS1619, which directly activates BK_{Ca} channels, anandamide must inhibit processes involved in the intracellular storage and release of calcium in the smooth muscle cells and in this way effect relaxation (Zygmunt *et al.*, 1997). This fits with evidence showing that the stable analogue of anandamide, R-methanandamide can deplete internal stores of calcium in astrocytes (Venance, Sagan and Giaume, 1997).

A possible explanation for the endothelium-dependent hyperpolarizing action of anandamide, is that this arachidonic acid derivative is metabolized by the endothelial cells (Pratt *et al.*, 1998). In the bovine coronary artery, anandamide induced an endothelium dependent relaxation of smooth muscle, which was resistant to the action of SR 141716A. Arterial segments with endothelium metabolized exogenous anandamide, as did endothelial cells in culture. These data are consistent with catabolism to arachidonic acid and then metabolism to form dilator eicosanoids, such as prostacylin or epoxyeicosatrienoic acids (EETs). However, a similar pathway does not operate in the mesenteric artery. In this blood vessel, arachidonic acid only had a slight relaxant effect, and a stable analogue of anandamide

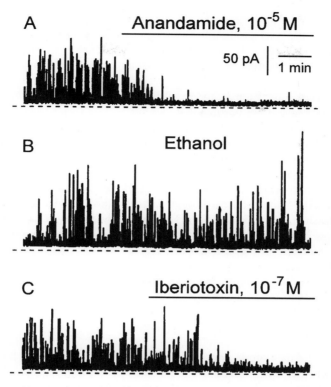

Figure 20–3. Effects of a. anandamide, b. 0.07% ethanol (the vehicle for anandamide) and c. the BK_{Ca} inhibitor iberiotoxin on spontaneous transient outward currents (STOC's). Hepatic cells were held at a potential of 0mV in a calcium-containing solution. Ethanol was present throughout the whole experimental period shown in (a) and (b). The records were obtained under voltage-clamp conditions from different cells and the dashed lines refer to the zero current level. Reproduced with the permission of the Stockton Press (Zygmunt *et al.*, 1997b).

was a more potent relaxant than anandamide itself (White and Hiley, 1998). In the coronary artery, SR 141716A attenuated the release of arachidonic acid from the endothelium, providing yet another action for this 'antagonist' (Pratt *et al.*, 1998).

Summary and Future Directions

The available evidence shows very clearly that anandamide is *not* EDHF. Perhaps the most compelling evidence for this conclusion is that the action of anandamide varies between vessels, and that hyperpolarization reflects an indirect effect exerted *throug*h the endothelial cells. Rather than causing direct hyperpolarization of vascular smooth muscle, and thus relaxation, anandamide acts to disrupt calcium handling in the smooth muscle cells and in this way affects relaxation independently of any cannabinoid receptor-linked mechanisms.

The moral of this story must surely be that to discover the identity of an endothelium-dependent hyperpolarizing factor, it is crucial to measure hyperpolarization of smooth

C.J. Garland

muscle directly. If a putative factor does directly stimulate hyperpolarization, then it must next be shown to display a similar pharmacology to EDHF. If the results are still positive at this stage, the next step is to investigate the universal nature of these observations. In the final analysis, it remains to be proven whether any putative EDHF is in fact the EDHF as opposed to an EDHF.

The most recent suggestion as to the identity of EDHF has followed precisely this experimental approach. It has been known for a long time that potassium ions efflux from endothelial cells upon agonist stimulation. It has now been shown that stimulation with acetylcholine causes an efflux of potassium ions which is blocked with a combination of apamin and charybdotoxin, a characteristic pharmacological feature of EDHF. The amount of potassium which is released from the endothelial cells can be measured with an ion sensitive microelectrode and, at around 6mM, is sufficient to cause both hyperpolarization and relaxation of the smooth muscle through pathways involving the activation of inwardly rectifying potassium channels and Na/K$^+$ ATPase (Edwards *et al.*, 1998).

Thus, modest increases in extracellular potassium ion concentration in the range of 5–10mM appear to fulfil all the criteria necessary to identify potassium as EDHF. But is potassium ion an EDHF or the EDHF? These data were obtained in rat hepatic and mesenteric arteries. In the former, the relaxation to EDHF and to potassium was blocked with a combination of ouabain and barium. In the mesenteric artery, ouabain and barium reduced but did not block relaxation. However, the extent to which potassium-induced relaxation was inhibited increased if the endothelium was destroyed. An explanation is that the variation could reflect more extensive coupling between endothelial and smooth muscle cells in the mesenteric artery. This might allow the spread of hyperpolarization from the endothelium to the smooth muscle cells. In this scenario, potassium is EDHF, but superimposed on the relaxation to potassium (EDHF) is a relaxation which follows the spread of agonist-induced hyperpolarization from the endothelium.

Whether or not potassium is a universal EDHF as opposed to an EDHF in some vascular beds remains to be resolved, as does the functional contribution from other endothelium-dependent mechanisms, such as gap junctional communication within the arterial wall.

ACKNOWLEDGEMENTS

This work was supported by the Wellcome Trust (UK).

21 Cannabinoid Receptors are Present in Arterial Smooth Muscle, are Coupled to the Inhibition of Adenylate Cyclase, but are not Involved in Relaxation

Michael Holland, R.A. John Challiss, Nicholas B. Standen and John P. Boyle

Department of Cell Physiology & Pharmacology, Maurice Shock Medical Sciences Building, University of Leicester, P.O. Box 138, Leicester, LE1 9HN, United Kingdom. Address for correspondence: Dr John P. Boyle, Department of Cell Physiology & Pharmacology, Maurice Shock Medical Sciences Building, University of Leicester, P.O. Box 138, Leicester, LE1 9HN, United Kingdom. Tel. (44) 116 2523074, Fax (44) 116 2525045, E-mail jpb3@le.ac.uk.

Endogenous production of the cannabinoid, anandamide, may explain the nitric oxide-independent (EDHF-mediated) vasorelaxation produced by acetylcholine. While anandamide repolarises/hyperpolarises and relaxes vascular smooth muscle, several studies have disputed whether these effects are mediated by CB_1 receptors. Cannabinoid receptors in the central nervous system are negatively coupled to adenylate cyclase via pertussis toxin sensitive G-proteins of the G_i sub-family. Experiments were designed to examine whether similar coupling occurs in the carotid artery and whether or not functional cannabinoid receptors exist on vascular smooth muscle cells.

Cyclic AMP accumulation was measured in rat carotid arteries using a binding protein assay. Relaxation was measured in carotid artery segments contracted with phenylephrine. Stimulation of carotid arteries with forskolin significantly increased the accumulation of cyclic AMP. However, in the presence of methanandamide, a stable analogue of anandamide, and HU-210, the forskolin-induced accumulation of cyclic AMP was reduced by up to 50%. This effect of methanandamide on cyclic AMP was inhibited by incubation of the arteries with pertussis toxin. These findings demonstrate that functional G_i-linked cannabinoid receptors are present in rat carotid artery smooth muscle. Methanandamide and anandamide both failed to relax carotid artery rings. Methanandamide inhibited forskolin-induced relaxation of the carotid artery, further illustrating the presence of cannabinoid receptors. Levcromakalim and pinacidil, vasodilators that act by inducing hyperpolarisation, almost fully relaxed this artery. These data demonstrate the presence of functional G protein-linked cannabinoid receptors in the carotid artery of the rat, but show that these receptors inhibit the accumulation of cyclic AMP rather than cause relaxation.

KEYWORDS: Anandamide, methanandamide, relaxation, cyclic AMP, rat carotid artery, SR141716A, pertussis toxin

INTRODUCTION

Endocannabinoids, such as anandamide, and stereochemically pure synthetic, cannabinoids, for example, (6aR)-*trans*-3-(1,1-Dimethylheptyl)-6a,7,10,10a-tetrahydro-1-hydroxy-6,6-dimethyl-6H-dibenzo[b,d]pyran-9-methanol (HU-210), produce bradycardia and hypotension *in vivo* and relaxation of vascular smooth muscle *in vitro* (Varga *et al.*, 1995, 1996; Randall *et al.*, 1996; Vidrio, Sanchezsalvatori and Medina, 1996; Calignano *et al.*, 1997; Plane *et al.*, 1997; Zygmunt *et al.*, 1997). Endogenous anandamide is synthesised by macrophages and platelets in response to lipopolysaccharide binding and its production may underlie the hypotension associated with haemorrhagic shock (Wagner *et al.*, 1997). However, the powerful effects of endocannabinoids on the cardiovascular system *in vivo* are complex. The intravenous administration of anandamide, for example, produces a combination of pressor and depressor effects (Varga *et al.*, 1996). The origin of the pressor effect is currently unclear, but it is not mediated by cannabinoid receptors (Varga *et al.*, 1996; Lake *et al.*, 1997). The depressor effects, which occur particularly in the resistance vessels, appear to involve both inhibition of noradrenaline-release from sympathetic nerve terminals (Varga *et al.*, 1996), possibly by inhibition of Ca^{2+} entry through N-type Ca^{2+} channels (Mackie and Hille, 1992), and direct effects on vascular smooth muscle (Randall *et al* 1996; Plane *et al.*, 1997; Zygmunt *et al.*, 1997). In most studies, the relaxation/ hypotension produced by endocannabinoids and synthetic cannabinoid receptor agonists is attenuated by the CB_1 receptor antagonist, N-(piperidin-1-yl)-5-(4-chlorophenyl)-1-(2,4-dichloropheny)-4-methyl-1 H-pyrazole-3-carboxamine HCl (SR141716A) (Varga *et al.*, 1995, 1996; Randall *et al.*, 1996; Vidrio, Sanchezsalvatori, and Medina, 1996; Zygmunt *et al.*, 1997), indicating that the hypotensive effects of endocannabinoids are likely to be mediated via CB_1 receptors. Little experimental evidence exists, however, to indicate that cannabinoid receptors are present in vascular smooth muscle, or to explain how the direct effects of endocannabinoids on vascular smooth muscle are mediated. One mechanism proposed to account for the direct relaxing effects of endocannabinoids, in hepatic artery smooth muscle cells, is anandamide-induced inhibition of caffeine-induced Ca^{2+}-release from intracellular stores (Zygmunt *et al.*, 1997). Whether this effect on Ca^{2+} homeostasis is CB_1 receptor-mediated was not determined but a similar pertussis toxin-sensitive effect of anandamide has previously been reported in astrocytes (Venance, Sagan and Giaume, 1997).

The present study addressed the questions whether or not there are functional cannabinoid receptors in rat carotid artery smooth muscle and whether or not methanandamide, a stable analogue of anandamide, causes inhibition of agonist-induced tone in this artery. The first question was answered by making use of the observation that cannabinoid receptors, at least in the spleen and nervous system, have been shown to be negatively coupled to adenylate cyclase. Therefore the effects of methanandamide, were assessed on the production of cyclic AMP and on forskolin-induced relaxations in the rat carotid artery.

METHODS

Tissue Preparation

Male Wistar rats were killed by stunning followed by exsanguination. The carotid arteries were carefully removed, cleaned of any attached nervous and connective tissues and placed in Krebs Henseleit buffer (4°C). The vessels were then either used in biochemical studies or in mechanical studies.

Cyclic AMP

The arteries were transferred into tubes containing 10 ml Krebs Henseleit buffer, cannabinoid receptor agonists added and the tubes transferred to a shaking waterbath and incubated at 37°C for 10–30 min. Forskolin was then added to the test arteries to stimulate the production of cyclic AMP, while control tissues were vehicle-treated (DMSO or ethanol <0.1% of the final concentration). After 5 min stimulation the arteries were removed, rapidly dried with damp filter paper and frozen in liquid nitrogen; 0.5 ml of cold (4°C) trichloroacetic acid (0.5M) was added to each sample, which was then homogenised (polytron, setting 6 for 5×15 sec). The homogenised samples were kept on ice for 45 min to allow total extraction of cyclic AMP.

Following the extraction of cyclic AMP, samples were centrifuged for 5 min at 6000 rpm at 4°C. The acid was removed from the supernatant by extraction with freon/tri-n-octylamine and the aqueous phase neutralised with $NaHCO_3$. Cyclic AMP levels were determined using a standard binding protein assay (Brown *et al.*, 1971) and the results expressed as pmol per mg of tissue protein.

To demonstrate that the cannabinoid receptors present in smooth muscle of the rat carotid artery are G protein-coupled, similarly to CB_1 and CB_2 receptors in other areas of the body, control arteries were incubated in minimum essential medium supplemented with 5% bovine serum albumin and 1% penicillin/streptomycin for 24 hours at 37°C in 5% CO_2. Test vessels were treated as control but with the addition of pertussis toxin at a concentration of 1 μg/ml for 24 hours.

Mechanical Responses

Segments of mesenteric and carotid artery were suspended in a Mulvany-Halpern wire myograph at 37°C for the recording of isometric tension in Krebs-Henseleit buffer. Artery segments were stretched to the equivalent of a transmural pressure of 100 mm Hg. The diameter was then calculated and set to 90% of this value and the tissue allowed to equilibrate for 45 min. Tone was induced in the segments by the addition of the α_1-adrenoceptor agonist, phenylephrine (10^{-5} M). After a stable increase in tone was established, forskolin was added cumulatively (10^{-9}–10^{-5} M) to the bath and relaxation recorded. When assessing the effects of cannabinoid receptor agonists, methanandamide, a stable analogue of anandamide was used, ensuring that any observed effects were not due to the breakdown of anandamide to other vasoactive metabolites. Identical procedures were used except methanandamide (10^{-5} M) was added to the bath 10 to 30 min before tone was produced.

Statistical Analysis

All data are expressed as the mean ± the standard error of the mean, while n is the number of observations for the mechanical responses and the number of observations made in triplicate for the biochemical studies. The significance of the difference between means was assessed using unpaired or paired Student's t-test. Results were accepted as significantly different when P was less than 0.05.

RESULTS

Conditions which produced a maximal accumulation of cyclic AMP in response to forskolin (10^{-5} M for 5 min) were used throughout and 10^{-6} and 10^{-5} M methanandamide and 10^{-6} M HU-210 were used to stimulate cannabinoid receptors.

Michael Holland et al.

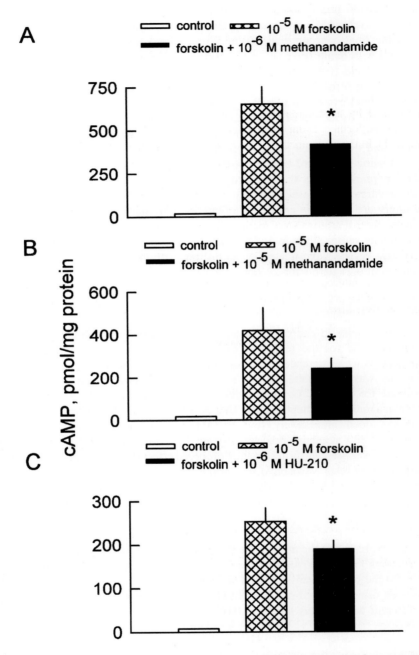

Figure 21-1. Effects of the cannabinoid agonists methanandamide, 10^{-6} M (A) and 10^{-5} M (B) and HU-210, 10^{-6} M (C) respectively on forskolin stimulated cyclic AMP accumulation in rat carotid arteries. All figures show, from left to right, the basal level of cyclic AMP, the increase in cyclic AMP following stimulation with forskolin and the significant reduction in cyclic AMP accumulation following incubation with the cannabinoid agonists. The asterisks denote statistically significant differences ($P < 0.05$). Data shown as means ± s.e.m.

Methanandamide and Forskolin-Induced Cyclic AMP Accumulation

Under control conditions, where carotid arteries were vehicle treated with ethanol or DMSO (0.1%), the concentration of cyclic AMP was 18.5 ± 2.5 pmol.mg^{-1} protein (n = 12). Following stimulation with forskolin (10^{-5} M) for 5 min the concentration of cyclic AMP was significantly increased 35 fold (n = 12) (Figure 21-1). Following incubation with methanandamide (10^{-6} M), the increase in cyclic AMP produced by forskolin was significantly reduced by approximately 28% (n = 12) (Figure 21-1). When identical experiments were performed using methanandamide (10^{-5} M) similar results were obtained (Figure 21-1). In these experiments the basal level of cyclic AMP was 17 ± 2.5 pmol cyclic AMP.mg^{-1} protein (n = 7), significantly increasing 25 fold (n = 8) following stimulation with forskolin. In the presence of 10^{-5} M methanandamide the increase in cyclic AMP induced by forskolin was significantly reduced by 42% (n = 7) (Figure 21-1). A lower concentration of methanandamide (10^{-7} M) also significantly attenuated forskolin-stimulated cyclic AMP accumulation.

HU-210 and Forskolin-Induced Accumulation of Cyclic AMP

Under control conditions where arteries were vehicle-treated, the concentration of cyclic AMP was 7.4 ± 0.6 pmol.mg^{-1} protein (n = 13). Cyclic AMP levels significantly increased 34 fold following stimulation with forskolin. This increase which was reduced significantly by 25% (n = 4) in the presence of 10^{-6} M HU-210.

SR141716A

Incubation with SR141716A (10^{-5} M) for 15 min or more significantly reduced forskolin-stimulated cyclic AMP-accumulation in rat carotid arteries (n = 8) from 295 ± 10 pmol.mg^{-1} protein to 154 ± 30 pmol.mg^{-1} protein, a decrease of 47%. Since this effect is similar, in magnitude and direction, to that of the cannabinoid agonists this antagonist could not be used to reverse the inhibitory effects of anandamide or methanandamide on the accumulation of cyclic AMP.

Pertussis Toxin

Blood vessels were incubated for 24 hours in tissue culture medium at 37°C. The basal level of cyclic AMP was 4.3 ± 0.8 pmol cyclic AMP.mg^{-1} protein (n = 3), somewhat lower than that observed in freshly-isolated vessels (Figure 21-2). Stimulation with forskolin significantly increased cyclic AMP levels (approximately 50 fold) (n = 4), while incubation with 10^{-5} M methanandamide significantly reduced cyclic AMP levels by approximately 50% (n = 4). When pertussis toxin, (1 μg/ml), was included in the tissue culture medium (minimum essential medium) and the vessels incubated at 37°C for 24 hours the basal level of cyclic AMP was 17.3 ± 2.8 pmol.mg^{-1} protein (n = 4). Stimulation with forskolin significantly increased the level of cyclic AMP while incubation with 10^{-5} M methanandamide failed to reduce cyclic AMP levels (n = 4) (Figure 21-2).

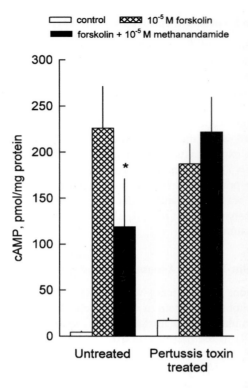

Figure 21-2.   Effects of pertussis toxin treatment (1 μg/ml for 24 hours) on methanandamide-induced attenuation of cyclic AMP-accumulation in rat carotid arteries. From left to right the first three bars illustrate; basal levels of cyclic AMP, the increase in cyclic AMP accumulation following stimulation with forskolin and the attenuation of cyclic AMP produced by the addition of methanandamide respectively, using carotid arteries maintained in minimum essential medium supplemented with 10% bovine serum albumin for 24 hours in the absence of pertussis toxin. Bars 4–6 illustrate data obtained using the same experimental conditions but including pertussis toxin. The asterisks denote statistically significant differences ($P < 0.05$). Data shown as means ± s.e.m.

Methanandamide and Vascular Responsiveness

Anandamide produced concentration-dependent relaxations of mesenteric arteries, yielding an EC_{50} value of 1.6×10^{-7} M (1.25 to 2.0, 95% confidence limits) (n = 16) (Figure 21-3). Concentrations of anandamide and methanandamide, twenty-fold larger than the EC_{50} value for anandamide in mesenteric arteries failed to relax phenylephrine-constricted carotid arteries. Pinacidil produced close to full relaxation of the rat carotid artery (81% relaxation, EC_{50} 2.5×10^{-7} M (2.23–2.81, 95% confidence limits) (n = 4). Similar results were obtained with a structurally dissimilar K_{ATP} channel opener, levcromakalim (data not shown).

Relaxation to forskolin (3×10^{-9} and 10^{-8} M) was significantly reduced by methanandamide (10^{-5} M) (Figure 21-4). Separate experiments were performed with a single dose of forskolin (10^{-8} M), rather than obtaining full concentration-effect curves. The use of a single concentration of forskolin, selected to lie on a linear portion of the concentration-

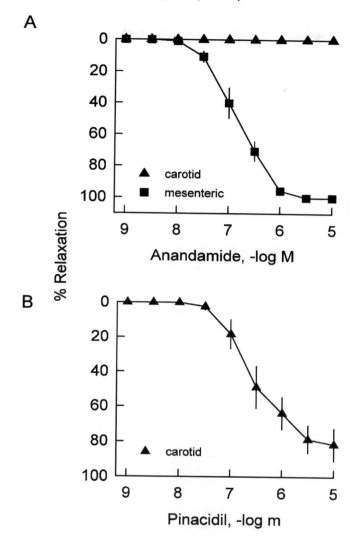

Figure 21-3. Relaxation produced by methanandamide and pinacidil. Panel A: Effects of anandamide or methanandamide on mesenteric (■) and carotid (▲) arteries. Panel B: The relaxant effect of pinacidil on phenylephrine-contracted carotid artery rings.

effect curve, allowed paired data to be obtained. Under control conditions forskolin produced a $57 \pm 7\%$ relaxation of arterial rings contracted with phenylephrine (10^{-5} M). This was significantly reduced to $35 \pm 2\%$ following their incubation with methanandamide (10–30 min), but unchanged ($54 \pm 6\%$) when the CB_1-selective antagonist SR141716A was included in combination with methanandamide. SR141716A alone did not influence the effects of forskolin (Figure 21-5).

Michael Holland et al.

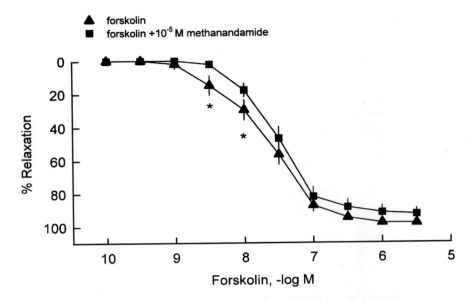

Figure 21-4. Inhibition by methanandamide of relaxation of the carotid artery by forskolin. Panel A: Concentration-effect curves for carotid artery rings to forskolin in the presence (■) and absence (▲) of methanandamide (10^{-6} M). The asterisks denote statistically significant differences (P < 0.05). Data shown as means ± s.e.m.

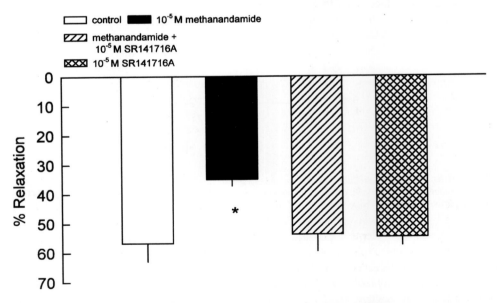

Figure 21-5. Relaxation of carotid artery rings to single doses of forskolin (10^{-9} M). From left to right the bars illustrate % relaxation under control conditions, in the presence of 10^{-5} M methanandamide, in the presence of 10^{-5} M methanandamide plus 10^{-5} M SR141716A and with SR141716A alone. The asterisks denote statistically significant differences (P < 0.05). Data shown as means ± s.e.m.

DISCUSSION

The precise mechanism by which the cannabinoids induce vasorelaxation remains unclear but a number of mechanisms have been proposed to account for their relaxing effects. The cannabinoids may produce hyperpolarisation/repolarisation (Randall *et al.*, 1996; Plane *et al.*, 1997), leading to relaxation by reducing Ca^{2+} entry through voltage-activated Ca^{2+} channels (Nelson and Quayle, 1995). Alternatively cannabinoids may interfere with Ca^{2+}-release from intracellular stores in smooth muscle (Zygmunt *et al.*, 1997), reducing the intracellular Ca^{2+} concentration. When the evidence from a number of studies is taken together (Randall *et al.*, 1996; Plane *et al.*, 1997; Zygmunt *et al.*, 1997; Chaitaigneau *et al.*, 1998) the proposal that cannabinoids produce relaxation by functioning as hyperpolarising factors appears less likely. In patch-clamp studies, for example, anandamide failed to activate whole cell currents in single hepatic artery smooth muscle cells, indicating that it was not producing K^+ channel opening, either directly or indirectly (Zygmunt *et al.*, 1997). Results from the present study support the idea that cannabinoids do not function as hyperpolarising factors, as anandamide, methanandamide and HU-210 all failed to relax the carotid artery, while the hyperpolarising vasodilators, pinacidil and levcromakalim produced close to 100% relaxation of constricted carotid arteries, illustrating that hyperpolarisation *per se* induces relaxation of this artery. The notion that anandamide produces relaxation by interfering with intracellular Ca^{2+} handling is therefore currently more attractive, although this mechanism of action would also be expected to cause relaxation of the smooth muscle of rat carotid artery.

One possible reason as to why cannabinoids fail to relax carotid arteries is the contractile agonist used to induce tone. Phenylephrine was used throughout the current study while methoxamine has been used by others. However, anandamide and methanandamide induce concentration-dependent relaxation of phenylephrine-induced tone in both mesenteric artery segments (Plane *et al.*, 1997) and isolated, perfused mesenteric beds (Gardner, Holland and Boyle, unpublished observation). Other studies have also reported that cannabinoids do not relax all vascular beds and ananadamide also failed to relax porcine coronary arteries (Chaitaigneau *et al.*, 1998). The effects of anandamide and other cannabinoids may therefore depend on the species and arterial bed used, with these compounds being more effective relaxants of resistance arteries.

The hypothesis that hyperpolarisation and relaxation to cannabinoids depends on activation of cannabinoid receptors, probably of the CB_1 subtype, is based on the ability of the selective cannabinoid CB_1 antagonist, SR141716A (10^{-6}–10^{-5} M), to antagonise the relaxing effects of methacholine and anandamide in rat mesenteric and hepatic arteries (Randall *et al.*, 1996; White and Hiley, 1997; Zygmunt *et al.*, 1997). Conversely, SR141716A did not significantly affect anandamide-induced relaxation in rat mesenteric arteries, and other cannabinoid receptor agonists were relatively ineffective relaxants (Plane *et al.*, 1997). The present study supports the idea that the interaction of cannabinoids with cannabinoid receptors does not lead to arterial relaxation.

The presence of membrane-delimited receptors is usually detected using radioligand binding methods. Cannabinoids, however, are extremely lipophilic and when used to determine specific binding to vascular smooth muscle cell membranes the level of non-specific binding is so high as to 'mask' specific binding (Kendall, personal communication). Because of these practical problems an alternative method was devised to assay for cannabinoid receptors in vascular smooth muscle. Both CB_1 and CB_2 receptors, although differentially distributed, share a similar signal transduction pathway such that occupation of either receptor, by a cannabinoid receptor agonist, leads (via the activation of G_i/G_o) to inhibition of adenylate cyclase (Howlett, 1996; Randall and Kendall, 1998). Thus the effects of cannabinoid receptor agonists were assessed on forskolin-induced accumulation of cyclic AMP in the rat carotid artery. Additionally the effects of several cannabinoid receptor agonists were examined to attempt to define which receptors, if any, are present in the rat carotid artery. The compounds used possessed varying degrees of specificity, but all three compounds are reported to have a higher affinity for CB_1 than CB_2 cannabinoid receptors (Felder *et al.*, 1995, 1998), indicating that the receptor present is more likely similar to the CB_1 sub-type.

Forskolin stimulated a large increase in the accumulation of cyclic AMP. If cannabinoid receptors were present on the smooth muscle cells of the rat carotid artery then, by analogy with results obtained using other tissue preparations, these increases would be expected to be attenuated by cannabinoid agonists. The significant inhibition of forskolin-induced increases in cyclic AMP accumulation by both methanandamide and HU-210 is in good agreement with data obtained from a variety of tissues, and from recombinant cannabinoid receptors expressed in a number of mammalian cells lines (Slipetz *et al.*, 1995; Rinaldi-Carmona *et al.*, 1996; Shire *et al.*, 1996). These data strongly suggest that cannabinoid receptors are not only present in carotid vascular smooth muscle but that they are functional, and negatively coupled to adenylate cyclase. Evidence that, in rat carotid artery, these receptors are coupled to adenylate cyclase via G proteins of the G_i/G_o family (in a similar way to that described in studies on other tissues and on cloned CB_1 and CB_2 receptors) comes from the sensitivity of the inhibitory effects of methanandamide to pertussis toxin. Cloned human and mouse CB_1 and CB_2 receptors expressed in a variety of mammalian cell lines are coupled to the inhibition of adenylate cyclase through the G_i/G_o family of G proteins (Slipetz *et al.*, 1995; Rinaldi-Carmona *et al.*, 1996; Shire *et al.*, 1996).

These results make it difficult to understand the mechanism by which activation of cannabinoid receptors by agonists such as anandamide could lead to relaxation of smooth muscle. It seems unlikely that the same receptor would be coupled, in the same cell, to a mechanism that causes relaxation and at the same time to another mechanism that impedes it. The data obtained concerning the effects of methanandamide on forskolin-induced relaxation in the rat carotid artery reinforces this point and illustrates that activation of the cannabinoid receptor should inhibit relaxation of smooth muscle, at least in this artery.

The sensitivity of the cannabinoid-induced inhibition of the accumulation of cyclic AMP to the selective CB_1 antagonist, SR141716A, was examined as were the effects of agonists which have been reported to show some selectivity between CB_1 and CB_2 receptors. The observation, that in functional studies at least, SR141716A reversed the inhibition of forskolin-induced relaxation, produced by methanandamide, suggests that the inhibitory effect of methanandamide occurs via the activation of CB_1 receptors. Further evidence supporting the role of CB_1 rather than CB_2 receptors in carotid artery smooth muscle cells came from the observation that the agonist HU 210 also caused inhibition of forskolin-induced cyclic AMP-accumulation. This agonist is reasonably selective for CB_1 cannabinoid receptors (Felder *et al.*, 1995; Slipetz *et al.*, 1995).

The observation that SR141716A has a direct effect on forskolin-stimulated cyclic AMP-accumulation, similar to that of cannabinoid agonists, suggests it may act as a low intrinsic activity agonist rather than a pure antagonist at CB_1 receptors. Further evidence in support of this idea can be found by examining data obtained on CB_1 receptors stably transfected into mammalian cell lines (Felder *et al.*, 1995) and myenteric neurones (López-Redondo, Lees and Pertwee, 1997).

ACKNOWLEDGEMENTS

This study was supported by the Medical Research Council.

22 Effects of SIN-1 on Potassium Channels of Vascular Smooth Muscle Cells of the Rabbit Aorta and Guinea-Pig Carotid Artery

J.-F. Quignard, T. Chataigneau, C. Corriu, J. Duhault, M. Félétou and P.M. Vanhoutte*

Dpt Diabétologie, Institut de Recherche Servier, 92150 Suresnes and
* Institut de Recherches Internationales Servier, 92410 Courbevoie, France.
Address for correspondence: Michel Félétou: Dpt Diabétologie, Institut de Recherche Servier, 92150 Suresnes

Experiments were designed to determine the subtype of potassium channels activated by the nitrovasodilator 3-morpholinosydnonimine (SIN-1). Membrane potential was recorded by means of intracellular microelectrodes inserted from the adventitial side, in the vascular smooth muscle cells of isolated segments of the carotid artery of the guinea-pig, superfused with thermostated modified Krebs-Ringer bicarbonate solution (containing N^{ω}-nitro-L-arginine and indomethacin). Potassium currents were recorded, at room temperature, in freshly dissociated smooth muscle cells from both preparations by the patch clamp technique (whole-cell, cell-attached, and excised configurations). SIN-1 caused a significant hyperpolarization of the smooth muscle cells of the guinea-pig carotid artery. The effects of SIN-1 were not affected by apamin, charybdotoxin, or the combination of the two toxins, but were abolished by glibenclamide. In freshly dissociated smooth muscle cells of the guinea-pig carotid artery, SIN-1 induced small or no modification of the whole-cell potassium current and of the iberiotoxin-sensitive potassium current, measured with the patch clamp technique (cell-attached configuration). However, the nitrovasodilator activated a small glibenclamide-sensitive current. In contrast, in freshly dissociated smooth muscle cells of the rabbit aorta, SIN-1 enhanced the activity of the global potassium current and of the iberiotoxin-sensitive potassium current. These findings suggest that the effects of NO in vascular smooth muscle are species-dependent. It activates ATP-sensitive potassium channels in smooth muscle cells of the guinea pig carotid artery and large conductance calcium-sensitive potassium channels in smooth muscle cells of the rabbit aorta. This results further confirm that NO and EDHF (which activates a glibenclamide-insensitive potassium conductance) are different in the guinea-pig carotid artery.

KEYWORDS: SIN-1, EDHF, calcium-activated potassium channel, ATP-sensitive potassium channel, electrophysiology

INTRODUCTION

NO, released by the endothelium or given exogenously relaxes vascular smooth muscle by increasing the cyclic GMP level. Intracellular microelectrode experiments indicate that in species such as the guinea-pig (Tare *et al.*, 1990; Parkington *et al.* 1995; Corriu *et al.*, 1996b) and the rabbit (Cohen *et al.*, 1997), NO causes hyperpolarization of the vascular smooth muscle cells. In the rabbit carotid artery, even in the presence of N^{ω}-nitro-L-arginine and indomethacin, an uncomplete blockade of NO synthase may permit sufficient

193

release of NO from the endothelial cells to account for the effects usually attributed to EDHF (Cohen *et al.*, 1997).

The purpose of the present study was to determine the type of potassium channel activated by NO donors in arterial smooth muscle cells of the guinea pig and the rabbit and to determine whether or not NO accounts for the endothelium-dependent hyperpolarization observed in the two species.

MATERIALS AND METHODS

Microelectrode Studies

Male Hartley guinea-pigs were anesthetized and the internal carotid arteries were dissected, cleaned of adherent connective tissues and pinned to the bottom of an organ chamber. The tissues were superfused with a thermostated modified Krebs-Ringer bicarbonate solution of the following composition (in mM): NaCl 118.3, KCl 4.7, $CaCl_2$ 2.5, $MgSO_4$ 1.2, KH_2PO_4 1.2, $NaHCO_3$ 25, EDTA 0.026 and glucose 11.1. Transmembrane potential was recorded by using glass capillary microelectrodes (tip resistance of 30 to 90 $M\Omega$) filled with KCl (3 M). Impalements of the smooth muscle cell were performed from the adventitial side. All the experiments were performed in the presence of N^ω-nitro-L-arginine (10^{-4} M) and indomethacin (5×10^{-6} M) to inhibit nitric oxide synthase and cyclooxygenase, respectively (Corriu *et al.*, 1996a).

Patch-Clamp Studies

The media of guinea-pig carotid artery was dissected from cleaned arteries. The smooth muscle cells were dissociated enzymatically (Quignard *et al.*, 1998). Whole-cell potassium currents were recorded at room temperature using the patch-clamp technique. In whole-cell experiments, to record voltage-sensitive potassium currents (K_V) a calcium-free intracellular solution was used with the following composition (in mM): KCl 130, $MgCl_2$ 2, ATP 3, GTP 0.5, HEPES 25, EGTA 10, Glucose 11. To record more accurately calcium-sensitive potassium currents (K_{Ca}), the concentration of EGTA was reduced to 1 mM and $CaCl_2$ (0.5 mM) was added. To record ATP-sensitive potassium currents (K_{ATP}), ATP was reduced to 0.1 mM, in the presence of oxaloacetic acid (5 mM) and pyruvic acid (2 mM). The cells were superfused with a solution containing (in mM): NaCl 125, KCl 5, $CaCl_2$ 2, $MgCl_2$ 1.2, HEPES 10 and glucose 11. Data were recorded with pClamp6 software (Axon Instruments, USA) through a RK-400 amplifier (Biologic, France).

Statistics

Data are shown as mean $\pm$ s.e. mean; n indicates the number of cells in which membrane potential was recorded. Statistical analysis was performed using Student's *t*-test for paired or unpaired observations. Differences were considered to be statistically significant when P was less than 0.05.

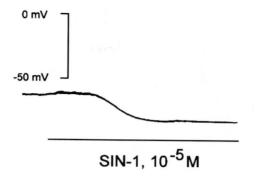

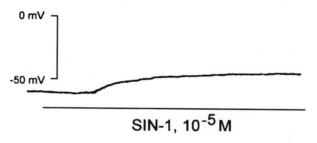

Figure 22-1. Effects of glibenclamide on the endothelium-dependent hyperpolarization evoked by SIN-1 (10^{-5} M), in the guinea-pig carotid artery in the presence of N$^\omega$-nitro-L-arginine (10^{-4} M) and indomethacin (5×10^{-6} M). In the presence of glibenclamide the hyperpolarization produced by SIN-1 was blocked and in some cells SIN-1 induced a depolarization.

RESULTS

Microelectrode

In guinea-pig carotid arteries, in the presence of N$^\omega$-nitro-L-arginine (10^{-4} M) and indomethacin (5×10^{-6} M), acetylcholine (10^{-6} M) induced an endothelium-dependent hyperpolarization (-18.1 ± 1.1 mV; n = 35); This hyperpolarization was reversed to a depolarization (4.4 ± 1.2 mV, n = 20) by the combination of charybdotoxin (10^{-7} M) plus apamin (5×10^{-7} M). Individually, these inhibitors were ineffective. Glibenclamide (10^{-6} M) did not affect significantly the acetylcholine-induced hyperpolarization (-18 ± 1.8 mV, n = 10). SIN–1 (10^{-5} M) induced an hyperpolarization of the vascular smooth muscle cells (-12.8 ± 2.0 mV, n = 6) (Figure 22-1). The effects of the nitrovasodilator were endothelium-independent and reversed by glibenclamide (10^{-6} M, 4.5 ± 4.6 mV, n = 4). The combination of charybdotoxin plus apamin did not alter the hyperpolarization to SIN-1 (data not shown).

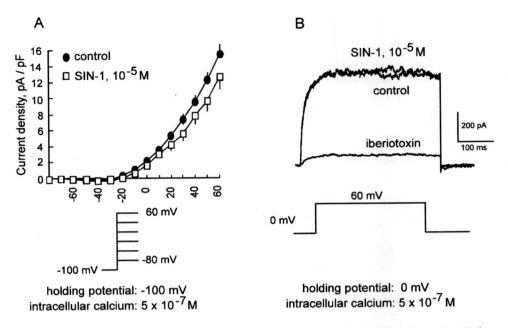

Figure 22-2. Effects of SIN-1 on whole-cell potassium currents in freshly isolated smooth muscle cells from the carotid artery of the guinea-pig (intracellular solution with 5×10^{-7} M free calcium). A) Current-voltage relationships of potassium current (holding potential -100 mV) before and after exposure to SIN-1 (10^{-5} M). B) Recording of potassium currents before and after exposure to SIN-1 (10^{-5} M). Currents were generated by stepping from a holding potential of 0 mV to the test potential of 60 mV.

Patch-Clamp

The freshly dissociated myocytes from guinea-pig carotid artery expressed different types of potassium channels (see Quignard *et al.*, Chapter 23 in this monograph). SIN-1 (up to 10^{-5} M) did not affect the current-voltage relationship of the current, either in the presence of high (n = 6) or low intracellular (n = 10) calcium concentrations (Figure 22-2). Dibutyril cyclic GMP (10^{-5} M), an analogue of cyclic GMP that permeates cell membrane, did not alter the amplitude of the potassium current (n = 5, data not shown). With the cell-attached configuration, large conductance K_{Ca} (BK_{Ca}) were recorded at the resting membrane potential. The unitary conductance was 280 ± 20 pS (n = 5). The unitary currents were inhibited by iberiotoxin (10^{-7} M) or by charybdotoxin (10^{-7} M) (outside-out configuration, n = 3). SIN-1 (10^{-5} M) did not modify significantly the open probability and the mean open time of these currents (Figure 22-3). Dibutyril cyclic GMP (10^{-5} M, n = 8) did not affect these parameters. However, in two cells out of ten, the open probability increased transiently ($+30 \pm 10\%$) during 3 minutes then returned to the basal values. By contrast, with the whole-cell configuration (low ATP concentration), SIN-1 induced, in three out of ten cells, an outward current which was inhibited by glibenclamide (10^{-6} M) at 0 mV (Figure 22-4). This current reversed near the equilibrium potential for potassium ions (-74 ± 5 mV, n = 3) and his density was of 0.8 ± 0.2 pA/pF.

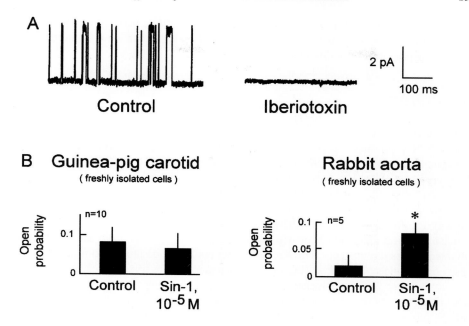

Figure 22-3. Effects of SIN-1 on unitary potassium currents in freshly isolated vascular smooth muscle cells of the guinea-pig carotid artery and rabbit aorta. A) Iberiotoxin (10^{-7} M) abolished BK_{Ca} in smooth mucle cells of the guinea-pig (outside-out configuration, holding potential: –20 mV). B) Effects of SIN-1 on the mean open probability of BK_{Ca}. Results were obtained in the cell-attached configuration (patch potential –20 mV). The asterisk indicates a statistical significant difference versus control values ($P < 0.05$); ns indicates no statistical difference; n represents the number of cells studied.

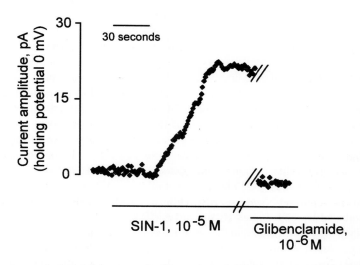

Figure 22-4. Effects of SIN-1 on K_{ATP} current amplitude, in function of time. SIN-1 induced a glibenclamide-sensitive, whole-cell potassium current (intracellular solution with a low ATP concentration, holding potential = 0 mV) in freshly isolated vascular smooth muscle cells of the guinea-pig carotid artery.

In freshly dissociated myocytes from the rabbit aorta, BK_{Ca} activity was also recorded in the cell-attached configuration. Application of SIN-1 induced a rapid increase in the mean open probability of this channel (Figure 22-3).

DISCUSSION

In the guinea pig carotid artery, NO can be considered as an hyperpolarizing factor. Indeed SIN-1 produces a glibenclamide-sensitive hyperpolarization, and patch-clamp experiments confirmed that SIN-1 activates an ATP-sensitive current. The characteristics of the current activated by SIN-1 (reversion of the current at the equilibrium potential for potassium ions and inhibition by the specific inhibitor of K_{ATP} channel, glibenclamide), are those of a K_{ATP} current. NO activates K_{ATP} currents through a cyclic GMP-dependent pathway in other vascular beds (Miyoshi et al., 1994).

In isolated vascular smooth muscle cells of the guinea-pig, SIN-1 did not affect calcium-activated potassium currents. This lack of effect could not be attributed to the absence of these conductances. Indeed, a whole-cell conductance is observed which was sensitive to iberiotoxin (specific BK_{Ca} blocker). With the cell-attached configuration, a large conductance unitary current was also recorded. The characteristics of this unitary current [conductance 280 pS, inhibition by iberiotoxin, activation by voltage and calcium (unpublished observations)] are similar to the BK_{Ca} unitary current (Kuriyama et al., 1995). The second messenger released by nitrovasodilatators, cyclic GMP, was without effect on the activity of the K_{Ca} current, confirming that in this tissue, nitrovasodilatators do not affect K_{Ca} (Corriu et al., 1996b).

In myocytes of the rabbit aorta, SIN-1 enhanced the calcium-activated potassium unitary current. Similar results have been described in other blood vessels (Archer et al., 1994). This effect may be related to either activation of soluble guanylate cyclase or a direct action of NO on the channel (Bolotina et al., 1994; Shin et al., 1997). The activation of K_{Ca} currents is involved in the NO-induced hyperpolarization in those tissues (Archer et al., 1994).

These finding suggest that NO activates different types of potassium channels, in different species. However, the anatomical origin of the tissues may be more important than the species *per se*. Thus, in the rabbit, NO activates a calcium-activated potassium current in the vascular smooth muscle cells of the carotid artery, (Plane et al., 1998) and an ATP-sensitive current in those of the mesenteric artery (Murphy & Brayden 1995a). By contrast, in some blood vessels such as the canine carotid and mesenteric arteries, NO does not produce hyperpolarization (Komori et al., 1988). This diversity may also depend on the source of exogenous NO. Thus, in the rabbit carotid artery, pure NO gas relaxes the blood vessel in a charybdotoxin-sensitive fashion, whereas the toxin has no effect on relaxation induced by NO donors (Plane et al., 1998).

In the guinea pig carotid artery, in the presence of N^{ω}-nitro-L-arginine and indomethacin, acetylcholine induced an endothelium-dependent hyperpolarization, which has been attributed to EDHF (Corriu et al., 1996a and b). In the same artery of the rabbit, an endothelial NO release persists, despite the presence of these inhibitors, and accounts for the acetylcholine-induced hyperpolarization. Thus, in this blood vessel (Cohen et al., 1997a), NO may be the endothelium-derived hyperpolarizing factor. However, in the carotid artery of the guinea pig, as in the rat hepatic artery (Zygmunt *et al.*, 1997a), the hyperpolarization caused by NO donors differs from EDHF-mediated hyperpolarization. As shown previously, a K_{ATP}, glibenclamide-sensitive current is involved in the SIN-1-induced hyperpolarization (Corriu *et al.*, 1996b). By contrast, EDHF-induced hyperpolarization is not altered by glibenclamide but is reversed by the combination of the K_{Ca} inhibitors apamin plus charybdotoxin. This combination of toxin does not affect the nitrovasodilatator-induced hyperpolarization. In addition, NO scavengers (8 carboxy PTIO or haemoglobin), do not alter the EDHF-mediated hyperpolarization (Garland and McPherson, 1992, unpublished observations). These results indicate that in the guinea-pig carotid artery, the acetylcholine-induced hyperpolarization in the presence of NO synthase inhibitors can not be attributed to a residual liberation of NO but must be caused by another factor.

23 Potassium Channels Involved in EDHF-Induced Hyperpolarization of the Smooth Muscle Cells of the Isolated Guinea-Pig Carotid Artery

J.-F. Quignard, T. Chataigneau, C. Corriu, J. Duhault, M. Félétou and P. M. Vanhoutte*

Dpt Diabétologie, Institut de Recherche Servier, 92150 Suresnes and
* Institut de Recherches Internationales Servier, 92410 Courbevoie, France.
Address for correspondence: Michel Félétou: Dpt Diabétologie, Institut de Recherche Servier, 92150 Suresnes

Experiments were designed to determine the nature of the potassium channels involved in the endothelium-dependent hyperpolarization caused by acetylcholine in the isolated carotid artery of the guinea-pig. The membrane potential of vascular smooth muscle cells was recorded by means of intracellular microelectrodes inserted in the adventitial side of isolated segments of this artery, in the presence of N^ω-nitro-L-arginine and indomethacin. Potassium currents were recorded, at room temperature, in freshly dissociated smooth muscle cells from the same preparation by the patch clamp technique (whole-cell configuration). The acetylcholine-induced hyperpolarization was absent in the presence of elevated potassium. Glibenclamide, apamin, scillatoxin, charybdotoxin and iberiotoxin individually produced no or only minor inhibition of the hyperpolarization. In the presence of the combination of charybdotoxin plus apamin or charybdotoxin plus scillatoxin, the effect of acetylcholine was abolished. The combination of iberiotoxin plus apamin was ineffective. 4-Aminopyridine alone induced a concentration-dependent inhibition of the hyperpolarization. Iberiotoxin, apamin and 4-aminopyridine-sensitive potassium currents could be recorded with the patch clamp technique. A glibenclamide-sensitive current was observed but only following stimulation with cromakalim. Addition of charybdotoxin after treatment with the combination of iberiotoxin plus apamin did not produce a further inhibition of the global potassium current. 4-Aminopyridine induced a concentration-dependent inhibition of the voltage-activated potassium channels in the same range of concentrations required for the inhibition of acetylcholine-induced hyperpolarization. These findings suggest that the potassium channel(s) involved in the endothelium-dependent hyperpolarization of the guinea-pig carotid artery do not correspond to an identified subtype. Neither ATP-sensitive nor large conductance calcium-sensitive channels seem to be involved. However the participation of voltage-activated potassium channels and small conductance calcium-sensitive channels or the association of these two channels should be considered.

KEYWORDS: potassium channels, 4-aminopyridine, apamin, charybdotoxin, iberiotoxin, electrophysiology

INTRODUCTION

Hyperpolarization of vascular smooth muscle cells mediated by EDHF can be attributed to an increase in cell membrane conductance to potassium ions as it is associated with an increased potassium efflux (Chen *et al.*, 1988). Potassium channels are involved in this mechanism as non selective potassium channel inhibitors (*e.g.* tetraethylamonium) prevent the hyperpolarization (Chen *et al.*, 1991). The exact nature of the potassium channel activated by EDHF is not clear and may be dependent of the vascular bed studied. ATP-sensitive potassium current (K_{ATP}) are not involved in the EDHF-mediated hyperpolarization (Vanhoutte and Félétou, 1996). However, activation of large (BK_{Ca}) or small (SK_{Ca}) conductance calcium-activated potassium currents (Cowan *et al.*, 1993; Murphy and Brayden, 1995b), or activation of an unknown conductance sensitive to the combination of apamin plus charybdotoxin have been described in various vascular beds (Chataigneau *et al.*, 1998a; Corriu *et al.*, 1996b; Zygmunt and Högestätt, 1996b). The present experiments were designed to define the different types of potassium channels present in isolated smooth muscle cells from the guinea pig carotid artery and, if possible, to identify the potassium channels involved in the N^{ω}-nitro-L-arginine/indomethacin-resistant endothelium-dependent hyperpolarizations produced by acetylcholine.

MATERIALS AND METHODS

Microelectrode Studies

Male Hartley guinea-pigs were anesthetized and the internal carotid arteries were dissected, cleaned of adherent connective tissues and pinned to the bottom of an organ chamber. The tissues were superfused with a thermostated modified Krebs-Ringer bicarbonate solution of the following composition (in mM)): NaCl 118.3, KCl 4.7, $CaCl_2$ 2.5, $MgSO_4$ 1.2, KH_2PO_4 1.2, $NaHCO_3$ 25, EDTA 0.026 and glucose 11.1. Transmembrane potential was recorded with glass capillary microelectrodes (tip resistance of 30 to 90 MΩ) filled with KCl (3 M). Impalements of the smooth muscle cell were performed from the adventitial side. All the experiments were performed in the presence of N^{ω}-nitro-L-arginine (10^{-4} M) and indomethacin (5×10^{-6} M) to inhibit nitric oxide synthase and cyclooxygenase, respectively (Corriu *et al.*, 1996a).

Patch-clamp Studies

The media of the guinea-pig carotid artery was dissected from cleaned arteries. The smooth muscle cells were dissociated enzymatically (Quignard *et al.*, 1998). Whole-cell potassium currents were recorded at room temperature using the patch-clamp technique. In whole-cell experiments, to record voltage sensitive potassium current (K_V) an intracellular calcium-free solution was used with the following composition (in mM): KCl 130, $MgCl_2$ 2, ATP 3, GTP 0.5, HEPES 25, EGTA 10, Glucose 11. To record more accurately K_{Ca} currents, the concentration of EGTA was reduced to 1 mM and $CaCl_2$ (0.5 mM) was added. To record K_{ATP} current, ATP was reduced to 0.1 mM, in the presence of oxaloacetic (5 mM) acid pyruvic acid (2 mM). The cells were superfused with a solution containing (in mM): NaCl 125, KCl 5, $CaCl_2$ 2, $MgCl_2$ 1.2, HEPES 10 and glucose 11. Data were recorded with pClamp6 software (Axon Instruments, USA) through a RK-400 amplifier (Biologic, France).

Statistics

Data are shown as mean ± s.e. mean; n indicates the number of cells in which membrane potential was recorded. Statistical analysis was performed using Student's *t*-test for paired or unpaired observations. Differences were considered to be statistically significant when P value was less than 0.05.

RESULTS

Intracellular Microelectrode

In isolated guinea-pig carotid arteries, experiments were performed in the presence of both N^{ω}-nitro-L-arginine (10^{-4} M) and indomethacin (5×10^{-6} M). Acetylcholine (10^{-6} M) induced an endothelium-dependent hyperpolarization (-18.1 ± 1.1 mV, n = 35; Figure 23-1). In the presence of potassium (35mM) in the superfusing solution, the cell membrane was depolarized to -32.8 ± 1.9 mV (n = 9) and acetylcholine did not produce any detectable changes in membrane potential (n = 3). By contrast, when potassium was removed from the solution, acetylcholine induced a significantly larger hyperpolarization (-37 ± 3 mV; n = 3).

Apamin (5×10^{-7} M), scillatoxin (5×10^{-8} M), charybdotoxin (10^{-7} M), iberiotoxin (10^{-7} M) or glibenclamide (10^{-6} M) induced no or minor changes in resting membrane potential and did not modify the acetylcholine-induced hyperpolarization (Figures 23-1 and 23-2). The combination of charybdotoxin plus apamin or charybdotoxin plus scillatoxin induced a small but significant depolarization of the resting cell membrane (nearly 5 mV) and the hyperpolarization induced by acetylcholine was abolished or converted to a depolarization (Figures 23-1 and 23-2). By contrast, the combination of iberiotoxin plus apamin did not affect the acetylcholine-induced hyperpolarization. 4-Aminopyridine (up to 5×10^{-3} M) by itself did not affect significantly the resting membrane potential but inhibited, in a concentration-dependent manner, the acetylcholine-induced hyperpolarization (Figure 23-3). The addition of charybdotoxin or apamin to 4-aminopyridine did not enhance the inhibitory effect of 4-aminopyridine alone. 4-Aminopyridine did not impair cromakalim-induced hyperpolarizations (data not shown).

Patch-Clamp

The cell capacity of carotid arterial myocytes was 28 ± 3 pA/pF (n = 40). Different types of potassium currents were recorded. With a holding potential of 0 mV (to inactivate K_V), and with a high concentration of intracellular calcium (5×10^{-7} M), depolarization of the myocytes induced a noisy outward current. The current activated at 10 mV and its density at +60 mV was 14 ± 2 pA/pF (n = 20, Figure 23-4). This current was inhibited partially by charybdotoxin (10^{-7} M) or iberiotoxin (10^{-7} M). The iberiotoxin-sensitive current was present in all cells and its density was 11 ± 2 pA/pF (n = 10). An apamin-sensitive potassium current could also be observed in 50% of the cells (10 out of 20, current density at 60 mV: 2 ± 2 pA/pF). In the presence of iberiotoxin to block BK_{Ca}, the apamin-sensitive current could still be recorded (1.8 ± 2 pA/pF, n = 6). In the presence of the combination of iberiotoxin plus apamin, charybdotoxin did not produce further inhibition (Figure 23-4).

With a holding potential of -100 mV, and with a low intracellular calcium, step depolarizations induced a slow inactivating current. The activation threshold of this current was -30 mV and the current density was 4 ± 1.5 pA/pF (n = 20) for a depolarization to

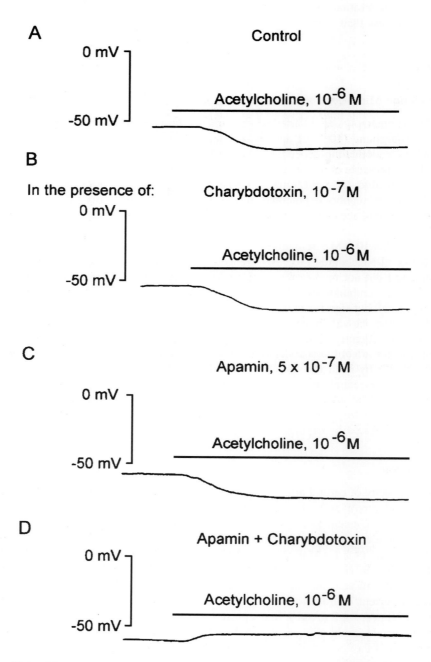

Figure 23-1. Effects of potassium channel blockers on the endothelium-dependent hyperpolarization evoked by acetylcholine (10^{-6} M), in the presence of N^{ω}-nitro-L-arginine (10^{-4} M) and indomethacin (5×10^{-6} M), in the guinea-pig isolated internal carotid artery. A) control. B) in the presence of charybdotoxin (10^{-7} M). C) in the presence of apamin (5×10^{-7} M) and D) in the presence of the combination of charybdotoxin plus apamin.

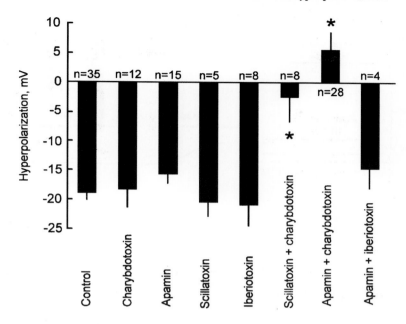

Figure 23-2. Effects of potassium channel blockers (apamin 5×10^{-7} M, charybdotoxin 10^{-7} M, scillatoxin 5×10^{-8} M, iberiotoxin 10^{-6} M) on the endothelium-dependent hyperpolarization evoked by acetylcholine (10^{-6} M), in the guinea-pig isolated internal carotid artery [presence of N^{ω}-nitro-L-arginine (10^{-4} M) and indomethacin (5×10^{-6} M)]. Data are shown as mean ± s.e.m. For the sake of clarity, the control values have been pooled. The asterisk indicates a statistically significant difference with control values ($P < 0.05$); n represents the number of cells studied.

20 mV. 4-Aminopyridine inhibited this current in a concentration-dependent manner (Figure 23-3). At higher depolarizations, an iberiotoxin-sensitive current could be recorded (data not shown).

With a low intracellular concentration of ATP, cromakalin (10^{-5} M) activated a potassium current. This current reversed near the equilibrium potential for potassium ions (-74 ± 5 mV, n = 5), and its density was 1 ± 0.2 pA/pF (n = 14). It was inhibited by glibenclamide (10^{-6} M, percent of inhibition: $90\% \pm 5$, n = 7).

An inwardly rectifying potassium current (K_{IR}) was recorded only in a few cells (3 out of 12). This current reversed at the equilibrium potential for potassium ions and was abolished by barium (5×10^{-5} M).

DISCUSSION

The present experiments indicate that the smooth muscle cells of the guinea-pig carotid artery express different types of functional potassium channels including a K_{Ca} channel inhibited by iberiotoxin and apamin, a K_V channel inhibited by 4-aminopyridine, and a K_{ATP} inhibited by glibenclamide.

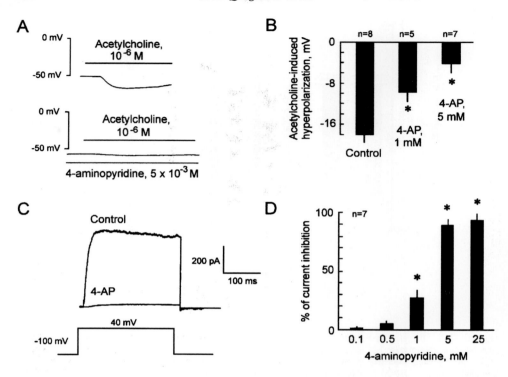

Figure 23-3. Effects of 4-aminopyridine (4-AP) on the endothelium-dependent hyperpolarization evoked by acetylcholine in the guinea-pig isolated internal carotid artery [presence of N^{ω}-nitro-L-arginine (10^{-4} M) and indomethacin (5×10^{-6} M)] and on the whole-cell potassium currents in isolated smooth muscle cells of the guinea-pig carotid artery. A) Effects of 4-aminopyridine (4-AP) on the endothelium-dependent hyperpolarization evoked by acetylcholine (10^{-6} M). B) Hyperpolarization (mV) shown as mean ± s.e.m. C) Effects of 4-aminopyridine on whole-cell potassium current under calcium free intracellular solution conditions (holding potential: -100 mV, step depolarization to 40 mV). D) Concentration-dependent inhibition of the potassium current produced by 4-aminopyridine (in percentage, shown as mean ± s.e.m). The asterisk indicates a statistical difference with control values ($P < 0.05$); n indicates the number of cells studied.

Two different types of K_{Ca} were observed in the present studies. The iberiotoxin-sensitive current shared the characteristic of the BK_{Ca} current (Kuriyama et al., 1995). Unitary potassium channels of large conductance were recorded in these cells (see Quignard et al., Chapter 22 this monograph). A charybdotoxin-sensitive current was recorded in the isolated smooth muscle cells. As this current was blocked by iberiotoxin, it also used BK_{Ca} channel. The apamin-sensitive current does not appear to be a classical SK_{Ca} current. In the present study, the current observed is an outward voltage-dependent current (threshold: -20 mV, unpublished observations) whereas classical SK_{Ca} are voltage-independent (Latorre et al., 1989). However, the current recorded in the present experiments is similar to the apamin-sensitive current recorded in renal arterioles of the rat (Gebremedhin et al., 1996). Under basal conditions, in isolated smooth muscle cells from the guinea-pig carotid artery, a specific charybdotoxin-apamin-sensitive current could not be recorded, confirming previous work in isolated smooth muscle cells from the rat hepatic artery (Zygmunt et al.,

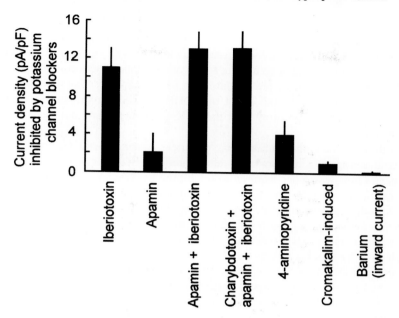

Figure 23-4.   Effects of potassium channel blockers on the whole-cell potassium current of isolated vascular smooth muscle cells from the guinea-pig carotid artery. Data are shown as mean current density sensitive to potassium channel blockers ± s.e.m. (iberiotoxin: 10^{-7} M, apamin: 5×10^{-7} M, charybdotoxin: 10^{-7} M, 4 aminopyridine: 5×10^{-3} M, cromakalim: 10^{-5} M, barium: 5×10^{-5} M).

1997a). A possible explanation is that this current is activated by EDHF and observed only in its presence. Alternatively, the combination of charybdotoxin plus apamin does not exert its effects at the level of vascular smooth muscle cell. K_{Ca} conductances have been also described in endothelial cells (Kuriyama *et al.*, 1995).

Two main classes of K_V are described in vascular smooth muscle cells: a rapidly (K_A) and a slowly inactivating current ($K_{V(s)}$; for review see Kuriyama *et al.*, 1995). In the guinea-pig carotid, none of the cells displayed a K_A current and only a $K_{V(s)}$ current was observed. This current is inhibited by 4-aminopyridine, a selective inhibitor of K_V (for review Faraci and Heistad, 1998), and is inactivated at a holding potential of 0 mV (unpublished observations), a characteristic of K_V channels. A K_{ATP} current (see also Quignard *et al.*, Chapter 22 same book) and a K_{IR} current are also present in the myocytes. In these cells, the K_{IR} current does not seem to be an important conductance in contrast to other blood vessels such as small porcine coronary arteries (for review see Quale *et al.*, 1997).

In the guinea pig carotid artery, acetylcholine induces an endothelium-dependent hyperpolarization attributed to EDHF. As glibenclamide and barium (Quignard *et al.*, 1999) do not inhibit the EDHF-mediated hyperpolarization, K_{ATP} and inwardly rectifying potassium channels, although present in the vascular smooth muscle cells, are probably not involved in the hyperpolarization. The inhibition of EDHF-induced hyperpolarization by the combination of apamin plus charybdotoxin or scillatoxin plus charybdotoxin suggests the role of a specific channel sensitive to the combination of the two toxins. Alternatively, two different potassium channels, each sensitive to one toxin, could be involved. As apamin and scillatoxin are two structurally different blockers of SK_{Ca} channel, this channel, or a subunit of this channel must contribute to the acetylcholine-induced hyperpolarization. Charybdotoxin and iberiotoxin are inhibitors of BK_{Ca} channels. Charybdotoxin, in association with apamin, reverse the effect of EDHF, but iberiotoxin, in association with apamin, has no effect, pointing out that BK_{Ca} channels are not involved in the hyperpolarization.

Charybdotoxin can also inhibit K_V channels (Chandy and Gutman, 1995). 4-Aminopyridine, a specific inhibitor of K_V, also inhibits the EDHF-induced hyperpolarization. The concentrations which are required to inhibit the hyperpolarization, are the same that the one required to inhibit the K_V currents recorded with the patch-clamp technique. Endothelium-dependent responses attributed to EDHF, sensitive to K_V inhibitors, have been described in other vascular beds. Thus, in porcine coronary arteries, 4 aminopyridine shortens the duration of EDHF mediated relaxations (Shimizu and Paul, 1998). Ciclazindol, a non-selective K_V blocker, inhibits partially the EDHF-induced hyperpolarization in the hepatic artery of the rat (Zygmunt *et al.*, 1997a). In association with apamin, ciclazindol abolishes the EDHF-mediated hyperpolarization, mimicking the effect of the combination of apamin plus charybdotoxin. Taken in conjunction, these observations could indicate that K_V channels, in association with SK_{Ca} channels, are involved.

However, the involvement of K_V channels sensitive to either 4-aminopyridine, ciclazindol or charybdotoxin is unlikely. In the carotid artery of the guinea-pig, 4-aminopyridine inhibited the endothelium-dependent hyperpolarization by itself while charybdotoxin alone was ineffective. The effects of 4-aminopyridine were not potentiated by apamin. Furthermore, although charybdotoxin may inhibit K_V currents (Chandy and Gutman, 1995), in the isolated smooth muscle cells of the guinea-pig carotid artery, at the concentration tested, 4-aminopyridine-sensitive currents were not influenced by charybdotoxin. Similarly, in the hepatic artery of the rat, ciclazindol did not inhibit K_V currents recorded with the patch-clamp technique (Zygmunt *et al.*, 1997a). Thus, it appears that 4-aminopyridine, ciclazindol and charybdotoxin are not acting on the same target. The present experiments do not permit to speculate further on the respective locations of the targets of charybdotoxin, apamin and 4-aminopyridine.

24 Potassium Channels Involved in Hyperpolarization and Relaxation induced by Endothelium-derived Hyperpolarizing Factor in Rat Mesenteric Arteries

Donald W. Cheung, Guifa Chen, Ethan Burnette and Guangdi Li

University of Ottawa Heart Institute, 40 Ruskin Street, Ottawa, K1Y 4W7, Canada.
Tel: (613) 761-5022. Email: dcheung@heartinst.on.ca.

The involvement of Ca^{2+}-activated K^+ channels has been implicated in the hyperpolarization induced by endothelium-derived hyperpolarizing factor (EDHF) since charybdotoxin and/or apamin are effective inhibitors of the response. In some arteries, 4-aminopyridine, a K^+ channel blocker better known for its action on delayed rectifier channels, is also effective. The effect of 4-aminopyridine on EDHF-induced hyperpolarizations and relaxations in the rat mesenteric artery was investigated by measuring the membrane potential and tension simultaneously. 4-Aminopyridine was effective in inhibiting the EDHF responses. To determine if 4-aminopyridine could be acting on Ca^{2+}-activated K^+ channels, patch-clamp recordings were made with freshly isolated cells from the same artery. 4-Aminopyridine inhibited whole-cell outward currents that were also sensitive to charybdotoxin. Single channel recordings from inside-out patches revealed two populations of Ca^{2+}-activated K^+ channels with relatively large conductances of 197 pS and 94 pS. These channels were inhibited by both charybdotoxin and 4-aminopyridine. Thus 4-aminopyridine's inhibition on Ca^{2+}-activated K^+ channels could in part account for its effect on the EDHF responses.

KEYWORDS: EDHF, 4-aminopyridine, Ca^{2+}-activated K^+ channels, vascular smooth muscle

INTRODUCTION

Endothelium-derived hyperpolarizing factor (EDHF) induces vasodilatation by hyperpolarizing vascular smooth muscle cells. Activation of K^+ channels is the most likely mechanism mediating this process. Thus it can be blocked by K^+ channel blockers such as tetraethylammonium and its amplitude is dependent on the K^+ concentration gradient (Chen, Suzuki and Weston, 1988). When more specific K^+ channel blockers were used to determine the nature of the K^+ channels being activated, the hyperpolarizations and the corresponding relaxations were inhibited to various extents either by charybdotoxin, a blocker of large conductance Ca^{2+}-activated K^+ channels (Chen and Cheung, 1997; Hashitani and Suzuki, 1997), or by apamin, a blocker of small conductance Ca^{2+}-activated K^+ channels (Murphy and Brayden, 1995b; Chen and Cheung, 1997; Hashitani and Suzuki, 1997). In the guinea-pig carotid artery, charybdotoxin and apamin in combination, but not alone were also effective (Corriu *et al.*, 1996b). These studies indicate that Ca^{2+}-activated K^+ channels are involved in EDHF-induced hyperpolarizations.

4-Aminopyridine ($<2 \times 10^{-3}$ M), a blocker generally recognized as more selective for delayed rectifier K^+ channels (Beech and Bolton, 1989; Bolzon, Xiong and Cheung, 1993), has also been tested. 4-Aminopyridine was not effective in inhibiting EDHF-induced hyperpolarizations in the rabbit mesenteric artery (Murphy and Brayden, 1995b) or guinea-pig submucosal arteriole (Hashitani and Suzuki, 1997). However, tension recordings from some other arteries indicate that 4-aminopyridine can inhibit the supposedly EDHF-component of the relaxation (Kitagawa *et al.*, 1994; Alonso *et al.*, 1993; Wu, Chen and Yen, 1997). It was not known in these cases whether or not the inhibition by 4-aminopyridine is due to a reduction in hyperpolarization, whether 4-aminopyridine is acting on the delayed rectifier or on Ca^{2+}-activated K^+ channels.

METHODS

Simultaneous Recording of Membrane Potential and Tension

Main superior mesenteric arteries were isolated from 10–12 week old male Wistar rats (Charles River) and immersed in physiological salt solution having the following composition ($\times 10^{-3}$ M): 120 NaCl, 25 $NaHCO_3$, 5 KCl, 2.5 $CaCl_2$, 1 NaH_2PO_4, 1 $MgSO_4$, and 11 glucose. Solutions were bubbled constantly with 5% CO_2–95%O_2, and maintained at 36°C. Rings (4 mm in length) were suspended by two tungsten wires and tension was recorded by a Narco F60 transducer. Glass microelectrodes of 40–60 Mohm resistance filled with 3M KCl were used for membrane potential recording (Chen and Cheung, 1996; 1997).

Patch-clamp Recordings

Conventional whole-cell and single channel recordings were made from freshly isolated cells of the rat mesenteric artery. Single cells were obtained from the arteries after treatment with collagenase (0.02%) and papain (0.1%) for two hours. The ionic composition of the extracellular solution was ($\times 10^{-3}$ M): NaCl 137, KCl 5.5, $CaCl_2$ 1.8, $MgCl_2$ 1.8, KH_2PO_4 0.4, $NaHCO_3$ 4.2, glucose 5.6, and the pH was buffered to 7.4 with 10 mM HEPES. The ionic composition for the intracellular solution was ($\times 10^{-3}$ M): KCl 150, $MgCl_2$ 1.0, Na_2 ATP 1.0, and the pH buffered to 7.2 with 10 mM HEPES. Free Ca^{2+} was adjusted to 1.55×10^{-7} M with EGTA. For single channel recording of inside-out patches, symmetrical K^+ concentrations of 1.5×10^{-1} M were used and the cytoplasmic Ca^{2+} concentration was adjusted to 5×10^{-7} M (Bolzon, Xiong and Cheung, 1993).

Results are expressed as means ± SE. Analysis of variance for repeated measures was used for statistical comparison. When a significance ($P < 0.05$) was indicated by the Bonferroni test, a paired Student's t-test was performed to compare responses at each acetylcholine concentrations.

RESULTS

4-Aminopyridine

The effects of 4-aminopyridine were determined by recording simultaneously the membrane potential and tension. 4-Aminopyridine ($1–2 \times 10^{-3}$ M) significantly depolarized the

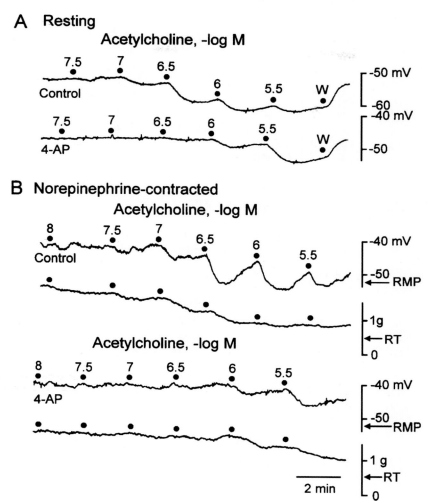

Figure 24-1.   A. Effect of 4-aminopyridine (4-AP, 1×10^{-3} M) on hyperpolarization induced by acetylcholine (ACh) in the rat mesenteric artery under resting conditions. B. Effect of 4-aminopyridine on the hyperpolarization (top traces) and relaxation (bottom traces) induced by acetylcholine in a rat mesenteric artery contracted by norepinephrine. RMP = resting membrane potential; RT = resting tension.

arteries from a resting membrane potential of -50.3 ± 0.5 mV to -45.8 ± 0.8 mV (n = 9). Acetylcholine caused dose-dependent hyperpolarization of the arteries in the presence of N^{ω}-nitro-L-arginine (3×10^{-5} M). The hyperpolarization was significantly reduced by 4-aminopyridine at all acetylcholine concentrations (Figure 24-1). At maximal stimulation with acetylcholine (3×10^{-6} M), the hyperpolarization was reduced significantly from 11.2 ± 0.6 mV to 7.8 ± 0.7 mV. The inhibition was more evident at lower concentrations of acetylcholine. Thus at 3×10^{-7}M acetylcholine, the reduction was from 7.7 ± 0.6 mV to 1.2 ± 0.2 mV.

In arteries contracted with norepinehrine, acetylcholine induced N^{ω}-nitro-L-arginine-resistant relaxation that could be correlated to membrane hyperpolarization (Chen and

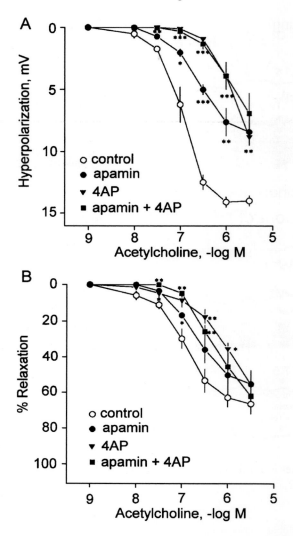

Figure 24-2. A. Effects of 4-aminopyridine (4AP, $1–2 \times 10^{-3}$ M), apamin (3×10^{-7} M), and a combination of 4AP and apamin on the hyperpolarization (A) and relaxation (B) induced by acetylcholine in rat mesenteric arteries contracted by norepinephrine. The asterisks indicate statistically significant ($P < 0.05$, n = 5–9) differences from control.

Cheung, 1997). 4-aminopyridine ($1–2 \times 10^{-3}$ M) inhibited both the hyperpolarization and relaxation (Figure 24-1). At 0.3 and 3×10^{-6} M of acetylcholine, the hyperpolarizations were reduced significantly to 0.9 ± 0.1 mV and 8.8 ± 0.8 mV from control values of 12.1 ± 0.9 mV and 13.4 ± 0.4 mV, respectively (n = 5). Similarly, the relaxation was reduced to $17.6 \pm 0.2\%$ and $55.4 \pm 3.5\%$ of the norepinephrine-induced tone from control values of $53.0 \pm 5.1\%$ and $61.3 \pm 0.9\%$, respectively (Figure 24-2). 4-Aminopyridine when used in combination with either charybdotoxin (1.5×10^{-7} M) or apamin (1.0×10^{-7} M)

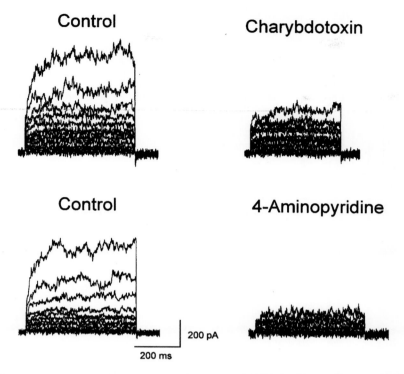

Figure 24-3. Effects of charybdotoxin (1.5×10^{-7} M) and 4-aminopyridine (2×10^{-3} M) on whole-cell currents recorded from freshly isolated cells of the rat mesenteric artery. Outward currents were elicited by depolarizing pulses from −80 mV to +60 mV. Holding potential = −80 mV.

did not cause further inhibition of the hyperpolarization or the relaxation than when given alone (Figure 24-2).

K⁺ Channel Activity

Patch-clamp recordings of whole-cell and single channel activities were made from freshly isolated mesenteric arterial cells. Whole-cell recordings showed that a large component of the outward currents could be blocked by both charybdotoxin (1×10^{-7} M) and 4-aminopyridine (1×10^{-3} M) (Figure 24-3). These currents were also blocked by tetraethylammonium (2×10^{-3} M) and were oscillatory. Single channel recording from inside-out patches showed two types of Ca²⁺-activated K⁺ channels with relatively large conductances of 197 pS and 94 pS in symmetrical K⁺ (1.5×10^{-1} M) solutions (Figure 24-4). These channels were sensitive to cytoplasmic Ca²⁺ concentrations (data not shown) and inhibited by charybdotoxin (1.5×10^{-7} M) (Figure 24-4). They were also effectively inhibited by 4-aminopyridine ($1–2 \times 10^{-3}$ M) (Figure 24-4). 4-Aminopyridine reduced significantly the probability of opening of the 197 pS channel from 0.048 to 0.003, and 0.014 to 0.002 for the 94 pS channel.

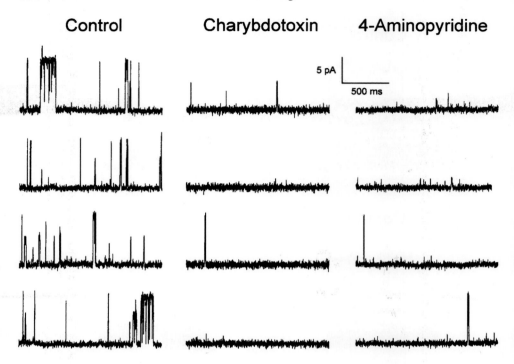

Figure 24-4. Effects of charybdotoxin (1.5×10^{-7} M) and 4-aminopyridine (2×10^{-3} M) on single channel current activities recorded in inside-out patches from freshly isolated cells of the rat mesenteric artery. Holding potential = +40 mV, $[Ca^{2+}]_{in} = 5 \times 10^{-7}$ M, $[K^+]_{in} = [K^+]_{out} = 1.5 \times 10^{-1}$ M.

DISCUSSION

EDHF-induced hyperpolarization has been attributed to be mediated by Ca^{2+}-activated K^+ channels. Thus blockers of Ca^{2+}-activated K^+ channels such as charybdotoxin and apamin are effective in inhibiting EDHF responses (Chen and Cheung, 1997; Hashitani and Suzuki, 1997). The combination of apamin and charybdotoxin was more effective than when either of the blockers was used alone (Chen and Cheung, 1997). The present study examined the mechanism of action of 4-aminopyridine on EDHF responses in rat mesenteric arteries. The inhibition of the relaxation by 4-aminopyridine was correlated to a reduction in the hyperpolarization. Patch clamp studies revealed that 4-aminopyridine could inhibit Ca^{2+}-activated K^+ channels, in addition to its recognized action on delayed rectifier K^+ channels.

There are similarities and differences in the effect of 4-aminopyridine on the rat mesenteric artery in comparison to other Ca^{2+}-activated K^+ channel blockers. The depolarization of the resting cell membrane caused by 4-aminopyridine was comparable to that induced by charybdotoxin and tetraethylammonium. This contrasts with apamin's lack of effect on the resting membrane potential (Chen and Cheung, 1997). 4-aminopyridine was more effective than charybdotoxin or apamin in inhibiting EDHF-induced hyperpolarization and relaxation (Chen and Cheung, 1997). Whereas charybdotoxin and apamin in combination were most effective (Corriu *et al.*, 1996; Chen and Cheung, 1997), addition of charybdotoxin or apamin produced no further effects than those obtained by 4-aminopyridine alone. Therefore, 4-aminopyridine possesses properties distinct from the other Ca^{2+}-activated K^+ channel blockers.

Whole-cell recordings of vascular smooth muscle cells usually reveal two major outward K^+ currents mediated by delayed rectifier and Ca^{2+}-activated K^+ channels (Beech and Bolton, 1989; Bolzon, Xiong and Cheung, 1993). These two channels are distinguishable by their sensitivity to 4-aminopyridine and charybdotoxin. There exist multiple charybdotoxin-sensitive Ca^{2+}-activated K^+ channels. For example, channels of 68 pS and 251 pS conductances are present in vascular smooth muscle cells from rat renal arterioles (Gebremedhin *et al.*, 1996). In the present study, two Ca^{2+}-activated K^+ channels of relatively large conductance were observed in smooth muscle cells of the rat mesenteric artery. These Ca^{2+}-activated K^+ channels of 197 pS and 94 pS conductances were sensitive to both charybdotoxin and 4-aminopyridine. The properties of these large conductance Ca^{2+}-activated K^+ channels in the rat mesenteric artery differ from those of the rat tail artery (Bolzon, Xiong and Cheung, 1993) or the rabbit portal vein (Beech and Bolton, 1989) which are not sensitive to 4-aminopyridine. In the rat tail artery, 4-aminopyridine inhibits only a small conductance delayed rectifier channel which is activated at a lower membrane potential threshold (Bolzon, Xiong and Cheung, 1993).

CONCLUSION

In rat mesenteric arteries, 4-aminopyridine inhibited EDHF-induced relaxations by reducing the hyperpolarization. In this artery, two Ca^{2+}-activated K^+ channels that are sensitive to 4-aminopyridine were identified. Thus the inhibition of EDHF-responses by 4-aminopyridine in the rat mesenteric artery and possibly other arteries could be accounted for, at least in part, by its effect on Ca^{2+}-activated K^+ channels. There are differences in the sensitivity of Ca^{2+}-activated K^+ channels to 4-aminopyridine amongst different vascular preparations.

ACKNOWLEDGEMENTS

This work was supported by the Heart and Stroke Foundation of Ontario.

25 Pharmacological Characterization of the Potassium Channels Targetted by Endothelium-Derived Hyperpolarizing Factor

Edward D. Högestätt, Jesper Petersson, David A. Andersson, Peter M. Zygmunt

Department of Clinical Pharmacology, Institute of Laboratory Medicine, University of Lund, Lund, Sweden
Address for correspondence: Edward D Högestätt; Phone: #46 46 17 33 59; Fax: #46 46 211 19 87; E-mail: Edward.Hogestatt@klinfarm.lu.se
Jesper Petersson; Phone: #46 40 33 10 00; Fax: #46 46 211 19 87; E-mail: Jesper.Petersson@neurolmas.lu.se
David A Andersson; Phone: #46 46 17 36 79; Fax: #46 46 211 19 87; E-mail: David.Andersson@klinfarm.lu.se
Peter M Zygmunt; Phone: #46 46 17 33 59; Fax: #46 46 211 19 87; E-mail: Peter.Zygmunt@klinfarm.lu.se

Endothelium-derived hyperpolarizing factor (EDHF) represents a new class of endogenous potassium (K^+) channel openers. However, the identity of the target K^+ channels for EDHF as well as the chemical identity of EDHF have not been established. After inhibition of cyclo-oxygenase with indomethacin and nitric oxide synthase with N^{ω}-nitro-L-arginine, acetylcholine and the calcium ionophore A23187 induce an endothelium-dependent hyperpolarization and relaxation, which are mediated entirely by EDHF in the rat hepatic and guinea-pig basilar artery. The effects of different K^+ channel inhibitors on these responses were examined with conventional tension recording and electrophysiological techniques. Apamin (SK_{Ca} inhibitor), iberiotoxin (BK_{Ca} inhibitor), 4-aminopyridine (inhibitor of K_V) and glibenclamide (K_{ATP} inhibitor) each did not affect the EDHF-mediated relaxation induced by acetylcholine in the rat hepatic and guinea-pig basilar artery. Charybdotoxin (inhibitor of BK_{Ca}, IK_{Ca} and certain K_V) partially inhibited the EDHF-mediated relaxation in the guinea-pig basilar artery, but had no effect in the rat hepatic artery. However, a combination of apamin and charybdotoxin abolished the EDHF-mediated hyperpolarization and relaxation in both arteries. In the rat hepatic artery, iberiotoxin and 4-aminopyridine were unable to replace charybdotoxin in this combination. Similar findings were obtained with iberiotoxin in the guinea-pig basilar artery, whereas a small inhibition of the EDHF-mediated relaxation with a maintained maximal response was observed when 4-aminopyridine was combined with apamin in this artery. Furthermore, apamin plus charybdotoxin did not affect the 4-aminopyridine-sensitive delayed rectifier K^+ current in isolated smooth muscle cells of the rat hepatic artery. The combination of apamin and charybdotoxin did not inhibit the responses to endogenous nitric oxide (NO), released by acetylcholine or A23187, and the endothelium-independent vasodilators 3-morpholino-sydnonimine (NO donor), levcromakalim (K_{ATP} opener), iloprost (prostacyclin mimetic) and anandamide (endocannabinoid) in the rat hepatic artery, indicating that the toxins do not impair the release of endothelium-derived relaxing factors or act as functional antagonists at the level of the smooth muscle cells. The results suggest that the target K^+ channel for EDHF in the rat hepatic and guinea-pig basilar artery is neither BK_{Ca}, K_{ATP} nor K_V. Rather, EDHF may activate a mixture of two different types of K^+ channel in each artery, possibly a "classical" apamin-sensitive SK_{Ca} and a charybdotoxin-sensitive IK_{Ca}.

KEYWORDS: Endothelium-derived hyperpolarizing factor, membrane potential, potassium channels, vascular endothelium

217

INTRODUCTION

Endothelium-derived hyperpolarizing factor (EDHF) is a recently discovered paracrine factor formed by the endothelial cells and of a yet unknown chemical identity and biological function. EDHF produces hyperpolarization and relaxation of vascular smooth muscle in both animal and man, effects which are considered to be mediated by activation of potassium (K^+) channels. Hence, EDHF may represent a new class of endogenous K^+ channel openers. However, the identity of the target K^+ channels for EDHF has not been established and may differ between different blood vessels and animal species.

In the rat hepatic and guinea-pig basilar artery, acetylcholine and the calcium ionophore A23187 induce an endothelium-dependent hyperpolarization and relaxation after inhibition of cyclooxygenase with indomethacin and the nitric oxide (NO) pathway with the NO synthase blocker N^{ω}-nitro-L-arginine, the guanylate cyclase inhibitors 1H-[1,2,4]oxadiazolo [4,3-a]quinoxaline-1-one (ODQ) or methylene blue, or the NO scavenger oxyhaemoglobin (Nishiye *et al.*, 1989; Zygmunt, Waldeck and Högestätt, 1994; Petersson *et al.*, 1998). Moreover, these responses are not associated with an increase of guanosine and adenosine 3':5'-cyclic monophosphates, and therefore considered to be mediated by EDHF alone (Zygmunt, Grundemar and Högestätt, 1994; Petersson *et al.*, 1998).

The present study was undertaken to examine the effects of different K^+ channel inhibitors on EDHF-mediated responses in the rat hepatic and guinea-pig basilar artery with conventional tension recording and electrophysiological techniques. To study EDHF-mediated responses, the synthesis of NO and prostanoids were inhibited with N^{ω}-nitro-L-arginine (3×10^{-4} M) and indomethacin (10^{-5} M).

METHODS

Experimental Procedure

Female Sprague-Dawley rats (200–250 g) and male guinea-pigs (300 g) were killed by CO_2 asphyxia followed by exsanguination. The rat hepatic and guinea-pig basilar artery were removed and divided into ring segments, 1 to 2 mm long, which were suspended between two metal pins in organ baths (2.5 to 5 ml), containing warmed (37°C) physiological salt solution of the following composition (mM): NaCl 119, $NaHCO_3$ 15, KCl 4.6, NaH_2PO_4 1.2, $MgCl_2$ 1.2, $CaCl_2$ 1.5 and (+)-glucose 6.0. The physiological salt solution was continuously bubbled with a mixture of 95% O_2 and 5% CO_2, resulting in a pH of 7.4. During an equilibration period of about one hour, the resting wall tension was adjusted to approximately 2 mN/mm vessel length. Isometric tension was measured by a force-displacement transducer (Grass Instruments FT03 C, USA), connected to a polygraph (for details, see Högestätt, Andersson and Edvinsson, 1983). In order to assess the contractile capacity, each vessel segment was contracted by an isosmolar 60 mM potassium solution (prepared as the physiological salt solution except for an equimolar substitution of NaCl with KCl).

Relaxations induced by acetylcholine, A23187, 3-morpholino-sydnonimine (SIN-1), iloprost, levcromakalim and anandamide were studied in blood vessels contracted by phenylephrine (rat hepatic artery) or prostaglandin $F_{2\alpha}$ (guinea-pig basilar artery). The concentration of phenylephrine or prostaglandin $F_{2\alpha}$ was titrated for each vascular segment to give a contraction equivalent to 70 to 90% of the initial response to the 60 mM potassium solution (Zygmunt, Grundemar and Högestätt, 1994). When stable contractions were obtained, the vasodilators were added cumulatively to determine concentration-response relationships. EDHF-mediated responses were always studied in the presence of N^{ω}-nitro-L-arginine (3×10^{-4} M) and indomethacin (10^{-5} M). The incubation time with the K^+ channel inhibitors was at least 20 min (unless specified otherwise). Each preparation was exposed to only one treatment.

Electrophysiology

Recording of membrane potential was made on intact ring segments of the rat hepatic artery suspended in a myograph at 37°C. Briefly, glass microelectrodes filled with 2 M KCl (80 to 120 MΩ) were advanced from the adventitial side of the artery at resting tension (Zygmunt *et al.*, 1998). Whole cell currents were measured on isolated smooth muscle cells obtained by enzymatic digestion of the rat hepatic artery. Currents were recorded under "calcium-free" conditions in amphotericin B-perforated membrane patches at 25°C by the patch clamp technique (Zygmunt *et al.*, 1997a).

Calculations and Statistics

Relaxations are expressed as the percentage reversal of the contraction induced by phenylephrine or prostaglandin $F_{2\alpha}$. The maximal relaxation induced by each concentration of vasodilator was recorded and used in subsequent calculations. The negative logarithm of the drug concentration eliciting half maximal relaxation (pEC$_{50}$) was determined by linear regression analysis, using the values immediately above and below half maximal response. E_{max} refers to the maximal relaxation achieved (100% denotes a complete reversal of the contraction). Values are presented as mean ± s.e. mean, and n indicates the number of vascular segments (animals) examined. Statistical analysis of pEC$_{50}$ and E_{max} values was performed by using Student's t-test (two-tailed) or multiple analysis of variance (MANOVA) followed by Bonferroni Dunn's *post hoc* test (Statview 4.12). Statistical significance was accepted when $P < 0.05$.

Drugs

Acetylcholine chloride, A23187, 4-aminopyridine, glibenclamide, N^{ω}-nitro-L-arginine and L-phenylephrine hydrochloride (Sigma, St Louis, MO, USA); anandamide (arachidonyl-ethanolamide; RBI, Natick, USA), apamin and iberiotoxin (Alomone labs, Jerusalem, Israel); synthetic charybdotoxin (Latoxan, Rosans, France); indomethacin (Dumex, Co-penhagen, Denmark); iloprost (Schering AG, Berlin, Germany); levcromakalim (SmithKline Beecham, Brentford, UK); 3-morpholino-sydnonimin hydrochloride (Biomol, Plymouth Meeting, PA, USA); Prostaglandin $F_{2\alpha}$ (Pharmacia Upjohn, Puurs, Belgium). A23187 was dissolved in absolute ethanol. Apamin and charybdotoxin were dissolved in saline. All other drugs were dissolved in distilled water. Stock solutions of the substances were stored at −70°C.

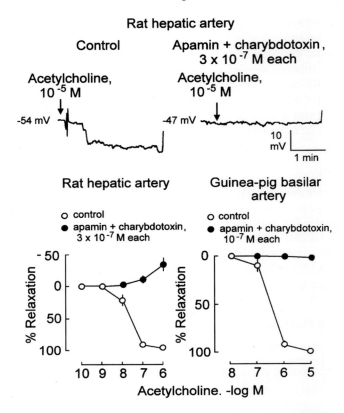

Figure 25-1. Effects of apamin plus charybdotoxin on acetylcholine-induced relaxations in rat hepatic (left, n = 7) and guinea-pig basilar (right, n = 6) arteries contracted by phenylephrine (hepatic) or prostaglandin $F_{2\alpha}$ (basilar). Upper traces show the effect of apamin plus charybdotoxin on EDHF-mediated hyperpolarization induced by acetylcholine in the rat hepatic artery. The membrane potential was recorded with glass microelectrodes filled with 2 M KCl (80 to 120 MΩ) in ring segments suspended in a myograph at 37°C (Zygmunt *et al.*, 1998).

RESULTS

In the presence of the NO synthase inhibitor N^{ω}-nitro-L-arginine (3×10^{-4} M) and the cyclooxygenase inhibitor indomethacin (10^{-5} M), acetylcholine or the calcium ionophore A23187 induced endothelium-dependent hyperpolarizations and relaxations in rat hepatic and guinea-pig basilar arteries contracted with phenylephrine and prostaglandin $F_{2\alpha}$, respectively (Figure 25-1; Zygmunt, Waldeck and Högestätt, 1994; Petersson, Zygmunt and Högestätt, 1997).

The combination of the K^+ channel inhibitors apamin and charybdotoxin abolished the EDHF-mediated hyperpolarization and relaxation in the rat hepatic artery, whereas each toxin alone had no effect (Figure 25-1, Table 25-1). This toxin combination also abolished the EDHF-mediated relaxation in the guinea-pig basilar artery (Figure 25-1). However,

Table 25-1 Effects of K$^+$ channel inhibitors on the acetylcholine-induced relaxation mediated by EDHF

	Control			Treatment		
	n	pEC_{50}	E_{max}	n	pEC_{50}	E_{max}
Rat hepatic artery						
Apamin (10^{-7} M)	6	7.8 ± 0.2	98 ± 1	6	7.6 ± 0.1	99 ± 1
Apamin (10^{-6} M)	6	7.6 ± 0.1	96 ± 1	6	7.6 ± 0.1	95 ± 2
Charybdotoxin (10^{-7} M)	7	7.8 ± 0.2	98 ± 1	7	7.5 ± 0.2	98 ± 1
Charybdotoxin (10^{-6} M)	4	7.5 ± 0.1	92 ± 5	4	7.6 ± 0.3	94 ± 3
Iberiotoxin (10^{-7} M)	4	7.7 ± 0.2	97 ± 3	4	8.0 ± 0.2	100 ± 1
Iberiotoxin (10^{-7} M) + apamin (10^{-7} M)	5	7.7 ± 0.2	97 ± 2	5	8.0 ± 0.1	99 ± 1
4-Aminopyridine (10^{-3} M)	6	7.8 ± 0.2	100 ± 1	6	7.4 ± 0.1	100 ± 1
4-Aminopyridine (10^{-3} M) + apamin (10^{-7} M)	4	7.5 ± 0.1	92 ± 1	4	7.5 ± 0.1	100 ± 1
Glibenclamide (10^{-5} M)	4	7.7 ± 0.2	96 ± 2	4	7.7 ± 0.3	97 ± 2
Guinea-pig basilar artery						
Apamin (10^{-7} M)	6	6.2 ± 0.2	93 ± 6	6	6.2 ± 0.1	95 ± 4
Charybdotoxin (10^{-7} M)	11	6.6 ± 0.1	99 ± 1	11	$4.9 \pm 0.5*$	$55 \pm 12*$
Iberiotoxin (10^{-7} M)	4	6.3 ± 0.2	99 ± 1	4	6.4 ± 0.3	99 ± 1
Iberiotoxin (10^{-7} M) + apamin (10^{-7} M)	5	6.5 ± 0.1	99 ± 1	5	6.4 ± 0.1	99 ± 1
4-Aminopyridine (10^{-3} M)	5	6.4 ± 0.2	99 ± 1	5	6.2 ± 0.2	100 ± 1
4-Aminopyridine (10^{-3} M) + apamin (10^{-7} M)	5	6.5 ± 0.1	96 ± 2	5	$5.5 \pm 0.1*$	98 ± 1
Glibenclamide (10^{-3} M)	4	6.4 ± 0.3	99 ± 1	4	6.6 ± 0.1	99 ± 1

*The asterisks indicate statistically significant (P < 0.05) difference from control

in this artery, charybdotoxin alone partially inhibited the EDHF-induced relaxation, while apamin was inactive (Table 25-1).

In contrast to charybdotoxin, iberiotoxin had no effect alone or when combined with apamin on the EDHF-mediated relaxation in both rat hepatic and guinea-pig basilar arteries (Table 25-1). Although both charybdotoxin and iberiotoxin were able to enhance phenylephrine-induced contractions in the presence of apamin, only the combination of apamin and charybdotoxin prevented the acetylcholine-induced relaxation mediated by EDHF in the rat hepatic artery (Figure 25-2). Glibenclamide was also without effect on EDHF-mediated responses in rat hepatic and guinea-pig basilar arteries (Table 25-1). The relaxation elicited by levcromakalim (K$_{ATP}$ opener) in the rat hepatic artery was unaffected by the combination of apamin and charybdotoxin (Figure 25-2, Table 25-2).

Apamin plus charybdotoxin did not inhibit the responses to endogenous NO, released by acetylcholine or A23187 in the presence of indomethacin alone, or to the endothelium-independent vasodilators SIN-1, iloprost, levcromakalim and anandamide in the rat hepatic artery (Table 25-2).

4-Aminopyridine (10^{-3} M) had no effect alone or when combined with apamin on the EDHF-mediated relaxation in the rat hepatic artery (Table 25-1). 4-Aminopyridine also did not affect this response in the guinea-pig basilar artery, but when combined with apamin, a small rightward shift of the acetylcholine concentration-response curve with a maintained E$_{max}$ was observed (Table 25-1). In isolated smooth muscle cells of the rat hepatic artery, apamin plus charybdotoxin did not affect the 4-aminopyridine-sensitive delayed rectifier K$^+$ current (Figure 25-3).

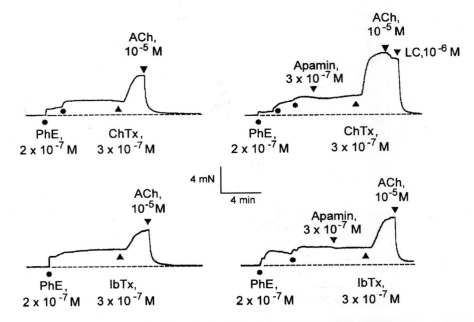

Figure 25-2. Traces showing the effects of charybdotoxin (ChTx) and iberiotoxin (IbTx) in the presence or absence of apamin on the acetylcholine (ACh)-induced relaxation mediated by EDHF in rat hepatic arteries contracted with phenylephrine (PhE). Levcromakalim (LC) was still able to induce a complete relaxation in the presence of apamin plus charybdotoxin when the acetylcholine-induced response was suppressed. Dashed lines indicate the baseline tension.

Table 25-2 Relaxation induced by acetylcholine, A23187, 3-morpholino-sydnonimin, iloprost, levcromakalim and anandamide in the absence and presence of apamin (3×10^{-7} M) plus charybdotoxin (3×10^{-7} M) in the rat hepatic artery

Agonist	Control			Apamin + charybdotoxin		
	n	pEC_{50}	E_{max} (%)	n	pEC_{50}	E_{max} (%)
Acetylcholine[a]	6	8.1 ± 0.1	100 ± 1	6	8.0 ± 0.1	99 ± 1
A23187[a]	6	5.8 ± 0.1	98 ± 1	10	6.0 ± 0.1	98 ± 1
SIN-1[b]	7	6.7 ± 0.2	99 ± 1	7	6.4 ± 0.2	99 ± 1
Iloprost[b]	5	7.1 ± 0.1	83 ± 7	5	6.9 ± 0.2	94 ± 2
Levcromakalim[b]	5	7.4 ± 0.1	99 ± 1	5	7.4 ± 0.1	100 ± 1
Anandamide[b]		6.1 ± 0.2	97 ± 2	8	5.9 ± 0.2	88 ± 10

[a]In the presence of indomethacin (10^{-5} M)
[b]In the presence of N^ω-nitro-L-arginine (3×10^{-4} M) and indomethacin (10^{-5} M)

DISCUSSION

Is The Target For EDHF a Single Type of K⁺ Channel?

There is a large controversy regarding which type of K⁺ channel is the target for EDHF. Some of this controversy could reside in the fact that EDHF may represent a family of compounds, each acting at different K⁺ channels and having a unique distribution in the

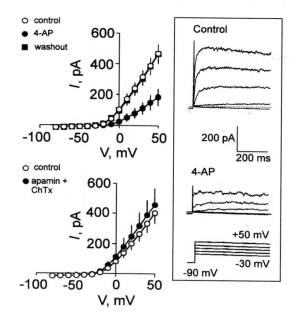

Figure 25-3.   Effect of 3×10^{-3} M 4-aminopyridine (4-AP; upper panel) and 3×10^{-7} M apamin plus 3×10^{-7} M charybdotoxin (ChTx; lower panel) on whole cell currents (I) under "calcium-free" conditions in isolated smooth muscle cells of the rat hepatic artery (n = 4–5). Currents at each test potential (V) were generated by stepping from a holding potential of –90 mV. Inset shows current traces before (upper) and after (middle) exposure to 4-AP, and the voltage protocol (lower). Dashed lines indicate zero current level.

vascular system. In a first attempt to resolve the identity of the target K⁺ channel for EDHF, it was hypothesized that EDHF is acting on a single type of K⁺ channel. The present study clearly rules out several types of K⁺ channels, since apamin, iberiotoxin, 4-aminopyridine and glibenclamide each did not affect the EDHF-mediated relaxation in the rat hepatic and guinea-pig basilar artery. These compounds are considered to selectively inhibit small (apamin) and large (iberiotoxin) conductance calcium-sensitive, voltage-dependent (4-aminopyridine) and ATP-sensitive (glibenclamide) K⁺ channels. However, charybdotoxin was able to partially inhibit the EDHF-mediated relaxation in the guinea-pig basilar artery, while this toxin alone was inactive in the rat hepatic artery, suggesting that EDHF may recruit charybdotoxin-sensitive K⁺ channels in the former. These results could be interpreted in at least two ways. Either the target for EDHF may be a new type of K⁺ channel, which is resistant to apamin, iberiotoxin, 4-aminopyridine and glibenclamide, or EDHF must activate more than one type of K⁺ channel in these arteries.

Does EDHF Activate Two Different Types of K⁺ Channel?

In the next series of experiments, the possibility that EDHF activates two populations of K⁺ channels with different pharmacological properties was examined. Several previous studies had reported an effect of either apamin or charybdotoxin on EDHF-mediated responses (Table 25-3). The effect of a combination of these K⁺ channel inhibitors was therefore tested. This combination completely abolished the EDHF-mediated hyper-

Table 25-3 Inhibitory effects of apamin and charybdotoxin each or combined on EDHF-mediated responses in different blood vessels

Vascular preparation (species)	Response	Apamin (conc.)	Charybdotoxin (conc.)	Apamin plus charybdotoxin (conc.)	References
Aorta (rabbit)	Relaxation	–	Complete (5×10^{-8} M)	–	a
Basilar artery (guinea-pig)	Relaxation	No (1×10^{-7} M)	Partial (10^{-7} M)	Complete (10^{-7} M each)	b, c
Carotid artery (guinea-pig)	Hyperpolarization	No (5×10^{-7} M)	No (10^{-7} M)	Complete (5×10^{-7} M + 10^{-7} M)	d
Carotid artery (rabbit)	Relaxation	–	Complete (5×10^{-8} M)	–	a
	Relaxation	–	Partial (10^{-8} M)	–	e
	Relaxation	Partial ($1{-}3 \times 10^{-6}$ M)	Complete (10^{-7} M)	–	f
Coronary artery (bovine)	Relaxation	Complete (10^{-6} M)	–	–	g
Coronary artery (porcine)	Relaxation	Complete (10^{-6} M)	No (10^{-8} M)	–	h
	Relaxation	No (10^{-6} M)	–	–	g
Coronary artery (guinea-pig)	Relaxation	No (10^{-6} M)	Partial (10^{-7} M)	Complete (10^{-7} M each)	i
	Relaxation	No (10^{-7} M)	Partial (10^{-7} M)	–	j
Gastroepiploic arteries (human)	Relaxation	No (10^{-6} M)	Partial (5×10^{-8} M)	–	k
	Hyperpolarization	No (10^{-6} M)	Partial (5×10^{-8} M)	–	k
Hepatic artery (rat)	Relaxation	No (10^{-6} M)	No (10^{-6} M)	Complete (3×10^{-7} M each)	l, m, n
	Hyperpolarization	–	–	Complete (3×10^{-7} M each)	n
Intestinal arterioles (guinea-pig)	Hyperpolarization	No (10^{-7} M)	Partial (5×10^{-8} M)	Complete (10^{-7} M + 5×10^{-8} M)	o
Mesenteric artery (rabbit)	Hyperpolarization	Complete (3×10^{-8} M)	–	–	p
Mesenteric artery (rat)	Relaxation	Partial (10^{-6} M)	No ($1-3 \times 10^{-7}$ M)	Complete (10^{-6} M + 10^{-7} M)	q
	Hyperpolarization	–	–	Complete (5×10^{-7} M + 10^{-7} M)	r
Mesenteric vascular bed (rat)	Relaxation	Complete (5×10^{-7} M)	–	–	s
	Relaxation	No (5×10^{-8} M)	No (10^{-7} M)	Complete (5×10^{-7} M +10^{-7} M)	t
Omental artery (human)	Relaxation	No (10^{-8} M)	No (5×10^{-8} M)	–	u
Oviductal artery (bovine)	Relaxation	Partial (10^{-6} M)	No (5×10^{-8} M)	–	v
Pial artery (human)	Relaxation	Partial (10^{-7} M)	No (10^{-7} M)	Complete (10^{-7} M each)	c
Renal artery (rabbit)	Relaxation	–	Partial (10^{-7} M)	–	c

a) Cowan et al., 1993
b) Petersson, Zygmunt and Högestätt, 1996
c) Petersson, Zygmunt and Högestätt, 1997
d) Corriu et al., 1996b
e) Lischke, Busse and Hecker, 1995a
f) Dong et al., 1997
g) Graier et al., 1996
h) Hecker et al., 1994
i) Eckman et al., 1998
j) Yamanaka, Ishikawa and Goto, 1998
k) Urakami-Harasawa et al., 1997
l) Zygmunt, 1995
m) Zygmunt and Högestätt, 1996b
n) Zygmunt et al., 1998
o) Hashitani and Suzuki, 1997
p) Murphy and Brayden, 1995b
q) White and Hiley, 1997
r) Chataigneau et al., 1998b
s) Adeagbo and Triggle, 1993
t) Randall and Kendall, 1998a
u) Ohlmann et al., 1997
v) Garcia-Pascual et al., 1995
x) Brandes et al., 1997

polarization and relaxation in the rat hepatic and guinea-pig basilar artery, clearly in favour of the two channel hypothesis.

Three different neuronal α-subunits of SK_{Ca} (SK1, SK2, SK3) have been cloned (Köhler *et al.*, 1996). When expressed in *Xenopus* oocytes, SK2 displays high sensitivity towards apamin (Köhler *et al.*, 1996), consistent with the presumed pharmacology of SK_{Ca} in non-neuronal tissue, whereas SK1 is apamin-resistant. However, K^+ currents recorded in cells transfected with both SK1 and SK2, or chimeric channel constructs have intermediate apamin sensitivity (Ishii, Maylie and Adelman, 1997). These findings suggest that apamin inhibits SK2 by binding to the α-subunit of the heterooligomeric channel protein. Apamin-sensitive K^+ currents also occur in vascular smooth muscle (Gebremedhin *et al.*, 1996). An apamin binding protein has been isolated and cloned from porcine vascular smooth muscle (Sokol *et al.*, 1994). However, the structure of this protein is distinct from known K^+ channel subunits, and the existence of SK_{Ca} similar to those observed in neuronal tissue has not yet been reported. Despite this uncertainty, the selectivity of apamin towards SK_{Ca} is undisputed, since apamin does not seem to interfere with other types of K^+ channels.

Based on our present knowledge of the pharmacology of apamin and charybdotoxin, the results of the present study could indicate the involvement of a "classical" apamin-sensitive SK_{Ca} and a charybdotoxin-sensitive K^+ channel in the EDHF-mediated relaxation. This latter channel does not seem to be a BK_{Ca}, since iberiotoxin was unable to replace charybdotoxin in combination with apamin. Although charybdotoxin is often used as a BK_{Ca} inhibitor, this toxin also affects certain K_V (Kv1.2, Kv1.3) in contrast to iberiotoxin (Chandy and Gutman, 1995). Both Kv1.2 and Kv1.3 channels are inhibited by submillimolar concentrations of 4-aminopyridine (Chandy and Gutman, 1995; Garcia *et al.*, 1995). However, 4-aminopyridine was inactive when combined with apamin in the rat hepatic artery, and this combination elicited only a small inhibition of the EDHF-mediated relaxation in the guinea-pig basilar artery without affecting the maximal relaxation. Furthermore, as shown in the patch-clamp experiments, the 4-aminopyridine-sensitive delayed rectifier current in isolated smooth muscle cells of the rat hepatic artery was unaffected by the combination of apamin and charybdotoxin, which abolished EDHF-mediated responses in intact arteries. Taken together, these findings indicate that the inhibitory effect of charybdotoxin on EDHF-mediated responses does not involve inhibition of K_V.

Two charybdotoxin-sensitive intermediate conductance calcium-sensitive K^+ channels (IK_{Ca}) were recently cloned from human pancreas and lymphocytes (Ishii *et al.*, 1997; Logsdon *et al.*, 1997). These channels have many electrophysiological and pharmacological properties in common with the Gardos channel in erythrocytes (Christophersen, 1991; Brugnara *et al.*, 1995). Both channels are inhibited by charybdotoxin, but insensitive to iberiotoxin and apamin. Although IK_{Ca} has not been demonstrated in vascular smooth muscle, the pharmacology of this channel clearly resembles the charybdotoxin-sensitive K^+ channel in the rat hepatic and guinea-pig basilar artery. Thus, the results of the present study could indicate that EDHF activates two different types of calcium-sensitive K^+ channels, an apamin-sensitive SK_{Ca} and a charybdotoxin-sensitive IK_{Ca}, which act in concert in rat hepatic and guinea-pig basilar arteries.

In the rat hepatic artery, N^ω-nitro-L-arginine and indomethacin as well as these inhibitors combined have no effect on the endothelium-dependent hyperpolarizations and relaxations induced by acetylcholine or the calcium ionophore A23187 (Zygmunt *et al.*, 1998). However, a significant N^ω-nitro-L-arginine-sensitive (NO-dependent) or indomethacin-sensitive (cyclooxygenase-dependent) response was disclosed when apamin plus charybdotoxin (inhibitors of EDHF) were combined with indomethacin or N^{ω}-nitro-L-arginine (Zygmunt and Högestätt, 1996b; Zygmunt *et al.*, 1998). Thus, lack of effect of cyclooxygenase (or NO synthase) inhibitors on endothelium-dependent responses does not provide solid proof to discard a role for prostanoids (or NO) unless the EDHF and NO (or cyclooxygenase) pathways are suppressed (see also Murphy and Brayden 1995b; Corriu *et al.*, 1996b). This also raises concern that these components may have been overlooked in previous studies.

The effects of apamin, charybdotoxin or a combination of these toxins on EDHF-mediated responses (in the presence of inhibitors of both NO synthase and cyclo-oxygenase) have been examined in several studies using different experimental models (Table 25-3). When applied separately, apamin and charybdotoxin produced variable inhibition, ranging from no effect to a complete inhibition of EDHF-mediated responses, depending mainly on the preparation examined. In the guinea-pig carotid artery (Corriu *et al.*, 1996b) and intestinal arterioles (Hashitani and Suzuki, 1997), rat mesenteric arteries (White and Hiley, 1997, Chataigneau *et al.*, 1998a,b) and the perfused mesenteric vascular bed (Randall and Kendall, 1998a), and human pial arteries (Petersson, Zygmunt and Högestätt, 1997), neither apamin nor charybdotoxin completely inhibited the responses mediated by EDHF. However, when these toxins were combined, the EDHF-mediated response was abolished. Iberiotoxin was unable to replace charybdotoxin in this combination in rat mesenteric and guinea-pig coronary and carotid arteries (White and Hiley, 1997; Chataigneau *et al.*, 1998a; Yamanaka, Ishikawa and Goto, 1998), which agrees well with the results obtained in the present study. The hypothesis that EDHF activates two different types of K^+ channels sensitive to either apamin or charybdotoxin could help explain much of the seemingly contradictory observations in the literature, assuming a heterogenous and species-dependent distribution of these channels in the vascular system.

Do Apamin and ChTx Act on the Endothelium?

K^+ channels sensitive to charybdotoxin or apamin have been demonstrated in the endothelium, and charybdotoxin or apamin prevents the hyperpolarization of the endothelium induced by endothelium-dependent vasodilators (Chen and Cheung, 1992; Marchenko and Sage, 1996; Wang, Chu and van Breemen, 1996b; Muraki *et al.*, 1997). Depolarization of the endothelial cell reduces calcium influx as a consequence of a decreased electrical driving force, thereby suppressing calcium-dependent formation of endothelium-derived relaxing factors (Suzuki and Chen, 1990; Lückhoff and Busse, 1990; Groschner, Graier and Kukovetz, 1992). A study using fura-2 microfluorometry and tension recordings of intact porcine coronary arteries provides strong evidence that the formation of EDHF is regulated by the intracellular calcium concentration (Higuchi *et al.*, 1996). This raises the question whether or not the K^+ channel inhibitors were acting at the level of the endothelium rather than on the smooth muscle cells in the present study. However, several pieces of evidence suggest that the effects of these compounds were confined to the smooth muscle cells. First, apamin plus charybdotoxin abolished the EDHF-mediated relaxation induced by the calcium ionophore A23187 in the rat hepatic and guinea-pig basilar artery (Zygmunt and Högestätt, 1996b; Petersson, Zygmunt and Högestätt, 1997). This mode of activation of the endothelium is receptor-independent and should be affected little by the membrane potential, since the transport of calcium by A23187 is electroneutral (Erdahl *et al.*, 1994). Furthermore, apamin plus charybdotoxin had no effect or only marginally affected the acetylcholine-induced relaxation in the absence of N^{ω}-nitro-L-arginine, when NO was released together with EDHF (present study, Petersson *et al.*, 1998; Zygmunt *et al.*, 1998). Although apamin plus charybdotoxin abolished the acetylcholine-induced relaxation mediated by EDHF in guinea-pig coronary arteries, this combination had no effect on acetylcholine-induced increases in intracellular free calcium concentration in endothelial cells (Yamanaka, Ishikawa and Goto, 1998). These findings strongly suggest that the calcium-dependent formation of endothelium-derived relaxing factors is preserved in the presence of apamin plus charybdotoxin.

Could Functional Antagonism Explain the Effect of Apamin and ChTx?

A number of precautions were taken to rule out functional antagonism of the toxin combination as a confounding factor. Since charybdotoxin enhanced the contraction induced by phenylephrine (rat hepatic artery) and prostaglandin $F_{2\alpha}$ (guinea-pig basilar artery), and occasionally increased the baseline tension in the guinea-pig basilar artery (Petersson, Zygmunt and Högestätt, 1997), the concentration of vasoconstrictor was titrated in each experiment to give a contraction equivalent to 70 to 90% of an initial reference response to a 60 mM potassium solution. Furthermore, the effect of apamin plus charybdotoxin was also tested on a number of endothelium-independent vasodilator agents with different mechanisms of action. The lack of effect of apamin plus charybdotoxin on vasodilator responses to SIN-1, iloprost, levcromakalim and anandamide in the rat hepatic artery strongly suggests that these toxins did not act as functional antagonists at the level of the smooth muscle cells. Finally, iberiotoxin and charybdotoxin both enhanced the phenylephrine-induced contractions in the rat hepatic artery, but only charybdotoxin was able to inhibit the EDHF-mediated relaxations when combined with apamin (Figure 25-2), further supporting this contension.

CONCLUSION

The results of the present study support the view that EDHF is an endogenous K^+ channel opener and suggest that EDHF activates two different types of K^+ channel in vascular smooth muscle, tentatively a "classical" apamin-sensitive SK_{Ca} and an IK_{Ca}, in the rat hepatic and guinea-pig basilar artery. Neither BK_{Ca}, K_{ATP} nor K_V seems to be the target K^+ channel for EDHF. Activation of either an apamin-sensitive SK_{Ca}, an IK_{Ca} or a mixture of these channels by EDHF could help explain much of the discrepant findings in the literature with regard to the effects of apamin and charybdotoxin on EDHF-mediated responses.

26 Mechanisms Underlying EDHF-Induced Relaxations

T. Nagao, S. Ibayashi, T. Kitazono, K. Fujii and M. Fujishima

Second Department of Internal Medicine, Faculty of Medicine, Kyushu University, Maidashi, Fukuoka 812-8582, Japan
Address correspondence to: Tetsuhiko Nagao, M.D., Ph.D., Second Department of Internal Medicine, Faculty of Medicine, Kyushu University, Maidashi 3-1-1, Fukuoka 812-82, Japan. Telephone: +81-92-642-5256; Fax: +81-92-642-5271; E-Mail Address: nagao@intmed2.med.kyushu-u.ac.jp

A major mechanism underlying relaxation induced by endothelium-derived hyperpolarizing factor (EDHF) is believed to be the inactivation of L-type voltage-dependent Ca^{2+} channels. However, if the inactivation of L-type voltage-dependent Ca^{2+} channels is the central mechanism for EDHF-induced relaxation, relaxing properties of EDHF would be quite limited from the physiological point of view. Evidence is accumulating that EDHF relaxes vascular smooth muscle cells through diverse mechanisms. One of the important relaxing mechanisms exerted by EDHF would be the inhibition of the production of inositol 1,4,5 trisphosphate and the subsequent reduction in Ca^{2+} release from intracellular stores. EDHF appears to inhibit receptor-operated Ca^{2+} influx. The influx may occur through nonselective cation channels or capacitative Ca^{2+} entry channels (calcium-release activated Ca^{2+} channels). Contractions induced by protein kinase C activation are also sensitive to EDHF, suggesting a possible interaction between EDHF and a process involved in the Ca^{2+}-sensitization of contractile elements by protein kinase C. The above mentioned relaxing properties of EDHF are shared by hyperpolarization induced by K^+ channel openers and activation of the Na^+-K^+ pump. Thus, such inhibitory effects of EDHF are most likely mediated by membrane hyperpolarization itself rather than by other unknown relaxing mechanisms.

KEYWORDS: capacitative Ca^{2+} entry, hyperpolarization, K^+ channel, protein kinase C, receptor-operated Ca^{2+} influx, voltage-dependent Ca^{2+} channel

INTRODUCTION

Endothelium-derived hyperpolarizing factor (EDHF) is a substance which is released from the endothelium upon activation of endothelial receptors by physiological vasoactive substances such as acetylcholine, bradykinin, histamine and serotonin (Nagao and Vanhoutte, 1993; Cohen and Vanhoutte, 1995). EDHF released from the endothelium diffuses to the underlying vascular smooth muscle cells and activates certain K^+ channels present on smooth muscle cells (Nagao and Vanhoutte, 1993; Cohen and Vanhoutte, 1995). The K^+ channel responsible for EDHF-induced hyperpolarization appears complex, and may not be the same under all conditions; previous studies have reported possible involvement of ATP-sensitive K^+ channels (Chen *et al.*, 1991), or of Ca^{2+} activated K^+ channels with large (Hwa *et al.*, 1994; Hansen and Olesen, 1997) or small (Adeagbo and Triggle, 1993) conductances. The efflux of K^+ leads to hyperpolarization of the smooth muscle plasma

229

membrane, and the change in membrane potential should inactivate voltage-dependent Ca^{2+} channels. Judging from the effectiveness of dihydropyridines on the blood pressure, L-type voltage-dependent Ca^{2+} channels appear to play an important role in the regulation of vascular tone.

Another possible mechanism for EDHF-induced relaxation is the inhibition of inositol 1,4,5-trisphosphate (IP_3) production (Itoh *et al.*, 1992; Yamagishi, Teshigawara and Taira, 1992). Decreased levels in IP_3 reduce Ca^{2+} release from intracellular store sites.

Although the activation of L-type voltage-dependent Ca^{2+} channels and IP_3-liberation are important in contractions of vascular smooth muscle cells, other mechanisms also play pivotal roles under certain circumstances, depending on the vascular bed considered and the stimulation applied. They include receptor-operated Ca^{2+} influx, activation of protein kinase C or tyrosine kinase, and production of arachidonic acid. Thus if the inhibition of L-type voltage-dependent Ca^{2+} channels and IP_3 mediates a major part of EDHF-induced relaxation, EDHF would play a limited role under physiological conditions. However, bradykinin, in the presence of blockers of nitric oxide synthase and cyclooxygenase, induces comparable relaxations in porcine coronary arteries contracted either with tetraethylammonium, prostaglandin $F_{2\alpha}$, phorbol 12,13-diacetate, or endothelin (Nagao and Vanhoutte, 1992b). Since contractions induced by the above substances depend on various contractile mechanisms, the relaxing potency of bradykinin should differ if EDHF relaxes vascular smooth muscle cells through the inhibition of L-type voltage-dependent Ca^{2+} channels and IP_3 production alone. The results suggest that EDHF relaxes vascular smooth muscle cells through more complex mechanisms.

TECHNICAL PITFALLS

A major problem with the study of EDHF is whether an observed relaxation is really mediated by EDHF alone. A standard technique to separate EDHF-mediated relaxation from nitric oxide-mediated one is to apply an endothelial agonist in the presence of blockers of nitric oxide synthase (Nagao and Vanhoutte, 1992a; Adeagbo and Triggle, 1993). The blockers are used at supramaximal concentrations as a rule. However, under certain conditions, production of nitric oxide is resistant to inhibitors of nitric oxide synthase. In particular, this holds true in arteries of the guinea-pig, a species often used for the study of EDHF. When carotid arteries of the guinea-pig were contracted with high K^+ solution (40 mM) in the combined presence of N^{ω}-nitro-L-arginine (3×10^{-5} M) and indomethacin (10^{-5} M) for more than 30 minutes, substance P induced considerable relaxations (Figure 26-1). EDHF never explains the relaxation induced by substance P since EDHF is incapable of hyperpolarizing vascular smooth muscle cells in high K^+ solution, which shifts the equilibrium potential for K^+ close to resting membrane potential. The relaxation induced by substance P declined with repeated application of the peptide and was abolished either by methylene blue (10^{-5} M) or 2-(4-carboxyphenyl)-4,4,5,5-tetramethylimidazoline-1-oxyl 3-oxide (carboxy-PTIO, 3×10^{-4} M), a scavenger of nitric oxide (Akaike *et al.*, 1993). The results are consistent with the view that the effect of N^{ω}-nitro-L-arginine, at the concentration used, was not potent enough to eliminate the production of nitric oxide in carotid arteries of the guinea-pig. Basilar arteries of the same

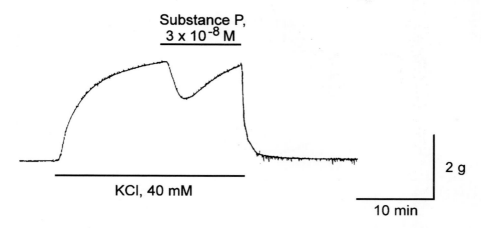

Figure 26-1.   Effect of substance P (3×10^{-8} M) on contraction induced by high K$^+$ (40 mM) in carotid artery of the guinea-pig. The experiment was performed in the presence of N$^{\omega}$-nitro-L-arginine (3×10^{-5} M) and indomethacin (10^{-5} M).

animal provide similar results when stimulated by acetylcholine (10^{-5} M). Hence, it is sometimes misleading to rely too much upon the effect of L-arginine analogs.

Another important aspect in the study of EDHF is that results obtained from tension recordings must be interpreted with caution. Isometric tension recordings provide a great deal of information and, in the study of the relaxing mechanism for EDHF, this experimental approach is essential. However, it must be remembered that pharmacological interventions may be accompanied by unexpected and undesired effects on the cell membrane which can be discovered only by measurements of membrane potential.

EFFECTS OF EDHF ON AGONIST-INDUCED CONTRACTIONS

Although the class of K$^+$ channels involved may differ, K$^+$ channel openers mimic relaxations induced by EDHF. Norepinephrine induces contractions in aortas of the rabbit, on which the Ca^{2+} antagonist nifedipine has little effect. Nevertheless, cromakalim, an opener of ATP-sensitive K$^+$ channels, is able to inhibit the norepinephrine-induced contractions (Quast and Baumlin, 1991). These results suggest that cromakalim relaxes rabbit aortas contracted with norepinephrine through a mechanism independent of the inactivation of L-type voltage-dependent Ca^{2+} channels. If this is the case, and if such an effect of cromakalim is mediated by membrane hyperpolarization, EDHF would show similar inhibition on contractions induced by mechanisms other than the activation of L-type voltage-dependent Ca^{2+} channels.

In basilar arteries of the guinea-pig, bradykinin (10^{-7} M) induced sustained contractions. The contractions were inhibited by a supramaximal concentration of diltiazem (10^{-5} M), a Ca^{2+} antagonist. However, there still remained sustained contractions in the presence of diltiazem. Acetylcholine (10^{-5} M) inhibited the diltiazem-resistant contractions induced by

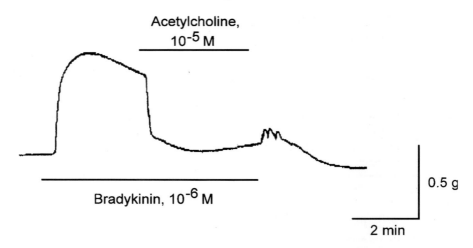

Figure 26-2.   Effect of acetylcholine (10^{-5} M) on basilar artery of the guinea-pig contracted by bradykinin (10^{-6} M). The experiment was performed in the presence of diltiazem (10^{-5} M), N^ω-nitro-L-arginine (NLA; 3×10^{-5} M), methylene blue (10^{-5} M), and indomethacin (10^{-5} M).

bradykinin (10^{-6} M) in the presence of N^ω-nitro-L-arginine (3×10^{-5} M), methylene blue (10^{-5} M) and indomethacin (10^{-5} M, Figure 26-2). Since the sustained contractions to bradykinin were abolished by Ca^{2+} free solution, they are dependent on Ca^{2+} influx from the extracellular space. Two possible routes for receptor-operated Ca^{2+} influx are nonselective cation channels and the capacitative Ca^{2+} entry channel (Iwasawa et al., 1997). The latter is a Ca^{2+} channel which is activated upon emptying of intracellular Ca^{2+} store sites and helps to refill them through Ca^{2+} influx from the extracellular space (Putney, 1986). Although it is not clear through which pathway Ca^{2+} entry takes place during receptor stimulation, K^+ channel openers did inhibit sustained Ca^{2+} increase induced by norepinephrine in mesenteric arteries of the rabbit (Itoh et al., 1992, Itoh et al., 1994). Concerning the effect of EDHF in other vascular beds, substance P relaxed porcine coronary arteries contracted with U46619 (a thromboxane A_2-mimetic) in the presence of nifedipine and N^ω-nitro-L-arginine (Kilpatrick and Cocks, 1994) although indomethacin was not present in this study.

Reports on voltage-dependency of nonselective cation channels are conflicting. Smooth muscle cells of rabbit aorta exhibited currents through nonselective cation channels upon stimulation with endothelin. The current did not show voltage-dependency (Enoki et al., 1995). In contrast to this observation, nonselective cation currents activated by agonists were voltage-dependent in the rabbit portal vein (Inoue and Kuriyama, 1993) and the guinea-pig ileum (Inoue and Isenberg, 1990).

Although experimental data regarding the relationship between membrane hyperpolarization and capacitative Ca^{2+} entry channel are limited, the current-voltage relationship for the Ca^{2+} entry does not show voltage-dependent activation (Zweifach and Lewis, 1993; Yao and Tsien, 1997). However, a possibility still remains that membrane hyperpolarization interferes with a Ca^{2+} refilling process which follows Ca^{2+} influx through capacitative Ca^{2+} entry channels.

EDHF AND PROTEIN KINASE C

Protein kinase C plays a central role in the contraction of vascular smooth muscle cells. The kinase, upon activation, potentiates vasoconstraction by sensitizing contractile protein to Ca^{2+} (Nishimura, Kolber and van Breemen, 1988). Receptor stimulation by agonists and the resulting production of diacylglycerol leads to the activation of protein kinase C in vascular smooth muscle cells.

Protein kinase C can be activated by phorbol esters such as phorbol 12,13-dibutyrate (PDB). PDB (10^{-7} M) induced sustained contractions, which develop slowly, in carotid arteries of the guinea-pig in the presence of diltiazem (10^{-5} M). Substance P inhibited the diltiazem-resistant contractions to PDB in the presence of N^{ω}-nitro-L-arginine (3×10^{-5} M), methylene blue (10^{-5} M) and indomethacin (10^{-5} M). Although a mechanism underlying such an inhibition remains to be elucidated, there should be an interaction between EDHF and the protein kinase C system. Although a most direct and simple explanation is that EDHF inhibits the activation of protein kinase C, so far there is no evidence supporting that concept. Another possible explanation would be the reduction of resting Ca^{2+} levels by EDHF. Pinacidil and Y-26763, two different K^+ channel openers, reduced resting Ca^{2+} levels in the mesenteric artery of the rabbit (Itoh *et al.*, 1992; Itoh *et al.*, 1994). Even a minor change in Ca^{2+} levels induced by hyperpolarization may result in a considerable change in tension if activation of protein kinase C takes place at the same time. The interaction between EDHF and the protein kinase C system provides another good explanation why EDHF inhibits diltiazem-resistant contractions induced by bradykinin in basilar arteries of the guinea-pig.

It may also be possible that the inhibitory effects of EDHF are mediated by its direct action on contractile protein to reduce Ca^{2+} sensitivity. As far as the effects of K^+ channel openers are concerned, most investigators agree that caffeine-induced contraction or Ca^{2+} release is not affected by cromakalim or lemakalim (Quast and Baumlin, 1991; Ito *et al.*, 1991; Yamagishi, Yanagisawa and Taira, 1992). By contrast, levcromakalim (L-cromakalim) shifts a Ca^{2+}-force relationship curve to the right, suggesting a decrease in Ca^{2+} sensitivity by the K^+ channel opener (Okada, Yanagisawa and Taira, 1993). Although, conflicting results are reported for cromakalim by using the same preparation in a similar experimental system (Yanagisawa, Teshigawara and Taira, 1990). In further support of the absence of Ca^{2+} desensitization by membrane hyperpolarization, activation of the Na^+-K^+ pump (by addition of K^+ to K^+ free solution) did not inhibit caffeine-induced contractions in basilar arteries of the rat (unpublished observation).

DOES EDHF POSSESS RELAXING PROPERTIES OTHER THAN MEMBRANE HYPERPOLARIZATION?

A question arises here as to whether EDHF inhibits L-type voltage-dependent Ca^{2+} channel-independent contractions through membrane hyperpolarization or through other unknown relaxing mechanism(s). As far as current knowledge on vascular physiology is concerned, there is no endogenous vasoactive substance which controls vascular tone as a pure K^+ channel opener. Thus, it is conceivable that EDHF possesses other relaxing mechanisms in addition to the opening of K^+ channels. However, the following considerations argue

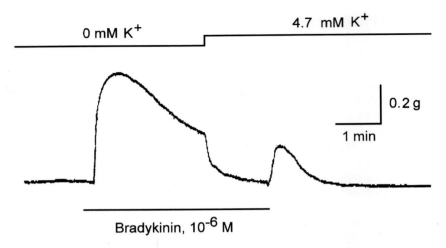

Figure 26-3. Effect of addition of K$^+$ (4.7 mM) to K$^+$ free solution (a procedure to activate the Na$^+$-K$^+$ pump) on basilar artery of the guinea-pig contracted with bradykinin (10^{-6} M). The experiment was performed in the presence of diltiazem (10^{-5} M).

against this possibility. First, the effects of EDHF on L-type voltage-dependent Ca^{2+} channel-independent contractions can be abolished by procedures known to inhibit the EDHF-action on membrane potential (high K$^+$ and tetrabutylammonium). Second, the effects of EDHF can be mimicked by membrane hyperpolarization induced by other procedures. Hyperpolarization induced by K$^+$ channel openers causes similar relaxations. In addition, the effects of K$^+$ channel openers were antagonized by their specific blockers, sulfonylureas. For further confirmation, another procedure to induce hyperpolarization, the activation of the Na$^+$-K$^+$ pump, was tested on diltiazem-resistant contractions induced by bradykinin in guinea-pig basilar arteries without endothelium. The pumps were activated by addition of K$^+$ (4.7 mM) to K$^+$ free solution. K$^+$ relaxed diltiazem-resistant contractions induced by bradykinin (10^{-6} M, Figure 26-3), and the effect was antagonized by ouabain (10^{-6} M). K$^+$ inhibited diltiazem-resistant contraction induced by PDB (10^{-7} M) in a similar fashion. Although the above considerations do not rule out the possibility that EDHF has other unknown relaxing mechanisms, the actions of EDHF can be explained, so far, by its property to hyperpolarize vascular smooth muscle cells.

CONCLUSIONS AND PERSPECTIVES

EDHF relaxes vascular smooth muscle cells through more diverse mechanisms than expected. In addition to the inactivation of L-type voltage-dependent Ca^{2+} channels, the endothelial factor inhibits the production of IP$_3$. The inhibition of receptor-operated Ca^{2+} influx and the interference with a process involved in the the protein kinase C-mediated Ca^{2+} sensitizing system are likely to play a crucial role. Reduction of basal Ca^{2+} levels may also be involved. Although EDHF may decrease Ca^{2+} sensitivity of the contractile protein through a protein kinase C-related mechanism, hyperpolarization does not appear

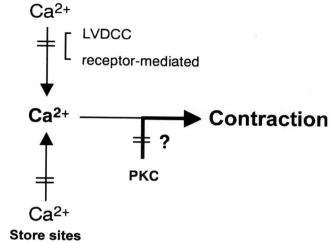

Figure 26-4. Putative mechanisms underlying EDHF-induced relaxation. EDHF inhibits Ca^{2+} influx through L-type voltage-dependent Ca^{2+} channel (LVDCC) and possibly receptor-operated Ca^{2+} influx. EDHF also inhibits Ca^{2+} mobilization from intracellular stores. EDHF interferes with a process involved in the Ca^{2+} sensitizing pathway mediated by protein kinase C (PKC) activation. The last mechanism amplifies the effect of decreased intracellular Ca^{2+} levels.

to exert a direct inhibition on Ca^{2+} sensitivity of the contractile machinery. Thus, EDHF reduces basal Ca^{2+} levels, inhibits Ca^{2+} influx through L-type voltage-dependent Ca^{2+} channels and possibly receptor-operated Ca^{2+} entry (Figure 26-4). As a direct consequence of reduced Ca^{2+} levels, contractile responses are inhibited. In addition, under activation of the protein kinase C pathway, even a minor decrease in intracellular Ca^{2+} would result in a larger inhibition of contractions than expected from the change in Ca^{2+} levels. These actions of EDHF are most likely to be mediated by membrane hyperpolarization itself.

An important but unsolved question regarding endothelium-derived relaxing factors is the differential role played by each factor under physiological conditions. It is of special importance to elucidate the respective roles of, and the possible interactions between, nitric oxide and EDHF. If the two endothelium-derived relaxing factors were to exert synergistic effects, this may seem desirable from a physiological point of view. However, this is not likely to be the case (Nagao and Vanhoutte, 1992b).

27 Heterocellular Calcium Signaling: A Novel Stimulus for the Release of EDHF?

K.A. Dora and C.J. Garland

Department of Pharmacology, University of Bristol, University Walk, Bristol BS8 1TD, UK. Ph: 44 (117) 928-8666; Fax: 44 (117) 929-3194; E-mail: kim.dora@bris.ac.uk; chris.garland@bris.ac.uk

Investigations into the identity and nature of endothelium-derived hyperpolarizing factor have generally focused on blockade of acetylcholine-stimulated hyperpolarization and/or relaxations that are independent of nitric oxide or prostacyclin synthesis. This form of hyperpolarization appears to be dependent on an increase in intracellular calcium concentration in the endothelial cells. However, arterial hyperpolarization and relaxation are also observed in response to stimulated rises in endothelial cell calcium concentration not due to agonists, suggesting that a rise in endothelial cell calcium *per se* is sufficient to modulate tone. The presence of myoendothelial junctions in arterial vessels offers the possibility for intercellular calcium signaling between smooth muscle and adjacent endothelial cells. These connections allow the transfer of current as well as small molecules between the two cell types. Intercellular calcium signaling secondary to α_1-adrenergic stimulation of vascular smooth muscle stimulates a rapid rise in endothelial cell calcium concentration and the release of endothelium-derived relaxing factors such as nitric oxide and possibly endothelium-derived hyperpolarizing factor, although the latter has not been investigated. Increases in contractions to α-adrenergic agonists after blockade of nitric oxide synthesis have been observed in multiple arterial beds. In small arteries, blockade of calcium-activated potassium channels (with the combination of apamin plus charybdotoxin) also augments the magnitude of α-adrenergic contraction, and appears to play a more significant role than inhibition of nitric oxide synthesis alone. This profile of blockade is consistent with that of hyperpolarization in response to acetylcholine in this tissue (which is abolished by the combination of apamin plus charybdotoxin). Thus, release of an endothelium derived hyperpolarizing factor can occur not only in response to direct activation of endothelial cells, but also secondary to the activation of the smooth muscle cells. These observations provide further support for the suggestion that endothelium-dependent hyperpolarization plays an important role in modulating constriction of small arteries.

KEYWORDS: K+-channel, nitric oxide, phenylephrine, modulation of contraction

INTRODUCTION

In small mesenteric arteries, dilatation mediated by the endothelium-dependent agonist acetylcholine is almost fully due to hyperpolarization of vascular smooth muscle. The blocker of prostacyclin synthesis, indomethacin, has no effect on responses to acetylcholine (Garland and McPherson, 1992), and the NO synthase inhibitor N^ω-nitro L-arginine methyl ester causes only a slight rightward shift in the acetylcholine concentration response curve for dilatation, and has no effect on repolarization (Garland and McPherson, 1992). The remaining dilatation and repolarization are abolished by raising the potassium concentration to 25 mM (Waldron and Garland, 1994) or by the combined application of apamin and charybdotoxin (Plane *et al.*, 1997). The exact target of these toxins is unclear, but the available evidence suggests that apamin is selective for small conductance, calcium-sensitive K-channels, whereas charybdotoxin acts mainly on large conductance, calcium-sensitive K^+-channels as well as other undefined K^+-channels (Zygmunt *et al.*, 1997). This 'non-selective' action of charybdotoxin is crucial to the blockade of EDHF-responses, since a combination of apamin and the more selective inhibitor of large conductance calcium-sensitive K^+-channels, iberiotoxin, does not block the EDHF response mediated by acetylcholine (Zygmunt *et al.*, 1997).

Apamin-sensitive K^+-channels appear to be confined to endothelial cells (Marchenko and Sage, 1996); whereas charybdotoxin-sensitive channels are found in both cell types (Edwards *et al.*, 1998; Edwards and Weston, 1998). Thus, hyperpolarization of smooth muscle mediated by acetylcholine appears to depend on stimulation of endothelial cell K^+-channels. Since acetylcholine causes an increase in endothelial intracellular calcium concentration (Busse, Pohl and Luckhoff, 1989), both via a release from intracellular stores and calcium influx mechanisms, the hyperpolarization may, at least in part, be associated with that rise in intracellular calcium concentration. Indeed, endothelial cell hyperpolarization caused by acetylcholine depends on an increase in endothelial cell intracellular calcium concentration (Chen and Suzuki, 1990), and arterial repolarization and relaxation is also observed in response to non-receptor-mediated increases in endothelial cell intracellular calcium concentration (Plane, Pearson and Garland, 1995). These data suggest, therefore, that a rise in intracellular calcium concentration in endothelial cells *per se* is sufficient alone to modulate vascular tone.

In order to study EDHF responses under isometric conditions, a constrictor agonist such as phenylephrine is necessary to stimulate the smooth muscle cells in the presence of NO synthase inhibitors. Under these conditions, the potency of phenylephrine is increased by the NO synthase inhibitor, an effect attributed to an inhibition of 'basal' NO release (Plane *et al.*, 1996). In contrast, in pressure myographs basal NO release is insufficient to affect myogenic tone, whereas in the presence of phenylephrine, NO release is stimulated as a result of hetero-cellular calcium signaling (Dora, Doyle and Duling, 1997). The resulting increase in endothelial intracellular calicum concentration may also stimulate the release of EDHF, so inhibition of one or both of these factors may underlie any increased contraction to phenylephrine. This possibility was tested by comparing contraction to phenylephrine in preparations with and without endothelium and after sequentially blocking the release of NO and EDHF.

METHODS

Small arterial segments (diameter 100–200 μm) from male Wistar rat mesenteric arcades were suspended in a Mulvany-Halpern myograph (10 ml) containing modified Krebs-Ringer bicarbonate solution (37°C, aerated with a 95% O_2:5% CO_2 gas mixture) and indomethacin (2.8×10^{-6} M). After normalization and equilibration, all arteries used maximally relaxed to acetylcholine (1×10^{-6} M, in pre-contracted vessels). In these studies, N^ω-nitro-L-arginine methyl ester (1×10^{-4} M, Sigma), apamin (5×10^{-9} M, Alamone Labs)

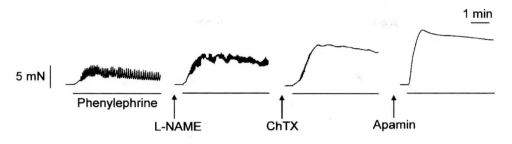

Figure 27-1. Augmentation of phenylephrine contraction by blockers of NO and EDHF synthesis. Under control conditions, contraction to phenylephrine tended to oscillate. Addition of N^{ω}-nitro-L-arginine methyl ester, followed by charybdotoxin (ChTx) and apamin each augmented the level of contraction stimulated by phenylephrine. This trace was obtained in a single experiment after repeated exposure to phenylephrine (3×10^{-6} M, indicated by bar). Average values obtained during the final 30 s of each exposure to phenylephrine were used for summary data (Figure 27-2).

and charybdotoxin (5×10^{-9} M, Alamone Labs) were added to the organ bath at least 20 min before addition of phenylephrine (Sigma). In each experiment, a concentration of phenylephrine ($1–3 \times 10^{-6}$ M) was chosen to evoke only a small increase in tension under control conditions, and the same concentration was then used throughout the remainder of the experiment. Phenylephrine was added for a period of 5 min. In some arteries, a hair was used to destroy the endothelium, which was assessed as a loss of relaxation to acetylcholine ($1–3 \times 10^{-6}$ M).

At the end of the experiment, the maximal contractile capability of the artery was determined by adding 1×10^{-5} M phenylephrine and 65 mM K^+. Data are means ± SEM and are expressed as a percentage of the maximal contraction; n indicates the number of animals.

RESULTS

Under control conditions, addition of phenylephrine caused a small increase in tension ($15.5 \pm 3.3\%$ of the maximal contraction, $n = 17$) which tended to oscillate (Figure 27-1). The magnitude of this control contraction did not vary with time (Figure 27-2). Addition of N^{ω}-nitro-L-arginine methyl ester (1×10^{-4} M) had no effect on basal tone (Table 27-1), but on average the magnitude of phenylephrine contraction was more than doubled ($34.0 + 4.9\%$ of the maximal contraction, $n = 17$). Similarly, the subsequent addition of either apamin or charybdotoxin did not affect basal tone (Table 27-1), but augmented the contraction to phenylephrine significantly (to $50.4 \pm 5.3\%$, $n = 11$ and $64.6 \pm 4.1\%$ of maximal contraction, $n = 4$, respectively). The combination of N^{ω}-nitro-L-arginine methyl ester with apamin plus charybdotoxin caused the greatest augmentation of the contraction to phenylephrine, more than five times greater than control ($80.8 \pm 3.7\%$ of maximum contraction, $n = 12$). After this treatment, oscillations in contraction were never observed.

In arteries without endothelium, the addition of N^{ω}-nitro-L-arginine methyl ester with and without toxins had no effect on basal tone. Control responses ($38.0 \pm 12.0\%$ of maximal contraction, $n = 6$) did not significantly increase with time. N^{ω}-nitro-L-arginine

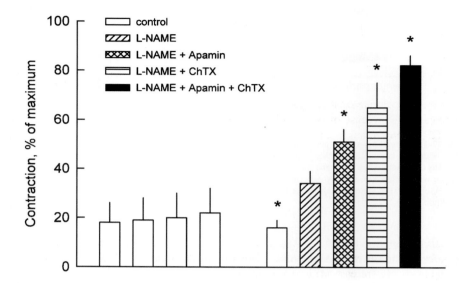

Figure 27-2. Summary of time course data. The level of phenylephrine contraction did not vary after repeated exposure under control conditions (Left). Addition of N$^{\omega}$-nitro-L arginine methyl ester significantly increased phenylephrine-mediated contraction. Further addition of blockers of EDHF production, apamin and charybdotoxin (ChTx) each significantly augmented the level of phenylephrine-induced contraction further, and the combination of N$^{\omega}$-nitro-L-arginine methyl ester plus both toxins had the most marked effect (Right). Values were averaged over the final 30 sec of phenylephrine addition. Mean values (± SEM) were analyzed with analysis of variance (ANOVA). The asterisks indicate statistically significant differences, P < 0.05; Dunnett Multiple Comparison Test. For each treatment, *n* values are given in Table 27-1.

Table 27-1 Effects of blocking release of NO and EDHF on basal tension in small mesenteric arteries of the rat

Treatment	Basal Tension (mN)	Maximal Tension (mN)	n
Endothelium Intact			
Control	1.2 ± 0.2	14.4 ± 1.0	17
N$^{\omega}$-nitro-L-arginine methyl ester	1.1 ± 0.1	14.4 ± 1.0	17
N$^{\omega}$-nitro-L-arginine methyl ester + Apamin	1.2 ± 0.2	14.5 ± 1.2	11
N$^{\omega}$-nitro-L-arginine methyl ester + Charybdotoxin	0.8 ± 0.3	15.7 ± 2.8	4
N$^{\omega}$-nitro-L-arginine methyl ester + Apamin + Charybdotoxin	1.1 ± 0.1	14.2 ± 1.2	12
Endothelium Denuded			
Control	0.9 ± 0.1	13.2 ± 2.5	6
N$^{\omega}$-nitro-L-arginine methyl ester	0.9 ± 0.1	13.2 ± 2.5	6
N$^{\omega}$-nitro-L-arginine methyl ester + Apamin	0.9 ± 0.1	12.3 ± 4.2	3
N$^{\omega}$-nitro-L-arginine methyl ester + Charybdotoxin	1.0 ± 0.2	14.0 ± 4.4	3
N$^{\omega}$-nitro-L-arginine methyl ester + Apamin + Charybdotoxin	0.9 ± 0.1·	13.2 ± 2.5	6

Data are means ± SEM of the number of animals indicated by *n*.

methyl ester alone (39.6 + 8.4%, $n = 6$) or together with apamin (46.3 ± 20.8%, $n = 3$) did not significantly increase the contraction to phenylephrine, whereas the addition of N^{ω}-nitro-L-arginine methyl ester plus charybdotoxin (66.5 ± 6.7%, $n = 3$) did.

DISCUSSION

This study demonstrates that contraction of vascular smooth muscle also activates endothelial cells to cause the release of NO and EDHF. Blockade of NO synthesis, alone or together with EDHF-release, augmented the contraction to phenylephrine. This observation demonstrates that even before the onset of contraction there are already signaling processes occurring between the endothelial and smooth muscle cells which lead to a modulation of the observed contraction.

Under basal conditions, the tension in the small mesenteric artery was unaffected after blockade of the NO and EDHF pathways. However, at this time there is no stimulus for myogenic (or other) tone in the preparation. Thus, it is not possible to assess whether either NO or EDHF were released by simply recording resting tension. As far as NO synthesis is concerned, measurements of cyclic GMP formation suggest that there is in fact a low basal level of NO release which is inhibited by a combination of N^{ω}-nitro-L-arginine methyl ester and the inhibitor of soluble guanylyl cyclase 1H-[1,2,4]oxadiazolo[4,3-a] quinoxalin-1-one (Plane *et al.*, 1998). There have been no reports of basal EDHF release in unstimulated preparations, however it has been shown that calcium-sensitive K^+-channels in smooth muscle cells are activated in myogencially-active cerebral arteries (Nelson *et al.*, 1995) and that EDHF may be released in perfused mesenteric arteries (Adeagbo and Triggle, 1993).

After supplying a stimulus for contraction by adding phenylephrine, the release of NO and EDHF became apparent. N^{ω}-nitro-L-arginine methyl ester significantly increased the contraction to phenylephrine, suggesting NO release was suppressing contraction. This finding has been observed in other tissues, including myogenically active arterioles (Dora, Doyle and Duling, 1997). Assays of cyclic GMP formation also support this observation. After the addition of phenylephrine to isolated arteries, the rate of cyclic GMP formation increases two to three fold in comparison to basal release (Plane *et al.*, 1996). Since α_1-adrenceptors are unlikely to be present on endothelial cells, the parsimonious explanation is that Ca^{2+} moves from the smooth muscle to adjacent endothelial cells and stimulates NO release, an effect which is blocked in the presence of N^{ω}-nitro-L-arginine methyl ester (Dora, Doyle and Duling, 1997).

Similarly, the augmentation of the contraction to phenylephrine with the toxins is also consistent with a phenylephrine-mediated increase in endothelial intracellular calicum concentration. Repeating the experiments in blood vessels without endothelium assessed the possibility that smooth muscle calcium-activated K^+-channels were also activated by phenylephrine, but both N^{ω}-nitro-L-arginine methyl ester and apamin failed to augment contraction to phenylephrine. The results are consistent with the presence of apamin-sensitive K^+-channels only in the endothelial cells (Marchenko and Sage, 1996; Mistry and Garland, 1998), but charybdotoxin-sensitive K^+-channels in both cell types (Edwards *et al.*, 1998; Edwards and Weston, 1998).

The present findings demonstrate another important stimulus for the release of NO and EDHF. The activation of the endothelium is not confined to endothelium dependent vasodilators or changes in wall shear stress. Stimulation of smooth muscle cells with phenylephrine leads to NO synthesis and K^+-channel opening, the latter causing the most pronounced modulation of contraction. This observation provides further evidence supporting an important role for endothelial hyperpolarization in controlling the diameter of small arteries.

ACKNOWLEDGEMENTS

This work was supported by the Wellcome Trust (UK).

28 The Role of Nonselective Cation Current in Regulation of Membrane Potential in Vascular Smooth Muscle Cells

Sergey I. Zakharov, Richard A. Cohen and Victoria M. Bolotina

Vascular Biology Unit, Department of Medicine, Boston Medical Center, Boston MA 02118, USA.
Address of correspondence: Dr. Victoria M. Bolotina, Vascular Biology Unit, R408, Boston University Medical Center, 88 East Newton Street, Boston MA 02118-2393. Tel: (617)-638-7118; Fax: (617)-638-7113; e-mail: vbolotina@med-med1.bu.edu

Activation of K^+ channels is believed to be a major mechanism underlying the endothelium-dependent hyperpolarization of smooth muscle cells which mediates vascular relaxation. Here alternative mechanisms are shown for hyperpolarization, which are not associated with activation of the major K^+ channels. Whole-cell currents and membrane potential were recorded using the perforated-patch clamp technique in smooth muscle cells from rabbit aorta. The role of Ca^{2+}-dependent K^+ (K^+_{Ca}), delayed rectifier K^+(K^+_{dr}), and different types of nonselective cation channels in the regulation of membrane potential was studied. K^+_{Ca} channels regulated resting membrane potential only in cells with elevated intracellular Ca^{2+}. The resting potential in such cells was from -60 to -45 mV and inhibition of K^+_{Ca} channels or loading cells with the Ca^{2+} chelator, BAPTA-AM, produced depolarization to -40 mV, which corresponded to the resting membrane potential of the majority of cells. In cells with resting membrane potential of -40 ± 2 mV, the inhibition of K^+_{Ca} channels had no effect, but inhibition of K^+_{dr} channels depolarized smooth muscle cells by 10–15 mV. In addition to K^+ channels, different nonselective cation channels could regulate the membrane potential. Activation of these channels caused depolarization, and their inhibition caused profound hyperpolarization (up to -70 mV) even without additional activation of K^+ channels. Thus, it is proposed that nonselective cation channels can be major regulators of membrane potential, and that a potential mechanism of endothelium-dependent hyperpolarization is inhibition of nonselective cation currents.

KEYWORDS: Nonselective cation channels, potassium channels, hyperpolarization, smooth muscle cells

INTRODUCTION

Membrane potential is an important determinant of the contractility of smooth muscle, and the role of different K^+ channels in its regulation is well established (Nelson and Quayle, 1995). It is also clear that other cation channels can regulate membrane potential. Typically resting smooth muscle cell freshly isolated from the rabbit aorta possess both outward and inward whole-cell currents as seen from the current-voltage relationship (Figure 28-1). A wide distribution of equilibrium membrane potentials for different cations, as well as depolarizing and hyperpolarizing forces of inward and outward currents are schematically shown in Figure 28-1. The resting membrane potential of smooth muscle cells is usually between -40 and -60 mV, and opening of any kind of K^+ channels (K^+_{Ca}, K^+_{dr}, K^+_{ir}, K^+_{ATP}) will increase the outward current and cause hyperpolarization. Opening of any nonselective

Sergey I. Zakharov et al.

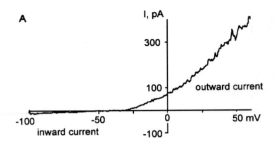

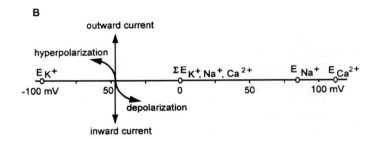

Figure 28-1. **A.** Typical whole-cell current-voltage relationship in resting rabbit aortic smooth muscle cell recorded during ramp depolarization. **B.** Schematic illustration of the wide distribution of equilibrium membrane potentials (E) for different cations (denoted by circles) along the voltage axes, and depolarizing and hyperpolarizing forces of inward and outward currents (denoted by arrows).

cation channels (which has an equilibrium potential around 0 mV) will necessarily produce an inward current and depolarization (Figure 28-1). The resting membrane potential, as well as membrane potential during activation of the smooth muscle cells, will result from the balance between the outward and inward currents. Depending on the kind of channel involved, a change in this balance can potentially depolarize or hyperpolarize the cell.

Several different K+ channels have been implicated in mediating endothelium-dependent smooth muscle cell hyperpolarization (for review see: Edwards and Weston, 1997). However, the role of nonselective cation channels remains obscure. Here it is shown that small changes in nonselective cation conductance cause large changes in membrane potential in rabbit aortic smooth muscle cells. Activation of several kinds of nonselective cation channels produces depolarization of the smooth muscle cell, and their inhibition causes cell hyperpolarization. Thus, regulation of different nonselective cation channels (those active under resting conditions, or those activated by contractile agonists) could be modulated by endothelium-derived hyperpolarizing factor(s) (EDHF's).

METHODS

Preparation of the Single Smooth Muscle Cells

New Zealand rabbits were killed after an intra venous dose of sodium pentobarbital (30 mg/kg). The thoracic aorta was excised and adhering fat tissue was removed. The aorta was opened longitudinally and 1 mm wide strips of the medial smooth muscle layer were separated and stored at 5°C in a low $CaCl_2$ (150 μM) dissociation medium containing glucose and 0.02% bovine serum albumin. After 1 hour the strips were transferred to dissociation media containing DL-dithiothreitol (2 mM) and papain (4 mg/ml) and placed at 37°C with gentle stirring for 30 minutes. The digested strips were washed in enzyme-free media and individual smooth muscle cells were obtained by trituration. Only relaxed cells were used for experiments.

Electrophysiological Recordings and Data Analysis

An aliquot of cells was added to the recording chamber (1 ml volume) which was mounted on the stage of an inverted microscope (Olympus IX70-S8F). Experimental solutions were applied to the individual cells through a tube that allowed quick (less than 3 sec) changes of solutions applied at a low perfusion rate. Experiments were carried out at room temperature (22 to 24°C). Whole-cell currents and membrane potential were studied with the perforated-patch technique using amphotericin B to permeabilize the membrane beneath the pipette. Data were recorded using an Axopatch 200A amplifier (Axon Instruments), filtered at 2 kHz and sampled at 5 kHz. Series resistance (26.2 ± 5.3 MΩ, n = 39) and cell capacitance (19 ± 1 pF, n = 39) were adjusted with the patch clamp amplifier. The whole-cell current was recorded at the holding potential of –80 or –30 mV, and voltage ramps (1 sec long) were applied every 5 or 30 sec to obtain current-voltage relationships. Membrane potential was measured in the current-clamp mode. Data were corrected for a junction potential which was determined at the end of each experiment.

Drugs and Solutions

For patch-clamp experiments the control bath solution contained (in mM): NaCl 140, KCl 2.8, $CaCl_2$ 2, $MgCl_2$ 1, HEPES 10, glucose 11 (pH 7.4 with NaOH). In a few experiments intracellular Ca^{2+} was buffered by treating cells with 2×10^{-5} M acetoxymethyl ester of 1,2-bis-(2-aminophenoxy) ethane-N,N,N′,N′-tetra acetic acid (BAPTA-AM) for 20 min. For whole-cell current and membrane potential recording, pipettes were filled with a solution containing (in mM): potassium aspartate 100, KCl 40, NaCl 4.5, EGTA 0.1, $MgCl_2$ 3, and HEPES 10, (pH 7.2 with KOH). In some experiments K^+ in the pipette was substituted with Cs^+ to inhibit outward K^+ currents. Amphotericin B was dissolved in dimethyl sulfoxide, sonicated, and added to the pipette solution (Rae *et al.*, 1991). Tetraethylammonium and 4-aminopyridine were prepared as aqueous solutions prior to use. All drugs used in this study were obtained from Sigma.

RESULTS AND DISCUSSION

In freshly isolated smooth muscle cells of the rabbit thoracic aorta both K^+ channels and nonselective cation channels were found to contribute to the regulation of resting membrane potential. These studies showed a strong correlation between the resting membrane potential and the relative magnitude of outward K^+ and inward nonselective cation currents.

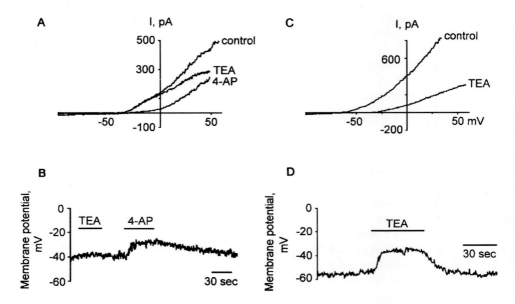

Figure 28-2. Typical whole-cell current-voltage relationships (**A,C**), and membrane potential recordings (**B,D**) in smooth muscle cells freshly isolated from rabbit thoracic aorta. **A** and **B**. Effects of tetraethylammonium (TEA, 5×10^{-3} M) and 4-aminopyridine (4-AP, 10^{-3} M) in the same smooth muscle cell with a resting potential of -40 mV. **C** and **D**. Effects of tetraethylammonium (5×10^{-3} M) in a "hyperpolarized" smooth muscle cell with high activity of K^+_{Ca} channels.

Role of K^+_{dr} and K^+_{Ca} Channels in Determining Resting Membrane Potential

The majority (75%) of freshly dispersed smooth muscle cells had a resting membrane potential of -40 ± 2 mV ($n = 32$) and had a typical whole-cell current-voltage (I-V) relationship (Figure 28-2) with: a) a pronounced outward, 4-aminopyridine sensitive K^+_{dr} current activated at -40 mV, b) an outward tetraethylammonium and charybdotoxin sensitive K^+_{Ca} current activated at membrane potentials above 0 mV, and c) a very small inward current seen at membrane potentials more negative than -40 mV. Inhibition of K^+_{dr} channels by 4-aminopyridine (10^{-3} M) caused about a 10 to 15 mV depolarization (Figure 28-2), indicating that K^+_{dr} current is involved in maintaining the resting membrane potential. Inhibition of K^+_{Ca} channels with tetraethylammonium (5×10^{-3} M) in these cells had no effect (Figure 28-2) showing that their role in regulation of resting membrane potential in such cells is negligible. Buffering of intracellular Ca^{2+} in such cells with BAPTA-AM (2×10^{-5} M applied for 20 min) did not affect membrane potential.

In contrast to these typical cells, about 20% of smooth muscle cell had a resting membrane potential which was more negative (between -40 and -60 mV). These cells had larger outward tetraethylammonium-sensitive K^+_{Ca} currents at negative membrane potentials (Figure 28-2). In such "hyperpolarized" smooth muscle cells, inhibition of K^+_{Ca} channels with tetraethylammonium (5×10^{-3} M) or charybdotoxin (10^{-7} M), or buffering of intracellular Ca^{2+} with BAPTA-AM depolarized the cells to about -40 mV (Figure 28-2). Thus, higher resting intracellular Ca^{2+} concentration likely accounts for basal activation of K^+_{Ca} channels leading to smooth muscle cell hyperpolarization to below -40 mV.

These results suggest that K^+_{dr} channels are capable of maintaining membrane potential in resting smooth muscle cells at about –40 mV (the potential above which K^+_{dr} are activated). Any rise in resting intracellular Ca^{2+} concentration can activate K^+_{Ca} channels that will allow additional outward K^+_{Ca} current and cause hyperpolarization of smooth muscle cells below –40 mV. In these studies perforated patch technique was used instead of the commonly used whole-cell dialysis in order to avoid Ca^{2+} buffering inside the cell. Thus, in resting smooth muscle cells the activation of K^+_{Ca} channels and hyperpolarization can be achieved not only by a direct effect on K^+_{Ca} channels, but also indirectly by a rise in intracellular Ca^{2+} concentration. Such a rise in Ca^{2+} (leading to activation of K^+_{Ca} channels, hyperpolarization, and relaxation) can occur in the sub-plasma membrane areas without a global rise in cytoplasmic Ca^{2+} (Nelson *et al.*, 1995). This possibility of indirect activation of K^+_{Ca} channels by a rise in intracellular Ca^{2+} should be addressed in further studies of the mechanisms of action of different EDHF's which are thought to produce K^+_{Ca} channel-dependent hyperpolarization and relaxation of the vascular smooth muscle cells.

Nonselective cation currents and resting membrane potential

In the physiological range of membrane potentials (from –60 to 0 mV) all K^+ channels (when opened) produce an outward current which tends to hyperpolarize smooth muscle cell towards the equilibrium potential for K^+ ions ($E_K = -100$ mV under normal ion gradients). Theoretically, if there are no other opposing inward currents, the resting membrane potential would reach E_K. Practically, though, membrane potential never reaches this point because of the presence of inward currents. These currents are not well understood and are often called "leakage" currents because of their low selectivity. "Leakage" currents in smooth muscle cells reverse around 0 mV and can be mediated by both nonselective cation channels in the plasma membrane or "simple leak" through the pipette-membrane interface which will allow the flux of both cations and anions.

In a majority of freshly isolated smooth muscle cells, a very small (less than 5 pA at –80 mV) inward nonselective cation current was found (Figure 28-3). Despite of its small density (less than 0.25 pA/pF) this inward current appeared to play a major role in determining the resting membrane potential. Elimination of this inward current by substitution of Na^+ with impermeable cations N-methyl-D-glucamine (Figure 28-3) or tetraethylammonium, significantly shifted the reversal potential of the whole-cell current (Figure 28-3), and caused rapid hyperpolarization of the smooth muscle cells (Figure 28-3) from –41 ± 1 mV to –55 ± 2 mV (n = 6). Substitution of extracellular Na^+ with Li^+ did not significantly affect this current (n = 3), showing that specific ion exchangers or pumps are not responsible for this inward current. High selectivity of this small inward current to cations, and not to anions, distinguish it from a simple "leak" which would be expected to be completely nonselective, conducting both cations and anions equally well. Thus, these results show that nonselective cation channels mediate this basal inward current in smooth muscle cells, and elimination of the current causes pronounced hyperpolarization. This suggests that small inward nonselective cation currents maintain the resting membrane potential by opposing the influence of outward K^+ currents and keeping the resting membrane potential more depolarized (Figure 28-1). EDHF's could hyperpolarize smooth muscle primarily by inhibiting this nonselective inward current, rather than by activating K^+ currents.

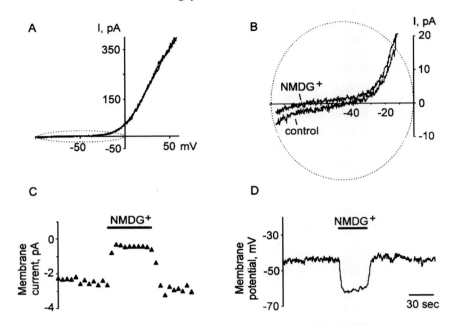

Figure 28-3.   Small nonselective cation current plays an important role in setting the resting membrane potential of smooth muscle cells. **A** and **B**. Whole-cell current-voltage relationship in control extracellular solution and when external Na^+ was completely substituted with N-methyl-D-glucamine ($NMDG^+$). **B** shows the same record as **A** but at higher resolution. **C**. Time course of the effect of $NMDG^+$ on the inward whole-cell current. Triangles represent values of current recorded at –80 mV every 5 sec. **D**. Hyperpolarizing effect of $NMDG^+$ in the same cell.

Activation of nonselective cation currents causes smooth muscle cell depolarization, and their inhibition produces hyperpolarization

There is another example of nonselective cation channels that can regulate membrane potential in smooth muscle (Figure 28-4). A novel monovalent cation current which is activated in smooth muscle cells by the removal of extracellular divalent cations has been recently described (Zakharov *et al.*, 1998). Activation of this current caused depolarization from -41 ± 1 mV to -28 ± 2 mV (n = 12). Blocking this inward current (together with the basal nonselective cation current) with N-methyl-D-glucamine, resulted in a large hyperpolarization to approximately –70 mV. This current is mediated by monovalent cation channels of very small (less than 0.4 pS) conductance which (in physiological range of membrane potential) allows Na^+ to go into the cell causing depolarization and could cause intracellular Na^+ to rise. Because this current is induced upon the removal of Ca^{2+}, these events could underlie the development of Ca^{2+} paradox-like phenomenon in vascular smooth muscle (Zakharov *et al.*, 1999).

A variety of contractile agonists is known to activate different nonselective cation channels and cause depolarization in smooth muscle cells (Isenberg, 1993). Recently, angiotensin II and thapsigargin (an inhibitor of sarcoplasmic-endoplasmic reticulum Ca^{2+}-ATPase) have been shown to activate a small (3.5 pS) nonselective cation channels which are regulated by the state of refilling of intracellular Ca^{2+} stores in smooth muscle from

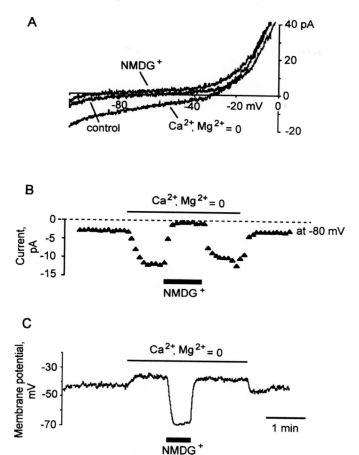

Figure 28-4. Regulation of membrane potential by nonselective cation current activated by removal of extracellular divalent cations. **A**. Current-voltage relationship (recorded during ramp depolarization) of the whole-cell current in control (contr), after removal of extracellular Ca^{2+} and Mg^{2+} (Ca^{2+}, Mg^{2+} = 0), and after all extracellular Na^+ was substituted with N-methyl-D-glucamine ($NMDG^+$) in the continuous absence of extracellular divalent cations. **B**. Time course of the changes in inward current (at −80 mV) in the same experiment. **C**. Changes in membrane potential of the same smooth muscle cell. Activation of inward current by removal of extracellular divalent cations produce cell depolarization, and inhibition of this inward current by substitution of Na^+ with $NMDG^+$ produces cell hyperpolarization.

rabbit aorta (Bolotina *et al.*, 1997; Gericke *et al.*, 1997). When a whole-cell current is activated at a holding potential of −30 mV by angiotensin II (2×10^{-6} M), it is inhibited by the nitric oxide donor, s-nitroso-n-acetyl penicillamine (SNAP, 10^{-4} M). This current could be also activated without a rise in intracellular Ca^{2+} caused by angiotensin II in cells pretreated with BAPTA-AM (2×10^{-5} M for 20 min) (Figure 28-5). The current could be inhibited by nickel (Ni^{2+}, 5×10^{-3} M), but not nifedipine (data not shown). The current-voltage relationship (Figure 28-5) obtained during ramp depolarizations applied every 30 sec (Figure 28-5) illustrates the nonselective cation nature of this current which conducted

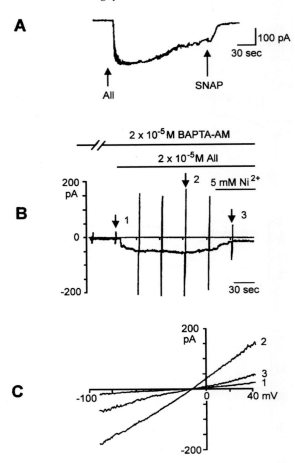

Figure 28-5. Store-operated nonselective cation current activated by angiotensin II in smooth muscle cells. **A.** Whole-cell inward current recorded at –30 mV. The current was activated with angiotensin II (AII, 2×10^{-5} M) and inhibition with SNAP (10^{-4} M). **B.** Whole-cell current was activated by angiotensin II (2×10^{-5} M) and inhibited by nickel (Ni^{2+}, 5×10^{-3} M) in a smooth muscle cell pretreated with BAPTA-AM (2×10^{-5} M) for 20 minutes. Current was recorded at a holding potential –30 mV, and ramp depolarizations were applied every 30 seconds. **C.** Current-voltage relationships of the whole-cell current during the corresponding ramps in the experiment shown in **B** : 1) control, 2) after activation with angiotensin II, 3) after inhibition with Ni^{2+}.

Na^+ (inward) and K^+ (Cs^+ in the outward direction) equally well. The reversal potential of this current in normal cells was about 0 mV, but it was shifted to about –10 mV in smooth muscle cells pretreated with BAPTA. Nonselective cation currents are also activated by endothelin-1 and inhibited by a nitric oxide donor in smooth muscle cells from the rat aorta (Minova *et al.*, 1997). As in the case of other nonselective cation currents, activation of such currents by contractile agonists will necessarily produce depolarization, and their inhibition will cause hyperpolarization. It is attractive to speculate that not only nitric oxide (Figure 28-5), but also other EDHF's can inhibit agonist-induced store-operated nonselective cation currents resulting in hyperpolarization of the smooth muscle cells and hence relaxation.

These few examples show that activation of nonselective cation channels of any kind will necessarily produce depolarization of smooth muscle cells, and their inhibition will cause hyperpolarization. The present results suggest that nonselective cation channels can be major regulators of membrane potential in resting and activated vascular smooth muscle cells, and that inhibition of these currents can cause hyperpolarization even without additional increase in the activity of K^+ channels. Under physiological conditions, non-selective cation channels, including those activated by contractile agonists, could be regulated by EDHF's.

29 Mechanisms of Action of Nitric Oxide and Metabolites of Cyclooxygenase and their Interaction with Endothelium-derived Hyperpolarizing Factor in the Rat Hepatic Artery

Peter M. Zygmunt, [1]Frances Plane, [1]Christopher J. Garland and Edward D. Högestätt

Department of Clinical Pharmacology, Institute of Laboratory Medicine, Lund University, S-221 85 Lund, Sweden. Tel: +46.(0)46.17.33.59, Fax: +46.(0)46.211.19.87, e-mail: Peter.Zygmunt@klinfarm.lu.se and Edward.Hogestatt@klinfarm.lu.se
[1]Department of Pharmacology, University of Bristol, University Walk, Bristol BS8 1TD, United Kingdom. Tel: +44.(0)171.928.7630. Fax: +44.(0)171.925.0168, e-mail: frances.plane@bristol.ac.uk and chris.garland@bristol.ac.uk

In the rat hepatic artery, acetylcholine and the calcium-ionophore A23187 induce endothelium-dependent relaxations mediated by endothelium-derived hyperpolarizing factor (EDHF), nitric oxide (NO) and a relaxing factor generated by cyclooxygenase. In contrast to EDHF, neither endothelium-derived NO nor the NO donor 3-morpholino-sydnonimine (SIN-1) hyperpolarized rat hepatic arteries. Furthermore, endothelium-derived NO and SIN-1 evoked relaxations in the presence of 30 mM KCl and in the presence of the potassium channel inhibitors apamin, charybdotoxin and glibenclamide. However, the soluble guanylate cyclase inhibitor 1H-[1,2,4]oxadiazolo[4,3-a]quinoxaline-1-one (ODQ) abolished relaxations evoked by SIN-1. The stable prostacyclin analogue iloprost and the endogenous relaxing factor produced by cyclooxygenase elicited hyperpolarization and relaxation which were unaffected by glibenclamide. Neither prostacyclin nor the product of cyclooxygenase could evoke relaxations when the extracellular K^+ concentration was raised to 30 mM. Continuous exposure of arteries to SIN-1 or iloprost did not affect the EDHF relaxation induced by acetylcholine in the presence of N^ω-nitro-L-arginine and indomethacin. Furthermore, replacement of N^ω-nitro-L-arginine with ODQ, which inhibited responses to endothelium-derived NO, did not affect the relaxation induced by acetylcholine in the presence of indomethacin indicating that an ongoing NO synthesis does not suppress responses mediated by EDHF. The results of the present study suggest that NO relaxes the rat hepatic artery via stimulation of guanylate cyclase without activation of potassium channels and subsequent membrane hyperpolarization. In contrast, iloprost and the endogenous metabolite of cyclooxygenase cause vasodilatation in rat hepatic arteries by hyperpolarizing the smooth muscle. EDHF is not regulated by NO or products of cyclooxygenase such as prostacyclin in the rat hepatic artery.

KEYWORDS: Endothelium-derived hyperpolarizing factor, iloprost, nitric oxide, potassium channels, prostanoids, vascular endothelium

INTRODUCTION

Endothelium-derived hyperpolarizing factor (EDHF), which is distinct from nitric oxide (NO) and products of cyclooxygenase, mediates vasodilatation in various vascular beds (see Cohen and Vanhoutte, 1995; Garland *et al.*, 1995) including rat hepatic, guinea-pig

253

basilar and human pial arteries (Zygmunt, Waldeck and Högestätt, 1994; Petersson *et al.*, 1995; Petersson *et al.*, 1998). The nature of EDHF is still unknown in the rat hepatic artery, but carbon monoxide (Zygmunt, Högestätt and Grundemar, 1994), cytochrome P450 monooxygeanse metabolites of arachidonic acid such as epoxyeicosatrienoic acids (Zygmunt *et al.*, 1996) and the endocannabinoid anandamide (Zygmunt *et al.*, 1997b) have been considered as EDHF candidates in this artery. Prostacyclin and NO induce relaxation of vascular smooth muscle by stimulation of adenylate and guanylate cyclase, respectively (Vane, Bunting and Moncada, 1982; Moncada and Higgs, 1993), whereas EDHF activates potassium channels without the involvement of cyclic AMP or cyclic GMP (see Cohen and Vanhoutte, 1995; Garland *et al.*, 1995). The identity of the target potassium channel(s) remains, however, to be established (Zygmunt, 1995; Corriu *et al.*, 1996b; Zygmunt and Högestätt, 1996b; Petersson, Zygmunt and Högestätt, 1997; Zygmunt *et al.*, 1997a. Part of the relaxation induced by NO and prostacyclin may involve activation of potassium channels either indirectly via cyclic nucleotide-dependent protein phosphorylation or, as suggested for NO, via a direct interaction with the channel protein (Archer *et al.*, 1994; Bolotina *et al.*, 1994; Murphy and Brayden, 1995a,b).

The contribution of EDHF to vasodilatation may be regulated by NO and EDHF may only operate when the action of NO is inhibited (Kilpatrick and Cocks, 1994; Kemp *et al.*, 1995; Bauersachs *et al.*, 1996b; Drummond and Cocks, 1996; McCulloch *et al.*, 1997). Relaxations mediated by EDHF are impaired by NO and 8-bromo cyclic GMP, possibly via an interaction with the target potassium channel(s) for EDHF or via inhibition of cytochrome P450 monooxygenase (Khatsenko *et al.*, 1993; Wink *et al.*, 1993; Olmos *et al.*, 1995; Bauersachs *et al.*, 1996b; Campbell *et al.*, 1996; Chen and Cheung, 1996; Popp *et al.*, 1996; McCulloch *et al.*, 1997). Even less is known about the effects of cyclooxygenase metabolites on relaxations mediated by EDHF.

The aim of the present study was to (a) investigate whether metabolites of cyclooxygenase contribute to endothelium-dependent relaxations, (b) to learn more about the mechanisms by which NO and metabolites of cyclooxygenase cause vasodilatation and (c) to examine the effects of these mediators on the action of EDHF in the rat hepatic artery.

METHODS

Experimental Procedure

Sprague-Dawley rats (250–300 g) were killed by CO_2 asphyxia followed by exsanguination. The hepatic artery was removed and divided into ring segments (1–2 mm long) which were suspended between two metal pins in organ chambers (2.5 ml), containing warmed (37°C) physiological salt solution of the following composition (mM): NaCl 119, $NaHCO_3$ 15, KCl 4.6, NaH_2PO_4 1.2, $MgCl_2$ 1.2, $CaCl_2$ 1.5 and (+)-glucose 6.0. The physiological salt solution was continuously bubbled with a mixture of 95% O_2 and 5% CO_2, resulting in a pH of 7.4. During an equilibration period of about one hour, the resting wall tension was adjusted to approximately 2 mN/mm vessel length. Isometric tension was measured by a force-displacement transducer (Grass Instruments FT03C, USA), connected to a polygraph (for details, see Högestätt, Andersson and Edvinsson, 1983). In order to assess the contractile capacity, each vessel ring was contracted twice, first by an isosmolar

60 mM potassium solution (prepared as the physiological salt solution except for an equimolar substitution of NaCl with KCl) and then by 10^{-5} M phenylephrine.

Relaxations induced by acetylcholine, A23187, SIN-1 and iloprost were studied in blood vessels contracted by phenylephrine, the concentration of which was titrated for each vascular ring to give a contraction equivalent to 70–90% of the initial response to 10^{-5} M phenylephrine (Zygmunt, Grundemar and Högestätt, 1994). When stable contractions were obtained, the vasodilators were added cumulatively to determine concentration-response relationships. The effects of SIN-1 and iloprost were studied on EDHF-mediated relaxations induced by acetylcholine as follows. When stable contractions by phenylephrine were obtained, either SIN-1 or iloprost was added to the organ bath until the contraction was reduced by more than 50%. The phenylephrine concentration was then increased to regain the initial level of tension before addition of acetylcholine. The incubation time with indomethacin, N^{ω}-nitro-L-arginine, apamin, charybdotoxin, glibenclamide and ODQ was at least 20 min. Each vessel ring was exposed to only one treatment.

Recording of smooth muscle membrane potantial was made on rings of the rat hepatic artery suspended in a myograph at 37°C (Waldron and Garland, 1994a). Glass microelectrodes filled with 2 M KCl and of a tip resistance between 80–120 MΩ were advanced from the adventitial side of the artery at resting tension.

Calculations and Statistics

Relaxations are expressed as the percentage reversal of the phenylephrine-induced contraction immediately before addition of the vasodilators. The maximal relaxation induced by each concentration of vasodilator was determined and used in subsequent calculations. The negative logarithm of the drug concentration eliciting half maximal relaxation (pEC_{50}) was determined by linear regression analysis, using the values immediately above and below half maximal response. E_{max} refers to the maximal relaxation achieved (100% denotes a complete reversal of the phenylephrine-induced contraction). Values are presented as mean ± s.e. mean, and n indicates the number of vascular rings (animals) examined. Statistical analysis of pEC_{50} and E_{max} values was performed by using Student's t-test (two-tailed) or multiple analysis of variance (MANOVA) followed by Bonferroni Dunn's *post hoc* test (Statview 4.12). Statistical significance was accepted when P was less than 0.05.

Drugs

Acetylcholine chloride (Aldrich, Steinheim, Germany); apamin (Alomone labs, Jerusalem, Israel); A23187, glibenclamide, N^{ω}-nitro-L-arginine, L-phenylephrine hydrochloride (Sigma, St Louis, MO, USA); synthetic charybdotoxin (Latoxan, Rosans, France); indomethacin (Confortid®, Dumex, Copenhagen, Denmark); 3-morpholino-sydnonimine hydrochloride (Biomol, Plymouth Meeting, PA, USA); iloprost (Ilomedin®, Schering AG, Berlin, Germany); ODQ (Tocris Cookson, Bristol, UK). A23187 and Glibenclamide were each dissolved in absolute ethanol, whereas ODQ was dissolved in dimethylsulfoxide. Apamin and charybdotoxin were dissolved in saline. All other drugs were dissolved in distilled water. Stock solutions of the substances were stored at −70°C.

Table 29-1 Effects of inhibitors of cyclooxygenase (indomethacin; 10^{-5} M), NO synthase (N^ω-nitro-L-arginine; 3×10^{-4} M) and the target potassium channel(s) for EDHF (apamin plus charybdotoxin; 3×10^{-7} M each) on endothelium-dependent relaxations induced by acetylcholine and A23187 in rat hepatic arteries contracted with phenylephrine

Inhibitors		Acetylcholine			A23187	
	n	pEC_{50}	E_{max} (%)	n	pEC_{50}	E_{max} (%)
Control	10	7.6 ± 0.2	99 ± 1	7	6.3 ± 0.1	92 ± 3
Indomethacin	8	7.7 ± 0.3	94 ± 1	6	5.8 ± 0.1	98 ± 1
Indomethacin + N^ω-nitro-L-arginine	11	7.6 ± 0.1	95 ± 2	11	6.0 ± 0.1	89 ± 2
Indomethacin + Apamin + Charybdotoxin	15	7.8 ± 0.1	98 ± 1	10	6.0 ± 0.1	98 ± 1
N^ω-nitro-L-arginine + Apamin + Charybdotoxin	11	6.3 ± 0.2^a	91 ± 1	9	6.2 ± 0.1	76 ± 8
Indomethacin + N^ω-nitro-L-arginine + Apamin + Charybdotoxin	12	–	-22 ± 11^a (contraction)	6	–	-44 ± 11^a (contraction)

Data are presented as mean $\pm$ s.e. mean. *n* denotes the number of vascular rings (animals).
[a]Significantly different from controls (P < 0.05).

RESULTS

Endothelium-Derived Relaxing Factors

Acetylcholine and A23187 elicited endothelium-dependent relaxations of rat hepatic arteries in the presence of N^ω-nitro-L-arginine plus indomethacin and indomethacin, apamin plus charybdotoxin, respectively (Table 29-1). Indomethacin alone or in combination with N^ω-nitro-L-arginine or with a mixture of apamin and charybdotoxin had no effect on relaxations induced by acetylcholine or A23187 (Table 29-1). However, indomethacin not only abolished the relaxations but converted them into contractions when the contribution of both EDHF and NO was prevented by apamin, charybdotoxin and N^ω-nitro-L-arginine (Table 29-1).

Membrane potential

The resting membrane potential of intact and endothelium-denuded vascular segments were -57 ± 1.5 mV (30 cells from 16 preparations) and -52 ± 2.5 mV (10 cells from 6 preparations), respectively. In intact arteries, the resting membrane potential was not affected by N^ω-nitro-L-arginine plus indomethacin (-54 ± 2.9 mV; nine cells from eight preparations), but reduced significantly to -47 ± 1.0 mV (10 cells from eight preparations) in the presence of charybdotoxin plus apamin.

Acetylcholine and A23187 (10^{-5} M each) elicited a hyperpolarization of 13 ± 1.0 mV (17 cells from 10 preparations) and 15 ± 1.3 mV (nine cells from eight preparations), respectively. The hyperpolarization induced by these agents was not affected by N^ω-nitro-L-arginine plus indomethacin (12 ± 1.2 mV for acetylcholine and 14 ± 1.5 mV for A23178; eight cells from eight preparations in each case), but reduced in the presence of apamin and charybdotoxin (7.0 ± 1.0 mV for acetylcholine and 10 ± 1.0 mV for A23187; five cells from five preparations in each case).

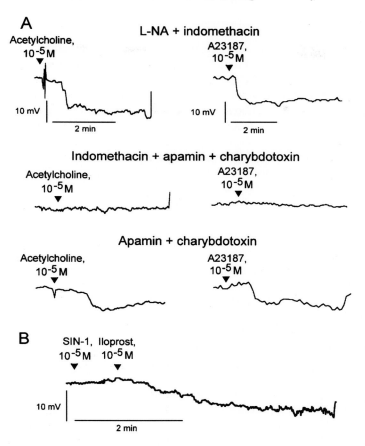

Figure 29-1. Traces showing the effects EDHF, NO and metabolites of cyclooxygenase on the membrane potential in segments of hepatic arteries at resting tension. (a) In arteries with endothelium, acetylcholine (10^{-5} M) and A23187 (10^{-5} M) evoked hyperpolarizations mediated by EDHF in the presence of indomethacin (10^{-5} M) plus N^{ω}-nitro-L-arginine (L-NA; 3×10^{-4} M) and by metabolites of cyclooxygenase in the presence of apamin (3×10^{-7} M) plus charybdotoxin (3×10^{-7} M). When apamin (3×10^{-7} M), charybdotoxin (3×10^{-7} M) and indomethacin (10^{-5} M) were present together, conditions under which endothelium-derived NO can be released and relax hepatic arteries, acetylcholine (10^{-5} M) and A23187 (10^{-5} M) failed to elicit membrane hyperpolarization. (b) Effects of SIN-1 (10^{-5} M) and iloprost (10^{-5} M) on membrane potential in a segment of hepatic artery without endothelium.

In the presence of indomethacin plus apamin plus charybdotoxin, the hyperpolarization to both acetylcholine and A23187 was abolished (four cells from four preparations in each case, Figure 29-1). SIN-1 (10^{-5} M) also did not alter the resting membrane potential of the smooth muscle cells of arterial segments without endothelium (Figure 29-1). The membrane potential before and after exposure to SIN-1 was −56 ± 2.0 mV and −57 ± 2.0 mV (five to six cells from three preparations in each case), respectively. In arterial segments without endothelium, iloprost (10^{-5} M) caused a hyperpolarization of 9.0 ± 1.5 mV (five cells from three preparations, Figure 29-1), which was unaffected by glibenclamide (8.5 ± 1.0 mV; three cells from two preparations).

Apamin and charybdotoxin

The SIN-1-induced relaxation was unaffected by the combination of apamin plus charybdotoxin (3×10^{-7} M each) in the presence of N^{ω}-nitro-L-arginine (3×10^{-4} M) and indomethacin (10^{-5} M); the pEC_{50} and E_{max} values were 6.4 ± 0.2 and $99 \pm 1\%$ in the presence and 6.7 ± 0.2 and $99 \pm 1\%$ in the absence of apamin plus charybdotoxin (n = 7). The toxin mixture was also without effect on SIN-1-induced relaxations in the absence of N^{ω}-nitro-L-arginine (n = 4, data not shown). The effect of apamin plus charybdotoxin was also tested on relaxations elicited by iloprost in the presence of N^{ω}-nitro-L-arginine and indomethacin. Again, this toxin combination was found to have no effect; the pEC_{50} and E_{max} values for iloprost were 6.9 ± 0.2 and $94 \pm 2\%$ in the presence and 7.1 ± 0.1 and $83 \pm 7\%$ in the absence of apamin and charybdotoxin (n = 5). Similar results were obtained when the experiments were performed in the absence of indomethacin (n = 4, data not shown).

Glibenclamide

Relaxations to acetylcholine (in the presence of indomethacin, apamin and charybdotoxin) or SIN-1 were unaffected by glibenclamide (10^{-5} M). The pEC_{50} and E_{max} values for acetylcholine were 7.5 ± 0.0 and $94 \pm 3\%$ in the presence and 7.5 ± 0.1 and $96 \pm 3\%$ in the absence of glibenclamide (n = 6). The pEC_{50} and E_{max} values for SIN-1 (in the presence of N^{ω}-nitro-L-arginine, indomethacin, apamin and charybdotoxin) were 6.2 ± 0.2 and $94 \pm 6\%$ in the presence and 6.5 ± 0.2 and $96 \pm 4\%$ in the absence of glibenclamide (n = 6).

   Glibenclamide (10^{-5} M) was also without significant effect on the indomethacin-sensitive relaxation induced by acetylcholine or the relaxation evoked by iloprost in the combined presence of N^{ω}-nitro-L-arginine, apamin and charybdotoxin; the pEC_{50} and E_{max} values for acetylcholine were 5.8 ± 0.1 and $84 \pm 4\%$ in the presence and 5.8 ± 0.2 and $84 \pm 4\%$ in the absence of glibenclamide (n = 7). The corresponding values for iloprost were 7.6 ± 0.2 and $94 \pm 3\%$ in the presence and 7.6 ± 0.1 and $99 \pm 1\%$ in the absence of glibenclamide (n = 5).

30 mM KCl

The indomethacin-sensitive relaxation induced by acetylcholine and the relaxation elicited by iloprost in the presence of N^{ω}-nitro-L-arginine, apamin and charybdotoxin together were abolished by 30 mM KCl (Figure 29-2), whereas acetylcholine evoked a relaxation in the presence of 30 mM KCl when N^{ω}-nitro-L-arginine was omitted (Figure 29-2). These responses were similar to those observed in normal physiological salt solution, containing indomethacin plus apamin plus charybdotoxin (Figure 29-2).

NO and Iloprost and Relaxations Mediated by EDHF

The acetylcholine-induced relaxation in the presence of N^{ω}-nitro-L-arginine (3×10^{-4} M) and indomethacin (10^{-5} M) was unaffected by the continuous presence of either SIN-1 or iloprost (Figure 29-3). Replacement of N^{ω}-nitro-L-arginine with ODQ (10^{-5} M), which significantly inhibited the relaxations induced by acetylcholine (in the presence of indomethacin plus apamin plus charybdotoxin) and those evoked by SIN-1, had no effect on the acetylcholine-induced relaxation in the presence of indomethacin (Figure 29-4).

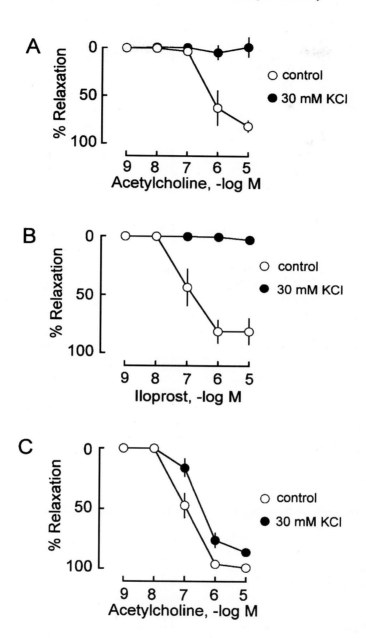

Figure 29-2. Effect of 30 mM KCl on relaxations mediated by metabolites of cyclooxygenase (a,b) and endothelium-derived NO (c) in hepatic arteries contracted with phenylephrine. Relaxations evoked by acetyl-choline (a) and iloprost (b) in the presence of N^{ω}-nitro-L-arginine (3×10^{-4} M), apamin (3×10^{-7} M) and charybdotoxin (3×10^{-7} M). (c) Acetylcholine-induced relaxations in the presence of indomethacin (10^{-5} M), apamin (3×10^{-7} M) and charybdotoxin (3×10^{-7} M) and in the additional presence of 30 mM KCl. Control experiments were performed in normal physiological salt solution (4.6 mM KCl). 100% on the y-axis denotes the amplitude of the phenylephrine-induced contraction before addition of acetylcholine or iloprost. Data are presented as means and vertical lines show s.e. mean of five to seven experiments.

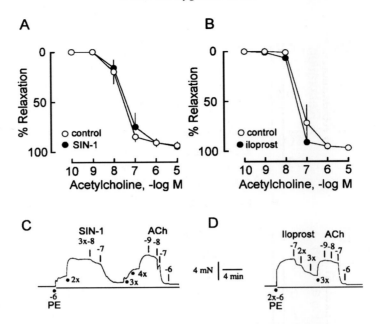

Figure 29-3.   Effects of SIN-1 (a) and iloprost (b) on EDHF-mediated relaxations induced by acetylcholine in hepatic arteries contracted with phenylephrine. SIN-1 or iloprost was given at a concentration which relaxed contracted preparations by more than 50%. The phenylephrine concentration was then increased to regain the initial level of tension. All experiments were perfomed in the presence of N^{ω}-nitro-L-arginine (3×10^{-4} M) and indomethacin (10^{-5} M). 100% on the y-axis denotes the amplitude of the phenylephrine-induced contraction before addition of acetylcholine. Data are presented as means and vertical lines show s.e. mean of five to six experiments. Traces showing the effects of SIN-1 (c) and iloprost (d) on EDHF-mediated relaxations evoked by acetylcholine (ACh) in arteries contracted by phenylephrine (PE) in the presence of N^{ω}-nitro-L-arginine (3×10^{-4} M) and indomethacin (10^{-5} M). Drug concentrations are shown as log molar concentrations. Dashed lines indicate the basal tension level before addition of phenylephrine.

DISCUSSION

Endothelium-Derived Relaxing Factors — Involvement of Cyclooxygenase

In the rat hepatic artery, acetylcholine and A23187 elicit endothelium-dependent relaxations (Zygmunt, Grundemar and Högestätt, 1994; Zygmunt, Ryman and Högestätt, 1995). The present study shows that such relaxations are mediated by at least three different mediators. The observation of a relaxing factor produced via cyclooxygenase in the rat hepatic artery is a novel finding, whereas responses mediated by NO and EDHF have been described previously in this blood vessel (Zygmunt, Grundemar and Högestätt, 1994; Zygmunt Waldeck and Högestätt, 1994; Zygmunt and Högestätt, 1996a,b).

The lack of effect of indomethacin either alone or when combined with the NO synthase inhibitor N^{ω}-nitro-L-arginine or apamin plus charybdotoxin, inhibitors of the target potassium channel(s) for EDHF (Zygmunt, 1995; Zygmunt and Högestätt, 1996b), illustrates that a role of prostanoids in endothelium-dependent relaxations cannot be ruled out unless all additional inhibitory pathways have been blocked (see also Murphy and Brayden, 1995b; Corriu *et al.*, 1996b).

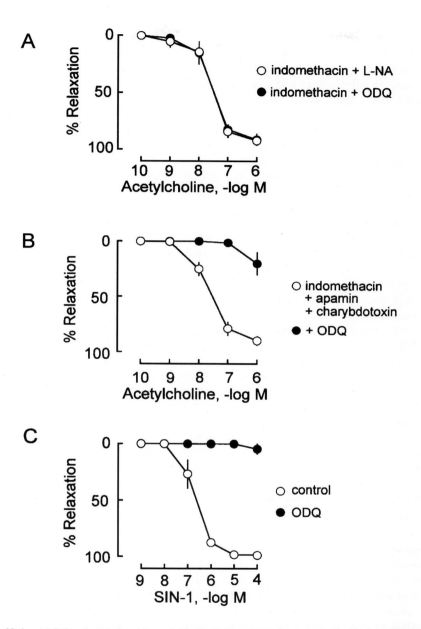

Figure 29-4. (a) Relaxations induced by acetylcholine in hepatic arteries contracted with phenylephrine in the presence of indomethacin (10^{-5} M) and either N^ω-nitro-L-arginine (L-NA; 3×10^{-4} M) or ODQ (10^{-5} M). (b) Effects of ODQ (10^{-5} M) on relaxations mediated by endogenous NO released by acetylcholine in the presence of indomethacin (10^{-5} M), apamin (3×10^{-7} M) and charybdotoxin (3×10^{-7} M) and (c) on relaxations evoked by SIN-1 in the presence of N^ω-nitro-L-arginine (L-NA; 3×10^{-4} M) and indomethacin (10^{-5} M). 100% on the y-axis denotes the amplitude of the phenylephrine-induced contraction before addition of either acetylcholine or SIN-1. Data are presented as means and vertical lines show s.e. mean of six to nine experiments.

The present study also shows that not only the relaxation but also the hyperpolarization mediated by EDHF is antagonized by apamin plus charybdotoxin in the rat hepatic artery, which further supports that hyperpolarization of the smooth muscle is the mechanism by which EDHF causes relaxation in this artery (Zygmunt, Waldeck and Högestätt, 1994). A similar effect of apamin and charybdotoxin on both relaxation and hyperpolarization mediated by EDHF has been found in small mesenteric arteries of the same species (White and Hiley, 1997; Chataigneau, *et al.*, 1998b).

Although acetylcholine and A23187 each activate the same mechanisms of relaxation in rat hepatic arteries, acetylcholine was less potent as an activator of relaxations dependent on cyclooxygenase activity than of those mediated by NO and EDHF. In bovine aortic endothelial cells higher concentrations of calcium are required for the production of prostacyclin than of NO (Parsaee *et al.*, 1992). Whether such a difference in calcium sensitivity could explain the present findings is unclear since the calcium ionophore A23187 was equally potent in activating all three types of responses.

When all mediators of relaxation were antagonized, acetylcholine and A23187 induced a contraction. The acetylcholine- but not A23187-induced response seems to be endothelium-dependent since rat hepatic arteries without endothelium do not contract when exposed to acetylcholine (Zygmunt, Ryman and Högestätt, 1995). This effect of acetylcholine could be due to the release of endothelium-derived contracting factors such as endothelin and cyclooxygenase-independent metabolites of arachidonic acid.

In the endothelium the main product of cyclooxygenase is prostacyclin, which affects both smooth muscle and platelets (Vane, Bunting and Moncada, 1982). In the present study, the stable prostacyclin analogue iloprost behaved like the relaxing factor(s) produced by cyclooxygenase. They both elicited a hyperpolarization and relaxation of similar amplitude. Furthermore, when the extracellular K^+ concentration was raised to 30 mM, conditions which prevent both hyperpolarization and relaxation mediated by EDHF (Zygmunt, Waldeck and Högestätt, 1994), neither factor could evoke a relaxation suggesting that the relaxations were indeed caused by hyperpolarization of the smooth muscle.

An indomethacin-sensitive hyperpolarization of vascular smooth muscle in response to acetylcholine has also been demonstrated in rabbit mesenteric and guinea-pig coronary and carotid arteries (Parkington *et al.*, 1993; Murphy and Brayden, 1995b; Corriu *et al.*, 1996b). In contrast to the present study, the cyclooxygenase-dependent hyperpolarization as well as the hyperpolarization induced by iloprost were considered to be of no importance for the relaxation in the guinea-pig coronary artery (Parkington *et al.*, 1993, 1995), whereas this aspect was not studied in rabbit mesenteric and guinea-pig carotid arteries (Murphy and Brayden, 1995b; Corriu *et al.*, 1996b).

In the rabbit mesenteric artery, both the endogenous hyperpolarization and that produced by iloprost were blocked by glibenclamide, an inhibitor of ATP-sensitive potassium channels (Murphy and Brayden, 1995b). However, the cyclooxygenase-dependent relaxation, and the hyperpolarization and relaxation elicited by iloprost in the rat hepatic artery were unaffected by glibenclamide at a concentration which completely prevents the hyperpolarization and relaxation induced by the potassium channel opener levcromakalim in this artery (Zygmunt, Waldeck and Högestätt, 1994). Furthermore, potassium channels sensitive to apamin and charybdotoxin also do not seem to be involved since these toxins combined did not affect the relaxation induced by iloprost. The present experiments do not allow to speculate further on the mechanism by which metabolites of cyclooxygenase induce hyperpolarization and relaxation in the rat hepatic artery.

The relaxation induced by the NO donor SIN-1 is inhibited by charybdotoxin in small mesenteric arteries of the rat (Plane *et al.*, 1996) and NO activates potassium channels sensitive to charybdotoxin in vascular smooth muscle (Archer *et al.*, 1994; Bolotina *et al.*, 1994). In other arteries the action of NO seems to involve activation of ATP-sensitive potassium channels, since glibenclamide inhibits the hyperpolarization caused by NO (Parkington *et al.*, 1995; Murphy and Brayden, 1995a,b; Corriu *et al.*, 1996b). However, in the present study, glibenclamide, apamin and charybdotoxin in combination did not affect the relaxation induced by SIN-1. Glibenclamide also had no effect on relaxations mediated by endogenous NO liberated by acetylcholine in the presence of indomethacin, charybdotoxin and apamin. Thus, in the rat hepatic artery, NO exerts its relaxant effect without the involvement of large-conductance calcium-activated potassium channels, ATP-sensitive potassium channels or the target potassium channel(s) for EDHF. Furthermore, neither endogenous NO nor SIN-1 elicited a hyperpolarization and 30 mM KCl had no effect on the acetylcholine-induced relaxation mediated by endogenous NO, indicating that potassium channel activation is not at all a mechanism by which NO relaxes this artery.

Do Nitric Oxide and Prostacyclin Inhibit the Action of Endothelium-Derived Hyperpolarizing Factor?

No evidence was found in the present study that the relaxing effect of EDHF is suppressed by NO or formed only in the presence of an inhibitor of NO synthase. Firstly, relaxations mediated by EDHF were unaffected by NO derived from SIN-1. A similar observation was also made in porcine coronary arteries (Nagao and Vanhoutte, 1992). Furthermore, the hyperpolarization caused by EDHF was unaffected by the NO donors SIN-1 in rabbit mesenteric arteries (Murphy and Brayden, 1995a) and sodium nitroprusside in rat small mesenteric arteries (Fukao *et al.*, 1997b). Secondly, relaxations mediated by EDHF in response to acetylcholine were unaltered when the action of NO but not its synthesis was prevented by the soluble guanylate cyclase inhibitor ODQ.

The activity of EDHF may, however, be down-regulated by NO and cyclic GMP in certain vascular beds (Olmos *et al.*, 1995; Bauersachs, *et al.*, 1996b; McCulloch *et al.*, 1997), possibly through an inhibitory effect of NO on cytochrome P450 monooxygenase (Khatsenko *et al.*, 1993; Wink *et al.*, 1993), the enzyme proposed to generate EDHF (Campbell *et al.*, 1996; Chen and Cheung, 1996; Popp *et al.*, 1996). Thus, the lack of effect of NO on EDHF in the present study may also support the conclusion that EDHF is not produced by cytochrome P450 monooxygenase in the rat hepatic artery (Zygmunt *et al.*, 1996). Similarly, the NO donor sodium nitroprusside was without effect on the hyperpolarization mediated by EDHF in rat small mesenteric arteries (Fukao *et al.*, 1997b).

EDHF also seems to act independently of the cyclooxygenase pathway, since iloprost was unable to affect the relaxation mediated by EDHF. The possibility that EDHF down-regulates the activity of cyclooxygenase and NO synthase has not been investigated.

30 Heterogeneity and Plasticity of Autacoid Release by the Endothelium of Resistance Arteries of the Rat

Guy. J.L. Lagaud, Peter L. Skarsgard, Ismail Laher and Cornelis van Breemen

Vancouver Vascular Biology Research Centre, for SPH-UBC and
Department of Pharmacology and Therapeutics, University of British Columbia,
Vancouver, Canada
Address for correspondence: Dr Guy Jean-Louis Lagaud, Department of Cardiovascular
Research, St. Paul's Hospital (Room 292), University of British Columbia,
1081 Burrard St Vancouver, B.C., V6T 1Z3 Canada. Tel: (604) 682–2344 ext 2722;
Fax: (604) 631-5351; e-mail: gjlagaud@prl.pulmonary.ubc.ca

In this study endothelial responses to calcium (Ca^{2+}) mobilizing agents in mesenteric arteries are compared to those in cerebral resistance arteries. Middle cerebral and small mesenteric arteries from Sprague-Dawley rats were harvested and mounted in a pressure myograph. The effects of substance P and cyclopiazonic acid were investigated in both resistance arteries with myogenic tone, in the absence or presence of the inhibitors of nitric oxide synthase (N^{ω}-Nitro-L-Arginine Methyl Ester) and cyclooxygenase (indomethacin). The effects of raised potassium (K^+), K^+ channel blockers, inhibitors of the metabolism of arachidonic acid, Na^+/K^+ ATPase inhibition and the selective cannabinoid receptor CB_1 antagonist, SR141716A were examined on the dilatation induced by cyclopiazonic acid, resistant to inhibitors of NO synthase and cyclooxygenase.

In pressurized cerebral arteries, only substance P was able to induce endothelium-dependent dilatation, while in small mesenteric arteries both cyclopiazonic acid and substance P induced endothelium-dependent dilatation. The dilatation observed in cerebral arteries was related solely to the activation of the NO pathway. In small mesenteric arteries, NO, cyclooxygenase-derived prostanoid(s) and another factor which possesses endothelium-derived hyperpolarizing factor (EDHF)-like properties were responsible for the observed vasodilatation. NO and cyclooxygenase derivative(s) were able to compensate for one another in the cyclopiazonic acid-induced endothelium-dependent vasodilatation when one of the two pathways was blocked. The EDHF-like component of dilatation induced by cyclopiazonic acid mainly activates the small Ca^{2+}-activated K^+ channels (SK_{Ca}) and to a lesser extend the large Ca^{2+}-activated K^+ channels (BK_{Ca}). Finally, the results of the present study suggest that the pathway by which agonists raise the intracellular calcium concentration ($[Ca^{2+}]_I$) may determine the nature of the endothelial secretory product.

The present investigation demonstrates that endothelium-dependent agonist responses differ between the small cerebral and mesenteric resistance arteries of the rat, and supports the concept of heterogeneity of the vascular endothelium.

KEYWORDS: endothelial heterogeneity, myogenic tone, EDHF, cerebral and mesenteric resistance arteries

INTRODUCTION

Heterogeneity in the responsiveness of blood vessels is to a large extend related to the variability in the types and densities of pharmacological receptors and ion transport mechanisms of smooth muscle (see, Mulvany and Aalkjaer, 1990). In addition there are a number of reports indicating that endothelial secretions may vary dependent on the size and location of the artery (Galle *et al.*, 1993; Hwa *et al.*, 1994; Archer *et al.*, 1996; Clark and Fuchs, 1997). The endothelium can release contracting factors (Luscher *et al.*, 1992) and relaxing factors such as nitric oxide (NO; Furchgott and Zawadzki, 1980; Palmer, Ferridge and Moncada, 1987), prostacyclin (Moncada *et al.*, 1977) and endothelium-derived hyperpolarizing factor (EDHF; Rubanyi and Vanhoutte, 1987; Chen, Suzuki and Weston, 1988; Taylor and Weston, 1988; Furchgott and Vanhoutte, 1989). The release of EDHF from the endothelium is triggered by agonists, such as acetylcholine, bradykinin, or substance P, and has been demonstrated in various arteries from different species (see Garland *et al.*, 1995; Cohen and Vanhoutte, 1995; Mombouli and Vanhoutte, 1997). The identity of EDHF has not yet been established but its action is believed to be the activation of K^+ channels in vascular smooth muscle cells. However the types of K^+-channels stimulated and also the endothelial secretion responsible for this stimulation seem to vary according to the preparation investigated. Indeed, EDHF activates ATP-activated K^+-channels (K_{ATP}) in rabbit middle cerebral and mesenteric arteries (Standen *et al.*, 1989), basilar artery (Plane and Garland, 1993), aorta and carotid artery (Cowan *et al.*, 1993) of rabbit. It activates large conductance Ca^{2+}-activated K^+-channels (BK_{Ca}) in guinea-pig carotid artery (Corriu *et al.*, 1996), rat and rabbit mesenteric arteries (Hwa *et al.*, 1994; Khan, Mathews and Meisheri, 1993; Asano, Masuzawa-Ito and Matsuda, 1993a), and dog middle cerebral artery and mesenteric arteries (Asano *et al.*, 1993). Furthermore, it has been found that EDHF activates small conductance Ca^{2+}-activated K^+-channels (SK_{Ca}) in guinea pig, bovine and porcine coronary arteries (Eckman, Frankovich and Keef, 1992; Hecker *et al.*, 1994), and in rat and rabbit mesenteric arterial beds (Adeagbo and Triggle, 1993; Murphy and Brayden, 1995b).

To further complicate matters, NO causes hyperpolarization in rabbit carotid artery (Cohen *et al.*, 1997) and activates BK_{Ca} in arterial smooth muscle cells (Bolotina *et al.*, 1994; Weidelt, Boldt and Markwardt, 1997) and K_{ATP} (Murphy and Brayden, 1995b; Weidelt, Boldt and Markwardt, 1997) in mesenteric arteries. In the mesenteric vascular bed, including the main mesenteric artery (Fujii *et al.*, 1992) and its smaller branches (McPherson and Angus, 1991; Garland and McPherson, 1992), the action of EDHF is not inhibited by K_{ATP} channels blockers. The inhibitory influence of charybdotoxin suggests rather the involvement of BK_{Ca} channels (Hwa *et al.*, 1994; Khan, Mathews and Meisheri, 1993). Studies performed in bovine, porcine and rat coronary arteries indicate that EDHF may be a metabolite of arachidonic acid, derived from cytochrome P450-dependent mono-oxygenase (Bauersachs, Hecker and Busse, 1994; Hecker *et al.*, 1994), but this does not seem to be the case in the rat mesenteric vascular bed (Vanheel and Van de Voorde, 1997).

The above emphasizes the complexity of vascular endothelial secretions as well as the need to explore further the fraction of the endothelial response due to EDHF when both the tissue and the agonist are varied. Another unresolved question remains whether the same EDHF released from different arteries targets different K^+ channels depending on the location of the smooth muscle, or that there are multiple EDHFs each targeting a specific K^+ channel.

Comparative studies on resistance arteries have revealed that cerebral arteries differ pharmacologically and electrophysiologically from arteries in other areas. For example, cerebral arteries are more sensitive than peripheral arteries to dihydropyridine Ca^{2+} antagonists and agonists (Cauvin, Loutzenhiser and Van Breemen, 1983; Asano *et al.*, 1987) and display very tight coupling between electrical and mechanical events (Harder and Waters, 1984). However to date no systematic studies comparing endothelial function in cerebral and peripheral resistance arteries have been carried out. Hence, experiments were designed to compare the endothelial responses of cerebral resistance arteries to a range of vasoactive agents with those of small mesenteric arteries, which contribute to peripheral resistance (Mulvany and Aalkjaer, 1990; Fenger-Gron, Mulvany and Christensen, 1995). The relative contribution of EDHF to the total endothelium mediated relaxation as a function of the mechanism whereby the agonist enhances the cytoplasmic calcium concentration is explored. A unique set of targets for EDHF released from the endothelium of small mesenteric arteries is identified. In order to recreate the most physiological relevant conditions, arteries were constricted by pressure rather than agonists. In the present study, the use of an agonist (substance P) and an inhibitor of endoplasmic reticulum Ca^{2+}-ATPase (cyclopiazonic acid), provide conclusive evidence for EDHF and identify its targets as small conductance Ca^{2+}-activated K^+-channels and large conductance Ca^{2+}-activated K^+-channels but not ATP-activated K^+-channels.

METHODS

Vessel Isolation and Cannulation

Male Sprague-Dawley rats (200–300g) were anesthetized with intraperitoneal injections of sodium pentobarbital (Somnotol, 30 mg/kg) and heparin (Hepalean, 500 U/kg), and then killed by decapitation. The brain or small intestine with attached mesentery was excised and transferred to a dissection dish filled with physiological salt solution at 4°C. Cerebral or mesenteric [third or fourth-generation arteries; Sun *et al.* (1992)] arteries were dissected from surrounding connective tissues, and transferred to the experimental chamber of an arteriograph filled with oxygenated physiological salt solution at 37°C.

Each blood vessel was tied onto a proximal glass microcannula with tip diameter of 60–80 μm using single strands (20 μm) of 4-0 braided nylon suture; the perfusion pressure was gently raised to clear the vessel of blood. The distal end of the artery was similarly mounted to the outflow microcannula. After several minutes of perfusion, the distal outflow cannula was closed, and the transmural pressure was slowly increased to 80 mm Hg by using an electronic pressure servo system (Living Systems, Burlington, VT).

The physiological salt solution in the vessel chamber was continuously recirculated by superfusion around the pressurized artery at a flow of 20–25 mL/min passing through an external reservoir that was bubbled with a gas mixture of 95% O_2, 5% CO_2. A heating pump connected to a glass heat exchanger warmed the physiological salt solution to 37°C, and a pH micro-probe was positioned in the bath; pH was maintained at 7.4 ± 0.04 by adjustment of the reservoir gassing rate.

The arteriograph containing a cannulated pressurized artery was placed on the stage of an inverted microscope with a monochrome video camera attached to a viewing tube and allowed to equilibrate for 60 min. Arterial dimensions were measured using a video system that provides automatic continuous readout measurements of luminal diameter and wall thickness (Halpern, Osol and Coy, 1984). The information is up-dated every 17 ms, and the precision of the diameter measurement is within 1%. Cerebral and mesenteric arterial myogenic tone developed spontaneously and consistently during equilibration, resulting in significantly reduced luminal diameter. Once attained, it remains stable for hours unless perturbed by changes in transmural pressure or the addition of vasoactive compounds (Sun *et al.*, 1992; Skarsgard, Van Breemen and Laher, 1997).

In some experiments, the endothelium was removed by intraluminal perfusion with 0.5% 3-[(3 Cholamidopropryl) dimethylammomio]-1-propanesulfonate (CHAPS) for 30 s (Andriantsitohaina *et al.*, 1996). The presence of functional endothelium was assessed in all preparations by the ability of 10^{-7} M substance P or 10^{-6} M acetylcholine to induce dilatation.

Experimental Procedure

Having established myogenic profile for these arteries, an investigation of endothelial function was undertaken in pressurized vessels.

To identify heterogeneity in receptor- and non receptor-mediated responsiveness between cerebral and mesenteric vascular beds, responses to substance P (10^{-8} M), cyclopiazonic acid (2.10^{-5}) and an exogenous NO donor, sodium nitroprusside (10^{-5} M) were investigated. A pilot study determined that the concentrations noted above yield a maximum response in cerebral and small mesenteric arteries. These vasoactive compounds were added to the circulating buffer at the final concentrations reported above, and the resulting changes in luminal diameter was measured.

Signaling in endothelial heterogeneity

To test the role of NO in the dilatation induced by various agents, a competitive inhibitor of nitric oxide synthases, N$^\omega$-Nitro-L-Arginine Methyl Ester (L-NAME; 2.10^{-4} M) was added to the superfusing buffer and allowed to circulate for 20 minutes until a new steady state diameter was reached. This was followed by reassessment of the vasodilatation provided by the agents.

Similarly, effects of cyclooxygenase blockade and depolarization (0 or 40 mM K$^+$ solution; i.e. in which NaCl was substituted for an equimolar concentration of KCl) were investigated in vessels challenged with cyclopiazonic acid. Increasing concentrations of cyclopiazonic acid (10^{-8}–10^{-3} M) were exposed to the arteries in the absence and in the presence of indomethacin, used at the maximally active concentration (10^{-5} M). Effects of cyclopiazonic acid were also investigated in normal physiological solution, in 0 mM or 40 Mm K$^+$ solution in the presence of indomethacin and N$^\omega$-Nitro-L-Arginine Methyl Ester. K$^+$ solution at 0 or 40 mM cause a fixed depolarization of the vascular smooth muscle; this treatment thus blocks cell hyperpolarization as a signalling event in cyclopiazonic acid induced vasodilatation.

Dilatation resistant to inhibitors of NO synthase and cyclooxygenase

The possible involvement of K$^+$ channels in the vasodilatation induced by cyclopiazonic acid in myogenically active vessels was assessed in the presence of indomethacin and N$^\omega$-Nitro-L-Arginine Methyl Ester. In these experiments, K$^+$ channel blockers were used at the concentrations known to selectively inhibit specific K$^+$ channels: tetraethylammonium (3 to 5.10^{-3} M), charybdotoxin (10 to 15.10^{-10} M), iberiotoxin (3.10^{-10} M) to inhibit large conductance Ca^{2+}-activated channel (SK_{Ca}), glibenclamide (10^{-5} M) to inhibit ATP-activated K^+-channels (K_{ATP}), 4-aminopyridine (10^{-4} M) to inhibit voltage gated K^+ channel (K_v) and barium chloride ($BaCl_2$; 3.10^{-5} M) to inhibit inward rectifier channels (K_{IR}).

Similarly, possible roles of metabolites of arachidonic acid, activation of Na^+/K^+ ATPase and activation of cannabinoid receptor CB_1 were investigated using the following inhibitors: quinacrine (1 to 5.10^{-5} M) and oleyloxyethylphosphorylcholine (OOPC; 10^{-6} to 10^{-5} M), inhibitors of phospholipase A_2; SKF 525a (10^{-5} to 10^{-4} M), clotrimazole (1 to 3.10^{-5} M) and 17-octadecynoid acid (17-ODYA; 2, 4 and 5.10^{-5} M), inhibitors of cytochrome P450; dihydroouabain (DHO; 3.10^{-5} M), a Na^+/K^+ATPase blocker and SR141617A (10^{-6} to 5.10^{-6} M), a cannabinoid receptor antagonist. The effect of the endogenous cannabinoid, anandamide (10^{-6} M) was also determined.

To obtain the same level of tone in myogenically active vessels phenylephrine (10^{-9} or 10^{-8} M) was added to some vessels when NO synthase inhibitor was used in combination with indomethacin plus one or several inhibitors. All the inhibitors were incubated with the myogenically active vessels for 20 minutes before cyclopiazonic acid was added.

At the conclusion of each experiment, the superfusion solution was changed to a calcium-free physiological salt solution that contained 2 mM EGTA and no $CaCl_2$. Vessels were incubated for 20 minutes and then the pressure steps were repeated to obtain the "passive" diameter of each vessel at each pressure in order to calculate the percentage of myogenic constriction

Expression of Results and Statistical Analysis

Myogenic tone at any given pressure was expressed as a percent decrease diameter from the "passive" diameter or

$$\text{percent constriction} = 100\% \times [(D_{Ca\text{-free}} - D_{PSS})/D_{Ca\text{-free}}]$$

where D is the arterial diameter in calcium free or normal physiological salt solution (PSS).

Vasodilator responses were expressed as percent increase in diameter from the initial diameter (due to myogenic tone) at the corresponding pressure.

$$\text{percent dilatation} = 100\% \times [(D_x - D_{mt})/(D_{Ca\text{-free}} - D_{mt})]$$

where D is the measured arterial diameter, and subscripts X, mt and Ca-free denote arterial diameters at each dose of agonists (X), initially (mt), and in Ca^{2+}-free buffer (Ca-free).

All results are expressed as mean ± S.E.M of n experiments. One vessel was taken from each animal. Statistical evaluation was done by analysis of variance (ANOVA). Means were considered significantly different when P was less than 0.05.

Drugs and Solutions

The ionic composition of the physiological salt solution was (in mM): NaCl 119, KCl 4.7, KH_2PO_4 1.18, $NaHCO_3$ 24, $MgSO_4.7H_2O$ 1.17, $CaCl_2$ 1.6, glucose 5.5. and EDTA 0.026. Acetylcholine chloride, A23187, 4-Amino-pyridine, $BaCl_2$, bradykinin, charybdotoxin, clotrimazole, CHAPS, dihydroouabain, glibenclamide, Hb, histamine, indomethacin, L-

NAME, phenylephrine, tetraethylammonium, 17-ODYA and SKF 525a were purchased from Sigma (Ontario, Canada). Anandamide, apamin, cyclopiazonic acid, iberiotoxin, OOPC and sodium nitroprussiate were purchased from Calbiochem (San Diego, CA, USA). Quinacrine was purchased from Research Biochemicals International (Boston, MA, USA). SR141716A was a generous gift from Sanofi (Montpellier, France). Stocks solution were diluted in deionized water (NANOpure). Cyclopiazonic acid, indomethacin, glibenclamide were prepared in dimethyl sulphoxide. Anandamide, clotrimazole, 17-ODYA, SKF 525a and SR141716A were dissolved in absolute ethanol. Reduced hemoglobin was prepared by treatment of hemoglobin solution with sodium dithionite (Martin *et al.*, 1985). The effects of ethanol and other solvents were tested at the dilution used, and none of the vehicle solutions altered the pressure-diameter relation or the vascular responses to norepinephrine and acetylcholine.

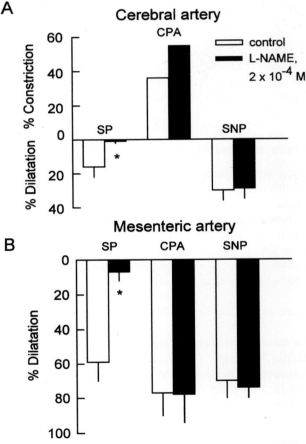

Figure 30-1. Comparative effects of vasoactive agents in rat distal middle cerebral and small mesenteric arteries with myogenic tone. Histograms showing the responses induced by substance P (SP; 10^{-8} M), cyclopiazonic acid (CPA; 2.10^{-5} M) and sodium nitroprusside (SNP; 10^{-5} M) in the absence (open columns) and in the presence (black columns) of N^{ω}-Nitro-L-Arginine Methyl Ester (L-NAME; 2.10^{-4} M). The values are mean ± S.E.M. (vertical bars) of 7 experiments. *P was less than 0.05 significantly different from the percentage of the dilatation obtained in the absence of L-NAME.

RESULTS

A pilot study performed initially in distal middle cerebral and small mesenteric arteries showed that cerebral arteries (range of 70–120 μm; mean 96.1 ± 18.2 μm; n = 7) developed graded myogenic constrictions over the physiologic pressure range. A maximal constriction ($30.12 \pm 5.5\%$) was obtained at 60 mm Hg. Mesenteric resistance-sized arteries (mean 100 ± 19.3 μm, n = 7) developed myogenic constrictions. Maximal constriction ($26.8 \pm 5.6\%$) was achieved at 80 mm Hg.

Vasomotion of cerebral and mesenteric arteries

In distal middle cerebral arteries (Figure 30-1), only substance P and sodium nitroprusside induced dilatation. Cyclopiazonic acid caused constriction. In the presence of NO synthase inhibitor, the dilatation induced by substance P was abolished and cyclopiazonic acid caused greater constrictions, while, sodium nitroprusside induced a dilatation that was not significantly different from control.

The effects of substance P, cyclopiazonic acid and sodium nitroprusside were investigated on myogenically active small mesenteric arteries (Figure 30-1). All the agents tested induced dilatation. N^{ω}-Nitro-L-Arginine Methyl Ester almost abolished the dilatation to substance P, but had no effect on the responses to cyclopiazonic acid and sodium nitroprusside. Removal of the endothelium completely abolished cyclopiazonic acid-induced dilatations (Table 30-1). Similar results were obtained with thapsigargin (10^{-5} M), another inhibitor of sarcoplasmic reticulum Ca^{2+}-ATPase (not shown).

Vasodilatation resistant to NO synthase inhibitor in mesenteric arteries

Since the NO synthase inhibitor-resistant components of cyclopiazonic acid dilatation could be due to endothelial production of prostacyclin, indomethacin was used to inhibit cyclooxygenase. Indomethacin or N^{ω}-Nitro-L-Arginine Methyl Ester alone did not affect the dilatation produced by cyclopiazonic acid (Figures 30-2, 30-3). The combination of indomethacin plus N^{ω}-Nitro-L-Arginine Methyl Ester partially affected cyclopiazonic acid-induced dilatation of small mesenteric arteries.

Table 30-1 Effects of physiological salt solution without potassium (K^+; 0 mM), with 40 mM K^+, the removal of endothelium or effects of the CB_1 antagonist SR141716A (5.10^{-6} M) on the response of cyclopiazonic acid (2.10^{-5} M) and anandamide (10^{-6} M) resistant to inhibitors of NO synthase and cyclooxygenase in small mesenteric resistance arteries of the rat pressurized at 80 mm Hg.

Treatment	Cyclopiazonic acid (percent dilatation)	Anandamide (percent dilatation)
(L-NAME + indomethacin)	74 ± 12	74 ± 1
0 mM K^+ solution	$2 \pm 1*$	$9 \pm 13*$
+ 40 mM K^+ solution	$-44 \pm 9*$	$18 \pm 3*$
endothelium removal	$-36 \pm 1*$	74 ± 1
+ SR141716A	72 ± 8	71 ± 6

Values are mean $\pm$ S.E.M of 3 to 5 experiments. *P was less than 0.05 significantly different as compared to control values. Negative values correspond to contraction.

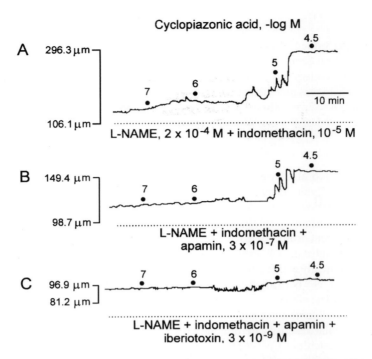

Figure 30-2. Dilatation independent of the NO synthase/cyclooxygenase pathways induced by cyclopiazonic acid in rat small mesenteric arteries pressurized at 80 mm Hg. Diameter representative traces of 5 to 7 experiments. Concentration-response curves to cyclopiazonic acid (10^{-7} M-3.10^{-5} M) in rat small mesenteric arteries with myogenic tone at 80 mm Hg in the presence of (A) N^{ω}-Nitro-L-Arginine Methyl Ester (L-NAME; 2.10^{-4} M) plus indomethacin (10^{-5} M), (B) L-NAME plus indomethacin plus apamin (3.10^{-7} M) and (C) L-NAME plus indomethacin plus apamin plus iberiotoxin (3.10^{-10} M).

Fixed vascular smooth muscle cell depolarization (0 or 40 mmol/L KCl) has been utilized to see if this could abolish the dilatation resistant to inhibitors of NO synthase and cyclooxygenase. The changes in K^+ concentration produced an increase in contraction of the arteries ($-27 \pm 7.6\%$; n = 4). Exposure to 0 or 40 mM K^+ solution abolished the dilatation induced by cyclopiazonic acid resistant to inhibitors of NO synthase and cyclooxygenase (Table 30-1). The dilatation produced by cyclopiazonic acid, resistant to N^{ω}-Nitro-L-Arginine Methyl Ester plus indomethacin and abolished by 0 or 40 mM K^+ solution, was abolished by endothelium removal (Table 30-1). The dilatation induced by sodium nitroprusside ($71 \pm 10\%$, n = 3 in the presence of endothelium) was preserved after exposure to 40 mM K^+ solution ($66 \pm 3\%$, n = 3) or in the absence of endothelium ($72 \pm 8\%$, n = 3).

EDHF signalling

Control experiments to determine the effects of the K^+ channel blockers on the pressurized small mesenteric arteries showed that charybdotoxin or iberiotoxin and tetraethylammonium induced slight but significant increases in tone (data not shown).

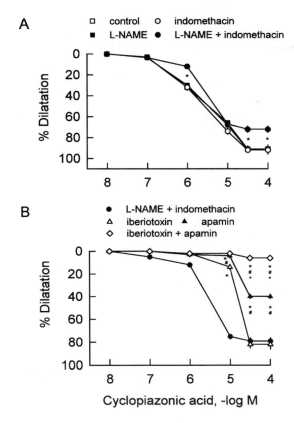

Figure 30-3. Mechanisms of cyclopiazonic acid-induced dilatation resistant to N$^\omega$-Nitro-L-Arginine Methyl Ester (L-NAME; 2.10^{-4} M) and indomethacin in rat small mesenteric arteries with myogenic tone at 80 mm Hg. (A) Concentration-response curves to cyclopiazonic acid in small mesenteric arteries of the rat pressurized at 80 mm Hg in the absence (open squares) and in the presence of the following drugs: 2.10^{-4} M N$^\omega$-Nitro-L-Arginine Methyl Ester (L-NAME; filled squares); 10^{-5} M indomethacin (open circles); L-NAME plus indomethacin (filled circles). (B) Concentration-response curves to cyclopiazonic acid in small mesenteric arteries of the rat pressurized at 80 mm Hg in the presence of the following drugs: 2.10^{-4} M N$^\omega$-Nitro-L-Arginine Methyl Ester (L-NAME) plus 10^{-5} M indomethacin (filled circles); L-NAME plus indomethacin plus 3.10^{-10} M iberiotoxin (open triangles); L-NAME plus indomethacin plus 3.10^{-7} M apamin (filled triangles); L-NAME plus indomethacin plus iberiotoxin plus apamin (open diamonds). The values are mean ± S.E.M. (vertical bars) of 5 to 7 experiments. The asterisk indicates a statistical significant difference (P was less than 0.05) as compared to the percentage of the dilatation induced by cyclopiazonic acid in the absence of any drugs. #P was less than 0.05 significantly different as compared to the percentage of the dilatation induced by cyclopiazonic acid in the presence of L-NAME plus indomethacin. +P was less than 0.05 significantly different as compared to the percentage of the dilatation induced by cyclopiazonic acid in the presence of L-NAME plus indomethacin plus iberiotoxin.

Blockade of the BK$_{Ca}$ channel with iberiotoxin did not abolish the dilatation to cyclopiazonic acid resistant to N$^\omega$-Nitro-L-Arginine Methyl Ester plus indomethacin (Figure 30-3). Similar results were obtained using other K$^+$ channel blockers (tetraethylammonium, charybdotoxin, BaCl$_2$, 4-amino-pyridine, glibenclamide) (Table 30-2). Only apamin (a blocker of SK$_{Ca}$) was effective in reducing the NO synthase inhibitor plus indomethacin-resistant responses to cyclopiazonic acid (Figures 30-2, 30-3). The combination of apamin

Table 30-2 Effects of K^+ channels blockers on cyclopiazonic acid (2.10^{-5} M) response resistant to inhibitors of NO synthase (N^ω-Nitro-L-Arginine Methyl Ester; L-NAME) and cyclooxygenase (indomethacin) in small mesenteric resistance arteries of the rat pressurized at 80 mm Hg.

Treatment	Cyclopiazonic acid (percent dilatation)
L-NAME + indomethacin	74 ± 12
+ glibenclamide (10^{-5} M)	74 ± 10
+ BaCl$_2$ (3.10^{-5} M)	74 ± 18
+ 4-amino-pyridine (10^{-4} M)	85 ± 7
+ tetraethylammonium (3.10^{-3} M)	71 ± 8
+ iberiotoxin (3.10^{-10} M)	81 ± 4
+ charybdotoxin (15.10^{-10} M)	83 ± 6
+ apamin (3.10^{-7} M)	16 ± 3*
+ apamin + iberiotoxin	2 ± 1* # +

Values are mean ± S.E.M. of 3 to 6 experiments. *P was less than 0.05 significantly different as compared the vasodilatation obtained in the presence of NO synthase inhibitor plus indomethacin. #P was less than 0.05 significantly different as compared to the dilatation obtained in the presence of NO synthase inhibitor plus indomethacin plus iberiotoxin.+P was less than 0.05 significantly different as compared to the dilatation obtained in the presence of NO synthase inhibitor plus indomethacin plus apamin plus iberiotoxin.

Table 30-3 Effects of inhibitors of cytochrome P450, phospholipase A$_2$, and Na$^+$/K$^+$ ATPase on cyclopiazonic acid (2.10^{-5} M) response resistant to inhibitors of NO synthase (N^ω-Nitro-L-Arginine Methyl Ester; L-NAME) and cyclooxygenase (indomethacin) in small mesenteric resistance arteries of the rat pressurized at 80 mm Hg.

Treatment	Cyclopiazonic acid (percent dilatation)
L-NAME + indomethacin	88 ± 15
+17-ODYA (2.10^{-5} M)	90 ± 10
+ SKF 525A (10^{-5} M)	90 ± 7
+ Clotrimazole (3.10^{-5} M)	90 ± 13
+ Quinacrine (10^{-5} M)	90 ± 12
+ OOPC (10^{-5} M)	91 ± 7
+ Dihydroouabain (3.10^{-5} M)	90 ± 12

Values are mean ± S.E.M of 3 to 5 experiments.

plus iberiotoxin resulted in near complete inhibition of the dilatation (Figures 30-2, 30-3).

Quinacrine, OOPC, 17-ODYA, SKF 525a, clotrimazole, and dihydroouabain had no effects on the N^ω-Nitro-L-Arginine Methyl Ester plus indomethacin-resistant responses to cyclopiazonic acid (Table 30-3). In small mesenteric artery 17-ODYA and OOPC blocked myogenic tone only at higher concentrations (data not shown). In this case, phenylephrine (10^{-8}–10^{-7} M) was added in the superfusion bath to boost the tone in order to study the mechanisms of the dilatation produced by cyclopiazonic acid.

Anandamide (10^{-6} M) induced endothelium-independent dilatation, resistant to NO synthase inhibitor plus indomethacin (Table 30-1). The CB_1 antagonist SR141716A (5.10^{-6} M) did not modify significantly anandamide-induced dilatation and had no effects on the N^ω-Nitro-L-Arginine Methyl Ester plus indomethacin-resistant responses to cyclopiazonic acid (Table 30-1). Exposure to 0 or 40 mM K^+ solution in the continuous presence of NO synthase inhibitor plus indomethacin inhibited the relaxation to anandamide (Table 30-1).

DISCUSSION

Comparison between cerebral and mesenteric resistance arteries has revealed profound differences in reactivity related to endothelium-mediated vasodilatation.

The present study indicates evidence for site specific adaptation in endothelial function. Indeed, in cerebral arteries substance P induced relaxation. The pattern was different in the mesenteric resistance arteries, where both substance P and cyclopiazonic acid induced endothelium-dependent relaxation.

A third type of variability was related to the nature of endothelial secretion stimulated by the different pharmacological agents. Thus substance P released mainly NO but no EDHF, while cyclopiazonic acid released mainly EDHF, but no NO. A small portion of the relaxation to cyclopiazonic acid could be attributed to the release of a prostaglandin probably prostacyclin I_2.

Since both the release of NO and EDHF are initiated by an increase in endothelial Ca^{2+} concentration, it is possible that differences exist in the pattern of $[Ca^{2+}]_I$ fluctuations which are important in determining the nature of the endothelial secretion. For example eNOS is associated with the plasmalemma (Pollock *et al.*, 1993; Forstermann *et al.*, 1994) and may be preferentially activated by agonist stimulated Ca^{2+} influx, while EDHF may be synthetized by an endoplasmic reticulum associated enzyme, which is more effectively activated by blockade of Ca^{2+}-dependent ATPase in sarcoplasmic/endoplasmic reticulum.

In myogenically active cerebral arteries, substance P produced endothelium dependent dilatation which was completely inhibited by NO synthase inhibitor indicating release of NO. The finding that distal middle cerebral arteries release mainly NO is consistent with previous studies performed in different types of cerebral arteries (Faraci and Brian, 1994; Ngai and Winn, 1995; Zimmermann *et al.*, 1997). However, cyclopiazonic acid caused vasoconstriction. The contractions observed were enhanced to the same extent by addition of NO synthase inhibitor or the removal of the endothelium. The observed discrepancy between the effects of substance P and cyclopiazonic acid may be related to differences in sensitivity of smooth muscle to the vasoactive agents.

For comparison with cerebral resistance arteries we chose small mesenteric arteries of similar size were used (Fenger-Gron, Mulvany and Christensen, 1995). The results showed that substance P and cyclopiazonic acid produced endothelium-dependent dilatations of the small mesenteric arteries, consistent with other studies in different vascular beds showing that these agents stimulate the release of EDRF (Garland *et al.*, 1995; Cohen and Vanhoutte, 1995). As in the cerebral arteries, the vasodilatation induced by substance P appeared to be mediated by NO, since the response was abolished by NO synthase inhibitor. However, dilatation in response to cyclopiazonic acid was due to NO-independent pathways.

The existence of a dilatation in small mesenteric resistance arteries which is resistant to inhibitors of NO synthase and cyclooxygenases suggests an alternative pathway, (see in Mombouli and Vanhoutte, 1997). Depolarizing solutions without K^+ or containing a high K^+ concentration have been used to prevent endothelium-dependent hyperpolarization (Adeagbo and Triggle, 1993; Fukao *et al.*, 1995a). In the present study, addition of 0 or 40 mM of K^+ solution abolished the cyclopiazonic acid-induced response resistant to inhibitors of NO synthase and cyclooxygenases. This finding indicates that hyperpolarization is likely to be responsible for the cyclopiazonic acid-induced dilatation. Exposure to high K^+ concentrations does not affect NO-mediated relaxations in a number of blood vessels including rat small mesenteric arteries (Plane and Garland, 1993; Fukao *et al.*, 1995a; Chen and Cheung, 1997) and does not impair the increase in content of cyclic GMP (Kuhberger *et al.*, 1994).

The hyperpolarization and the subsequent relaxation produced by EDHF involve the activation of K^+ channels in vascular smooth muscle cells (Chen, Suzuki and Weston, 1988; Taylor and Weston, 1988). A good correlation between acetylcholine-induced hyperpolarization and relaxation (measured simultaneously) exists in the rat mesenteric vascular bed (McPherson and Angus, 1991). The present results indicate that the cyclopiazonic acid-induced dilatation of small mesenteric arteries which is resistant to inhibitors of NO synthase and cyclooxygenases does not involve ATP-dependent K^+ channels, voltage-gated K^+ channels or inwardly-rectifying K^+ channels. These results are in contrast to data indicating that EDHF can activate K_{ATP}, K_v, K_{ir} in different vascular preparations (Cowan *et al.*, 1993; Plane and Garland, 1993; Weidelt, Boldt and Markwardt, 1997). These different results may correspond to the presence of different K^+ channels or differential mechanisms of activation by EDHF. Overall, these studies indicate that Ca^{2+}-activated K^+ channels play a major role mediating to EDHF responses although there are differences as to the relative involvement of the large and small conductance K^+ channels among arteries.

In the rat mesenteric preparation, cyclopiazonic acid and thapsigargin induce release of EDHF by increasing the intracellular calcium concentration ($[Ca^{2+}]_i$; Fukao *et al.*, 1995a). The present results suggest a major contribution by the small conductance, Ca^{2+}-activated K^+ channels since apamin was most effective in blocking the cyclopiazonic acid-induced relaxation resistant to inhibitors of NO synthase and cyclooxygenases. Abolition of the dilatation of arteries could be achieved only when the two types of Ca^{2+}-activated K^+ channels were blocked by apamin and iberiotoxin. Thus, apamin-sensitive channels seem to be dominant in the dilator response, while the contribution of iberiotoxin-sensitive channels is unmasked when the apamin channels are blocked. Another possibility is that the release of EDHF is modified by the depolarization of endothelial cells by iberiotoxin (Fukao *et al.*, 1995a), thus decreasing the electrical driving force for endothelial Ca^{2+}-entry. The inhibition could thus be due to a combination of the action of iberiotoxin on endothelial cells and apamin on smooth muscle. The current study provides evidence that both types of Ca^{2+}-activated K^+ channels (large and small) participate in the hyperpolarization and dilatation induced by cyclopiazonic acid in myogenically active small mesenteric resistance arteries of the rat.

EDHF could be a cytochrome P450-derived metabolite of arachidonic acid (Hecker *et al.*, 1994; Campbell *et al.*, 1996a). In the present study, three approaches were adopted to test this proposition. Firstly, tissues were exposed to the suicide-substrate inhibitor 17-ODYA, which is a potent inhibitor of the cytochrome P450-dependent epoxygenase responsible for the formation of EETs (Zou *et al.*, 1994). Experiments were also conducted in the presence of SKF 525a and clotrimazole, two mechanistically-different inhibitors of a large number of cytochrome P450-dependent systems (see Murray and Reidy, 1990). Finally, the effects of inhibitors of phospholipase A_2 on the endothelium-dependent dilatation were also investigated. The lack of effect of 17-ODYA in the present study is a strong indication that EDHF is unlikely to be an EET in the small mesenteric artery. This finding contrasts, however, with observations made in the rat isolated perfused heart, where the compound (at a concentration of only 2.10^{-6} M) inhibited the bradykinin-induced vasodilator responses possibly mediated by EDHF (Fulton *et al.*, 1995). EETs are reported to relax blood vessels from several vascular regions (Gebremedhin *et al.*, 1992; Hecker *et al.*, 1994; Campbell *et al.*, 1996a). The discrepancy between these findings and those of the present study could indicate that "EDHF" is not the same in all tissues. The finding against the proposal that EDHF is a cytochrome P450-derived metabolite, at least in the small mesenteric artery of the rat is in accordance with findings reported in rat hepatic and main mesenteric and in the guinea-pig carotid arteries (Zygmunt *et al.*, 1996; Corriu *et al.*, 1996; Vanheel and Van de Voorde, 1997). The apparent contradictions in the literature may be due to liberation of different EDHFs as well as different distribution of various K^+ channels in various smooth muscle preparations.

The EDHF-induced relaxation of the isolated mesenteric bed of the rat is inhibited by CB_1 cannabinoid receptor antagonist SR141716A, and anandamide, the endogenous cannabinoid, may be an EDHF in mesenteric arteries (Randall *et al.*, 1996). However, in the present study the selective CB_1 antagonist SR141716A did not alter anandamide- and cyclopiazonic acid-elicited dilatation of myogenically active mesenteric arteries resistant to inhibitors of NO synthase and cyclopoxygenases and abolished by high K^+ solution. The present observations confirm that anandamide is not an EDHF in the small mesenteric artery of the rat (Plane *et al.*, 1997). The reason for the discrepancy between these results and those previously reported (Randall *et al.*, 1996) is not known.

The $[Ca^{2+}]_i$ of the endothelial cells is an important stimulus not only for the formation of NO (Busse *et al.*, 1993) but also for the activation of the NO-independent pathways (Suzuki and Chen, 1990; Nagao and Vanhoutte, 1993; Fukao *et al.*, 1997b). This seems to be true also in myogenically active small mesenteric arteries, since the agonists used in this study activate NO-dependent and NO-independent pathways, and increase $[Ca^{2+}]_i$ in endothelial cells (Dinerman, Lowenstein and Snyder, 1993). The increase in $[Ca^{2+}]_i$ induced for example by acetylcholine, bradykinin or substance P is due to release from the endoplasmic reticulum by IP_3, and Ca^{2+} influx (Lambert, Kent and Whorton, 1986; Dinerman, Lowenstein and Snyder, 1993). A23187 increases Ca^{2+} permeability of both cell and endoplasmic reticulum membranes, whereas cyclopiazonic acid depletes intracellular Ca^{2+} stores by selectively inhibiting Ca^{2+}-dependent ATPase in sarcoplasmic/endoplasmic reticulum in a variety of tissues (Seidler *et al.*, 1989; Lagaud, Stoclet and Andriantsitohaina, 1996). The striking difference between cyclopiazonic acid which releases EDHF but no NO and substance P which releases NO but no EDHF strongly suggests that the mode of Ca^{2+} elevation determines the secretary product. Hence, this may mean that cyclopiazonic acid, by slowly releasing Ca^{2+} from the endoplasmic reticulum through inhibition of

Ca^{2+} re-uptake by Ca^{2+}-dependent ATPase in sarcoplasmic/endoplasmic reticulum, stimulates the synthesis of EDHF. On the other hand, substance P may increase $[Ca^{2+}]_i$ mainly by activating Ca^{2+} influx and perhaps Ca^{2+} release from the superficial endoplasmic reticulum (Sharma and Davis, 1995) thus, stimulating eNOS localized near the cell membrane (Pollock *et al.*, 1993; Forstermann *et al.*, 1994).

In summary, the present results show that endothelium-dependent agonist responses differ between the distal middle cerebral and small mesenteric resistance arteries of the rat. In myogenically active cerebral arteries, substance P produced endothelium-dependent dilatation. However, in small mesenteric arteries both substance P and cyclopiazonic acid caused endothelium-dependent dilatation. In cerebral arteries the observed dilatations were related solely to the production of NO. In small mesenteric arteries, NO, prostanoid(s) and another factor with properties consistent with EDHF were involved in the response. The NO and cyclooxygenase pathways might compensate for one another in cyclopiazonic acid-induced endothelium-dependent vasodilatation. In addition, the small Ca^{2+}-activated K^+ channel and to a lesser extend large Ca^{2+}-activated K^+ channels are implicated in the EDHF-like component of the dilatation to cyclopiazonic acid. The data presented support the concept of heterogeneity among endothelial cells.

ACKNOWLEDGEMENTS

This research was support by the Heart and Stroke Foundation of British Columbia and the St. Paul's Hospital Foundation. The authors thank Mrs. K. Park for carefully reading the manuscript.

31 The Endothelium-derived Hyperpolarizing Factor Released from Porcine Coronary Arteries Enhances Tyrosine Phosphorylation and Activates Erk1/2 in Cultured Human Coronary Smooth Muscle Cells

Ingrid Fleming and Rudi Busse

Institut für Kardiovaskuläre Physiologie, Klinikum der J.W.G.-Universität, Theodor-Stern-Kai 7, D-60590 Frankfurt am Main, Germany
Address for correspondence to: Ingrid Fleming, Institut für Kardiovaskuläre Physiologie, Klinikum der J. W. Goethe-Universität, Theodor-Stern-Kai 7, D-60590 Frankfurt/Main, Germany. Telephone: (49) 69 6301 6052; Telefax: (49) 69 6301 7668; E.mail: fleming@em.uni-frankfurt.de

In porcine coronary arteries, endothelium-derived hyperpolarizing factor (EDHF) is transiently produced in response to receptor-dependent agonists (e.g., bradykinin) and continuously produced in response to cyclic strain. In this study, it was investigated whether or not EDHF affects intracellular signalling pathways which might be involved in the regulation of gene expression. EDHF was generated in porcine coronary arteries by applying rhythmic changes in intraluminal pressure, under conditions of combined blockade of nitric oxide synthase and cyclooxygenase. Thereafter the luminal incubate was removed and applied to cultured smooth muscle cells, also treated with N^{ω}-nitro-L-arginine and diclofenac. Under these conditions, EDHF elicited the tyrosine phosphorylation of several cellular proteins, the most prominent of which were the extracellular signal-regulated protein kinases, Erk1 and Erk2. The increase in the phosphorylation of Erk1/2 was time-dependent; it appeared first after 2 min and reached a maximum five to ten min after the addition of the EDHF-containing solution. This increase in Erk1/2 phosphorylation was not observed following the application of the luminal incubate from porcine coronary arteries treated with miconazole or 17-ODYA, which attenuate the production of EDHF. 11,12-Epoxyeicosatrienoic acid, a candidate for EDHF, evoked a time-dependent activation of Erk1/2, similar to that induced by authentic EDHF. These observations indicate that EDHF may not solely be a vasodilator and that its continuous release, especially in arteries with decreased levels of bioactive nitric oxide, may play a role in the regulation of cellular signalling and possibly of gene regulation.

KEYWORDS: porcine coronary artery, tyrosine phosphorylation, extracellular signal-regulated protein kinases, rhythmic vessel distension

INTRODUCTION

The existence of a humoral hyperpolarizing factor from native endothelial cells was originally proposed on the basis of an endothelium-dependent hyperpolarization of vascular smooth muscle cells in an arterial segment (Chen *et al.*, 1988; Feletou and Vanhoutte, 1988). However, it proved difficult to substantiate the release of a diffusible hyperpolarizing factor using classical bioassay techniques. Evidence that EDHF is indeed a transferable factor was obtained by monitoring the membrane potential of detector vascular smooth muscle cells situated downstream, and electrically isolated from donor endothelial cells

(Bauersachs *et al.*, 1996; Popp *et al.*, 1996; Gebremedhin *et al.*, 1998). Receptor-dependent agonists have mainly been used to elicit the synthesis of, as well as to characterise endothelium-derived vasodilator and -constrictor substances although mechanical stimuli such as fluid shear stress and pulsatile stretch are physiologically more important for the continuous generation of the endothelium-derived autacoids. The release of EDHF, as assessed by the hyperpolarization of detector smooth muscle cells, is greater following pulsatile stretch than in response to supra-physiological concentrations of bradykinin (Popp *et al.*, 1998). Moreover, unlike the responses to receptor-dependent agonists, the release of EDHF into the luminal incubate of rhythmically stretched coronary segments was not subject to tachyphylaxis. Thus, the continuous mechanical stimulation of arteries may ensure the maintained release of this endothelium-derived vasodilator.

Although the exact chemical nature of EDHF is still unknown, the hyperpolarizing factor produced by human and porcine, carotid, coronary and renal arteries displays characteristics similar to those of a cytochrome P450-derived metabolite of arachidonic acid (for reviews see Mombouli and Vanhoutte, 1997; Quilley *et al.*, 1997). Inhibitors of cytochrome P450 enzymes, such as miconazole and 17-ODYA, markedly attenuate EDHF-mediated relaxation in a number of vessels. Moreover, induction of cytochrome P450 enzymes using β-naphthoflavone enhances the release of EDHF from cultured porcine and human endothelial cells (Popp *et al.*, 1996). Although concern has been expressed regarding the non-selective effects of some P450 inhibitors (i.e., clotrimazole and econazole) on K^+_{Ca} channel activity and endothelial Ca^{2+} signalling, it should be stressed that 17-ODYA exhibits neither of these effects and selectively inhibits the production of EDHF rather than interfering with its action on vascular smooth muscle cells (Popp *et al.*, 1996). These observations along with others led to the proposal that the EDHF released from coronary artery endothelium is most probably an epoxyeicosatrienoic acid (EET) synthesised by a cytochrome P450 epoxygenase. Indeed, EETs are generated by endothelial cells and mediate part of the dilator effect of arachidonic acid (Harder *et al.*, 1995). Moreover, the 11,12-EET- and 5,6-EET-induced relaxation of arteries without endothelium is sensitive to K^+_{Ca} channel inhibitors, and both EETs activate large conductance K^+_{Ca} channels in native and cultured smooth muscle cells (Gebremedhin *et al.*, 1992; Rosolowsky and Campbell, 1993; Campbell *et al.*, 1996).

Under conditions in which the bioavailability of nitric oxide is decreased, the endothelium is able to maintain much of its vasodilator producing capacity due to the removal of the nitric oxide-mediated inhibition of the EDHF synthase activity and expression (Olmos *et al.*, 1995; Popp *et al.*, 1996). This supportive role of EDHF may well extend beyond its function as a vasodilator. Therefore, the aim of the present investigation was to determine whether or not EDHF derived from the porcine coronary artery activates cellular targets other than K^+_{Ca} channels, which may result in alterations of gene expression in the vascular wall.

METHODS

Vessel Preparation

Porcine hearts were obtained from a local abattoir and placed immediately into cold Hank's solution (GIBCO, Eggenstein, Germany). Segments of coronary epicardial arteries (~50 mm length; mean external diameter 2.4–2.8 mm) were excised and cleaned of fat and connective tissue. Side branches were tied using surgical silk. The blood vessels were incubated in cold physiological salt solution (mM: 140 NaCl; 4.7 KCl; 1 $MgCl_2$; 1.3 $CaCl_2$; 10 HEPES; 5 D-glucose, pH 7.4, 37°C, pO_2 140 mmHg) until use.

Smooth Muscle and Endothelial Cells

Human coronary smooth muscle and endothelial cells were obtained from CellSystems (St. Katharinen, Germany) and cultured in SmGM-2 (CellSystems) to confluence. Experiments were performed using confluent cells between passages 4 and 16.

Rhythmic Vessel Distension and Cell Stimulation

Segments of porcine coronary artery with endothelium were attached to steel cannulae in an organ chamber, stretched to their *in situ* length and perfused with physiological salt solution (1 mL/min) at an intraluminal pressure of 30 mmHg. Rhythmic changes (1 Hz) in circumferential wall stretch were induced in these segments, following blockade of the outflow, by applying sinusoidal volume oscillations in a magnitude which produced corresponding pressure oscillations between 30, 60 and 90 mmHg. After two min of rhythmic vessel distension, 200 μL of the intraluminal fluid was withdrawn and applied to cultured smooth muscle cells (supernatant 300 μL) and incubated as described in Results. Thereafter, cells were washed with ice-cold HEPES buffer containing NaF (10 mM), $Na_4P_2O_7$ (15 mM), and Na_3VO_4 (2 mM) and harvested by scraping.

Immunoblotting

Endothelial cell suspensions were centrifuged at 13,000g for 60 sec, cells contained in the pellet were then lysed in buffer containing leupeptin (2 μg/mL), pepstatin A (2 μg/mL), trypsin inhibitor (10 μg/mL), phenylmethylsulfonyl fluoride (PMSF; 44 μg/mL) and Triton X-100 (1% v/v), left on ice for 10 min and centrifuged at 10 000g for 10 min. Proteins in the resulting supernatant were eluted by heating with SDS-PAGE sample buffer and were separated by 10% SDS-PAGE, as described (Fleming *et al.*, 1995). Proteins were detected using their respective antibodies as described in the results section and were visualised by enhanced chemiluminescence using a commercially available kit (Amersham, Germany). Prestained molecular weight marker proteins (BioRad) were used as standards for the SDS-PAGE.

RESULTS

Effect of the Luminal Incubate From Rhythmically Stretched Porcine Coronary Arteries on Cultured Human Coronary Smooth Muscle Cells

Rhythmic vessel distension of porcine coronary arteries was used to generate EDHF without using Ca^{2+}-elevating receptor-dependent agonists (Popp *et al.*, 1998). Immunoblotting of protein from confluent cultures of human smooth muscle cells using a monoclonal anti-phosphotyrosine antibody revealed a relatively high basal level of apparently consti-

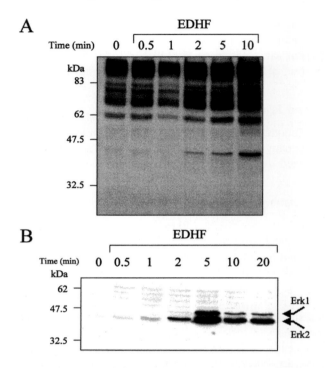

Figure 31-1. EDHF induces the tyrosine phosphorylation of cellular proteins and the activation of Erk1/2. EDHF production was stimulated in porcine coronary arteries by applying rhythmic changes in intraluminal pressure ($\Delta 90$ mmHg, 1 Hz, 2 min) under conditions of combined nitric oxide synthase/cyclo-oxygenase blockade (N^ω-nitro-L-arginine, 3×10^{-4}M; diclofenac 10^{-5} μM). Thereafter the luminal incubate was removed and applied to cultured human coronary smooth muscle cells also pretreated with N^ω-nitro-L-arginine and diclofenac for 30 sec to 20 min. Incubation was stopped, the Triton X-100-soluble cell fraction was prepared and proteins separated by SDS-PAGE and Western blotting was performed using antibodies which selectively recognised either (A) phosphotyrosine or (B) the threonine/tyrosine phosphorylated forms of Erk1 and Erk 2. The Western blots presented are representative of results obtained in four independent experiments.

tutively tyrosine-phosphorylated proteins (Figure 31-1). The application of luminal incubate from rhythmically stretched porcine coronary arteries resulted in the time-dependent phosphorylation of several Triton X-100-soluble proteins, the most prominent of which were two proteins of 42 and 44 kDa (Figure 31-1). A significant increase in tyrosine phosphorylation of the 42/44 kDa doublet was detected two min after the addition of the luminal incubate, peaked between five and ten min and returned towards baseline values within 20 to 30 min. This protein doublet was identified as the extracellular signal-regulated kinases Erk 1 and Erk 2 using an antibody which selectively recognises the threonine/tyrosine phosphorylated forms of Erk1 and Erk 2 (Figure 31-1). Incubation of detector smooth muscle cells with either the tyrosine kinase inhibitor genistein or, PD 98059, an inhibitor of the mitogen-activated protein kinase kinase, prevented the activation of Erk1/2 by the coronary artery luminal incubate.

Similar effects were observed when human coronary endothelial cells were used as detector cells (data not shown).

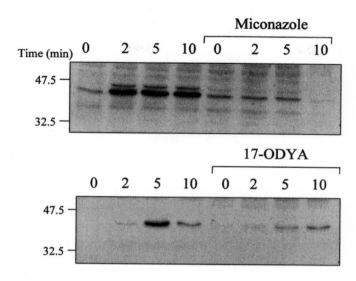

Figure 31-2. The cytochrome P450 inhibitors miconazole and 17-ODYA, which inhibit EDHF production, prevent the activation of Erk1/2 by the luminal incubate from porcine coronary arteries. Luminal effluate from a rhythmically distended porcine coronary artery, treated with either miconazole (3×10^{-6}M, 40 min) or 17-ODYA (3×10^{-6}M), was applied to cultured human coronary smooth muscle cells. Both the donor artery and the detector cells were continuously treated with N^{ω}-nitro-L-arginine (3×10^{-4}M) and diclofenac (10^{-5}M), and the detector smooth muscle cells were incubated in the presence of miconazole and 17-ODYA throughout. At the time points indicated the incubation was stopped, the Triton X-100-soluble cell fraction was prepared and proteins separated by SDS-PAGE. Western blotting was performed using an antibody which selectively recognises the threonine/tyrosine phosphorylated forms of Erk1 and Erk 2. The Western blots presented are representative of results obtained in four independent experiments.

Pharmacological Characterisation of the Erk1/2-Activating Factor

To characterise the factor contained within the luminal incubate from mechanically stimulated arteries which was responsible for the phosphorylation and activation of Erk1/2, experiments were performed using the cytochrome P450 inhibitors miconazole and 17-ODYA. Neither of these compounds affected basal tyrosine phosphorylation in cultured smooth muscle cells, or the increase in tyrosine phosphorylation induced by thrombin (data not shown). However, incubation of donor segments with either miconazole (3×10^{-6}M) or 17-ODYA (3×10^{-6}M) inhibited the luminal incubate-induced activation of Erk1/2 in detector cells (Figure 31-2).

As CB1 cannabinoid receptor agonists inhibit EDHF-mediated relaxation in a number of blood vessels (Figure 31-3) via a mechanism independent of cytochrome P450 inhibition (Fleming *et al.*, 1999) the effect of HU 210 on the "EDHF-induced" activation of Erk1/2 was determined. In this case, cultured human coronary smooth muscle cells of passage 9 and higher were used as detector, as these cells no longer expressed CB1 receptor mRNA or exhibited a biological response to HU 210. Under these experimental conditions, the incubation of the EDHF-donor artery with HU 210 abolished the stimulatory effect of the luminal incubate from rhythmically stretched arteries on Erk1/2.

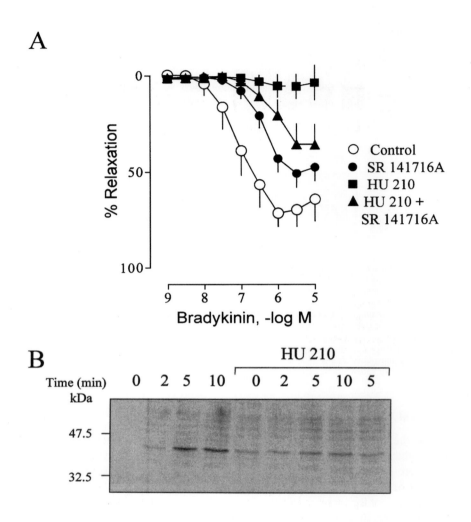

Figure 31-3. The cannabinoid receptor agonist HU 210 abolishes EDHF production in porcine coronary arteries and the activation of Erk1/2 by the luminal incubate from rhythmically stretched coronary arteries. (A) Concentration-response curves to bradykinin were obtained using U46619 (1–3×10^{-7}M) precontracted porcine coronary artery rings in the presence of either solvent (0.06% ethanol), SR 141716A (3×10^{-5}M), HU 210 (3×10^{-5}M; A), or the combination of HU 210 and SR 141716A. All experiments were performed in the continuous presence of N$^{\omega}$-nitro-L-arginine (3×10^{-4}M) and diclofenac (10^{-5}M) and the results are presented as the mean ± SEM of six separate experiments. (B) Luminal effluate from a rhythmically distended porcine coronary artery, treated with HU 210 (3×10^{-5}M), was applied to cultured human coronary smooth muscle cells. Both the donor artery and the detector cells were continuously treated with N$^{\omega}$-nitro-L-arginine (3×10^{-4}M) and diclofenac (10^{-5}M), and the detector smooth muscle cells were incubated with HU 210 (3×10^{-5}M) throughout. At the time indicated the incubation was stopped, the Triton X-100-soluble cell fraction was prepared and proteins separated by SDS-PAGE. Western blotting was performed using an antibody which selectively recognises the threonine/tyrosine phosphorylated forms of Erk1 and Erk 2. The Western blots presented are representative of results obtained in three independent experiments.

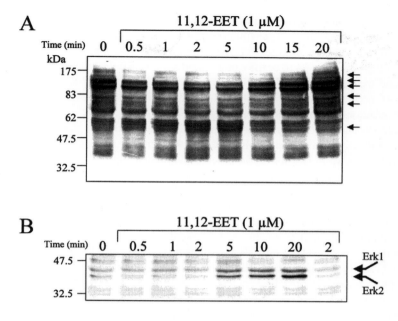

Figure 31-4. 11,12-EET induces a time-dependent increase in the tyrosine phosphorylation of smooth muscle proteins including the activation of Erk1/2. 11,12-EET was applied to cultured human coronary artery smooth muscle cells. At the time points indicated the incubation was stopped and the Triton X-100-soluble cell fraction prepared. Western blotting was performed using either an antibody which selectively recognises either (A) phosphotyrosine or (B) the threonine/tyrosine phosphorylated forms of Erk1 and Erk 2. The Western blots presented are representative of results obtained in six independent experiments.

Comparison with the Effects of 11,12 Epoxyeicosatrienoic Acid

The effects of 11,12 EET, an EDHF known to be produced by coronary arteries, was determined on cellular tyrosine phosphorylation and the activation of Erk1/2. 11,12-EET (10^{-6}M) induced the transient phosphorylation of several Triton X-100-soluble proteins, most notably those corresponding to 42, 44 and ~50 kDa. The EET also induced the transient tyrosine dephosphorylation of several proteins ranging between 59 and 130 kDa (see arrows on Figure 31-4), suggesting the time-dependent activation of both tyrosine kinases and tyrosine phosphatases. Western blot analysis using an antibody against the threonine/tyrosine phosphorylated forms of Erk1 and Erk 2 revealed that 11,12 EET, like the EDHF-containing luminal incubated from rhythmically stretched arteries, elicited the time-dependent activation of these kinases (Figure 31-4). Activation of Erk1/2 could be observed in response of 11,12-EET at concentrations as low as 10^{-9}M (data not shown).

DISCUSSION

The results of this investigation indicate that the luminal incubate from porcine coronary arteries exposed to pulsatile stretch enhances the tyrosine phosphorylation of a number of Triton X–100-soluble proteins and activates Erk1/2 in human coronary artery smooth muscle cells. The substance responsible for this activation was identified as EDHF, as the activation of Erk1/2 was inhibited following treatment of the donor segments with either inhibitors of cytochrome P450 or the cannabinoid agonist HU 210, which have been shown to attenuate the production of EDHF in a bioassay system. 11,12-EET, a candidate for EDHF also modulated cellular levels of phosphotyrosine and activated Erk1/2 in a time- and concentration-dependent manner.

Pulsatile stretch elicits the release of an EDHF which exhibits pharmacological char- acteristics and a half-life in physiological salt solution identical to those of the EDHF released following stimulation with the receptor-dependent agonist bradykinin (Popp *et al.*, 1998). As Ca^{2+}-elevating receptor-dependent agonists may exert effects on both donor and detector cells, such compounds are unsuitable to determine secondary targets of EDHF in vascular cells. For this reason rhythmic distension was used in the present study to generate EDHF in porcine coronary arteries. The results presented demonstrate that the EDHF-containing luminal incubate of rhythmically stretched porcine coronary arteries enhances the cellular levels of phosphotyrosine and activates Erk1/2. Comparison with the effects of 11,12-EET suggests that EDHF may not only activate tyrosine kinases but also tyrosine phosphatases, as the EET induced a pronounced tyrosine-dephosphorylation of a number of smooth muscle cell proteins. However, as the amount of EDHF produced within the coronary artery varies from donor to donor, quantification of the effects of EDHF on protein tyrosine phosphorylation was not possible.

At this stage it is uncertain how EDHF elicits the activation of tyrosine kinases and/ or phosphatases, as upstream signalling events may prove to be sensitive to the activation of K^+_{Ca} channels and the subsequent changes in membrane potential, direct activation of G proteins, or the activation of an as yet unidentified membrane receptor. Indeed, since EDHF activated K^+_{Ca} channels in cell-attached membrane patches (Popp *et al.*, 1996), it would appear that this factor activates K^+ channels in an indirect manner possibly involving membrane-associated second messenger pathways. EDHF however, activates Erk1/2 partly by a Ca^{2+}-dependent mechanism, as the intracellular Ca^{2+}chelator, BAPTA, attenuated activation of the kinases in response to the luminal incubate from porcine coronary arteries (unpublished observation).

The activation of Erk1/2 is essential for both migration and proliferation of smooth muscle cells and a number of arachidonic acid metabolites can affect cellular kinase activity and proliferation. For example, arachidonic acid increases Erk1/2 activity and promotes mitogenesis in aortic smooth muscle cells of the rat via its conversion into 15-hydroxyeicosatetraenoic acid (Rao *et al.*, 1994). Additional cytochrome P450 and lipoxy- genase metabolites of arachidonic acid, including EETs, 20-hydroxyeicosatetraenoic acid (20-HETE), and 12(S)-HETE are also known to increase [^{3}H]thymidine incorporation in vascular smooth muscle cells (Chen *et al.*, 1998; Uddin *et al.*, 1998). Distinct temporal phases of kinase activation are required to stimulate smooth muscle migration and pro- liferation, with activation of Erk1/2 within 15 minutes being associated with effects on cell migration, and activation between one and four hours associated with smooth muscle proliferation (Nelson *et al.*, 1998). Activation of Erk1/2 frequently results in the activation and nuclear translocation of c-Fos and c-Jun, two components of the activator protein-1 (AP-1) transcription factor complex and directly affect the function of ETS-domain tran- scription factors (Sharrocks *et al.*, 1997).

Exactly which genes are responsive to the transcription factors activated, is again determined by the time course of activation, and the spectrum of additional transcription factors which are co-activated by a given stimulus. However, the response observed varies with the cell type under investigation, as activation of the Erk1/2 signalling cascade can either stimulate or inhibit DNA synthesis depending upon whether its activation is acute, phasic or chronic (Tombes *et al.*, 1998). The model used in the present study to generate EDHF, enabled the immediate response of detector cells to be monitored but as the EDHF produced has a relatively short half-life (approximately 70 seconds) (Popp *et al.*, 1998), this system is unsuitable for more chronic investigations to determine whether EDHF stimulates or inhibits smooth muscle cell proliferation. Without such experiments it is impossible to state whether EDHF, the production of which is enhanced in the absence of nitric oxide, would be expected to exert a pro- or anti-atherogenic effect on the vasculature.

An increase in the oxidative stress in endothelial and smooth muscle cells is thought to occur in conditions associated with endothelial dysfunction, such as hypertension and atherosclerosis. In such a situation, where the bioavailability of nitric oxide is decreased, the production of EDHF together with numerous additional eicosanoids and iso-eicosanoids, would however be expected to further contribute to oxidant stress. Thus, while EDHF may partially maintain endothelium-dependent vasodilator responses in states associated with a manifest "endothelial dysfunction", the hyperpolarizing factor is not necessarily able to compensate for the loss of the anti-atherogenic effects of nitric oxide on gene regulation and proliferation of smooth muscle (for review see Busse and Fleming, 1996).

In summary, rhythmic distension of the endothelium-intact porcine coronary artery results in the generation of a cytochrome P450-dependent factor which exhibits the pharmacological properties of EDHF and results in the activation of Erk1/2 in vascular smooth muscle cells. These observations indicate that EDHF may not be solely a vasodilator and that its continuous release, especially in arteries with decreased levels of bioactive nitric oxide, may play a role in the regulation of cellular signalling and possibly gene regulation.

32 Depression of Endothelium-dependent Relaxation Despite Normal Release of Nitric Oxide in the Aorta of Spontaneously Hypertensive Rats: Possible Role of Protein Kinase C

Anatoly Soloviev, Sergei Tishkin, Alexander Parshikov, Alexander Stefanov and Irina Mosse

Institute of Phamacology & Toxicology, Academy of Medical Sciences, 14 Eugene Pottier Str., 252057, Kiev, Ukraine Tel: +38 044 446 02 88, Fax: +38 044 241 88 85/446 30 31

This study was designed to clarify the mechanisms responsible for the depression of endothelium-dependent relaxation in hypertension. The relaxations of norepinephrine-contracted aortic rings to acetylcholine were compared in spontaneously hypertensive (SHR) and normotensive (WKY) rats. Acetylcholine-induced relaxations were smaller in tissues obtained from hypertensive animals than in those from healthy rats. The protein kinase C blockers, chelerythrine chloride and calphostin C, normalized acetylcholine-induced relaxations. The relaxation in WKY rats aortas was reduced significantly in the presence of indomethacin and N^{ω}-nitro-L-arginine. The remaining relaxation elicited by acetylcholine probably is due to the release of endothelium-derived hyperpolarizing factor. A similar intervention in aortas from SHR abolished almost the acetylcholine-induced relaxation. The measurement of acetylcholine-stimulated nitric oxide release using a chemoluminescence assay showed no difference in the two strains. The sensitivity of tissues from SHR to authentic nitric oxide (NO) was lower than those from WKY rats and increased when calphostin C was added to the bath solution. Electrophysiological studies demonstrated that application of acetylcholine hyperpolarized control aortas. In aortas from SHR the hyperpolarization observed upon exposure to acetylcholine was smaller than that seen in WKY rats. Chelerythrine chloride increased the amplitude of hyperpolarization to acetylcholine in aortas from SHR close to the degree observed in WKY rats. These results are consistent to hypothesis that impairment of vascular endothelium-dependent relaxation in hypertension may be due to: a) decreased sensitivity of smooth muscle cells to nitric oxide and endothelium derived hyperpolarizing factor resulting from an increased protein kinase C activity; b) decreased synthesis and/or release of endothelium-derived hyperpolarizing factor; and c) protein kinase C mediated inactivation of Ca^{2+}-dependent potassium channels.

KEYWORDS: hypertension, endothelium, nitric oxide (NO), endothelium-derived hyperpolarizing factor (EDHF), protein kinase C

INTRODUCTION

Endothelium-dependent relaxations are depressed in arteries from spontaneously hypertensive rats (SHR). A similar reduction has been documented in various models of experimental hypertension including genetic, renovascular, deoxycorticosterone-salt, and coarctio-induced hypertension (Konishi and Su, 1983; Lockette *et al.*, 1986; Vanheel and Van de Voorde, 1996). The cellular mechanism underlying these changes in hypertension are not yet understood. The loss of endothelium-dependent relaxation in vascular tissue might have been due to a decrease in synthesis, release, metabolism of endothelium-derived relaxing factor (EDRF/NO), to changes in sensitivity of smooth muscle cells to this factor (Ignarro, 1989) or to the release of endothelium-derived contracting factor (Lüscher and Vanhoutte, 1986). Endothelial cells also release endothelium derived hyperpolarizing factor (EDHF), which induces relaxation by opening potassium channels (Feletou and Vanhoutte, 1988; Chen *et al.*, 1991). It is uncertain whether or not decreased endothelium-dependent hyperpolarization underlies the alteration of endothelium-dependent relaxation observed in the vascular tissues from SHR, although endothelium-dependent hyperpolarization is impaired in arteries from that strain (Fujii *et al.*, 1996).

An elevated protein kinase C activity my be involved in the maintenance of hypertension in the SHR (Bazan, Campbell and Rapoport, 1992; Soloviev and Bershtein, 1992). It is unknown whether or not this participates in the depression of endothelium-dependent responses in hypertension, in general, and endothelium-dependent hyperpolarization, in particular. Thus, experiments were designed to compare the endothelium-dependent relaxations and hyperpolarizations, as well as release of NO in aortas of SHR and normotensive Wistar-Kyoto (WKY) rats, in the presence or absence of inhibitors of protein kinase C, chelerytrine chloride and calphostin C.

METHODS

Experimental Preparation

Experiments were performed on rings of thoracic aortas obtained from adult (18–20 weeks) SHR and WKY rats. Systolic blood pressure measured using a cuff tail sphygmomanometer S-2 (Hugo Sachs Electronik, Germany) was 165 ± 6 and 114 ± 4 mm Hg in SHR and WKY rats, respectively. The rats weighing approximately 250–300 g were anesthetized with pentobarbital sodium (80 mg/kg, intraperitoneally) and sacrificed by cervical dislocation. The rings were prepared with special care in order to keep the endothelium intact. When appropriate, the endothelium was removed by gentle rubbing of the inner surface of the aortas with a cotton swab. These tissues contracted normally to norepinephrine, but did not relax in response to acetylcholine (data not shown).

Tension Measurements

Aortic rings (1 mm width) were suspended isometrically in an organ chamber between a stationary stainless steel hook and an isometric transducer (type DY1, Ugo Basile, Comerio-Varese, Italy) coupled to a multipurpose polygraph (Nihon Cohden, Tokyo, Japan). The rings were equilibrated for one hour with a resting tension of 15 mN for muscles from WKY rats. Preliminary studies had indicated that 20–24 mN of tension was needed to achieve maximal contractile force in aortas from SHR. Experiments were performed at 37°C in modified Krebs bicarbonate buffer of the following composition (in mM): NaCl, 133; KCl, 4.7; $NaHCO_3$, 16.3; NaH_2PO_4, 1.38; $CaCl_2$, 2.5; $MgCl_2$, 1.2; glucose, 7.8; pH, 7.4.

Sucrose Gap Experiments

Spiral strips (0.5 mm) were dissected and suspended in a sucrose gap apparatus for simultaneous recording of mechanical and electrical activity. One end of the strip was fixed mechanically in K-high solution whereas the other end [in normal or test solution (1.5 ml/min, flow rate)] was connected to a isometric force transducer and stretched to a passive tension of about 4–5 mN (WKY) or 5–10 mN (SHR). The middle part of the tissue (5 to 7 mm length) was superfused by deionized isotonic sucrose solution. The potential difference between two the ends of the muscle was registered using a MZ-4 amplifier (Nihon Kohden, Japan) and two Ag-AgCl nonpolarized electrodes. The signals were recorded on a two-channel pen recorder (model 2210, LKB, Bromma, Sweden).

Microelectrode Studies

Aortic rings were inverted and suspended between two stainless steel hooks for the recording of membrane potential. Glass electrodes filled with 3 M KCl (tip resistance 60–80 MΩ) were inserted into smooth muscle cells from the intimal side. A microelectrode amplifier MZ-4 (Nihon Kohden, Japan) was used for measurement of transmembrane potentials. Membrane potentials were displayed continuously on an oscilloscope (ATAC-250, Nihon, Cohden, Japan) and recorded on a pen recorder (model 2210, LKB, Bromma, Sweden).

Chemiluminescence

The detection of NO was based upon the chemiluminescent reaction between nitric oxide (NO) and the luminol (5 amino-2,3-dihydro-1,4-phthalazinedine) — H_2O_2. The reaction solution contained 0.2 mM luminol, 250 μM 1,10–phenantroline, 50 mM H_2O_2 and 4 mM potassium carbonate. For the chemiluminescence assay, the effluent of the organ chamber (2 ml/min, flow rate) was mixed with a luminol-containing solution (0.8 ml/min) in a special reactor cell connected to a flow cell-type chemiluminescence detector with a photomultiplier tube (model R 2693, Hamamatsu). The limit of NO determination was approximately 5×10^{-11} M (linear range of 10^{-11} M $- 3 \times 10^{-9}$ M). NO concentrations were detected within five seconds after the superfusate exited from the organ chamber.

Drugs

The following compounds were used: acetylcholine iodide, 1,10-phenantroline (Serva, Heidelberg, Germany); chelerythrine chloride (Sigma-Aldrich, Saint-Quentin-Fallavier, France); luminol sodium salt, norepinephrine, calphostin C (Sigma Chemical, St. Louis, USA). Authentic NO was prepared according to the following procedure. At first, NO was prepared by flushing 99% NO gas through a sealed bottle for 30 minutes. Further, deoxygenated water was prepared by flushing O_2-free argon for 3 hours through a sealed bottle containing HPLS-grade H_2O. Then NO gas was injected into the deoxygenated water. A saturated solution of NO (3 mM) was obtained by adding 5 ml NO gas to 20 ml deoxygenated water.

Anatoly Soloviev et al.

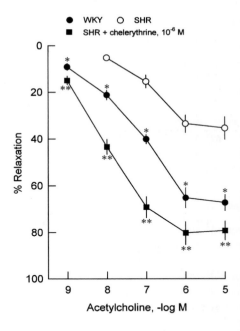

Figure 32-1. Concentration-relaxation curves to acetylcholine obtained in isolated rat thoracic aortas from WKY (closed circles) and SHR before (open circles) and after (closed squares) treatment of SHR aortas with chelerythrine (10^{-6} M). Results are expressed in percent of the contraction evoked by norepinephrine (10^{-6} M) and are presented as means ± SEM (n = 12). Asterisk indicates that difference between closed circles and open circles is statistically significant. Double asterisks indicates that difference between closed squares and open circles is statistically significant (* P < 0.05).

Statistical Analysis

Data are reported as mean values ± standard errors of the mean (SEM). Comparisons were made using Student's t-test. P values less than 0.05 were considered to be statistically significant. ED_{50s} were calculated after the approximation of concentration-response data to the sigmoidal fit curve based on the Boltzman equation.

RESULTS

Acetylcholine caused relaxations of aortic rings from both SHR and WKY rats during contractions to norepinephrine. Aortas from SHR exibited significantly decreased endothelium-dependent relaxantions compared to WKY (Figure 32-1). The maximal amplitude of acetylcholine-induced relaxation in WKY was 67.3 ± 3.6% of the contraction to norepinephrine but only 36.1 ± 4.4% in SHR (n = 12).

Treatment of SHR tissues with the C-kinase inhibitor, chelerythrine chloride (10^{-6} M), augmented relaxations to acetylcholine (Figure 32-1).

The acetylcholine-induced relaxation in WKY rats aortas was reduced by 52.2 ± 6.5% in the presence of indomethacin (5×10^{-6} M) and N^{ω}-nitro-L-arginine (10^{-5} M). In aortas

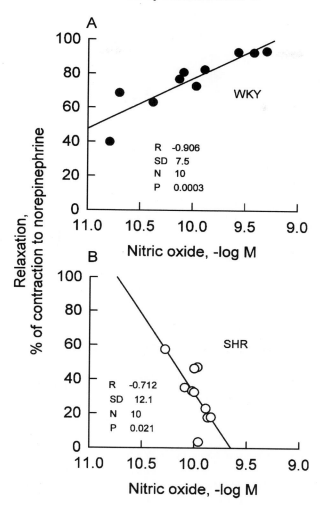

Figure 32-2. Relationship between acetylcholine-stimulated increment of NO concentration in the bath so-lution and the amplitude of relaxation in thoracic aortas from WKY rats (A) and SHR (B). Acetylcholine was given in a final concentration of 10^{-6} M.

from SHR, indomethacin plus N^{ω}-nitro-L-arginine nearly abolished the acetylcholine-induced relaxation (from $36.1 \pm 4.4\%$ to $7.1 \pm 2.3\%$, n = 11).

The measurement of NO release by chemiluminescence showed no statistically signifi-cant differences in the amount of NO released per 1 mg aortic tissue per minute, in the presence of acetylcholine (10^{-5} M), between WKY and SHR aortas ($1.0 \pm 0.1 \times 10^{-10}$ M and $1.0 \pm 0.3 \times 10^{-10}$ M, respectively, n = 10).

A positive correlation existed between the relaxation to acetylcholine and the release of NO in WKY aortas (Figure 32-2). By contrast, a negative correlation was observed in the SHR aorta (Figure 32-2).

Anatoly Soloviev et al.

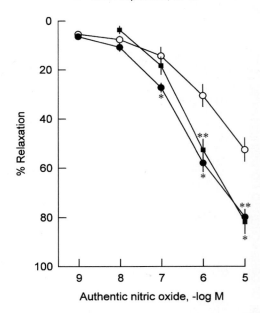

Figure 32-3. Concentration-relaxation curves to authentic NO obtained using thoracic aortas from WKY rats (closed circles) and SHR in the absencewithout (open circles) and presence (squares) of calphostin C (10^{-7} M). Data shown as means ± SEM (n = 12). Asterisk indicates that difference between closed circles and open circles is statistically significant. Double asterisks indicates that difference between closed squares and open circles is statistically significant (* P < 0.05).

Aortas without endothelium from both WKY and SHR relaxed to authentic NO (Figure 32-3). The EC_{50}s for WKY and SHR aortas were significantly different (4.6 ± 0.3 × 10^{-7} M and 2.8 ± 0.15 × 10^{-6} M, respectively; n = 12). After treatment of SHR aortas with calphostin C (10^{-7} M), the sensitivity of smooth muscle cells to NO increased significantly (EC_{50} – 7.1 ± 0.4 × 10^{-7} M) and became even significantly greater than that observed in the WKY.

Sucrose gap experiments demonstrated that the maximal amplitude of hyperpolarization evoked by acetylcholine (10^{-5} M) in WKY aortas was 5.7 ± 0.27 mV (n = 17). N^{ω}-nitro-L-arginine (10^{-5} M) and indomethacin (5 × 10^{-5} M) did not significantly reduce the acetylcholine-induced hyperpolarization. In SHR aortas the hyperpolarization was significantly smaller — 2.4 ± 0.29 mV (n = 10). Calphostin C (10^{-7}M), had no effect on the acetylcholine-induced hyperpolarization in the WKY aorta (Figure 32-4). By contrast, it significantly increased acetylcholine-induced hyperpolarizations in the SHR aorta [to 4.8 ± 0.46 mV (n = 10)] (Figure 32-4).

The microelectrode studies confirmed that acetylcholine (10^{-5} M) hyperpolarizes smooth muscle cells of WKY aortas. A peak amplitude of 14.2 ± 2.4 mV (n = 11) from resting membrane potential was reached within one minute. In aortas from SHR, the hyperpolarization to acetylcholine was significantly smaller (4.1 ± 1.1 mV, n = 10). Calphostine C

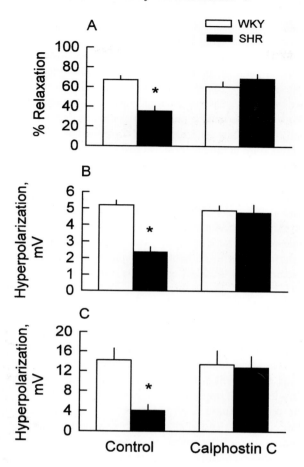

Figure 32-4.   A, B — amplitude of cell membrane hyperpolarizations, measured in the sucrose gap apparatus, and relaxations induced by acetylcholine (10^{-5} M) in rat thoracic aortas from SHR and WKY rats, in the absence (left) and presence (right) of the protein kinase C blocker, calphostin C (10^{-7} M). C — hyperpolarization, measured with microelectrodes, induced by acetylcholine (10^{-5} M) in rat thoracic aorta from SHR and WKY rats incubated with (left) and without (right) calphostin C (10^{-7} M) registered with microelectrodes. Data are shown as means ± SEM (n = 11). The asterisk indicates that the difference between WKY and SHR parameters is statistically significant (* $P < 0.05$).

(10^{-7} M) increased the amplitude of the hyperpolarization in SHR aortas to a level similar to that was observed in WKY tissues (12.9 ± 2.2 mV, Figure 32-4).

DISCUSSION

The present findings demonstrate that the blunted acetylcholine-induced relaxation in SHR may be attributed, at least partially, to a diminished release of EDHF. This conclusion is based on two main observations. First, the acetylcholine-stimulated NO release was not different in SHR and WKY aortas. Second, acetylcholine-induced hyperpolarization was significantly smaller in aortas from SHR.

The hyperpolarization produced by acetylcholine when the endothelium is present is most likely caused by an increase in the potassium permeability of the smooth muscle cells membrane brought about by EDHF (Bolton, Lang and Takewaki, 1996). EDHF can activate at least two types of potassium channels in vascular smooth muscle cells (Brayden and Murphy, 1996; Vanhoutte and Feletou,1996), calcium-dependent potassium channels of large and/or small conductance. Alternatively, the acetylcholine-induced hyperpolarization may contribute to relaxation in the rat aorta by closing voltage-dependent calcium channels and reducing the intracellular calcium concentration (Chen and Suzuki, 1989). The single-channel potassium currents measurements performed in cell-attached and inside-out patches from aortic membranes demonstrated that the Ca^{2+}-activated potassium channel function is altered in arterial muscle of SHR compared to the WKY (England et al., 1993).

The present experiments performed with protein kinase C blockers demonstrated an increased amplitude of vascular responses (contractile and electrical) to acetylcholine in the SHR. The experiments with authentic NO indicated that the sensitivity of smooth muscle to NO is decreased in the SHR, and that a normal responsiveness can be restored by a protein kinase C blockers. These observations suggest that protein kinase C is involved in the depression of endothelium-dependent relaxations and hyperpolarizations to acetylcholine in the aorta of the SHR.

The reason for the high protein kinase C activity in vascular tissues of the SHR is unknown. An alteration of the phospholipid profile of smooth muscle cell membranes of the thoracic aorta has been demonstrated in this strain, in particular a reduced phosphatidylcholine content (Cox and Tulenko, 1991). Liposomes prepared from phosphatidylcholine restore endothelium-dependent responses in the SHR (Soloviev et al., 1992). Diacylglycerol, which can activate protein kinase C, is produced not only through phosphatidylinositol 4,5-biphosphate but can be generated also from phosphatidylcholine (Cabot et al., 1988). The latter is a major source of the sustained accumulation of diacylglycerol observed in agonist-stimulated vascular smooth muscle (Lee and Severson, 1994). Activation of protein kinase C inhibits Ca^{2+}-dependent potassium channels and may cause membrane depolarization and vascular contraction, implying that it antagonizes EDHF (Minami et al., 1993).

The results obtained in the present study support the idea that the depression of endothelium-dependent vascular responses in genetic hypertension is a consequence of an increased protein kinase C activity in the vascular smooth muscle cells. Together with a decreased sensitivity of these smooth muscle cells to NO and EDHF, the blunted endothelium-dependent relaxations observed in SHR aorta may be due to either a protein kinase C — mediated decrease in the synthesis of EDHF or a protein kinase C — mediated inactivation of potassium channels.

33 Diminished Cyclic GMP- and Hyperpolarization-dependent Relaxations to Acetylcholine in Rabbit Carotid Arteries with a Neointima

Cor E. Van Hove, Katelijne E. Matthys, Arnold G. Herman and Hidde Bult

Div. of Pharmacology, Univ. of Antwerp (UIA), Universiteitsplein 1, B-2610 Wilrijk, Belgium. Tel: (+32) 3 820 26 36; Fax: (+32) 3 820 25 67; Email: corvhove@uia.ua.ac.be

In rabbit carotid arteries with a collar-induced neointima (collar-arteries), the relaxation to acetylcholine remains unchanged. The role of the L-arginine-NO pathway and of endothelium-derived hyperpolarizing factor (EDHF) in control and collar-arteries was investigated.

Collars were placed around both carotid arteries of anaesthetised rabbits and 14 days later rings were cut from the collar region and the proximal region outside the collar (control). Relaxations to acetylcholine were measured in the presence of N^ω-nitro-L-arginine, which suppresses the L-arginine-NO pathway, and in high K^+-contracted rings, to prevent the hyperpolarization.

N^ω-nitro-L-arginine shifted the concentration-relaxation curve to acetylcholine during contractions to phenylephrine (EC_{50}) in control rings to the right and reduced the amplitude of the relaxation from 100 to 50%. In collar-rings similar changes were seen in ten of 16 rings whereas the N^ω-nitro-L-arginine-resistant relaxation was completely blocked in the remaining six collar-rings. In control rings, contraction in a depolarizing solution (30 mM K^+) reduced the sensitivity to acetylcholine and the maximal relaxation was only 60%. High K^+ reduced the sensitivity of collar-rings to acetylcholine to a larger extent. In collar-arteries the maximal relaxations obtained through the L-arginine-NO pathway and hyperpolarization showed a positive correlation, which suggests that NO not only acts as guanylate cyclase activator but also as an EDHF in the rabbit carotid artery.

These experiments suggest that endothelium-dependent relaxations mediated by cyclic GMP and hyperpolarization are less pronounced when studied individually in rabbit carotid arteries with a neointima. Since the overall response to acetylcholine is not diminished, either mechanism must compensate for the other one if it is inhibited.

KEYWORDS: nitric oxide, endothelium-derived hyperpolarizing factor, intima, atherosclerosis

INTRODUCTION

The release of nitric oxide (NO), endothelium-derived relaxing factor (EDHF) and prostacyclin by endothelial cells contributes to the local control of vascular tone (Lüscher *et al.*, 1993; Cohen and Vanhoutte, 1995). NO diffuses to the underlying smooth muscle cells where it stimulates soluble guanylate cyclase. Subsequent production of cyclic GMP induces relaxation by mechanisms which are still not completely understood (Lincoln *et al.*, 1996). The identity of EDHF is still under debate but its mechanism of action involves the opening of potassium channels.

297

Atherosclerotic arteries display diminished endothelium-dependent relaxations, probably because of a defective release of nitric oxide (NO) (Verbeuren *et al.*, 1986; Shimokawa and Vanhoutte, 1989c; Harrison, 1996; Matthys and Bult, 1997). However, other endothelium-derived relaxing factors may be affected as well. Thus, a diminished release of prostacyclin has been demonstrated in hypercholesterolemic rabbits (Beetens *et al.*, 1986). Intimal thickenings are predilection sites for atherosclerosis (Schwartz, de Blois and O'Brien, 1995). They can be induced in the rabbit carotid artery by the positioning of a soft silicone collar around the vessel (Kockx *et al.*, 1992). The formation of the neointima does not affect the endothelium-dependent relaxation to acetylcholine (Van Put *et al.*, 1995), an agonist which promotes the release of NO and prostacyclin as well as EDHF. The aim of the present study was to assess whether or not the relative contributions of the L-arginine-NO pathway and hyperpolarization in the endothelium-dependent relaxations of blood vessels with a neointima are altered in such arteries.

MATERIALS AND METHODS

Experimental Model of Intimal Thickening

Male New Zealand white rabbits (2.5 to 3.5 kg) were anaesthetised with sodium pentobarbital (30 mg/kg intravenously) and the common carotid arteries were exposed surgically. Non-occlusive, flexible silicone collars (inlet/outlet diameter 1.8 mm) were placed around the carotid arteries and closed with silicone glue (Kockx *et al.*, 1992). After 14 days the rabbits were anaesthetised and both arteries were removed and placed in cold Krebs-Ringer solution (NaCl 118, KCl 4.7, $CaCl_2$ 2.5, KH_2PO_4 1.2, $MgSO_4$ 1.2, $NaHCO_3$ 25, CaEDTA 0.025 and glucose 11.1 mM).

Vascular Reactivity

After careful removal of the collar and loose connective tissue, rings (2mm length) were cut from the collared region as well as from the region proximal to the collar (control). Collar- and control rings were suspended in organ chambers filled with 25 ml Krebs-Ringer solution maintained at 37°C and continuously aerated with 95% O_2–5% CO_2. Indomethacin (10^{-6}M) was always added to the bath solution to prevent possible interference due to the release of vasoactive prostanoids. Tension was measured isometrically with a Statham UC_2 force transducer (Gould Cleveland, OH, USA) connected to a data acquisition system (Moise 3, EMKA technologies, France). The rings were stretched gradually to a tension of 6 g, which had been determined in preliminary experiments to bring both types of rings to their optimal length-tension relationship. The rings were allowed to equilibrate for 45 min. A cumulative concentration-response curve to phenylephrine ($3 \times 10^{-9} - 3 \times 10^{-5}$ M) was obtained to determine the concentration giving a half-maximal contraction for each individual ring. Subsequently, cumulative concentration-response curves to acetylcholine ($10^{-9} - 10^{-5}$ M) were obtained after half-maximal contraction with phenylephrine or KCl (30 mM). The depolarizing potassium solution was used to preclude hyperpolarization and was prepared by equimolar substitution of NaCl by KCl. Incubation with N^{ω}-nitro-L-arginine (3×10^{-4} M, added 30 min before acetylcholine) was used to suppress the L-arginine-NO pathway. Relaxations are expressed as % of the initial contraction. As collar-rings develop less force in response to contractile agonists, the initial contraction differed among both groups studied. Different levels of phenylephrine-induced contractions do not influence the characteristics of the relaxation to acetylcholine (Van Put *et al.*, 1995). Between concentration-response curves, rings were allowed to equilibrate for 30 min, during which the bath solution was changed 3 times.

Histological Evaluation of Artery Rings

At the end of the vascular reactivity protocol, the physiological salt solution was replaced by formalin (4%) for 15 min. The rings were then removed from the organ chamber, further formalin-fixed overnight and subsequently paraffin-embedded. Transverse sections were stained with hematoxylin/eosin and the intimal thickness was measured at 400× magnification. Immunohistochemical detection of endothelial cells (CD31) was done using a specific monoclonal antibody visualised by the indirect peroxidase antibody conjugate technique. The length of the lumen perimeter covered by CD31-positive cells was expressed as a percentage of the total lumen perimeter. In these measurements the small area damaged by the two organ bath hooks was not taken into account.

Statistical Analysis

All data are expressed as the mean ± standard error of the mean; n refers to the number of arterial rings (obtained from four different rabbits). In the vascular reactivity studies, the raw data (g contraction) were fitted to a logistic function to estimate pD_2 values (Van Put *et al.*, 1995). The pD_2 is the negative logarithm of the molar concentration (EC_{50}) producing half of the maximum response (E_{max}). Statistical comparison of the maximal relaxations and the pD_2 values in the control and collar rings was performed by paired Student t-tests. When P was less than 0.05, differences were considered to be statistically significant.

Materials

Drugs used were from Sigma except for acetylcholine (Sterop, Brussels, Belgium), indomethacin (Merck, Sharp and Dohme) and sodium pentobarbital (Sanofi, Libourne, France). Silicone (MED-4211) was obtained from Nusil Technology (CA,USA) and JC/70A (anti-CD31) was purchased from Dako (Glostrup, Denmark).

RESULTS

Collar-induced Intimal Thickening

Control rings proximal to the collars retained their normal appearance. They did not develop intimal thickening and were covered by a continuous layer of flat endothelial cells resting directly on the internal elastic lamina. The positioning of a collar for 14 days resulted in discrete intimal thickening (Table 33-1). A continuous layer of CD31-positive cells lined the lumen of the collar-arteries (Table 33-1), confirming the presence of an intact endothelium.

Table 33-1 Intimal thickness and length of the lumen covered by endothelial cells (%) in control and collar rings

	Intimal thickness (μm)	Endothelium (%)
Control	2 ± 1	98 ± 3
Collar	13 ± 3	97 ± 4

Table 33-2 Sensitivity to acetylcholine in control and collar rings, expressed as pD_2 ($-\log EC_{50}$)

		rings contracted with	
	phenylephrine (EC_{50})	+ L-NA (3×10^{-4}M)	30 mM K^+
Control	7.60 ± 0.06 (n = 16)	6.55 ± 0.04 [a] (n = 16)	6.93 ± 0.04 [a] (n = 7)
Collar	7.46 ± 0.13 (n = 16)	6.49 ± 0.07 [a] (n = 10)	6.49 ± 0.20 [a,b] (n = 5)

[a] P < 0.05 compared with phenylephrine contracted rings without L-NA.
[b] P < 0.05 compared with 30mM K^+ contracted control rings.

Endothelium-dependent Relaxations to Acetylcholine

Relaxations to acetylcholine were studied in rings constricted with the EC_{50} of phenylephrine. In control rings, complete relaxation ($99 \pm 1\%$) was obtained with 3×10^{-6} M acetylcholine and the pD_2 was 7.60 ± 0.06 (Table 33-2). Collar-rings with a neointima had a comparable sensitivity to acetylcholine (Table 33-2) and complete relaxation ($98 \pm 1\%$) was attained at the same acetylcholine concentration as in control rings (Figure 33-1).

N^ω-nitro-L-arginine

Treatment of control rings with N^ω-nitro-L-arginine (3×10^{-4} M) significantly shifted the relaxation curve to the right and reduced the maximal relaxation to $52 \pm 5\%$ (n = 16) of the initial contraction. In collar-rings, N^ω-nitro-L-arginine abolished the relaxation to acetylcholine in six rings, whereas the shift of the relaxation curve in the ten remaining rings was comparable to the controls (Figure 33-1, Table 33-2).

30 mM K^+

Control rings constricted with a high K^+ solution displayed a decreased sensitivity to acetylcholine (Table 33-2) as well as a diminished maximal relaxation ($61 \pm 3\%$, n = 7, Figure 33-2). In collar-rings the high K^+ concentration resulted in a more pronounced loss of sensitivity to acetylcholine than in control rings (Table 33-2). The maximal relaxation varied between 30 and 60% (n = 5), while the two remaining rings no longer dilated in response to acetylcholine. Repeating the concentration-relaxation curves in the presence of N^ω-nitro-L-arginine (3×10^{-4} M) abolished the relaxation in control as well as in collar rings. Plotting the maximal relaxations obtained in the presence of N^ω-nitro-L-arginine alone vs. those obtained in high K^+ alone revealed a positive correlation (Spearman's correlation coefficient: 0.829; significant at the 0.01 level (2-tailed).

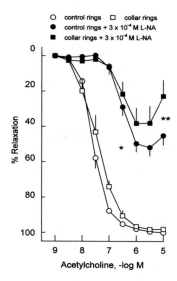

Figure 33-1. Concentration-relaxation curves to acetylcholine obtained in isolated control (circles) and collar-rings (squares) of rabbit carotid arteries. Results are expressed in percent of the contraction evoked by phenylephrine (EC$_{50}$), and are presented as means $\pm$ SEM (n = 16). The data were collected in the absence (open symbols), and in the presence of 3×10^{-4}M N$^{\omega}$-nitro-L-arginine (closed symbols). Asterisk indicates that difference in pD$_2$ between open and closed symbols is statistically significant. Double asterisks indicates that difference in maximal relaxation between open and closed symbols is statistically significant.

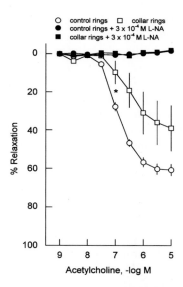

Figure 33-2. Concentration-relaxation curves to acetylcholine obtained in isolated control (circles) and collar-rings (squares) of rabbit carotid arteries. Results are expressed in percent of the contraction evoked by K$^+$ (30mM) and are presented as means $\pm$ SEM (n = 7). The data were collected in the absence (open symbols), and in the presence of 3×10^{-4}M N$^{\omega}$-nitro-L-arginine (closed symbols). Asterisk indicates that difference in pD$_2$ between open circles and open squares is statistically significant.

DISCUSSION

Positioning of a non-occlusive collar around the rabbit carotid artery results in the appearance of a discrete neointima after two weeks, which is composed mainly of smooth muscle cells (Kockx *et al.*, 1992) and covered by an intact endothelial layer as shown by CD31 staining. The lesions resemble the human intimal thickening that constitutes a predilection site for the occurrence of atherosclerosis (Schwartz, de Blois and O'Brien, 1995). Endothelium-dependent relaxation is impaired in atherosclerotic arteries (Verbeuren *et al.*, 1986; Shimokawa and Vanhoutte,1989c; Harrison, 1996; Matthys and Bult, 1997). Less is known on possible functional changes during the stage of intimal thickening and this constituted the aim of the present study.

Besides prostacyclin, two different mechanisms contribute to the relaxation elicited by endothelium-dependent agonists i.e. a cyclic GMP-dependent pathway and a cyclic GMP-independent component associated with hyperpolarization (Cohen and Vanhoutte, 1995). The latter action could be exerted by nitric oxide itself or by another endothelium-dependent factor. The two pathways can be pharmacologically separated by studying the relaxations in the presence of N^{ω}-nitro-L-arginine and in the presence of high potassium which precludes hyperpolarization (Cowan *et al.*, 1993; Adeagbo and Triggle, 1993; Najibi *et al.*, 1994). Potassium channels play a crucial role in the endothelium-dependent hyperpolarization of vascular smooth muscle. The identity of the K^{+}-channels involved is still not fully solved. Therefore a Krebs-Ringer solution with a high potassium concentration rather than specific K^{+}-channel blockers was used to inhibit hyperpolarization in the present experiments.

In the absence of blockade of nitric oxide synthase and hyperpolarization, control and collar-rings displayed complete relaxation with comparable sensitivity to acetylcholine.

In control rings, the exclusion of hyperpolarization with potassium (30 mM) diminished not only the maximal amplitude of the relaxation but also the sensitivity to acetylcholine. These results confirm the existence of a hyperpolarization component in the endothelium-dependent relaxation to acetylcholine. That the L-arginine-NO pathway was the sole mediator of the relaxations in high potassium was shown by the lack of relaxation to acetylcholine by all rings upon further administration of N^{ω}-nitro-L-arginine. Compared to control rings, all collar-rings displayed an additional shift to the right of the concentration-relaxation curve to acetylcholine in high potassium. This suggests that the L-arginine-NO pathway became less active in carotid rings with a neointima. In some collar-rings the L-arginine-NO pathway was reduced to such an extent that acetylcholine no longer induced relaxation when hyperpolarization was excluded. This reduction of the endogenous pathway could also explain the supersensitivity of collar-rings to exogenous nitric oxide (Matthys *et al.*, 1998).

In the presence of N^{ω}-nitro-L-arginine the maximal relaxation of control rings still reached about 50% of the phenylephrine-induced contraction, which is in accordance with published results (Cowan *et al.*, 1993; Najibi *et al.*, 1994; Lischke, Busse and Hecker, 1995c; Cohen *et al.*, 1997; Plane *et al.*, 1998). In more than half of the collar-rings, the inhibitory effect of N^{ω}-nitro-L-arginine was comparable to that in control rings. However, the N^{ω}-nitro-L-arginine-resistant relaxation which is generally attributed to endothelium-dependent hyperpolarization, was absent in six collar-rings suggesting that hyperpolarization was defective in those rings. Collar-rings that failed to relax in the presence of N^{ω}-nitro-

L-arginine also failed to relax in the presence of high potassium. This suggests that the hyperpolarizing-dependent and -independent mechanisms involve the same endothelium-derived factor which is in accordance with Cohen *et al.* (1997). They demonstrated by direct measurements of NO release and membrane potentials that nitric oxide itself is the main EDHF candidate in the rabbit carotid artery.

Cohen *et al.* (1997) also demonstrated that N^{ω}-nitro-L-arginine, even in combination with other L-arginine analogues, could not completely eliminate the release of NO and that these traces are sufficient to induce relaxation via hyperpolarization and independent of cyclic GMP. In view of their findings the N^{ω}-nitro-L-arginine resistant, acetylcholine-induced relaxations observed in control rings in the present experiments are presumably due to smooth muscle hyperpolarization by residual nitric oxide release. Disturbance of the L-arginine-NO pathway in collar-rings could reduce and even eliminate this residual nitric oxide which might explain the lack of relaxation of some collar-rings in the presence of N^{ω}-nitro-L-arginine. Although in some carotid arteries with a neointima both mechanisms were apparently not active anymore when studied individually, a paradoxical undisturbed relaxation was observed when both pathways were operative, suggesting a functional synergism between the two mechanisms.

The present experiments in rabbit carotid arteries with a neointima demonstrated defects in the endothelium-dependent relaxation mechanisms which could only be revealed when studied individually. A functional compensation mechanism is suggested which provides the vessel with sufficient "reserve capacity" to overcome diminished nitric oxide production or release in an early stage of atherosclerosis.

34 Antihypertensive Therapy Improves Endothelium-dependent Hyperpolarization

Uran Onaka, Koji Fujii, Isao Abe and Masatoshi Fujishima

Second Department of Internal Medicine, Faculty of Medicine, Kyushu University, Fukuoka, Japan
Address for correspondence: Koji Fujii, M.D., Second Department of Internal Medicine, Faculty of Medicine, Kyushu University, Maidashi 3-1-1, Higashi-ku, Fukuoka, 812-8582, Japan, Tel. +81-92-642-5256; Fax. +81-92-642-5271; E-mail fujii@intmed2.med.kyushu-u.ac.jp

Acetylcholine-induced hyperpolarization of vascular smooth muscle, mediated by endothelium-derived hyperpolarizing factor (EDHF), is impaired in the mesenteric arteries of adult spontaneously hypertensive rats (SHR), a model of genetic hypertension. The studies summarized in this chapter were designed to determine: a) whether the EDHF-mediated hyperpolarization is altered in other types of experimental hypertension; and b) whether antihypertensive treatment can improve EDHF-mediated responses in the SHR. Membrane potentials were recorded from mesenteric arteries of SHR, Dahl salt-sensitive rats and 2-kidney, 1-clip renal hypertensive rats using conventional microelectrodes. The endothelium-dependent hyperpolarization to acetylcholine was significantly impaired in SHR, Dahl salt-sensitive rats fed a high salt-diet and 2-kidney, 1-clip hypertensive rats compared with their respective normotensive control rats. Adult SHR were treated with either enalapril or a combination of hydralazine plus hydrochlorothiazide for three months. Both treatments reduced blood pressure similarly. Acetylcholine-induced hyperpolarizations were significantly improved in treated SHR. Furthermore, the acetylcholine-induced relaxation of norepinephrine-contracted rings (in the presence of both indomethacin and N$^\omega$-nitro-L-arginine) was improved in treated SHR. On the other hand, levcromakalim-induced hyperpolarization and relaxation were similar among the different experimental groups. These findings suggest that EDHF-mediated hyperpolarization is impaired in experimental hypertension regardless of its etiology, and that antihypertensive treatment restores the impaired EDHF-mediated responses in hypertensive blood vessels.

KEYWORDS: endothelium-derived hyperpolarizing factor, hypertension, arteries, antihypertensive treatment, acetylcholine

INTRODUCTION

Endothelial cells modulate vascular tone by the release of relaxing factors, such as nitric oxide (NO) (Furchgott and Zawadzki, 1980; Palmer, Ferrige and Moncada, 1987; Ignarro *et al.*, 1987b), prostacyclin, and endothelium-derived hyperpolarizing factor (EDHF) (Félétou and Vanhoutte, 1988; Chen, Suzuki and Weston, 1988). EDHF hyperpolarizes the underlying vascular smooth muscle cells by opening K$^+$ channels, thereby leading to relaxation (Chen and Suzuki, 1989; Kauser *et al.*, 1989; Cohen and Vanhoutte, 1995). Endothelium-dependent relaxation is impaired by hypertension both in humans and animals (Lüscher and Vanhoutte, 1986; Vanhoutte, 1989; Panza *et al.*, 1990; Linder *et al.*, 1991). The mechanisms for this appears to be heterogeneous (Lüscher and Vanhoutte, 1986; Koga *et al.*, 1989; Van de Voorde, Vanheel and Leusen, 1992). The acetylcholine-induced hyperpolarization via EDHF is attenuated in mesenteric arteries of adult spontaneously hypertensive rats (SHR), and this partly accounts for the impaired endothelium-dependent relaxation in this model (Fujii *et al.*, 1992, 1993). Although several studies examined the effects of antihypertensive treatment on endothelial fuction in experimental hypertension (Lüscher, Vanhoutte and Raij, 1987; Clozel, Kuhn and Hefti, 1990; Kähönen *et al.*, 1995; Dohi, Kojima and Sato, 1996; Takase *et al.*, 1996), none focused on EDHF-mediated hyperpolarization itself. The experiments, summarized here, were designed to determine: a) whether or not endothelium-dependent hyperpolarizations to acetylcholine are impaired in experimental hypertension with different etiology; and b) whether or not chronic antihypertensive treatment in the SHR can improve EDHF-mediated hyperpolarizations and relaxations.

METHODS

Animals

Male SHR and age-matched Wistar Kyoto rats (WKY), Dahl salt-sensitive and resistant rats, and Wistar rats were used. Dahl salt-sensitive and resistant rats were fed either a 8% NaCl diet or a 0.3% NaCl diet for 8 weeks. Wistar rats were anesthetized with pentobarbital sodium, and through an incision in the left flank, an U-shaped silver clip was applied to the left renal artery. The right kidney remained intact. Two-kidney, 1-clip (2-K, 1-C) rats were used three months after the clipping procedure. SHR, WKY and 2-K, 1-C rats were fed a standard rat chow. At the age of eight to nine months, SHR were assigned to one control (SHR-C) and two treatment groups. The SHR-E group was treated with enalapril 40 mg/kg/day, whereas SHR-H group was treated with a combination of hydralazine 25 mg/kg/day and hydrochlorothiazide 7.5 mg/kg/day. All drugs were given in the drinking water. Drugs were withdrawn several days prior to the experiment.

Systolic blood pressure was measured in conscious rats by the tail-cuff method. Rats were anesthetized with ether and killed by decapitation. The main branch of the mesenteric artery was excised and bathed in cold Krebs solution [composition (in mM): Na^+, 137.4; K^+, 5.9; Mg^{2+}, 1.2; Ca^{2+}, 2.5; HCO_3^-, 15.5; $H_2PO_4^-$, 1.2; Cl^-, 134; and glucose, 11.5]. The artery was cut into rings of 3 mm and 1.2 mm for the electrophysiological and tension experiments, respectively.

Membrane Potential Recordings

Transverse strips cut along the longitudinal axis of the rings were placed in 2-ml experimental chambers with the endothelial layer up. Tissues were superfused with Krebs solution (36°C) aerated with 95% O_2–5% CO_2 (pH 7.3–7.4) at the rate of 3 ml/min. After equilibration for at least 60 minutes, glass capillary microelectrodes filled with 3 M KCl and with tip resistance of 50–80 MΩ were inserted into the smooth muscle cell from the endothelial side. Electrical signals were amplified through an amplifier (MEZ-7200, Nihon Koden, Tokyo, Japan), monitored on an oscilloscope (VC-11, Nihon Koden), and recorded with a pen recorder (RJG-4002, Nihon Koden).

Isometric Tension Recordings

Rings with intact endothelium were placed in 5-ml organ chambers filled with Krebs solution (36°C) aerated with 93% O_2–7% CO_2 (pH 7.4). Two fine, stainless steel wires were placed through the lumen of the ring; one was anchored, and the other was attached to the mechano-transducer (UM-203, Kishimoto, Kyoto, Japan). The rings were allowed to equilibrate for 60 minutes at an optimal resting tension of 1.0 g, before challenging them with contractile agonists.

The rings were contracted with norepinephrine (10^{-5} M) or 77 mM KCl solution in the presence or absence of indomethacin (10^{-5} M) and N^ω-nitro-L-arginine (10^{-4} M). After the contraction had reached a steady level, relaxing agents were applied to the organ chamber in a cumulative manner. The extent of relaxation was expressed as percentage of the initial contraction evoked by the contractile agonist.

Drugs and Solutions

Acetylcholine chloride, indomethacin, N^ω-nitro-L-arginine, norepinephrine, sodium nitroprusside, enalapril, hydralazine and hydrochlorothiazide were obtained from Sigma Chemical Co (St. Louis, MO, USA). Levcromakalim was a gift from SmithKline Beecham pharmaceuticals (Worthing, UK). The solutions containing 20 mM or 77 mM KCl were obtained by equimolar replacement of NaCl by KCl in Krebs solution. Indomethacin was dissolved in 10 mM Na_2CO_3, L-NNA in 0.2 M HCl, and levcromakalim in 90% ethanol. Other agents were dissolved in distilled water.

Statistics

Results are given as mean ± SEM. Concentration-response curves describing hyperpolarization and relaxation were analyzed by two-way analysis of variance followed by Scheffé's test for multiple comparisons. Other variables were analyzed by one-way analysis of variance followed by Scheffé's test for multiple comparisons or paired Student's *t*-test. When *P* was less than 0.05, differences were considered to be statistically significant.

RESULTS

Endothelium-dependent Hyperpolarization in SHR, Dahl Salt-sensitive Rats and 2-K, 1-C Renal Hypertensive Rats

Hyperpolarizations in response to acetylcholine, applied under basal conditions, were significantly reduced in SHR, Dahl salt-sensitive rats fed a high salt diet, and 2-K, 1-C renal hypertensive rats compared with their respective normotensive controls (Figure 34-1). When acetylcholine was applied to the preparations depolarized with norepinephrine (10^{-5} M), the acetylcholine-induced hyperpolarization was also significantly less in SHR, salt-loaded Dahl salt-sensitive rats and 2-K, 1-C rats compared with control rats (data not shown).

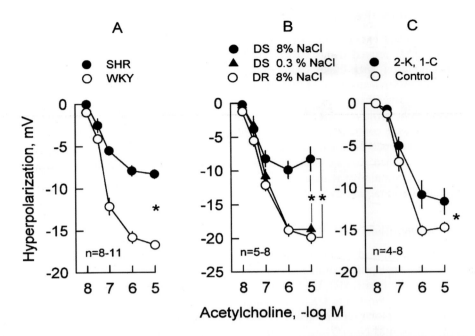

Figure 34-1. Concentration-hyperpolarization curves to acetylcholine in mesenteric arteries (with endothelium) of spontaneously hypertensive rats (SHR) and Wistar Kyoto rats (WKY) (A), Dahl salt-sensitive rats (DS) and Dahl salt-resistant rats (DR) (B), and 2-kidney, 1-clip renal hypertensive rats (2-K, 1-C) and control rats (C). Acetylcholine was applied under resting conditions. Values are mean ± SEM. The asterisk indicates that differences are statistically significant ($P < 0.05$, by two-way analysis of variance).

Effects of Antihypertensive Treatments on Endothelium-dependent Hyperpolarization in SHR

Systolic blood pressure

Systolic blood pressure of SHR was lowered to a level comparable to that of WKY by the antihypertensive treatment with either enalapril or the combination of hydralazine plus hydrochlorothiazide.

Resting membrane potential

The resting membrane potential was significantly less negative in control SHR (–45 ± 2 mV) than in WKY (–50 ± 1 mV). Antihypertensive treatments led to a partial repolarization of the membrane (SHR with enalapril: –49 ± 1 mV; SHR with the combination: –46 ± 1 mV).

Endothelium-dependent hyperpolarization

Acetylcholine-induced hyperpolarization was impaired significantly in control SHRs compared to WKY (Figure 34-2) in preparations contracted with norepinephrine. Treatment with enalapril significantly improved the acetylcholine-induced hyperpolarization.

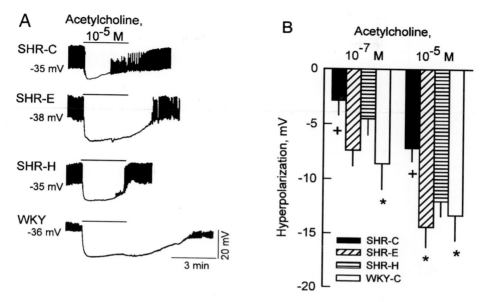

Figure 34-2. (A), Recordings showing hyperpolarization to acetylcholine (ACh) (10^{-5} M) during depolarization with norepinephrine (10^{-5} M) in the presence of indomethacin (10^{-5} M) in mesenteric arteries with endothelium of untreated spontaneously hypertensive rats (SHR-C), enalapril-treated SHR (SHR-E), hydralazine plus hydrochlorothiazide-treated SHR (SHR-H), and Wistar Kyoto rats (WKY). (B), hyperpolarizations to 10^{-7} M and 10^{-5} M acetylcholine. Values are mean ± SEM. There were 7 to 8 rats in each group. The asterisk indicates that values are statistically significantly different from those of control SHR ($P < 0.05$, by two-way analysis of variance). The dagger indicates that values are statistically significantly different from those of WKY ($P < 0.05$, by two-way analysis of variance). (Modified from Onaka *et al.*, Circulation 1998, with permission of the American Heart Association)

Treatment with hydralazine and hydrochlorothiazide improved the acetylcholine-induced hyperpolarization to a level comparable to that observed in WKY. When acetylcholine was applied under resting conditions, the two treatments also improved the hyperpolarization in response to acetylcholine (data not shown).

Endothelium-dependent relaxation

In mesenteric arterial rings contracted with norepinephrine (10^{-5} M), the acetylcholine-induced relaxation was significantly less in control SHR than in WKY, in the absence or presence of indomethacin (10^{-5} M). The two treatments tended to improve these relaxations. The EDHF-mediated component was assessed by determining the relaxation to acetylcholine in the presence of both indomethacin and N^{ω}-nitro-L-arginine (Figure 34-3). The EDHF-mediated relaxation was virtually absent in untreated SHR. The two treatments markedly restored the EDHF-mediated relaxation, which became comparable to that observed in WKY.

When rings were contracted with KCl (77 mM) to eliminate EDHF-mediated responses in the presence of indomethacin, the relaxation to acetylcholine was almost identical among the four groups. This relaxation was abolished by N^{ω}-nitro-L-arginine.

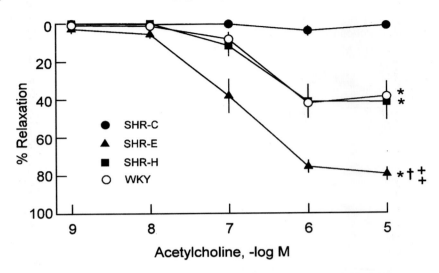

Figure 34-3. Concentration-relaxation curves to acetylcholine in mesenteric arterial rings with endothelium contracted with norepinephrine (10^{-5} M) in the presence of indomethacin (10^{-5} M) and N$^\omega$-nitro-L-arginine (10^{-4} M) in untreated spontaneously hypertensive rats (SHR-C), enalapril-treated SHR (SHR-E), hydralazine plus hydrochlorothiazide-treated SHR (SHR-H), and Wistar Kyoto rats (WKY). Values are mean ± SEM. There were 7 to 9 rats in each group. The asterisk indicates that values are statistically significantly different from those of control SHR ($P < 0.05$, by two-way analysis of variance). The dagger indicates that values are statistically significantly different from those of WKY ($P < 0.05$, by two-way analysis of variance). The double dagger indicates that values are statistically significantly different from hydralazine plus hydrochlorothiazide-treated SHR ($P < 0.05$, by two-way analysis of variance). (Modified from Onaka *et al.*, Circulation 1998, with permission of the American Heart Association)

Endothelium-independent hyperpolarization and relaxation

The hyperpolarization in response to levcromakalim was comparable in the mesenteric arteries among the four groups (Figure 34-4). The relaxation in response to levcromakalim in rings contracted with norepinephrine (10^{-5} M) was also similar among the four groups (Figure 34-4). The relaxation to sodium nitroprusside, a NO donor, in rings contracted with norepinephrine (10^{-5} M) was comparable among the four groups.

DISCUSSION

The studies described here demonstrate that endothelium-dependent hyperpolarizations of mesenteric arteries are impaired in various models of experimental hypertension, and that antihypertensive treatments can restore the normal endothelium-dependent hyperpolarizations and relaxations in the SHR.

Endothelium-dependent Hyperpolarization in Experimental Hypertension

The SHR is the most widely used animal model of essential hypertension (Yamori and Swales, 1994). Dahl salt-sensitive rats develop hypertension only when they are exposed to a high salt-diet, and thus represent a salt-sensitive form of hypertension with low-renin

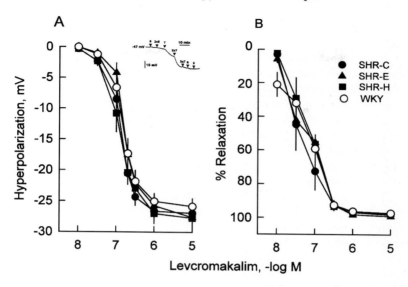

Figure 34-4. (A), concentration-hyperpolarization curves to levcromakalim in mesenteric arteries with endothelium of untreated spontaneously hypertensive rats (SHR-C), enalapril-treated SHR (SHR-E), hydralazine plus hydrochlorothiazide-treated SHR (SHR-H), and Wistar Kyoto rats (WKY). Inset, records showing hyperpolarization to increasing concentrations of levcromakalim in SHR-C. Hyperpolarizations were obtained under control condition. There were 6 to 8 rats in each group. (B), concentration-relaxation curves to levcromakalim in mesenteric arteries of SHR-C, SHR-H, SHR-E and WKY. Relaxations were obtained in rings contracted with norepinephrine (10^{-5} M) in the presence of indomethacin (10^{-5} M). There were 7 to 9 rats in each group. Values are mean ± SEM.

profile (Rapp, 1994). Two-K, 1-C renovascular hypertensive rats develop hypertension that depends al least in part on renin-angiotensin system (Thurston, 1994). In the present study, endothelium-dependent hyperpolarizations to acetylcholine in mesenteric arteries were reduced in these three models of experimental hypertension with different etiology. Endothelium-dependent hyperpolarization to carbachol is also depressed in the aorta of 2-K, 1-C hypertensive rats (Van de Voorde, Vanheel and Leusen, 1992). Preliminary evidence indicates that the acetylcholine-induced hyperpolarization is attenuated in rats made hypertensive by chronic inhibition of NO synthase (Onaka *et al.*, 1996). Furthermore, endothelium-dependent hyperpolarization is normal in prehypertensive SHR but impaired in SHR with established hypertension (Fujii *et al.*, 1993). These findings, taken in conjunction, suggest that endothelium-dependent hyperpolarizations are impaired as a consequence of the chronic exposure to hypertension. Then, the question arises as to whether or not the deleterious influence of hypertension on endothelium-dependent hyperpolarization is reversible.

Effects of Antihypertensive Treatment on Endothelium-dependent Hyperpolarization

The data summarized here demonstrates that antihypertensive treatments improved endothelium-dependent hyperpolarization and relaxation to acetylcholine in the mesenteric arteries of SHR. On the other hand, the NO-mediated relaxation was preserved and was not influenced by antihypertensive treatments. These observations suggest that the EDHF system may play a pivotal role in the endothelial dysfunction and its improvement by drug therapy in SHR, the model of essential hypertension.

Because two different regimens improved EDHF-mediated responses, the blood pressure reduction *per se* may be important in reversing the impairment in endothelial function. However, enalapril treatment tended to improve EDHF-mediated responses to a greater extent than did the combination of hydralazine and hydrochlorothiazide. This raises the possibility that the converting enzyme inhibitor improved endothelial function in part through a mechanism unrelated to blood pressure. In vitro treatment with perindoprilat, another converting enzyme inhibitor, augmented bradykinin-induced hyperpolarization in the isolated canine and human coronary arteries (Mombouli and Vanhoutte, 1989; Nakashima *et al.*, 1993), probably by inhibiting the breakdown of bradykinin. However, it is unlikely that such an effect of converting enzyme inhibitors accounts for the present findings, since the incubation of isolated blood vessels of SHR with converting enzyme inhibitors did not affect acetylcholine-induced hyperpolarization (Onaka *et al.*, 1998). Hyperpolarization and relaxation to levcromakalim, a direct activator of K_{ATP} channels (Taylor *et al.*, 1988), were not different between SHR and WKY and were not modified by antihypertensive therapy. Therefore, the beneficial effect of drug therapy can not be attributed to alterations of the vascular smooth muscle function.

The existence of EDHF-mediated, endothelium-dependent hyperpolarization has been demonstrated in human blood vessels (Nakashima *et al.*, 1993; Petersson *et al.*, 1996; Urakami-Harasawa *et al.*, 1997). Furthermore, in human small gastroepiploic arteries the contribution of EDHF to relaxations elicited by bradykinin even exceeds that of NO (Urakami-Harasawa *et al.*, 1997). It remains to be tested whether or not these EDHF-mediated hyperpolarization are altered or modified by drug therapy in blood vessels from patients with essential hypertension.

NO-Mediated Relaxation in Experimental Hypertension

In the present study, NO-mediated relaxations, as assessed by the relaxation to acetylcholine in rings contracted with high K solution, were not impaired in the SHR and not modulated by drug therapy. NO-mediated relaxations have been reported to be reduced in certain forms of experimental hypertension, e.g., Dahl salt-sensitive rats, 2-K, 1-C rats and stroke-prone SHR (Hayakawa *et al.*, 1993). Furthermore, several studies have demonstrated beneficial effects of antihypertensive therapy on NO-mediated relaxation (Tschudi *et al.*, 1994; Dohi, Kojima and Sato, 1996). Obviously the effects of drug therapy on NO-mediated relaxation in hypertension may differ depending on the type of hypertension or the vascular bed studied.

35 Chronic Exposure to Lead Inhibits Endothelium-dependent Hyperpolarization in the Mesenteric Artery of the Rat

Hirotaka Oishi, *Mikio Nakashima, *Tadahide Totoki and Katsumaro Tomokuni

Department of Community Health Science and
*Department of Anesthesiology and Critical Care Medicine, Saga Medical School, Saga, Japan
Address for correspondence: Mikio Nakashima, M.D., Ph.D., Surgical Center, Saga Medical School, 1-1 5-chome, Nabeshima, Saga 849, Japan.
Tel: (81) 952 31 6511; Fax: (81) 952 33 2518

Experiments were designed to determine the effect of chronic lead exposure on endothelium-dependent responses to acetylcholine in isolated blood vessels of the rat. Male Wistar rats were maintained for one or three months, with or without oral administration of lead. Membrane potential and isometric tension were measured in mesenteric arteries. Acetylcholine caused concentration-dependent and endothelium-dependent relaxations in rings with endothelium contracted with phenylephrine. There was no significant difference in relaxation between lead-exposed and control animals. In the presence of N^ω-nitro-L-arginine methyl ester, both the endothelium-dependent hyperpolarizations and relaxations to acetylcholine were significantly reduced in those from the group for three month exposed to lead. In the aortas from the lead-exposed groups, endothelium-dependent relaxations to acetylcholine were not significantly different from age-matched controls, both of which were inhibited in the presence of N^ω-nitro-L-arginine methyl ester. The basal levels of cyclic GMP in the aorta were not affected by the exposure to lead regardless of its duration. These data indicate that both endothelium-dependent hyperpolarizations and N^ω-nitro-L-arginine methyl ester resistant relaxations decrease with chronic exposure to lead in the rat mesenteric arteries. They may suggest that lead is an inhibitor of endothelium-derived hyperpolarizing factor (EDHF).

KEYWORDS: lead exposure, EDHF, EDRF/NO, cytochrome P450, cyclic GMP, rat artery

INTRODUCTION

Chronic exposure to lead, resulting from pollution or occupational hazards can cause various health problems in humans, including cardiovascular disorders such as hypertension (Harlan *et al.*, 1985; Beevers *et al.*, 1976; Neri *et al.*, 1988; Sharp *et al.*, 1988), although some studies in humans show no correlation between lead and high blood pressure (Elwood *et al.*, 1988; Grandjean *et al.*, 1989). However, the influence of the exposure to heavy metal on the cardiovascular system still remains unclear. In hypertensive rats exposed to lead, endothelial function might be impaired, since some endothelial factors, such as increased endothelin-3 or decreased plasma cyclic GMP (as a reflection of the production of nitric oxide) are observed (Khalil *et al.*, 1993). The endothelial monolayer cell is a source of vasodilator mediators, including endothelium-derived relaxing factor

313

(EDRF; most likely nitric oxide), prostacyclin and an as-yet-unidentified endothelium-derived hyperpolarizing factor [EDHF; (Chen *et al.*, 1988; Félétou and Vanhoutte, 1988; Beny and Brunet, 1988b)]. The purpose of the present study was to investigate the effect of chronic exposure to lead on endothelium-dependent responses by detecting mechanical and electrophysiological responses as well as measuring tissue levels of cyclic GMP changes in rat arteries.

MATERIALS AND METHODS

Preparations

Male Wistar rats (7 weeks, weighing 260.1 ± 5.4 g; Kyudo, Kumamoto, Japan) were assigned randomly to two different protocols. Animals in lead-exposed groups were given drinking water containing lead acetate (50 ppm), while those in controls were given distilled water. Both groups were fed with a standard diet (CE-2, Nihon CLEA, Tokyo, Japan), and were housed for one month or three months in a chronic care facility prior to sacrifice. Systolic blood pressure and heart rate were measured indirectly by the tail cuff method in unanesthetized animals (MK-1000, Muromachi Kikai, Tokyo, Japan).

The rats were anesthetized by sodium pentobarbital (50 mg/kg, intraperitoneally; Abbott Laboratories, North Chicago, IL). The main branches of the superior mesenteric artery and thoracic aorta were excised and put in modified Krebs-Ringer bicarbonate solution (4°C) [millimolar composition; NaCl 118.3, KCl 4.7, $CaCl_2$ 2.5, $MgSO_4$ 1.2, KH_2PO_4 1.2, $NaHCO_3$ 25.0, Ca-EDTA 0.026, and glucose 11.1, aerated with 95% O_2 and 5% CO_2 (control solution)]. The blood vessels were cleaned of surrounding tissues, and cut into rings 4–5 mm long. In some rings, the endothelium was removed mechanically by inserting the tip of a forceps into the lumen of the blood vessel and rolling the rings back and forth on a paper towel wetted with control solution. Whole blood from each animal was stored at −80°C for determination of lead content. Plasma was separated from red blood cells by centrifuging at 2,500 rpm for 20 minutes at 4°C and stored at −80°C for determination of delta-aminolevulinic acid (δ-ALA). The concentration of lead in the blood was measured using a flameless atomic absorption spectrometer (Z-9000, Hitachi, Tokyo, Japan) after a ten-fold dilution of whole blood with 0.1 N HNO_3 containing 1% Triton X-100. The plasma δ-ALA was determined according to the method of fluorimetric high performance liquid chromatography (Shimadzu, Kyoto, Japan) (Oishi *et al.*, 1996). All experiments were performed in the presence of indomethacin (10^{-5} M) to inhibit the formation of vasoactive prostanoids.

Organ Chamber Studies

Rings were suspended between two stirrups in organ chambers (60 ml) filled with control solution (gassed with 95% O_2 plus 5% CO_2 and maintained at 37°C). One of the stirrups was anchored to the inside of the organ chamber and the other was connected to a force transducer (TB-611T, Nihon Kohden, Tokyo, Japan) to record changes in isometric tension. In all experiments, the rings were stretched in a stepwise manner to the optimal point of their length-tension curve and were allowed to equilibrate for 45 minutes. In preliminary experiments, concentration-contraction curves to phenylephrine were performed to determine the concentration inducing 80% of the maximal contraction (ED_{80}). The preparation were contracted with ED_{80} of phenylephrine, and a concentration-response curve to acetylcholine (10^{-8} to 10^{-6} M) was obtained in rings equilibrated in the presence or absence of N^{ω}-nitro-L-arginine methyl ester (10^{-4} M).

Electrophysiological Studies

Rings of the superior mesenteric artery were cut open along their longitudinal axis, and pinned down on the bottom of an organ chamber (2 ml in capacity) with the endothelial side facing upward. The arteries were superfused with control solution (37°C, 3 ml/min) for 45 minutes before starting the recording. The membrane potential of arterial smooth muscle cells were measured with glass microelectrodes (FHC, ME, U.S.A) filled with 3 M-KCl solution (tip resistance, 40–80 megaohms). The electrical signal was recorded from the intimal side of the arteries (Nakashima and Vanhoutte, 1993b), and was amplified by means of a recording amplifier (MEZ-7101, Nihon Kohden, Tokyo, Japan). The membrane potential was monitored continuously on an oscilloscope (VC-10, Nihon Kohden, Tokyo, Japan). Experiments were performed in the presence of N^{ω}-nitro-L-arginine methyl ester (10^{-4} M).

Tissue Content of Cyclic GMP

A radioimmunoassay technique was used to determine the tissue content of cyclic GMP. The rings was frozen quickly by liquid nitrogen and stored −80°C. At the time of assay, rings were homogenized in 1 ml of 6% trichloroacetic acid at 4°C and centrifuged three times at 3, 000 rpm for 10 min. The supernatant was removed from each sample and placed into a glass tube, extracted three times with 6 ml of water-saturated ethyl ether. Then the samples were assayed using a cyclic GMP Radioimmunoassay kit (Yamasa Shoyu Co, Chiba, Japan). Levels of cyclic GMP in tissues were expressed in femtomoles per mg wet weight of tissue.

Drugs

The drugs used were phenylephrine, acetylcholine, indomethacin, N^{ω}-nitro-L-arginine methyl ester (all of Sigma, St. Louis, MO, U.S.A), and lead acetate (Wako pure chemicals, Osaka, Japan). Indomethacin was dissolved in distilled water together with equimolar concentration of Na_2CO_3. The other drugs were prepared with distilled water.

Statistical Analysis

The data were presented as mean ± S.E.M., and were analyzed with Student's t test (paired or unpaired), or with analysis of variance followed by Fisher's PLSD test. "n" represents the number of animals examined. P values less than 0.05 were regarded as statistically significant.

Table 35-1   Body weight, blood pressure, heart rate, blood lead, and plasma δ-aminolevulinic acid (δ-ALA) in control and lead-exposed rats at 1- and 3-months

	1-month		3-month	
	Control	*Lead-exposed*	*Control*	*Lead-exposed*
Body weight (g)	397.0 ± 8.4 (12)	392.7 ± 5.9 (24)	516.1 ± 12.2 (7)	534.4 ± 10.3 (18)
Systolic blood pressure (mmHg)	135.8 ± 3.8 (9)	143.3 ± 2.4 (24)	154.6 ± 8.6 (7)	159.1 ± 4.6 (18)
Heart rate (bpm)	451.9 ± 9.4 (9)	474.9 ± 9.0 (24)	414.7 ± 12.1 (7)	426.4 ± 8.9 (18)
Blood lead (μg/dll)	0.34 ± 0.1 (7)	13.3 ± 3.6* (16)	N.D. (5)	10.8 ± 3.1* (10)
Plasma δ-ALA (μg/l)	6.6 ± 0.4 (6)	9.7 ± 1.3* (16)	7.5 ± 0.0 (5)	7.5 ± 0.5 (10)

Data shown as mean ± S.E.M. Number in parentheses indicates the number of animals. N.D. shown as not detectable
The asterisks indicate a statistically significant difference between control and lead-exposed groups (P < 0.05).

RESULTS

There was no statistical significance in body weight, heart rate, and systolic blood pressure between lead-exposed and age-matched control groups. After both one and three month exposure to lead, the blood levels of lead were higher than those in age-matched controls. Plasma levels of δ-ALA were significantly higher after one month exposure to lead in control animals (Table 35-1).

Organ Chamber Studies

Thoracic aorta

In rings with endothelium, there was no significant difference in the contraction to phenylephrine between one month lead-exposed and age-matched control groups either in the absence or presence of N^ω-nitro-L-arginine methyl ester (Figure 35-1). The contraction to phenylephrine was increased significantly in rings from three month lead-exposed in both in the presence and absence of N^ω-nitro-L-arginine methyl ester (Figure 35-1). In tissues without endothelium, there was no significant difference in contraction to phenylephrine between the groups, regardless of the duration of the exposure to lead (Figure 35-1).

Acetylcholine (10^{-8} to 10^{-6} M) caused concentration-dependent relaxations in rings with endothelium from either lead-exposed or control rats. There was no significant difference in relaxations between lead-exposed and age-matched control groups, regardless of the duration of the exposure (Figure 35-1).

In rings with endothelium plus N^ω-nitro-L-arginine methyl ester or in rings without endothelium, the relaxations to acetylcholine were almost abolished (less than 3% of maximal relaxation expressed as percentage of the contraction of phenylephrine), and no significant difference was observed between lead-exposed and age-matched control animals (Figure 35-1).

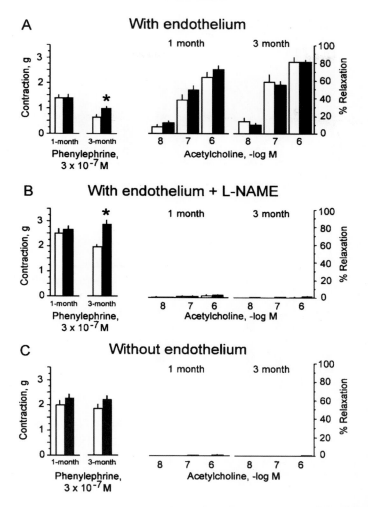

Figure 35-1. Relaxation induced by acetylcholine in control (open bars) and lead-exposed groups (filled bars) in rat aorta. Responses were obtained from rings with endothelium (A) or with endothelium plus N^{ω}-nitro-L-arginine methyl ester (L-NAME;10^{-4} M) (B) or without endothelium (C). Relaxation were expressed as percentage of the contraction to phenylephrine (3×10^{-7} M). Data shown as means $\pm$ S.E.M. n = 5–15.

Mesenteric artery

In rings with endothelium, there was no significant difference in contraction to phenylephrine (10^{-6} M), between lead-exposed and control groups, regardless of the duration of the exposure (Figure 35-2A). In the presence of N^{ω}-nitro-L-arginine methyl ester, the contraction to phenylephrine was increased significantly in rings from the exposed groups of either duration (Figure 35-2B).

Hirotaka Oishi et al.

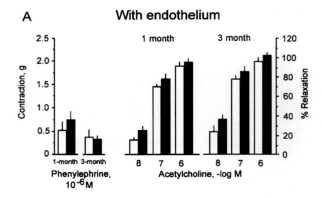

A **With endothelium**

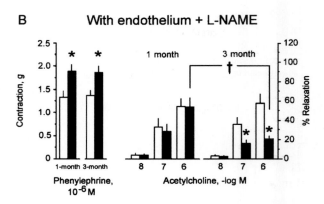

B **With endothelium + L-NAME**

Figure 35-2. Relaxation induced by acetylcholine in control (open bars) and lead-exposed groups (filled bars) in the rat mesenteric artery. Responses were obtained from rings with endothelium (A) or with endothelium plus N^ω-nitro-L-arginine methyl ester (L-NAME ;10^{-4} M) (B). Relaxations were expressed as percentage of the contraction to phenylephrine (10^{-6} M). Data shown as means ± S.E.M. n = 5–7. The asterisks indicate a statistically significant difference between control and lead-exposed groups (P < 0.05). The dagger indicates a statistically significant difference between one- and three-month duration in control or lead-exposed groups (P < 0.05).

In the absence of N^ω-nitro-L-arginine methyl ester, acetylcholine (10^{-8} to 10^{-6} M) produced concentration-dependent relaxations in rings with endothelium, contracted with phenylephrine. There was no significant difference in relaxation between lead-exposed and control preparation (Figure 35-2).

In the presence of N^ω-nitro-L-arginine methyl ester, the concentration-dependent relaxations to acetylcholine were decreased significantly compared to those in its absence (Figure 35-2). The relaxation to acetylcholine (10^{-7} and 10^{-6} M) was decreased significantly in rings from the three-month exposure compared to those from age-matched controls, in the presence of N^ω-nitro-L-arginine methyl ester (Figure 35-2). A significant decrease in the relaxation to acetylcholine (10^{-6} M) was also observed in rings after three month, compared to one-month exposure to lead (Figure 35-2).

Table 35-2 Resting membrane potentials in vascular smooth muscle of the rat mesenteric arteries with endothelium

	Membrane Potential (mV)	
	1-month	*3-month*
Control	-55.7 ± 2.1 (6)	-54.4 ± 1.2 (5)
Lead-exposed	-52.4 ± 1.2 (14)	-53.6 ± 0.8 (10)

Data shown as mean $\pm$ S.E.M.
Number in parentheses indicates the number of animals.
All data were obtained in the presence of indomethacin (10^{-5} M) and N^{ω}-nitro-L-arginine (10^{-4} M).

Electrophysiological Studies

There was no significant difference in resting membrane potentials of the arteries among the four protocol groups (Table 35-2).

In the presence of N^{ω}-nitro-L-arginine methyl ester, acetylcholine (10^{-8} to 10^{-6} M; in cumulative application) produced concentration-dependent hyperpolarizations of the vascular smooth muscle cells. (Figure 35-3). There was no significant difference in hyperpolarization between tissues from one-month exposed and control rats, while in tissues from those given lead for three months, hyperpolarizations to acetylcholine (10^{-6} M) were decreased significantly in lead-exposed animals (Figure 35-3).

Production of Cyclic GMP

There was no significant difference in the basic levels of cyclic GMP in the aortas of lead-exposed versus age-matched controls (Figure 35-4).

DISCUSSION

The present experiments confirm that N^{ω}-nitro-L-arginine methyl ester, an inhibitor of nitric oxide synthase, completely blocks endothelium-dependent relaxations to acetylcholine in the rat aorta, but only partly in the mesenteric artery of the same species (Nagao *et al.*, 1992a). They further demonstrate that both endothelium-dependent hyperpolarizations and relaxations resistant to N^{ω}-nitro-L-arginine methyl ester decrease with chronic exposure to lead in the mesenteric artery, while the N^{ω}-nitro-L-arginine methyl ester sensitive relaxation does not in either the aorta or the mesenteric artery. Relaxations resistant to N^{ω}-nitro-L-arginine methyl ester could be best explained by EDHF (Chen *et al.*, 1988; Félétou and Vanhoutte, 1988; Beny and Brunet, 1988b). EDHF increases the K$^+$ conductance of the vascular smooth muscle cell membrane by opening K$^+$ channels (Chen *et al.*, 1988; Beny and Brunet, 1988b; Taylor *et al.*, 1998). Thus, the present results suggest that inhibition of EDHF may, at least in part, account for the impaired relaxations resistant to N^{ω}-nitro-L-arginine methyl ester in lead-treated animals.

One of the possible mechanisms underlying the impairment of endothelium-dependent hyperpolarization with chronic exposure to lead includes its cytotoxicity at the endothelium.

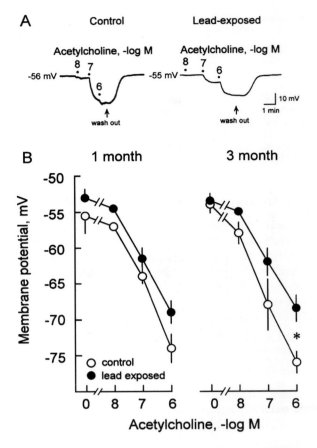

Figure 35-3. Resting membrane potentials and hyperpolarizations to acetylcholine in rat mesenteric arteries with endothelium in the presence of N^{ω}-nitro-L-arginine methyl ester (10^{-4} M). (A), typical traces of membrane potential changes to acetylcholine in the three-month lead-exposed and age-matched control groups. (B), concentration-dependent hyperpolarizations to acetylcholine in mesenteric arteries of lead-exposed (closed circle) and control (open circle) groups (closed circle) at one- or three-month. The amplitudes of hyperpolarization to acetylcholine (10^{-8}, 10^{-7}, and 10^{-6} M) in one-month rat averaged 2.1 ± 0.3, 9.2 ± 1.1, 18.9 ± 1.9 mV and 2.0 ± 0.5, 9.7 ± 1.3, 16.8 ± 1.1 mV, in control and lead-exposed groups , respectively. Data shown as means $\pm$ S.E.M. n = 5–10. The asterisk indicates a statistically significant difference between the control and lead-exposed groups ($P < 0.05$).

If chronic lead-exposure were to inhibit functions of endothelial cells, in particular the production of EDRF via alteration of cell membrane receptors, or changes in the release of the the factor (Illiano *et al.*, 1992), the relaxation of, or the increase in cyclic GMP in the vascular smooth muscle cells should be reduced. However, there was no significant decrease either in the relaxation to acetylcholine or in the levels of in cyclic GMP in the aorta after chronic exposure to lead.

A decreased release of nitric oxide could also explain a decrease in the electrical response, since the concentration of cyclic GMP in plasma and urine (as a reflection of

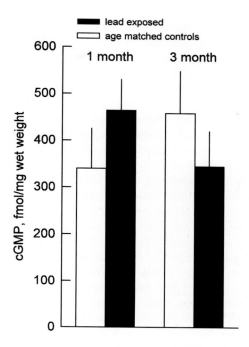

Figure 35-4. Basal levels of cyclic GMP in rings with lead-exposed (filled bars) and age-matched controls (open bars) of rat aorta, after one- and three month. The values were expressed as femtomoles of cyclic GMP per mg wet weight of aorta. Data shown as means ± S.E.M. (n = 6–16).

production of nitric oxide) decrease in the rats with chronic exposure to lead (Khalil *et al.*, 1993). Nitric oxide can cause membrane hyperpolarization in the mesenteric artery of this species (Garland and McPherson, 1992). However, in the present experiments, it is not likely that a reduced release and/or production of EDRF/NO is responsible for the decreased hyperpolarization with exposure to lead, since the part of relaxation sensitive to N^ω-nitro-L-arginine methyl ester, and the plasma levels of cyclic GMP are not decreased in the groups exposed to the metal.

Lead could inhibit the production or release of EDHF. One of the endothelium-derived cytochrome P450 metabolites of arachidonic acid, 5, 6-epoxyeicosatrienoic acid, displays the characteristics of EDHF (Bauersachs *et al.*, 1994). Indeed, cytochrome P450 inhibitors (e.g. clotrimazole, metyrapone, and SKF525a) abolishes the EDHF-mediated relaxation elicited by acetylcholine (Lischke *et al.*, 1995b). Exposure to lead causes a disturbance of cytochrome P450 in the rat liver (Degawa *et al.*, 1994). In rat liver microsomes, treatment with ionic lead decreases total cytochrome P450 to 60–80% of control, and inhibits the expression of cytochrome $P450_{1A2}$ (Degawa *et al.*, 1994). Taken together, these findings suggest that the production of EDHF, possibly a cytochrome P450-metabolite, is inhibited by exposure to lead. The present study also indicates that exposure of more than one month is required to cause a decrease in the acetylcholine-induced hyperpolarization.

In addition to changing endothelial function, chronic lead exposure may also affect that of vascular smooth muscle cell function. In particular, sensitivity to EDHF of the smooth muscle might be changed in the metal-treated animals. Since EDHF is believed to cause hyperpolarization by activating either the Na-K pump (Félétou and Vanhoutte, 1988) or potassium channels (Chen *et al.*, 1988) in vascular smooth muscle, lead might affect these targets. This hypothesis is supported by observations that lead can inhibit sodium-potassium-activated adenosine triphosphatase (Na-K-ATPase) in the dog cerebral cortex (Weiler *et al.*, 1988), voltage-dependent potassium channels in the rat brain (Madeja *et al.*, 1995) and voltage-activated calcium channels in the dorsal root ganglion neuron of the rat (Büsselberg *et al.*, 1994). However, it remains unclear whether or not these direct effects of lead ion explain the decreased electrical response in vascular smooth muscle from animals exposed chronically to lead.

In earlier observations in lead-treated rats, a positive correlation was obtained between blood lead levels and systemic arterial blood pressure (Khalil *et al.*, 1993; Piccinini *et al.*, 1977) although plasma cyclic GMP was reduced (Khalil *et al.*, 1993). By contrast, in the present study, despite the increased blood lead levels of the treated rats, systolic blood pressure did not increase, and the basal levels of cyclic GMP in the aorta were not changed. These discrepancies might be due to the concentration of lead used, since in the former study, higher blood level of lead was achieved (Khalil *et al.*, 1993).

The present study shows that aortic rings from low lead-treated rats, at least after a shorter exposure exhibit an increased force-generating ability to phenylephrine, as is the case in the rat tail artery (Webb *et al.*, 1981). The mechanisms underlying this increased reactivity with chronic lead treatment remains unknown. It is not likely that endothelium-derived contracting factor (Lüscher and Vanhoutte, 1990) is responsible for the phenomenon, since all experiments were performed in the presence of indomethacin. The increased vascular contractility might be due to a direct action of lead on smooth muscle cells, since the tissue content of radioactive calcium is increased following exposure to the metal (Piccinini *et al.*, 1977), and since lead induces endothelium-independent and calcium-dependent contractions resulting from protein kinase C activation in the rabbit mesenteric artery (Watts *et al.*, 1993).

In summary, the present study indicates that chronic exposure to lead inhibits endothelium-dependent hyperpolarizations and relaxations sensitive to N^{ω}-nitro-L-arginine methyl ester in the rat mesenteric artery without affecting EDRF/NO induced relaxation. Lead, if given chronically, may act as an inhibitor of endothelium-derived hyperpolarizing factor in the rat mesenteric artery.

ACKNOWLEDGMENT

The authors thank Dr. Masayoshi Ichiba and Kaori Takahashi for technical help.

36 Effect of Aging on the Endothelium-dependent Relaxation Induced by Acetylcholine in Mesenteric Resistance Arteries of the Rat

Maria Alvarez de Sotomayor, Jean-Claude Stoclet and Ramaroson Andriantsitohaina

Pharmacologie et Physico-Chimie des Interactions Cellulaires et Moléculaires,
UMR CNRS (ex ERS 653, ex URA 491), Université Louis Pasteur de Strasbourg,
Faculté de Pharmacie, BP 24, 67401 Illkirch-Cedex, France
Address for correspondence: Dr Ramaroson Andriantsitohaina,
Pharmacologie et Physico-Chimie des Interactions Cellulaires et Moléculaires,
UMR CNRS (ex ERS 653, ex URA 491), Université Louis Pasteur de Strasbourg,
Faculté de Pharmacie, BP 24, 67401 Illkirch-Cedex, France.
Phone: 33-3-88-67-69-67; Fax: 33-3-88-66-46-33; E-mail: nain@pharma.u-strasbg.fr

Experiments were designed to investigate the effect of aging on the contribution of endothelium-derived hyperpolarizing factor (EDHF) to the acetylcholine-induced relaxation. Responses were compared in branch II or III of superior mesenteric arteries from 14 (adult) or 70–90 (old) week male rats. Organ chamber experiments were performed to record the relaxation induced by acetylcholine in arteries contracted with norepinephrine. Acetylcholine produced relaxations in a concentration-dependent manner in blood vessels from both adult and old rats. The maximal relaxation to acetylcholine was reduced in vessels from old as compared to adult rats. In arteries from adult rats, the nitric oxide synthase inhibitor, N^{ω}-nitro-L-arginine but not the cyclooxygenase inhibitor, indomethacin, significantly attenuated the response to acetylcholine; the combination of N^{ω}-nitro-L-arginine plus indomethacin did not produce further inhibition. In arteries from old rats, N^{ω}-nitro-L-arginine alone had no effect on the response to acetylcholine. Indomethacin significantly potentiated the response to acetylcholine in the small mesenteric arteries from old rats. The combination of N^{ω}-nitro-L-arginine plus indomethacin did not affect the acetylcholine-response in arteries from old rats. Thus, the N^{ω}-nitro-L-arginine plus indomethacin-insensitive component of acetylcholine-induced relaxation, which probably reflects the EDHF-component of the response, was not significantly different in arteries from adult and old rats. These results suggest that the decrease in acetylcholine-induced relaxation with aging involves decreased NO-mediated dilatation and increased generation of vasoconstrictor product(s) from cyclo-oxygenase. The EDHF-component of the relaxation to acetylcholine appears not affected by aging in small mesenteric arteries of the rat.

KEYWORDS: acetylcholine, endothelium, EDHF, aging, resistance artery

INTRODUCTION

In addition to structural changes of the vascular wall, aging is associated with generalized dysfunction of both the endothelium and vascular smooth muscle. Thus, endothelium-dependent relaxations to different vasodilator agonists, including acetylcholine, are reduced with aging (Hongo *et al.*, 1988; Koga *et al.*, 1989; Taddei *et al.*, 1995). Several mechanisms may account for this dysfunction of the endothelium such as a decreased release of endothelium-derived relaxing factors or an increased production of endothelium-derived contracting factors. Among the relaxing factors, the endothelium can produce at least three substances which cause relaxation of vascular smooth muscle: endothelium-derived nitric oxide (NO) (Furchgott, 1983; Palmer, Ferrige and Moncada, 1987), prostanoid derivatives mainly prostacyclin and endothelium-derived hyperpolarizing factor (EDHF) (Bolton, Lang and Takewaki, 1984; Chen, Suzuki and Weston, 1988). NO can activate different targets to induce relaxation. However, it acts primarily via activation of the soluble guanylate cyclase in the underlying smooth muscle cells with a resulting increase in the content of cyclic GMP (Rapoport and Murad, 1983). Prostanoid derivatives (mainly prostacyclin) produce relaxation by the activation of adenylate cyclase or a direct activation of K^+ channels. Finally, a non NO and non prostanoid endothelial factor has been characterized but its nature is still under debate (Félétou and Vanhoutte, 1988; Taylor *et al.*, 1988; Mombouli and Vanhoutte, 1997). The latter can hyperpolarize vascular smooth muscle cells and induces relaxation. Due to the number of vasodilator substances released by the endothelium, the nature of age-related endothelial dysfunction is complex and not fully elucidated. The contribution of NO is reduced with aging because of an increased breakdown resulting from the augmented production of superoxide anions (Gryglewski, Palmer and Moncada, 1986), the decreased expression of endothelial NO synthase (Barton *et al.*, 1997) or both. With respect to EDHF, endothelium-dependent hyperpolarizations and relaxations to acetylcholine are impaired in the mesenteric artery with aging (Fujii *et al.*, 1992, 1993). However, it is unclear whether or not EDHF-mediated relaxation is altered with aging in resistance arteries. The latter plays a major role in the regulation of peripheral resistance. Therefore, the present study was aimed to determine the age-related changes in endothelium-dependent relaxation to acetylcholine in a resistance artery, the small mesenteric artery of the rat. The experiments were designed to study the relative contribution of the relaxing factors, especially EDHF, in response to acetylcholine with aging using different pharmacological tools such as the NO synthase inhibitor, N^ω-nitro-L-arginine and the cyclooxygenase inhibitor, indomethacin.

METHODS

Animals

Male Wistar rats, 14 (adult) or 70–90 (old) weeks old were bred from genitors provided by Iffa-Credo (Lyon, France). All rats were maintained in a colony room with fixed dark: light cycles and constant humidity and temperature, and they were provided with rodent chow n° A04 from U.A.R. (Villemoisson, France) and tap water *ad libitum*. This investigation conforms with the authorization for the use of laboratory animals given by the French government (Department of Agriculture).

Arterial Preparation

The animals were killed by cervical dislocation and exsanguinated. The viscera were exposed, and a proximal segment of the small bowel was removed and pinned in a dissecting dish containing physiological salt solution of the following composition (in mM): NaCl 119, KCl 4.7, KH_2PO_4 0.4, $NaHCO_3$ 14.9, $MgSO_4$ 1.17, $CaCl_2$ 2.5, glucose 5.5.

Branch II or III resistance arteries were cleaned of fat and connective tissue, and a segment 1.6 to 2.0 mm in length was removed. The segment was then suspended in a myograph using two tungsten wires (20 μm in diameter) inserted through the lumen of the vessel. Mechanical activity was recorded isometrically by a force transducer (Kistler-Morse, DSG BE4). The blood vessels were placed in physiological salt solution, kept at 37°C and gassed continuously with 95% O_2 and 5% CO_2 (pH 7.4). After an equilibration period of 30 min, the blood vessel was stretched to a length that yields a circumference equivalent to 90% of that given by an internal pressure of 100 mm Hg; this required a load of about 200 mg. After setting the vessel to its working length, it was challenged twice with norepinephrine (10^{-5} M) to elicit reproducible contractile responses. The presence of functional endothelium was assessed by the ability of acetylcholine (10^{-6} M) to induce more than 50% relaxation of vessels contracted with norepinephrine (10^{-5} M).

Experimental Protocols

Arteries with functional endothelium were contracted at 80% of maximal contraction with norepinephrine. The concentration of norepinephrine was adjusted for each preparation (0.3 to 1×10^{-6} M). The contractions produced by norepinephrine were 3.2 ± 0.35 ($n = 24$) and 3.2 ± 0.32 mN mm^{-1} ($n = 22$) in small mesenteric arteries from adult and old rats, respectively (not significantly different with age). When the contraction reached a plateau, cumulative addition of acetylcholine (from 10^{-9} to 3×10^{-6} M) was performed.

The involvement of the endothelial release of cyclooxygenase products and NO in the relaxation produced by acetylcholine was tested using the cyclooxygenase inhibitor, indomethacin and the NO synthase inhibitor, N$^{\omega}$-nitro-L-arginine. Concentration-response curves to acetylcholine were performed in the absence and in the presence of indomethacin, N$^{\omega}$-nitro-L-arginine or indomethacin plus N$^{\omega}$-nitro-L-arginine. Indomethacin and N$^{\omega}$-nitro-L-arginine were used at maximally active concentrations which were 10^{-5} M and 3×10^{-5} M respectively. When indomethacin, N$^{\omega}$-nitro-L-arginine or indomethacin plus N$^{\omega}$-nitro-L-arginine were used, the concentration of norepinephrine was adjusted in order to obtain the same level of contraction, the incubation period was 20 min prior to contraction with norepinephrine.

The involvement of EDHF in the relaxation produced by acetylcholine was characterized by obtaining concentration-response curves in normal physiological salt solution containing 25 mM KCl in the presence of indomethacin plus N$^{\omega}$-nitro-L-arginine.

Chemicals

Acetylcholine, indomethacin and N$^{\omega}$-nitro-L-arginine were purchased from Sigma Chemical (Grenoble, France).

Expression of Results and Statistical Analysis

Preliminary experiments had shown that two consecutive concentration-response curves of acetylcholine were not significantly different in the absence of treatment. The first curve was thus taken as control. Relaxation was expressed as a percentage of the level of contraction. All results are expressed as mean $\pm$ S.E.M. of n experiments. The differences

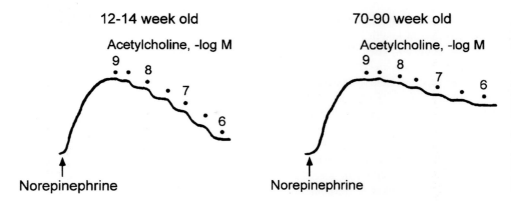

Figure 36-1. Concentration-dependent relaxations to acetylcholine in small mesenteric arteries [contracted with norepinephrine (NE)] taken from 12–14 week old (left) and 70–90 week old (right) rats.

between the responses to acetylcholine were tested for significance by use of analysis of variance (ANOVA). Differences were considered statistically significant when P was less than 0.05.

RESULTS

In arteries with functional endothelium contracted with norepinephrine taken from both adult and old rats, acetylcholine (from 10^{-9} to 1×10^{-6} M) produced relaxations in a concentration-dependent manner (Figure 36-1). Acetylcholine failed to produce relaxation of arteries without endothelium (not shown). However, arteries taken from old rats displayed reduced endothelium-dependent relaxation. The maximal relaxation to acetylcholine was significantly decreased from $79 \pm 6\%$ to $28 \pm 6.8\%$ in blood vessels from old compared to adult rats.

Neither indomethacin alone, N^{ω}-nitro-L-arginine alone or the combination of both inhibitors changed the basal tone of the arteries with functional endothelium taken from either adult and old rats (data not shown).

In arteries taken from adult rats (Figure 36-2), indomethacin did not alter response to acetylcholine. Blockade of endothelial NO using N^{ω}-nitro-L-arginine reduced significantly the relaxant effect of acetylcholine. The combination of indomethacin and N^{ω}-nitro-L-arginine did not produce further inhibition of the response to acetylcholine. The component of acetylcholine relaxation resistant to indomethacin plus N^{ω}-nitro-L-arginine relaxation was almost abolished in vessels partially depolarized with 25 mM KCl physiological salt solution.

In arteries taken from old rats (Figure 36-2), indomethacin significantly enhanced the response to acetylcholine. N^{ω}-nitro-L-arginine failed to reduce the response to acetylcholine. Also, the combination of indomethacin plus N^{ω}-nitro-L-arginine did not affect the relaxing response to acetylcholine. The indomethacin plus N^{ω}-nitro-L-arginine resistant component of the response to acetylcholine was abolished in vessels partially depolarized with 25 mM KCl physiological salt solution.

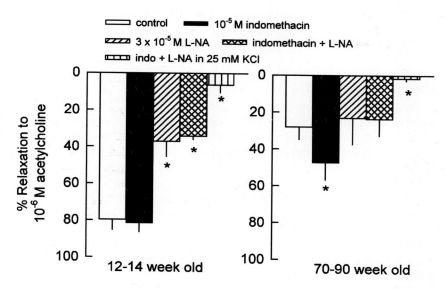

Figure 36-2. Relaxations to acetylcholine (10^{-6} M) obtained in the absence (open columns) or the presence of indomethacin (filled columns, 10^{-5} M), N$^\omega$-nitro-L-arginine (hatched columns, 3×10^{-5} M), indomethacin plus N$^\omega$-nitro-L-arginine (gray columns) or indomethacin plus N$^\omega$-nitro-L-arginine in physiological salt solution containing 25 mM KCl (horizontal columns). Results obtained in vessels from 12–14 week old (left) and 70–90 week old (right) rats. Results are mean ± S.E.M. of 5 to 7 experiments. The asteriks indicate statistically significant differences ($P < 0.05$) compared to controls.

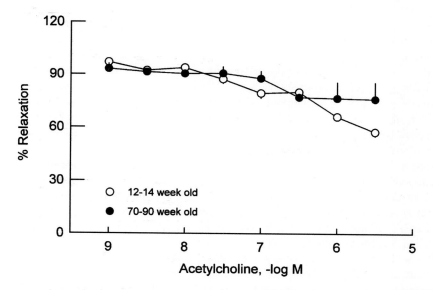

Figure 36-3. Concentration-response curves to acetylcholine obtained in the presence of indomethacin (10^{-5} M) plus N$^\omega$-nitro-L-arginine (3×10^{-5} M) in arteries taken from 12–14 week old (open circles) and 70–90 week old (filled circles) rats. Results are mean ± S.E.M. of 5 to 7 experiments.

The concentration-response curves to acetylcholine were not significantly different in the presence of the combination of indomethacin and N^{ω}-nitro-L-arginine between arteries taken from adult and old rats (Figure 36-3).

DISCUSSION

The major finding of the present study is that aging impairs endothelium-dependent relaxation in resistance arteries. The latter involves decreased NO-mediated dilatation and increased generation of vasoconstrictor product(s) from cyclooxygenase. The EDHF-component of acetylcholine relaxation appears not to be affected by aging in the small mesenteric arteries studied.

In resistance arteries from adult rats, the endothelium-dependent relaxation induced by acetylcholine comprised two components, an endothelium-derived NO (abolished by N^{ω}-nitro-L-arginine) and a N^{ω}-nitro-L-arginine-resistant component which was abolished by partial depolarization. The latter component is involved in the hyperpolarization produced by acetylcholine in the mesenteric resistance artery and can be attributed to EDHF (Garland and McPherson, 1992; Fukao *et al.*, 1995). EDHF contributes to the hyperpolarization and relaxation, and opens K^+ channels different from glibenclamide-sensitive K^+ channels activated by NO in this resistance artery (Siegel *et al.*, 1989b). Products of cyclooxygenase appear not to play a role in the endothelium-dependent relaxation to acetylcholine in adult resistance arteries.

In resistance arteries from old rats, the relaxation to acetylcholine was markedly attenuated, in line with previous observations reported in different vessels such as the aorta, pulmonary and superior mesenteric arteries (Fujii *et al.*, 1993; Tschudi *et al.*, 1996; Barton *et al.*, 1997). An increase in contractility to norepinephrine of the blood vessels cannot account for the reduced endothelium-dependent relaxation because the contractions induced by the catecholamine were matched between adult and old rats. There are several possible explanations for the reduced endothelium-dependent relaxation in aged rats. These include a decreased release of endothelium-derived-NO, EDHF or an increased production of endothelium-derived contracting factors. With respect to NO, the N^{ω}-nitro-L-arginine-sensitive component of the response to acetylcholine was almost abolished in resistance arteries from old rats suggesting that the NO production is dramatically reduced with aging. This result is in accordance with observations suggesting a reduced participation of NO pathway with aging (Dohi *et al.*, 1990; Tschudi *et al.*, 1996; Barton *et al.*, 1997). The latter could be the consequence of an increased NO breakdown due to an augmented production of superoxide anions (Gryglewski, Palmer and Moncada, 1986; Rubanyi and Vanhoutte, 1986), an alterations of antioxidant defense systems and an increased production of reactive oxygen species (Beckman and Ames, 1998). Alternatively a decrease in the expression and activity of endothelial NO synthase may explain it (Barton *et al.*, 1997; Chou *et al.*, 1998). As regards endothelium-derived vasoconstrictor products, the response to acetylcholine was enhanced in the presence of indomethacin in arteries from old but not adult rats. Thus, aging was associated with an increased participation of endothelial vasoconstrictor products of cyclooxygenase, the nature of which remains to be determined (Dohi *et al.*, 1995; Koga *et al.*, 1989; Mombouli and Vanhoutte, 1993). Also, whether the endothelial production of vasoconstrictor prostaglandins or its effect was enhanced in mesenteric resistance arteries from old rats is not clear. The participation of the endothelial vasoconstrictor prostaglandins was abolished after blockade of NO-synthase in arteries from old rats. This suggests that endothelial factors including NO and vasoconstrictor products interact with each other and that blockade of one pathway masked the participation of the other. Such an interaction between NO and products of cyclooxygenase has been reported in the literature (Auch-Schwelk *et al.*, 1992; Ohlmann *et al.*, 1997; Salvemini *et al.*, 1993; Sautebin *et al.*, 1995).

The relaxation insensitive to N^{ω}-nitro-L-arginine plus indomethacin, which probably reflects the EDHF component of the response, was not different between arteries from adult and old rats. This suggests that the relaxation caused by EDHF is preserved with aging. Since membrane potential was not measured, the possible role of hyperpolarization in the relaxation could not be evaluated. In the superior mesenteric artery of the rat, Fujii *et al.* (1993) found that both hyperpolarizations and relaxations to acetylcholine declined with age. They suggested that these alterations are due to reduced synthesis, release or diffusion of EDHF. Possible reason for these different results could be anatomical heterogeneity. Small mesenteric resistance arteries were used here whereas Fujii *et al.* (1993) studied the main superior mesenteric arteries. A similar anatomical heterogeneity has been reported with aging between the aorta and the femoral artery (Barton *et al.*, 1997).

37 Gender Differences in, and Contribution of Estrogen to the Functions of Endothelium-derived Hyperpolarizing Factor in Middle-aged Rats

Ichiro Sakuma, Atsushi Sato*, Ming-yue Liu*, Morio Kanno* and Akira Kitabatake

*Department of Cardiovascular Medicine and *Department of Pharmacology, Hokkaido University School of Medicine, N-15, W-7, Kita-ku, Sapporo 060-8638, Japan*
Author for correspondence: Ichiro Sakuma, Department of Cardiovascular, Medicine, Hokkaido University School of Medicine, N-15, W-7, Kita-ku, Sapporo 060-8638, Japan.
Tel: +81-11-716-1161 Ext. 6973; Fax: +81-11-706-7874;
e-mail: cvstaf@med.hokudai.ac.jp

Experiments were designed to examine whether or not gender differences exist in relaxation of smooth muscle and the function of endothelium-derived hyperpolarizing factor in the middle-aged rat. The effects of oophorectomy and those of estrogen-replacement were also studied. Fifty-week-old male and female Wistar rats were studied. The female rats were divided into three groups, control (sham-operated), oophorectomy group, and estrogen-replacement group in which an osmotic pump delivering 3 μg/day of estradiol was implanted at oophorectomy. Four weeks after the operation, the cell membrane potential of the mesenteric artery with intact endothelium and changes in its isometric tension were monitored. The resting membrane potential was more shallow in the oophorectomy group than in controls, and estrogen-replacement partially restored it. The acetylcholine-induced membrane hyperpolarization was depressed in male compared to control females. It was markedly depressed in the oophorectomy group and restored in estrogen replacement group. Pinacidil-induced hyperpolarizations were similar among the four groups. Thus, gender differences exist in the importance of endothelium-derived hyperpolarizing factor in the mesenteric artery of middle-aged rats. Estrogen probably causes this difference and regulates the relaxing properties of the artery.

KEYWORDS: endothelium-derived hyperpolarization, estrogen, oophorectomy, resting membrane potential, acetylcholine

INTRODUCTION

Women are about ten years older than men at the initial manifestation of coronary heart disease and about twenty years older at the occurrence of myocardial infarction. This phenomenon is thought to be caused by the production of estrogens in women during the normal menstruation cycle. In support of this notion there is accumulating evidence that estrogen replacement therapy protects against the development of coronary atherosclerosis and ischemic heart disease in postmenopausal women (Rossouw, 1996). This cardiac protection has been attributed to the effects of estrogens on lipid profile, its anti-oxidant and fibrinolytic actions. Among these there are vascular relaxing effects of estrogens after acute or chronic administration which result from their calcium channel blocking action and the enhanced production of endothelial nitric oxide (Gisclard, Miller and Vanhoutte,

1988; Williams *et al.*, 1988; Weiner *et al.*, 1994). However, the effects of estrogens on membrane potential of vascular smooth muscle cells and on endothelium-derived hyperpolarizing factor (EDHF) have not been studied. The purpose of the present study was to determine whether or not a gender difference exists in the resting membrane potential and endothelium-dependent hyperpolarization in the rat. Furthermore, in order to detect the role of estrogens on those parameters, the effects of oophorectomy and those of estrogen-replacement were investigated in female rats. Since surgical or spontaneous menopause, and consequent estrogen-deficiency are events usually encountered by women in their middle age, middle-aged female rats which had a past history of pregnancy and delivery were studied.

METHODS

Fifty-weeks-old male and female Wistar rats were studied. The female rats (at normal cycling) were divided into three groups: control (sham-operated), oophorectomy group, and estrogen-replacement group in which an osmotic pump (Alza Corporation, 2ML4, Palo Alto, CA, U.S.A.) delivering 3 μg/day of water soluble estradiol was implanted at oophorectomy. Four weeks after the operation the rats were anaesthetized with diethyl ether, and the mesenteric arteries were removed carefully and cleaned of surrounding fat and connective tissue in oxygenated physiological salt solution at room temperature. The arteries were cut into rings (3-mm in length). Care was taken to ensure that the endothelial layer was not damaged during the preparation.

Recording of Membrane Potential

The arteries were opened longitudinally and suspended in an organ chamber (capacity 3 ml), endothelial side facing up using tiny pins. The tissues were superfused with warm (37°C) aerated (95% O_2 and 5% CO_2) physiological salt solution at a constant flow rate of 7 ml/min and allowed to equilibrate for at least 60 min before starting the recording. By means of glass capillary microelectrodes filled with 3M KCl (tip resistances 40–80 MΩ), smooth muscle cells were impaled from the intimal side. The microelectrode was coupled by a Ag/AgCl junction to a high impedance capacitance neutralizing amplifier (Nihon Kohden, MEZ-8201, Tokyo, Japan). An agar bridge containing 3M KCl was used as a reference electrode. Electrical signals were continuously monitored on an oscilloscope (Nihon Kohden, VC-10, Tokyo, Japan) and recorded on a chart recorder (Watanabe Sokki WR3101, Tokyo, Japan). After a stable membrane potential had been measured for at least 2 min, application of acetylcholine (ACh) was commenced (Fukao *et al.*, 1995). The composition of physiological salt solution was as follows (in mM); NaCl 118.2, KCl 4.7, $CaCl_2$ 2.5, $MgCl_2$ 1.2, KH_2PO_4 1.2, $NaHCO_3$ 25.0 and glucose 10.0.

Drugs

The following drugs were used; Acetylcholine chloride (Wako, Osaka, Japan) and water-soluble 17β-estradiol (Sigma Chemical, St. Louis, MO, U.S.A.). Acetylcholine chloride was dissolved in distilled water. PSS was used for further dilution to the proper concentrations. Water-soluble 17β-estradiol was dissolved in saline.

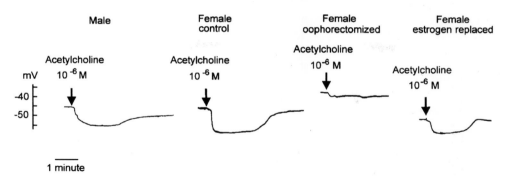

Figure 37.1.   Membrane potential recordings and effects of acethylcholine (10^{-6} M) on the membrane potential of the mesenteric arteries from middle-aged rats with intact endothelium. Arrow heads denote the application of acetylcholine.

RESULTS

Resting membrane potential of the mesenteric arterial muscle was significantly shallower in arteries from oophorectomy group (–37 mV) compared with those from male (–45 mV), control female (–46 mV) and estrogen replaced female (–49 mV) (Figure 37-1).

The acetylcholine-induced hyperpolarization in the mesenteric artery with endothelium was significantly depressed in the artery from male (–6 mV from baseline) in comparison with the control female group (–13 mV). It was depressed in arteries from oophorectomy group (–2 mV) and was partially restored in those from the estrogen-replacement group (–9 mV) (Figure 37-1).

DISCUSSION

Alterations in functions of EDHF have been reported in various pathophysiological conditions including aging (Fujii *et al.*, 1993; Nakashima and Vanhoutte, 1993a), hypertension (Fujii *et al.*, 1992) and diabetes mellitus (Fukao *et al.*, 1997c). In the present study, the resting membrane potential of rat mesenteric arterial smooth muscle was similar between male and female, but became shallower following ovariectomy and was restored by estrogen replacement. Thus, these observations suggest that estrogens physiologically functions to increase the membrane potential and that deprivation of estrogen renders vascular smooth muscle prone to constrict. Possible mechanisms underlying these effects of estrogen include a direct effect on potassium channels of the smooth muscle cell membrane or an indirect effect through activation of the release of EDHF from the endothelium.

The acetylcholine-induced hyperpolarization of the mesenteric artery was also modulated by estrogen. Estrogen enhanced the functions of EDHF released by acetylcholine from the artery in a resting state.

These effects of estrogen may facilitate dilatation of the peripheral resistance arteries (Popp, Fleming and Busse, 1998) and contribute to the adjustment of adequate vascular compliance and to the control of peripheral blood flow.

38 Contribution of Nitric Oxide and Endothelium-derived Hyperpolarizing Factor to the Relaxation Evoked by Bradykinin in Porcine Coronary Arteries Four Weeks After Removal of the Endothelium

Catherine Thollon, Marie P. Fournet-Bourguignon, Jean P. Bidouard, Christine Cambarrat, Delphine Saboureau, Isabelle Delescluse, Paul M. Vanhoutte* and Jean P. Vilaine

*Division Pathologies Cardiaques et Vasculaires, Institut de Recherches SERVIER, 11 rue des Moulineaux, 92150 Suresnes, France; and * Institut de Recherches internationales SERVIER, 6 place des pléiades, 92415 Courbevoie, France. Tel: (33) 01 41 18 23 41; Fax: (33) 01 41 18 24 30*

Four weeks after balloon denudation, the regenerated endothelium selectively loses the pertussis-toxin sensitive G-protein coupled relaxation to serotonin while that to bradykinin is maintained. The aim of the present study was to investigate the relative contribution of nitric oxide (NO) and endothelium-derived hyperpolarizing factor in the relaxation induced by bradykinin in porcine coronary arteries with regenerated endothelium. Transmembrane potential and isometric tension were evaluated simultaneously in each strip with regenerated endothelium and compared with those of the corresponding control coronary artery. Measurements were performed at the end of the resting period, when the contraction to prostaglandin $F_{2\alpha}$ had reached a stable level and 1.5 min after adding bradykinin. In parallel, levels of cyclic GMP and cyclic AMP were determined by radio-immunoassay. Under basal conditions, the smooth muscle cells from coronary arteries with regenerated endothelium were depolarized in comparison with those from control arteries. The level of cyclic AMP was normal while that of cyclic GMP level was reduced significantly in arteries with regenerated endothelium. Prostaglandin $F_{2\alpha}$ induced comparable contractions of both types of arteries, associated with depolarization. In the arteries with regenerated endothelium, the relaxation to bradykinin in the presence of the inhibitor of cyclooxygenase, indomethacin, was not modified. Furthermore, exposure to bradykinin induced identical hyperpolarizations in both types of vessels while the production of cyclic GMP was decreased in arteries with regenerated endothelium. Thus, the unaltered relaxation to bradykinin, despite the reduced production of NO, suggests that the unchanged amplitude of endothelium-dependent hyperpolarization is sufficient to maintain a normal relaxation in coronary arteries with regenerated endothelium.

KEYWORDS: regenerated endothelium, endothelium-derived hyperpolarizing factor, cyclic GMP, cyclic AMP, relaxation, bradykinin

INTRODUCTION

All major cardiovascular risk factors cause endothelial dysfunction, which precedes clini-cally apparent vascular disease and its complications (Cohen, 1995). Thus, hypertension (Lüscher and Vanhoutte, 1986), diabetes mellitus (Cohen, 1993), hypercholesterolemia (Creager et al., 1990) and atherosclerosis (Förstermann et al., 1988) are associated with decreased responses to endothelium-dependent vasodilators. Experiments in animal models suggest that the occurrence of abnormal endothelium — dependent relaxations is an early event in the development of vascular disease (Yamamoto *et al.*, 1987; Cohen *et al.*, 1988; Shimokawa *et al.*, 1987). Since the turnover of endothelial cells is accelerated under most of these conditions (Schwartz and Benditt, 1977), the functional alterations in the regenerated endothelium could play an important role in the genesis of coronary disease. Previous studies (Shimokawa *et al.*, 1987) demonstrated that, in porcine coronary arteries with regenerated endothelium after balloon denudation, endothelium-dependent relaxations are normalized eight days after the procedure. However, four weeks after such denudation, the pertussis-toxin sensitive G-protein coupled relaxation to serotonin is reduced while that induced by bradykinin is normal. In the porcine coronary artery, it is unlikely that NO fully explains the endothelium-dependent relaxation induced by bradykinin (Richard *et al.*, 1990; Cowan and Cohen, 1991; Nagao and Vanhoutte, 1992). The aim of the present study was to investigate the involvement of nitric oxide (NO) and endothelium-derived hyperpolarizing factor (EDHF) in the relaxation induced by bradykinin in porcine coronary arteries with regenerated endothelium.

MATERIALS AND METHODS

Endothelial Denudation

Seven Large White pigs (8 weeks, 18–23 kg), were anesthetized with an intramuscular injection of Tiletamine plus Zolazepam, (20 mg/kg). Additional doses of anesthetic (sodium thiopental) were given intravenously, as necessary. Animals were intubated and ventilated with a respirator. Heparin (250 I.U./kg) and lysine acetylsalicylate (10 mg/kg) were administrated. Using a guide catheter, the balloon (3.5 mm diameter × 20 mm length) was introduced via the femoral artery. Under fluoroscopic guidance, the balloon was advanced into the coronary artery and the endothelium was removed by inflating the balloon three times. The animals were sacrified 28 days after the endothelial denudation. Each coronary artery with regenerated endothelium was compared with one coronary artery with native endothelium from the same heart.

Simultaneous Recording of Membrane Potential and Mechanical Activity

The coronary arteries were dissected free, cleaned of adherent fat and connective tissue and maintained in an oxygenated Krebs-Ringer solution at room temperature. Rings of coronary arteries (approximately 4 mm long) were cut open along the longitudinal axis. One end of the segment was pinned down to the bottom of an experimental chamber, the endothelial side upward. At the other end three ligatures with micro-surgical silk thread (8/0) were performed to attach the tissue to a wire connected to a force transducer in order to measure contraction. The strips were superfused continuously at 5 ml/min with an oxygenated, modified Krebs-Ringer solution of the following composition (mmol/L): NaCl (118), KCl (4.7), $MgSO_4$ (1.2), KH_2PO_4 (1.2), $CaCl_2$ (2.5), $NaHCO_3$ (25), EDTA (0.026) and glucose (11). The preparations were allowed to equilibrate for 30 to 45 min. During this period a basal tension of about 4 g was applied progressively. The membrane potential was measured with conventional glass microelectrodes (30–40 MegΩ), filled with 3M KCl.

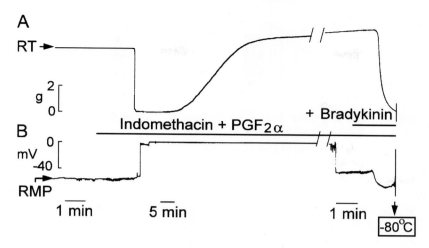

Figure 38-1.   Illustration of simultaneous recordings of the endothelium-dependent relaxation (A) and hyperpolarization (B) induced by bradykinin (3×10^{-8} M) in a control coronary artery taken from a pig not subjected to a surgical procedure. For animals subjected to angioplasty four weeks before, the coronary arteries with regenerated endothelium were compared to control arteries with native endothelium from the same heart. Coronary artery segments were contracted with prostaglandin $F_{2\alpha}$ (3×10^{-6} M), in presence of indomethacin (10^{-5} M).

The membrane potential was measured: a) at the end of the resting period (RT = resting tension and RMP = resting membrane potential); b) when the contraction to prostaglandin $F_{2\alpha}$ (3×10^{-6} M) in presence of indomethacin (10^{-5} M) had reached a stable level; and c) after adding bradykinin (3×10^{-8} M). The test segment was frozen at the time of maximal relaxation to bradykinin (1.5 to 2 min). In parallel, two other strips from the same coronary artery were frozen just before the addition of bradykinin: one under basal conditions (Krebs-Ringer solution) and the other superfused simultaneously with the same solution as the test segment (except bradykinin). Then the vascular segments were maintained at $-80°C$ until the content of cyclic nucleotides was measured.

Recording of membrane potential was performed: a) at the end of the resting period (control values), b) after contraction with prostaglandin $F_{2\alpha}$ (3×10^{-6} M) in the presence of indomethacin (10^{-5} M) and c) during the endothelium-dependent relaxation to bradykinin (3×10^{-8} M) (Figure 38-1).

Measurement of Cyclic Nucleotides

For each experiment, three tissue segments were rapidly frozen at the different steps of the protocol: under basal conditions, after contraction with prostaglandin $F_{2\alpha}$ and at the maximal relaxation induced by bradykinin (Figure 38-1). Frozen tissue was placed in medium containing acetate buffer (0.05 M, pH = 5.8) and a phosphodiesterase inhibitor, 3-isobutyl-1-1-methylxanthine (IBMX 10^{-4} M). The frozen tissue was rapidly homogenized and sonicated. Total protein content was determined in each homogenate using the Biorad method. Then, the homogenate was centrifugated at 2000 g for 20 minutes at 4°C. The cyclic GMP and cyclic AMP levels were determined in the supernatant by radioimmunoassay using the Amerlex method. The results were expressed in pmol per mg of total protein content.

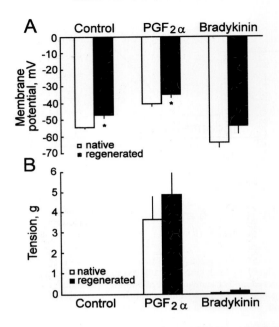

Figure 38-2. Changes in membrane potential of smooth muscle cells (A) and tension (B) during contractions to prostaglandin $F_{2\alpha}$ (3×10^{-6} M) and relaxations to bradykinin (3×10^{-8} M) in porcine coronary arteries with regenerated endothelium. Data are mean $\pm$ SEM from 7 experiments. All experiments were performed in the presence of indomethacin (10^{-5} M). The asterisks indicate statistically significant differences from the corresponding control coronary arteries with native endothelium ($P < 0.05$).

Drugs

The following drugs were used: indomethacin, prostaglandin $F_{2\alpha}$, bradykinin and IBMX (all from Sigma Chemical Co.). Indomethacin (10^{-2} M, in ethanol), prostaglandin $F_{2\alpha}$ (3×10^{-3} M, in ethanol) and bradykinin (10^{-3} M, in H_2O) were prepared as stock solutions ($-20°C$) and diluted in Krebs-Ringer solution to reach the final concentrations reported.

Statistical Analysis

Data are expressed as the mean $\pm$ SEM from n experiments. Student's t-test for paired observations was used. Differences were considered to be statistically significant when P was less than 0.05.

RESULTS

Membrane Potential and Tension

Twenty-eight days after surgery, in coronary arteries with regenerated endothelium, before any pharmacological intervention, the smooth muscle cells were significantly less polarized than in the corresponding control arteries (Figure 38-2A).

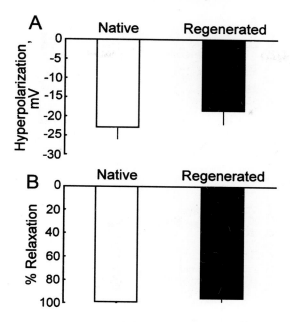

Figure 38-3. Maximal values of hyperpolarization (A) and relaxation (B) induced by bradykinin (3×10^{-8} M) in porcine coronary arteries with regenerated endothelium contracted with prostaglandin $F_{2\alpha}$ (3×10^{-6} M). Data are mean ± SEM from 7 experiments. All experiments were performed in the presence of indomethacin (10^{-5} M).

Prostaglandin $F_{2\alpha}$ (3×10^{-6} M) in the presence of indomethacin (10^{-5} M) induced similar contractions in both types of coronary arteries. Furthermore, the development of tension was associated with a depolarization of vascular smooth muscle cells of similar amplitude: 12.1 ± 1.6 and 13.7 ± 1.2 mV, for coronary arteries with regenerated and with native endothelium, respectively (Figure 38-2). In two coronary arteries with regenerated endothelium among the seven studied, membrane potential fluctuations were observed under exposure to prostaglandin $F_{2\alpha}$, while they were quiescent under resting conditions.

Addition of bradykinin (3×10^{-8} M) produced hyperpolarization of the cell membrane which was not significantly different in vessels with native or regenerated endothelium (Figure 38-3A). Moreover, the resultant endothelium-dependent relaxation was not changed in the coronary arteries with regenerated endothelium (Figure 38-3B).

Cyclic GMP and Cyclic AMP Levels

Under basal conditions the cyclic AMP levels were identical in coronary arteries with native and regenerated endothelium, while the cyclic GMP level was significantly reduced (by $73.0 \pm 6.0\%$) in blood vessels with regenerated endothelium (Figure 38-4). When both were contracted with prostaglandin $F_{2\alpha}$ (3×10^{-6} M), a comparable difference was observed ($60.0 \pm 18.0\%$ less cyclic GMP in arteries with regenerated endothelium compared to arteries with native endothelium). The cyclic GMP production induced by bradykinin

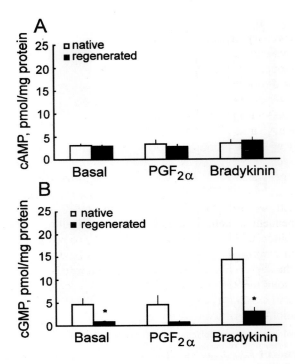

Figure 38-4. (A) Levels of cyclic AMP in porcine coronary arteries with native or regenerated endothelium, under basal conditions (without indomethacin), after contraction with prostaglandin $F_{2\alpha}$ (3×10^{-6} M) and during relaxation with bradykinin (3×10^{-8} M), in the presence of indomethacin (10^{-5} M). (B) Levels of cyclic GMP in porcine coronary arteries with native or regenerated endothelium, under basal conditions (without indomethacin), after contraction with prostaglandin $F_{2\alpha}$ (3×10^{-6} M) and during relaxation with bradykinin (3×10^{-8} M), in the presence of indomethacin (10^{-5} M). The asterisks indicate statistically significant differences from the corresponding control coronary arteries with native endothelium ($P < 0.05$).

was significantly lower in coronary arteries with regenerated than in those with native endothelium (Figure 38-4B). The cyclic GMP content of vessels with regenerated endothelium amounted to only $33.3 \pm 13.0\%$ of that present in arteries with native endothelium. However, stimulation with bradykinin increased the cyclic GMP content by about four-fold in both types of blood vessels.

DISCUSSION

Resting Membrane Potential and Basal Levels of Cyclic GMP

The present study demonstrates that the smooth muscle cells from coronary arteries with regenerated endothelium are depolarized in comparison to those from corresponding control arteries. The change in resting membrane potential in these coronary arteries is relevant as it could modify the myogenic tone of these blood vessels. This interpretation is confirmed by the observation of spontaneous electrical activity in two arteries with regenerated endothelium while such membrane potential instability was never observed in the corresponding control arteries. Rhythmic spontaneous activity has been demonstrated in human coronary arteries and, with higher frequency in blood vessels from patients with cardiovascular diseases (e.g. Ross *et al.*, 1980; Kalsner, 1985). Hence, the depolarization of smooth muscle cells observed four weeks after balloon denudation may be a key factor in the development of alterations in vasomotion of these coronary arteries. In coronary arteries with regenerated endothelium, such electrophysiological modifications always were associated with a decrease in the levels of cyclic GMP, although the levels of cyclic AMP were unchanged. Thus, in this dysfunctional endothelium, the NO pathway is altered selectively while the prostacyclin pathway is normal under basal conditions. The reduced levels of cyclic GMP could be involved in the depolarization observed in the preparations with regenerated endothelium. Nitric oxide could directly or indirectly (via cyclic GMP) inhibit an inward depolarizing current and/or activate a potassium current. Nitric oxide could activate cyclic GMP-stimulated phosphodiesterases (Mery *et al.*, 1993) or cyclic GMP-dependent protein kinases (Wahler and Dollinger, 1995) resulting for example in an inhibition of I_{Ca} (an inward depolarizing current). NO also can directly activate a potassium current (Bolotina *et al.*, 1994). All of these effects would favor a depolarization and the appearence of membrane potential instability under reduced levels of NO and of its second messenger, cyclic GMP. The two segments showing membrane potential oscillations in the presence of prostaglandin $F_{2\alpha}$ had the lowest increase of cyclic GMP during relaxation to bradykinin (less than 1 pmol/mg protein).

Endothelium-dependent Hyperpolarization and Cyclic GMP Levels during Relaxation to Bradykinin

In the same porcine model, the endothelium-dependent relaxation to serotonin is decreased while that induced by bradykinin remains normal (Shimokawa *et al.*, 1987; 1989; Borg-Capra *et al.*, 1997). In the presence of KCl, a greater shift in the concentration response curve to bradykinin was noted for coronary arteries with regenerated endothelium compared to those with native endothelium (Borg-Capra *et al.*, 1997), suggesting an increased involvement of EDHF in mediating the relaxation. Moreover, in conditions of inhibition of nitric oxide synthase, four weeks after denudation, the endothelium-dependent hyperpolarizations induced by bradykinin are related to the initial membrane potential values (Thollon *et al.*, 1997). Since it is conceivable that the presence of contracting agents could modify the resting membrane potential, the endothelium-dependent hyperpolarization induced by bradykinin was measured under the same conditions as those used for the study of relaxations. Furthermore, the endothelium-dependent hyperpolarizations and the increase in cyclic GMP content, as a reflection of the NO production during relaxation to bradykinin, were evaluated simultaneously. During similar contraction with prostaglandin $F_{2\alpha}$, identical depolarizations were observed in both types of blood vessels. Hence, although the cyclic GMP level was reduced in coronary arteries with regenerated endothelium, the hyperpolarization in response to bradykinin was normal in these preparations. Since the resulting endothelium-dependent relaxation to bradykinin was identical despite the alteration of the NO pathway, this suggests that the unchanged amplitude of hyperpolarizations is sufficient to maintain normal relaxation in coronary arteries with regenerated endothelium. This may imply that in blood vessels with native endothelium, NO curtails the relaxing effect of EDHF rather than inhibiting its production (Olmos *et al.*, 1995).

39 Endothelium-dependent Responses to Bradykinin in Porcine Coronary Arteries after Heart Transplantation: Reduced Relaxations Resistant to N$^{\omega}$-Nitro-L-Arginine Despite Preserved Hyperpolarization

Louis P. Perrault*, Catherine Thollon, Nicole Villeneuve, Jean-Pierre Bidouard, Jean-Paul Vilaine and Paul M. Vanhoutte

Cardiovascular Division, Institut de Recherches Servier, 11 rue des Moulineaux, Suresnes, 92150, France
*Current address: Research Center and Department of Surgery, Montreal Heart Institute, 5000 Belanger east, Montreal, Quebec, Canada, H1T 1C8
Address for correspondence to: Paul Vanhoutte MD, Ph.D. Institut de Recherches Internationales Servier, 6 place des Pléiades, 92415 Courbevoie, France.
Tel: 33 (1) 46 41 61 23 Fax: 33 (1) 46 41 72 73

Thirty days after heart transplantation the endothelial function of the porcine coronary artery is altered selectively, since the endothelium-dependent relaxations to serotonin are reduced while those to bradykinin are preserved. One month later, this dysfunction becomes more generalized, leading to an alteration in the relaxation to bradykinin and to nitric oxide (NO) donors. In parallel, the basal levels of cyclic GMP are decreased. The present study was designed to evaluate the alteration in the L-Arginine NO pathway associated with modifications in endothelium-dependent hyperpolarizations.

Vascular reactivity and membrane potential measurements were made in rings of native and allograft epicardial coronary arteries six weeks after transplantation, using standard isometric force transducers and intracellular microelectrodes, respectively. All experiments were performed in the presence of indomethacin, an inhibitor of cyclooxygenase. In both types of experiments, the preparations were contracted with prostaglandin F2α.

In the presence of N$^{\omega}$-Nitro-L-Arginine, endothelium-dependent relaxations to bradykinin were reduced in allograft coronary arteries compared to native coronary rings, while there was no significant change in the maximal relaxation to cromakalim (an opener of ATP-dependent potassium channels). Simultaneous electrical and mechanical recordings showed no alteration in resting membrane potential of smooth muscle cells, depolarization during contraction to prostaglandin $F_{2\alpha}$ or hyperpolarization in the presence of bradykinin. Despite the normal hyperpolarization to bradykinin, the relaxation was reduced in coronary strips from allograft hearts.

In this model of endothelial dysfunction, part of the reduction in the endothelium-dependent relaxation to bradykinin may be attributed to an alteration in the activity of endothelium-derived hyperpolarizing factor. In the presence of an inhibitor of NO synthase, a dissociation exists between the preserved endothelium-dependent hyperpolarization to bradykinin and the reduced resistant relaxation, while the capacity of smooth muscle cells to relax in response to an opener of ATP-dependent potassium channels is intact. The reason why EDHF is unable to completely relax the smooth muscle cells in allograft coronary arteries is unclear.

KEYWORDS: endothelium, coronary artery, transplantation, EDHF

INTRODUCTION

The contribution of hyperpolarization in endothelium-dependent vascular relaxation varies as a function of the size of the arteries and is prominent in resistance vessels (Nakashima *et al.*, 1993; Campbell *et al.*, 1996a). In large arteries, both EDHF and NO can contribute to endothelium-dependent relaxations, but the role of NO predominates under normal circumstances. Nevertheless, in these arteries, EDHF can mediate near normal endothelium-dependent relaxations when the synthesis of NO is inhibited (Kilpatrick and Cocks, 1994). This preservation may be important for the regulation of vascular tone in disease when the production and/or the activity of NO are reduced. Bradykinin causes endothelium-dependent relaxations by binding to B_2-kinin membrane receptors coupled to Gq-proteins. This results in activation of endothelial nitric oxide synthase and of NO production. Release of the endothelium-derived hyperpolarizing factor (EDHF) also follows, causing endothelium-dependent relaxation by hyperpolarizing the vascular smooth muscle (Cohen and Vanhoutte, 1995; Félétou and Vanhoutte, 1988; Corriu *et al.*, 1996). In some models of atherosclerosis, a decreased production of EDHF could contribute to the reduced efficacy of bradykinin, although under certain conditions, such as hypercholesterolemia, potassium channels play a greater role than under normal conditions as if EDHF compensates for reduced NO mediated relaxations (Najibi *et al.*, 1994).

Although heart transplantation remains the treatment of choice for medically unresponsive terminal heart disease and is associated with a five year survival of 70%, coronary graft vasculopathy develops in a majority of transplant recipients and is the main cause of death beyond the first year after transplantation (Billingham, 1987). The appearance of accelerated atherosclerosis is preceded by reduced dilatations of the coronary artery to endothelium-dependent agonists suggesting the presence of an early endothelial dysfunction. When identified, this is predictive of the development of graft coronary disease one year after graft implantation and of the occurence of cardiac events and death (Davis *et al.*, 1996). The exact mechanisms underlying the coronary endothelial dysfunction after transplantation remain elusive (Perrault, 1997a). Clinical evidence has hinted at a selective endothelial dysfunction which occurs after heart transplantation even in patients without overt acute rejection. Patients studied one to five months after graft implantation demonstrate a selective impairment of endothelium-dependent dilatation to acetylcholine while the dilatation to bradykinin is maintained (Aptecar *et al.*, 1997). Several noxious stimuli may impair endothelial function in the course of transplantation including exposure to depolarizing cardioplegic solutions [which are known to impair the EDHF mediated relaxations (He, Yang and Yang, 1997)], preservation solutions, warm ischemia-reperfusion injury during implantation, treatment with cyclosporine A and cytomegalovirus infections (Drinkwater *et al.*, 1995).

The present study was designed to evaluate the integrity of relaxations due to endothelium-derived hyperpolarizing factor (EDHF) after heart transplantation.

MATERIAL AND METHODS

Animals and Immunological Studies

Thirty large-white swines (EARL de Fresnelles, Boisemont, France) of either sex, aged 10 ± 0.3 weeks and weighing 23 ± 0.8 kg were used. The experiments were performed in compliance with the "Guide for the Care and Use of Laboratory Animals" published by the National Institutes of Health (NIH publication no. 85–23, revised 1985). All procedures used in this study were approved by the local institutional committee on animal care. Preoperative blood samples were drawn for determination of blood type and the class I antigen of the Swine Lymphocytes Alloantigen system (SLA) by the microlymphocytoxicity technique (Perrault *et al.*, 1997b) on swines from the same litter to ensure a rate of recombination between the class I and class II region of the major histocompatibility complex of less than 1%. The transplantations were performed between animals compatible for blood type and for the SLA class I antigen. The serum typing was performed at the Institut National de Recherches Agronomiques (INRA, Jouy-en Josas, France).

Anesthesia, Cardioplegia and Surgical Technique

Heterotopic retroperitoneal heart transplantation was performed with normothermic blood cardioplegic arrest and immediate reimplantation (ischemic time less than 60 minutes) to avoid endothelial injury (Perrault *et al.*, 1997b). After standard ventilatory weaning, the animals were left to recover in temperature controlled quarters and fed standard piglet chow (number 8, Pietrement, Provins, France) and water *ad libitum*. No immunosuppressive drugs were used. The recipients were sacrificed electively six weeks (n = 15) after transplantation.

Explantation Protocol

Allograft hearts (n = 13): After induction of anesthesia, venous access, volume replacement and ventilatory support were established. The abdomen was reentered and complete mobilization of the allograft was performed taking care not to injure the epicardial surface of the heart. No intraarterial injection of potassium chloride was used prior to harvest to avoid the effect of the cardioplegic solution on the signaling pathway (He, Yang and Yang, 1997). The heart was excised rapidly. Native hearts (n = 13) were excised through a median sternotomy and arrested in the same fashion. Control hearts (n = 4) : Hearts from normal swines of similar weight and age which received the same tranquillizing drug without transplantation, were used as controls.

HISTOLOGY

Myocardium

Myocardial biopsies of all allografts and selected native hearts were taken from the septum, the right and left ventricles of fresh specimens and fixed in formaldehyde (10%). Hematoxylin eosin-safran staining was performed and the biopsies were evaluated for rejection grade (Billigham, 1987), extent of necrosis and ischemic changes.

VASCULAR REACTIVITY

Native, allograft and control hearts were placed in a modified Krebs-bicarbonate solution (composition in mM: NaCl 118.3, KCl 4.7, $MgSO_4$ 1.2, KH_2PO_4 1.2, glucose 11.1, $CaCl_2$ 2.5, $NaHCO_3$ 25 and calcium ethylenediaminotetraacetic acid 0.026; control solution). Oxygenation was insured using a 95% O_2/5% CO_2 gas mixture. The epicardial coronary arteries of the control, native and allograft hearts were dissected free from the epicardium, myocardium and from adventitial tissue and divided into rings (4 mm wide; 20 rings from the allograft and 20 rings from the native heart). Rings from the left anterior descending, left circumflex and right coronary arteries were used randomly but matched between native and allograft preparations in all experiments.

The reactivity of control, native and transplanted arteries was studied in organ chambers filled with control solution (20 ml) at 37°C. The rings were suspended between two metal stirrups, one of which was connected to an isometric force transducer. Data were collected using data acquisition software (IOS3, Emka Inc., Paris, France). All experiments were performed in the presence of indomethacin (10^{-5} M, to exclude production of endogenous prostanoids) and propranolol (10^{-7} M, to prevent the activation of beta-adrenergic receptors).

Each preparation was stretched to the optimal point of its active length-tension curve (approximately 4 grams) as determined by measuring the contraction to potassium chloride (30 mM) at different levels of stretch and then allowed to stabilize for 90 minutes. A maximal contraction was determined with potassium chloride (60 mM). Rings were excluded if they failed to contract to potassium chloride (exclusion rate less than 5%).

After washing and thirty minutes of stabilization, endothelium-dependent relaxations were studied in preparations contracted with prostaglandin $F_{2\alpha}$ (range 2×10^{-6} to 10^{-5} M) to achieve a contraction approximating 50% of the maximal contraction to KCl (60 mM). Responses to serotonin (10^{-10} to 10^{-5} M; in the presence of 10^{-6} M ketanserin; incubated 40 minutes before the addition of serotonin to block serotonin $5HT_2$ receptors), bradykinin (10^{-10} to 10^{-5} M), bradykinin (10^{-10} to 10^{-5} M) in the presence of N^{ω}-Nitro-L-Arginine (10^{-4} M) and bradykinin (10^{-10} to 10^{-5} M) in the presence of N^{ω}-Nitro-L-Arginine plus a high potassium concentration (20 mM) were compared between control and native and allograft coronary rings six weeks after transplantation.

In separate experiments, in the presence of N^{ω}-Nitro-L-Arginine, relaxations to a single injection of bradykinin (3×10^{-8} M) were compared in native and allograft coronary arteries to avoid the possible desensitizing effect of incremental doses of bradykinin (Olmos *et al.*, 1995). In other experiments, the response to cromakalim (a K_{ATP}-dependent potassium channel opener, 10^{-9} to 10^{-6} M) in the presence of N^{ω}-Nitro-L-Arginine were compared in native and allograft coronary arteries. No rings were exposed to more than one endothelium-dependent agonist in the course of the experiments.

At the end of the experiments, endothelium-independent responses were studied in rings with endothelium by comparing relaxations to Sin-1 (10^{-5} M; 3-morpholino sydnonimine), a nitric oxide donor.

MEMBRANE POTENTIAL RECORDING

Rings of coronary arteries were cut open along the longitudinal axis. One side of the segment was pinned down to the bottom of an experiment chamber (2.5 ml) covered with Sylgard, with the endothelial side facing upward. On the other side, three sutures of microsurgical silk thread (8/0) were used to attach the tissue to a wire connected to a force transducer in order to measure contraction. The strips were perfused continuously at 5 ml/min with an oxygenated modified Krebs-Ringer solution ($36.5 \pm 0.5°C$, pH 7.4). The preparations were allowed to equilibrate for 30 to 45 minutes. During this period, a basal tension of 4 to 5 g was applied progressively to the preparation. The membrane potential was measured with a conventional glass microelectrode (30–40 MΩ), filled with 3M KCl. The microelectrode was connected to an amplifier and the membrane potential was monitored simultaneously on a digital storage oscilloscope and a pen-chart recorder. In all experiments, the response of the coronary arteries from the allograft heart were compared with those from the native heart. All experiments were performed in the presence of indomethacin (10^{-5} M); and N^{ω}-Nitro-L-Arginine (10^{-4} M, to inhibit nitric oxide synthase). The membrane potential was measured at the end of the resting period (control values), after the developed tension in response to prostaglandin $F_{2\alpha}$ (2×10^{-6} M) had reached a stable plateau and during the administration of bradykinin (3×10^{-8} M).

Drugs

All solutions were prepared daily. Bradykinin, cromakalim, 5-Hydroxytryptamine creatinine sulfate (serotonin), indomethacin, ketanserin, N^{ω}-Nitro-L-Arginine, propranolol, prostaglandin $F_{2\alpha}$, were purchased from Sigma Chemical Co. (St-Quentin Falavier, France). Sin-1 was synthesized at the Servier Research Institute (Suresnes, France).

Statistical Analysis

Relaxations and contractions are expressed as a percentage of the maximal contraction to potassium chloride for each group and expressed as means $\pm$ standard error of the mean (SEM); n refers to the number of animals studied. Student's t-test for paired/unpaired observations was used for statistical analysis. Differences were considered to be statistically significant when P was less than 0.05. ANOVA studies were performed to compare dose-response curves. The Newman-Keuls test was used as the *post-hoc* test.

RESULTS

HISTOLOGY

Myocardial Biopsies

The myocardium from native hearts was normal. Allograft hearts six weeks after transplantation showed extensive lymphocytic infiltration with a moderate amount of necrosis compatible with ISHLT classification grade III to IV and limited subendocardial fibrosis compatible with perioperative ischemic injury.

VASCULAR REACTIVITY

Contractions

In the rejection group, the amplitude of the contraction to potassium chloride was significantly lower in allograft coronary rings compared with native coronary arteries. There were no statistically significant differences in the amplitude of contractions to prostaglandin $F_{2\alpha}$ between native and allograft coronary rings.

Endothelium-dependent Relaxations

Serotonin

There were no significant differences in the endothelium-dependent relaxations to serotonin between rings from control and native hearts. Relaxations to serotonin were decreased significantly six weeks after transplantation in allograft coronary rings compared to native coronary arteries.

Bradykinin

Control coronary arteries

The relaxation-concentration curves were shifted significantly to the right in the bradykinin plus N^{ω}-Nitro-L-Arginine group, and shifted further to the right in arteries exposed to bradykinin, N^{ω}-Nitro-L-Arginine and potassium chloride, compared to those receiving bradykinin alone.

Native coronary arteries

The relaxation-concentration curves were significantly shifted to the right in the bradykinin plus N^{ω}-Nitro-L-Arginine group and further shifted to right in the bradykinin, N^{ω}-Nitro-L-Arginine and potassium chloride group compared to the group receiving braykinin alone.

Allograft coronary arteries

The relaxation-concentration curve to bradykinin was shifted significantly to the right six weeks after transplantation compared to native arteries but there were no statistically significant differences in maximal relaxations. In allograft coronary rings, there were no statistically significant differences between concentration-response curves to N^{ω}-Nitro-L-Arginine plus bradykinin and bradykinin plus N^{ω}-Nitro-L-Arginine and potassium chloride. The response of both groups was statistically different from the response to bradykinin alone.

The relaxation to a single injection of bradykinin was significantly lower in the allograft coronary rings compared to native coronary arteries despite similar levels of contraction to prostaglandin $F_{2\alpha}$.

Endothelium-independent Relaxations

There was no significant difference in the maximal relaxations to cromakalim between native and transplanted coronary arteries six weeks after transplantation. There were no statistically significant differences in maximal relaxations between native and allograft rings to Sin-1 six weeks after transplantation (100% relaxation).

RESTING MEMBRANE POTENTIAL

Before any pharmacological intervention, the resting membrane potential of the coronary smooth muscle cells from allograft hearts was not significantly different from that of the corresponding native coronary arteries (-55 ± -1.7 mV versus -58 ± 2 mV; $n = 4$). The applied resting tension was about 4 g (4.2 g ± 0.25 and 4.2 ± 0.1 g for the native and the allograft hearts, respectively). Prostaglandin $F_{2\alpha}$ (2×10^{-6} M) in the presence of indomethacin and N^{ω}-Nitro-L-Arginine induced similar levels of contraction (7.4 ± 2 and 9.1 ± 1.8 g) and similar levels of depolarization of vascular smooth muscle cells (16.8 ± 1.5 and 21.5 ± 2.4 mV in the allograft and the native coronary strips, respectively). The addition of bradykinin (3×10^{-8} M) induced the same extent of hyperpolarization of the cell membrane (-27.2 ± 1.25 in allograft and -28.8 ± 2.0 mV in native arteries) but the simultaneous relaxation resistant to N^{ω}-Nitro-L-Arginine was reduced in allograft coronary strips (mean relaxation 50%) compared to coronary strips from the corresponding native hearts (mean relaxation 96%).

DISCUSSION

The major finding of the present study is that rejection causes an impairment of the endothelium-dependent relaxation resistant to a NO synthase inhibitor, which can be attributed to EDHF. EDHF relaxes vascular smooth mucle cell by causing hyperpolarization. The early endothelial dysfunction due to rejection after heart transplantation causes an alteration of Gi-protein mediated endothelium-dependent relaxations of coronary arteries without major endothelial cell loss. It is associated with a decreased basal release of NO as evidenced by decreased basal cyclic GMP content (Perrault *et al.*, 1997b). In the same model, sixty days after transplantation, relaxations to agonists using non Gi-mediated pathways (adenosine diphosphate, bradykinin) are also reduced as well as the sensitivity to Sin-1 (Perrault *et al.*, 1997c). This model permits the study of the effect of the rejection process on endothelial function since it does not induce initial ischemia-reperfusion injury. Furthermore the absence of immunosuppressive drugs eliminates their potential confounding effects on coronary endothelial dysfunction (Perrault *et al.*, 1997a). The alterations studied are due to a functional impairment since there is no destruction of the endothelial cell lining up to sixty days after graft implantation.

Both cardioplegic depolarizing solutions and cardiac preservation cause alterations of the EDHF mediated responses which are due to calcium-dependent potassium channels in the porcine coronary artery (He Yang and Yang, 1997; Nagao and Vanhoutte, 1992). This may be important since the effect of different preservation solutions on the endothelium has been incriminated in the subsequent development of graft coronary vasculopathy (Drinkwater *et al.*, 1995). Indeed, ischemia and reperfusion causes a selective decrease in endothelial G-protein mediated relaxations evidenced by the decreased endothelium-dependent relaxations evoked by aggregating platelets (Pearson, Schaff and Vanhoutte, 1990) and reduced cyclic AMP and cyclic GMP mediated-vasodilatation. Persistence of this selective dysfunction has been documented three months after severe ischemia-reperfusion injury but the long term evolution, as well as its contribution to graft coronary vasculopathy after heart transplantation, is unknown (Perrault *et al.*, 1996).

Endothelium-dependent relaxations not mediated by Gi-proteins are also affected by the rejection process, as shown previously by the rightward shift in the concentration-responses curves to the calcium ionophore A23187, bradykinin and ADP sixty days after heart transplantation. However, at that time the maximal vasodilator capacity was maintained. The rightward shift could be explained by a decreased sensitivity of smooth muscle cells to nitric oxide, as found in earlier results obtained with the nitric oxide donor Sin-1 (Perrault *et al.*, 1997b). One hypothesis to explain the shift of the bradykinin concentration-response curve is that a decreased production of EDHF contributes to the decrease in efficacy of the agonist, although in some models of atherosclerosis, EDHF compensates for reduced NO mediated relaxations (Najibi *et al.*, 1994).

In aging animals and in atherosclerotic human coronary arteries (Nakashima *et al.*, 1993a), endothelium-dependent hyperpolarizations are reduced, suggesting that the absence of endothelium-dependent hyperpolarizations may contribute to the abnormal vascular responses characteristic of these conditions. The endothelial dysfunction in rejection may represent such a pathological state where there is a functional alteration of multiple endothelial cell signaling pathways including the NO and EDHF pathway. The potential effect of desensitization to bradykinin (Olmos *et al.*, 1995) due to increasing concentrations of the agonist and mediated by cyclic GMP, was evaluated in the experiments using a single injection of bradykinin in the presence of a blocker of NO synthesis. These studies confirmed in the allograft rings the reduction in endothelium-dependent relaxations which one, resistant to N^ω-Nitro-L-Arginine.

In the endothelial dysfunction due to balloon denudation of the porcine coronary endothelium followed by regeneration, electrophysiological studies showed that endothelium-dependent hyperpolarization may maintain a normal relaxation to bradykinin (Thollon *et al.*, present issue). In the present study, there is a dissociation between the reduced endothelium-dependent relaxation to bradykinin resistant to NO synthase inhibitors and preservation of smooth muscle cell hyperpolarization to bradykinin. Beny *et al.* (1997) demonstrated that a reduced relaxation may occur despite a normal endothelial cell hyperpolarization as a result of an electrotonic dissociation between endothelial cells and smooth muscle cells. Limitation of the diffusion of EDHF (Kauser and Rubanyi, 1992) to the vascular smooth muscle may occur in intimal thickening and atherosclerosis and account for the reduction in endothelium-dependent relaxations. Dissociation between the hyperpolarization and relaxant effect may occur in this model as smooth muscle cells conserve the capacity to relax to the ATP-dependent potassium channel opener cromakalim. The abnormal response mediated by EDHF identified here may involve calcium-activated potassium channels which are believed to be the main effectors of EDHF in the porcine coronary artery (He, Yang and Yang, 1997). Induction of iNOS by the rejection process could decrease the effectiveness of EDHF (Kessler *et al.*, 1997). However, in the present model, no evidence has been found for the presence of a functional iNOS in vascular reactivity studies performed both in the setting of acute and chronic rejection (Perrault *et al.*, 1997a, 1997c), although this induction occurs in humans and rats after heart transplantation (Worrall *et al.*, 1996).

Since EDHF can cause near normal relaxation when the synthesis of NO is inhibited, the alteration of this pathway may further compound the coronary endothelial dysfunction and accelerate the pathological events occuring at the level of the vascular wall. Preservation of both NO and EDHF may be important to prevent development of graft coronary atherosclerosis.

ACKNOWLEDGEMENT

We would like to thank of Isabelle Delescluse, Christine Cambarrat, Christine Jacquemin, Ludovic Lesage, Aline Pillon and Hélène Reure for their technical assistance; Lisa Maiofiss for the statistical work; Roselyne Prioux for aid in the preparation of the manuscript as well as the contribution of Christine Renard and Marcel Vaiman for the immunologic typing.

Dr Perrault is supported by the Clinician-Scientist program of the Medical Research Council of Canada.

40 Role of Endothelium-derived Hyperpolarizing Factor in Relaxation of Pregnant Rat Uterine Artery by Corticotropin-releasing Factor

Venu Jain, Yuri P. Vedernikov, George R. Saade and Robert E. Garfield

The University of Texas Medical Branch, Reproductive Sciences, Department of Obstetrics and Gynecology, 301 University Boulevard, Route J-62, Galveston, TX 77555-1062, USA Address for correspondence: Venu Jain, The University of Texas Medical Branch, Reproductive Sciences, Department of Obstetrics and Gynecology, 301 University Boulevard, Route J-62, Galveston, TX 77555-1062, USA. Phone: (409) 772-7590; Fax: (409) 772-2261; E-mail: vjain@mspo2.med.utmb.edu
Yuri P. Vedernikov, The University of Texas Medical Branch, Reproductive Sciences, Department of Obstetrics and Gynecology, 301 University Boulevard, Route J-62, Galveston, TX 77555-1062, USA. Phone: (409) 747-0483; Fax: (409) 772-2261; E-mail: yvederni@utmb.edu
George R. Saade, The University of Texas Medical Branch, Reproductive Sciences, Department of Obstetrics and Gynecology, 301 University Boulevard, Route J-62, Galveston, TX 77555-1062, USA. Phone: (409) 747-0482; Fax: (409) 772-2261; E-mail: gsaade@utmb.edu
Robert E. Garfield, The University of Texas Medical Branch, Reproductive Sciences, Department of Obstetrics and Gynecology, 301 University Boulevard, Route J-62, Galveston, TX 77555-1062, USA. Phone: (409) 772-7590; Fax: (409) 772-2261; E-mail: rgarfiel@utmb.edu

Corticotropin-releasing factor (CRF) a vasorelaxant peptide hormone, is increased during human pregnancy. It is thought to be important in modulating uteroplacental blood flow. CRF causes relaxation by endothelium-dependent as well as -independent mechanisms in the rat aorta. Experiments were designed to evaluate the role of endothelium-derived hyperpolarizing factor, nitric oxide and prostacyclin in the relaxation of the pregnant rat uterine artery by CRF. Segments of the uterine artery (outer diameter 300–400 μm) from day 18 pregnant rats were suspended in a small vessel myograph in physiological salt solution, contracted with norepinephrine or depolarizing solution, and the relaxations to cumulative concentrations of CRF were studied. CRF relaxed the uterine artery in a concentration dependent manner. Mechanical removal of the endothelium decreased the CRF-induced relaxation. Use of depolarizing solution for contraction as well as incubation with thiopental sodium (a cytochrome P-450 inhibitor) prior to contraction with norepinephrine significantly inhibited the relaxation to CRF. Similarly, incubation with N^ω-nitro-L-arginine methyl ester (a nitric oxide synthase inhibitor) decreased CRF responses. In contrast, indomethacin (a cyclooxygenase inhibitor) failed to inhibit the relaxation by CRF. Therefore, these data support a role for endothelium-derived hyperpolarizing factor and nitric oxide but not prostacyclin in relaxation of the pregnant rat uterine artery by CRF.

KEYWORDS: corticotropin-releasing factor, endothelium-dependent relaxation, nitric oxide, endothelium-derived hyperpolarizing factor, prostacyclin, rat uterine artery

INTRODUCTION

Corticotropin-releasing factor (CRF), a 41 amino acid peptide, is produced by the hypothalamus (Perkins and Linton, 1995). It acts on the anterior pituitary to release adrenocorticotropin which in turn stimulates cortisol production by the adrenal glands (Perkins and Linton, 1995). In humans, the levels of CRF are normally undetectable but increase exponentially during pregnancy, particularly during the final weeks, to decline rapidly in the immediate postpartum period (Campbell *et al.*, 1987). The placenta is the main source of this increased production of CRF during pregnancy (Riley *et al.*, 1991). Intravascular administration of CRF lowers blood pressure in rats (Gardiner, Compton and Bennet, 1988) as well as humans (Hermus *et al.*, 1987). CRF is a potent vasodilator (Lei *et al.*, 1993; Clifton *et al.*, 1995; Jain *et al.*, 1997, 1998b). Therefore, CRF may have a role in the modulation of peripheral vascular resistance during pregnancy, especially in the uteroplacental vascular bed.

The mechanism of the vasodilator effect of CRF is not well characterized. Both the endothelium and vascular smooth muscle possess CRF binding sites (Dashwood, Andrews and Wei, 1987). Endothelium-dependent relaxation was first demonstrated by Furchgott and Zawadzki (1980). The proposed endothelium-derived relaxing factor (EDRF) has been identified as nitric oxide (Palmer, Ferridge and Moncada, 1987; Ignarro *et al.*, 1987; Furchgott, 1988). However, pharmacological studies using various inhibitors of the metabolism of arachidonic acid showed that endothelium-dependent relaxation involves at least three different pathways (De Mey, Claeys and Vanhoutte, 1982). Prostacyclin is one of these endothelial relaxing factors (Moncada and Vane, 1979; Lüscher and Vanhoutte, 1990). Part of the vascular response to acetylcholine is due to an endothelium-dependent hyperpolarization of the smooth muscle (Félétou and Vanhoutte, 1988). The endothelial factor mediating this effect has been termed endothelium-derived hyperpolarizing factor (EDHF) (Félétou and Vanhoutte, 1988; Huang, Busse and Bassange, 1988; Chen, Suzuki and Weston, 1988; Kauser *et al.*, 1989). The relative importance of each of these three factors may be variable depending on the specific vascular bed and the species. Nitric oxide is synthesized under normal physiological conditions by the constitutive endothelial nitric oxide synthase (Wu, 1995), whereas cyclooxygenase is the responsible enzyme for the synthesis of prostacyclin (Moncada *et al.*, 1977). EDHF may be a product of cytochrome P450 (Rubanyi and Vanhoutte, 1987; Hecker, Fleming and Busse, 1995; Campbell *et al.*, 1996).

The relaxing effect of CRF in the rat aorta is predominantly endothelium dependent and mediated by the L-arginine/nitric oxide-cyclic GMP pathway (Jain *et al.*, 1997). These findings are in agreement with observations that the CRF-induced vasodilatation in the human feto-placental circulation is mediated by nitric oxide and cyclic GMP (Clifton *et al.*, 1995). However, responses to CRF in the mesenteric artery of the male rat are endothelium independent (Lei *et al.*, 1993). The uterine vasculature may be an important target for CRF during pregnancy. The present experiments were designed to evaluate the role of EDHF, nitric oxide and prostacyclin as endothelial factors involved in relaxation of the pregnant rat uterine artery by CRF.

METHODS

Animals

Timed pregnant female Sprague-Dawley rats were obtained from Charles River Laboratories (Wilmington, MA). They were housed separately in temperature and humidity controlled quarters with constant light:dark cycles of 12 hr:12 hr and were provided with food and water *ad libitum*. The pregnant rats used in these studies have a gestational period of 22 days, day 1 being the day the sperm plug was observed. The animals were sacrificed on day 18 of gestation by CO_2 inhalation.

Drugs And Solutions

Acetylcholine hydrochloride, corticotropin-releasing factor, indomethacin, norepinephrine bitartrate, N^{ω}-nitro-L-arginine methyl ester and thiopental sodium were purchased from Sigma, St. Louis, MO. Stock solutions of all the drugs were prepared in deionized water with the exception of indomethacin (10^{-2} M) which was prepared in a 150 mM solution of $NaHCO_3$ (pH 8.3). Stock solutions for norepinephrine were freshly prepared for each experiment. The composition of the physiological salt solution was as follows (mM): NaCl, 115; KCl, 5; NaH_2PO_4, 1.2; $NaHCO_3$, 25; $MgCl_2$, 1.2; $CaCl_2$, 2.5; EDTA, .026; glucose, 11. The depolarizing solution (high-K^+ physiological salt solution) was made by replacing NaCl with equimolar KCl, the final K^+ concentration in that solution being 120 mM.

In Vitro Experiments

Two millimeter segments of the uterine artery (outer diameter 300–400 μm) were suspended in the jaws of a wire myograph (Model 410A, J.P. Trading I/S, Aarhus, Denmark) over 25 μm tungsten wires. The endothelium was removed in some preparations by rubbing their luminal surface with a human hair (Osol, Cipolla and Knutson, 1989). The preparations were bathed in physiological salt solution maintained at 37°C, pH 7.4. A mixture of 95% O_2 and 5% CO_2 was bubbled continuously through the solution. The force was recorded by an isometric force transducer and analyzed using Windaq data acquisition and playback software (Dataq Instruments, Inc., Akron, OH). The blood vessels were given a preload based on the length-tension curve for each vessel. Myosight software (J.P. Trading I/S, Aarhus, Denmark) was used for estimating the circumference that each vessel would have had under a transmural pressure of 100 mmHg *in situ* and the circumference of the preparation was adjusted to 90% of the estimated circumference (Mulvany and Halpern, 1977). The arteries were allowed to equilibrated for 1 hr. Then two successive stimulations of 15 min duration were given with high-K^+ physiological salt solution, separated by a 15 min equilibration in physiological salt solution. The presence or absence of endothelium in the preparations was confirmed by contracting them with norepinephrine (10^{-6} M) and eliciting a relaxation with acetylcholine (10^{-6} M). After washing, the preparations were contracted with norepinephrine (10^{-6} M), and the relaxations to cumulative concentrations of CRF (10^{-10} to 10^{-6} M) were studied.

To assess the role of endothelium, the responses to CRF were studied in arteries without endothelium. In blood vessels with endothelium, responses to CRF during contraction with depolarizing solution (120 mM K^+) were compared to those during contraction with norepinephrine to determine the involvement of EDHF. Responses were also studied in arteries incubated with thiopental (cytochrome P-450 inhibitor, 3×10^{-4} M, for 30 min) to inhibit EDHF production (Lischke, Busse and Hecker, 1995). To examine the role of nitric oxide and prostacyclin in the response to CRF, studies were performed in arteries incubated with N^{ω}-nitro-L-arginine methyl ester (nitric oxide synthase inhibitor, 10^{-4} M, for 30 min) or with indomethacin (cyclooxygenase inhibitor, 10^{-5} M, for 45 min).

Data Analysis

The data is expressed as mean $\pm$ S.E.M., n represents the number of vessels from different rats used in the experiments. The effect of CRF on the uterine artery was quantified as percent relaxation of the preexisting tone in preparations contracted with norepinephrine or depolarizing solution. Concentration-response curves were generated based on responses to cumulative concentrations of CRF. The ED_{50} values for CRF (concentration of CRF producing 50% of the maximal relaxation, in log M) were calculated. The area under the concentration-response curves was calculated and expressed in arbitrary units. For statistical analysis, Student's t-test or one-way analysis of variance followed by Newman-Keuls Multiple Comparisons Test were used as appropriate. When p was less than 0.05, differences were considered to be statistically significant.

RESULTS

CRF relaxed the uterine artery of pregnant rats in a concentration-dependent manner (Figure 40-1). The relaxation to CRF was significantly inhibited in the absence of endothelium (Figure 40-1). The ED_{50} value as well as the area under the concentration response curve were decreased significantly by the removal of endothelium (Table 40-1).

When depolarizing solution (120 mM K^+) was used to contract the arteries, the relaxation to CRF was reduced compared to that observed in uterine arteries contracted with norepinephrine (Figure 40-2). In addition, in rings contracted with norepinephrine, incubation with thiopental significantly decreased the relaxation to CRF (Figure 40-2). The ED_{50} values were not significantly decreased by 120 mM K^+ or thiopental, but the areas under the curves were significantly decreased by both (Table 40-1). The relaxation during norepinephrine-induced contractions by CRF was decreased in the presence of inhibitor of nitric oxide synthase N^{ω}-nitro-L-arginine methyl ester (Figure 40-3). Both the ED_{50} value and the area under the curve were decreased significantly by N^{ω}-nitro-L-arginine methyl ester (Table 40-1). Indomethacin did not affect the relaxation to CRF in the uterine artery contracted with norepinephrine (Figure 40-4). The ED_{50} value or the area under the curve were not significantly different from controls (Table 40-1).

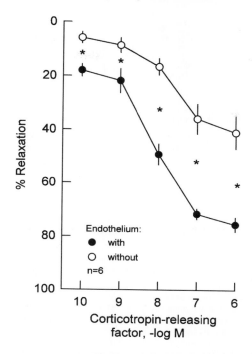

Figure 40-1. Effect of removal of the endothelium on responses of the uterine artery of pregnant rats (day 18) to cumulative concentrations of CRF. Uterine arteries were contracted with norepinephrine (10^{-6} M). The 100% contraction induced by norepinephrine was 1.95 ± 0.15 g and 2.14 ± 0.15 g in preparations with or without endothelium, respectively. Each point represents mean $\pm$ S.E.M. ($n = 6$). Students's t-test was used; control versus deendothelized. The asterisks indicate statistically significant differences ($P < 0.05$).

Table 40-1 Relaxations to CRF in pregnant rat uterine artery

	ED_{50}	Area under the curve
Effect of removal of endothelium		
With endothelium	-8.32 ± 0.1	189.26 ± 8.29
Without endothelium	$-7.83 \pm 0.09*$	$84.44 \pm 13.75*$
Effect of inhibition of EDHF		
Norepinephrine	-7.73 ± 0.07	129.74 ± 13.54
120 mM K^+	-7.56 ± 0.07	$70.22 \pm 6.06*$
norepinephrine/thiopental	-7.61 ± 0.05	$83.86 \pm 4.45*$
Effect of inhibition of nitric oxide synthase		
Control	-8.09 ± 0.18	165.35 ± 19.40
N^ω-nitro-L-arginine methyl ester	$-7.40 \pm 0.16*$	$41.13 \pm 13.36*$
Effect of inhibition of cyclooxygenase		
Control	-7.73 ± 0.07	129.74 ± 13.54
Indomethacin	-7.63 ± 0.04	129.42 ± 10.26

The asterisks indicate statistically significant differences ($P < 0.05$).

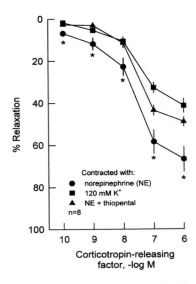

Figure 40-2. Effect of membrane depolarization (120 mM K$^+$) or inhibition of cytochrome P-450 (thiopental) on responses of the uterine artery of pregnant rats (day 18) to cumulative concentrations of CRF. The vessels were contracted with depolarizing solution (120 mM K$^+$) or incubated with thiopental (3×10^{-4} M) for 30 min and contracted with norepinephrine (10^{-6} M). The 100% contraction induced by norepinephrine was 1.96 ± 0.11 g and 1.88 ± 0.08 g in the absence and presence of thiopental, respectively, and the 100% contraction induced by depolarizing solution was 1.41 ± 0.08 g. Each point represents mean $\pm$ S.E.M. ($n = 8$). One-way analysis of variance followed by Newman Keuls Multiple Comparisons Test was used; norepinephrine versus 120 mM K$^+$ and norepinephrine/thiopental. The asterisks indicate statistically significant differences ($P < 0.05$).

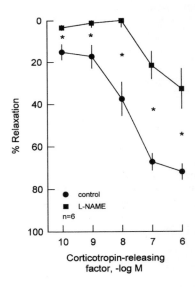

Figure 40-3. Effect of inhibition of nitric oxide synthase on responses of the uterine artery of pregnant rats (day 18) to cumulative concentrations of CRF. Uterine arteries were incubated with N$^\omega$-nitro-L-arginine methyl ester (10^{-4} M) for 30 min and contracted with norepinephrine (10^{-6} M). The 100% contraction induced by norepinephrine was 1.92 ± 0.09 g and 2.17 ± 0.16 g in the absence and presence of N$^\omega$-nitro-L-arginine methyl ester, respectively. Each point represents mean $\pm$ S.E.M. ($n = 6$). Students's t-test was used; control versus L-NAME. The asterisks indicate statistically significant differences ($P < 0.05$).

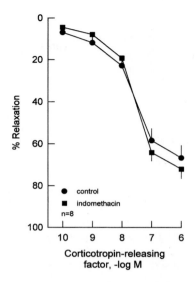

Figure 40-4. Effect of inhibition of cyclooxygenase on responses of the uterine artery of pregnant rats (day 18) to cumulative concentrations of CRF. Uterine arteries were incubated with indomethacin (10^{-5} M) for 45 min and contracted with norepinephrine (10^{-6} M). The 100% contraction induced by norepinephrine was 1.98 ± 0.10 g and 2.00 ± 0.13 g in the absence or presence of indomethacin, respectively. Each point represents mean $\pm$ S.E.M. ($n = 8$). Students's t-test was used; control versus indomethacin. The asterisks indicate statistically significant differences ($P < 0.05$).

DISCUSSION

CRF when administered chronically in pregnant rats causes a decrease in blood pressure (Jain *et al.*, 1998b). In addition, it is a relaxant of the rat aorta *in vitro* and its relaxing effect is endothelium-dependent and mediated by the nitric oxide-cyclic GMP pathway (Jain *et al.*, 1997). The present study demonstrates that CRF is a potent dilator of the uterine artery of pregnant rats. The relaxation is predominantly endothelium-dependent and is mediated by EDHF and nitric oxide, but not prostacyclin. There is also an endothelium-independent component at higher concentrations of CRF. Since in humans the production of CRF increases during the final weeks of pregnancy (Campbell *et al.*, 1987), the factor may play a significant role in increasing the uteroplacental blood flow.

Two subtypes of the CRF receptor have been identified in humans and rats, CRFR$_1$ and CRFR$_2$ (Grigoriadis *et al.*, 1996). Both receptor subtypes have been cloned and sequenced. CRFR$_2$ has two splice variants i.e. CRFR$_{2\alpha}$ and CRFR$_{2\beta}$. The CRF receptor isoform expressed in the rat vasculature is CRFR$_{2\beta}$ (Jain *et al.*, 1998a). CRF has binding sites on the vascular endothelium as well as smooth muscle (Dashwood, Andrews and Wei, 1987). However, immunohistochemical studies show that the CRF receptor is present predominantly in the endothelium and to a lesser extent in the vascular smooth muscle (Jain *et al.*, 1998a). In the present study, the removal of endothelium prevented a major component of the relaxation of the uterine artery to CRF, especially at lower (and more physiological) concentrations of the hormone. This prompts the conclusion that in the uterine artery also, the relaxation to CRF is endothelium-dependent. The endothelium regulates the tone of the underlying vascular smooth muscle by releasing various contracting and relaxing factors (Furchgott and Vanhoutte, 1989; Rubanyi, 1993). The release of the relaxing factors is mediated by an increase in free cytoplasmic Ca^{2+} levels in the endothelial cells (Furchgott, 1983). Indeed, the constitutive endothelial nitric oxide synthase as well as phospholipase A_2, which liberates arachidonic acid from membrane phospholipids to be metabolized into prostacyclin, are Ca^{2+}-dependent enzymes (Lückoff *et al.*, 1988; Busse and Mülsch, 1990; Hong and Deykin, 1982). The increase in intracellular Ca^{2+} is effected by various mechanisms including the influx of extracellular Ca^{2+}, Na^+-Ca^{2+} exchange and liberation of Ca^{2+} from intracellular stores (Singer and Peach, 1982; Winquist, Bunting and Schofield, 1985; Lückoff and Busse, 1986). CRF can cause an increase in intracellular calcium through receptor-operated Ca^{2+} channels (Kiang, 1994). Hence, it may enhance the production of endothelium-derived relaxing factors by increasing the intracellular Ca^{2+} in the endothelial cells.

EDHF causes relaxation of smooth muscle by activation of Ca^{2+}-activated K^+-channels (Hecker *et al.*, 1994; Fulton, McGiff and Quilley, 1994). At least one class of EDHFs has been shown to be produced by cytochrome P450 (Fulton, McGiff and Quilley, 1994; Lischke, Busse and Hecker, 1995). Since EDHF acts by causing hyperpolarization of the cell membrane, the use of depolarizing solution for the contraction of vessels *in vitro* eliminates its effect. Contraction of the uterine artery with depolarizing solution as well as inhibition of cytochrome P-450 with thiopental inhibited the relaxation caused by CRF in the uterine artery. This supports the role of a cytochrome P450-dependent EDHF in the relaxation to CRF.

Nitric oxide, a product of the endothelial nitric oxide synthase (see Wu, 1995) activates soluble guanylate cyclase of vascular smooth muscle and increases cyclic GMP, which in turn causes relaxation of the smooth muscle (Holzmann, 1982). The relaxing effect of CRF in the perfused fetal placental circulation is mediated by the nitric oxide-cyclic GMP system (Clifton *et al.*, 1995). Similar findings were reported in the isolated aorta of nonpregnant and pregnant rats (Jain *et al.*, 1997). In the present study, blockade of nitric oxide synthase with N^ω-nitro-L-arginine methyl ester reduced the relaxations to CRF. This finding indicates that relaxation caused by CRF in the rat uterine artery is mediated in part by nitric oxide. Nitric oxide is an inhibitor of cytochrome P450 enzymes (Wink *et al.*, 1993). Therefore, inhibition of nitric oxide production by N^ω-nitro-L-arginine methyl ester may increase the production of EDHF. This interaction may also be functionally relevant; in vascular beds or physiological/pathological states where nitric production is high, cytochrome P450-dependent EDHF production may be lower and vice versa.

Prostacyclin, the first endothelial relaxing factor to be identified, is produced by cycloxygenase (Moncada, 1977). However, inhibition of cyclooxygenase with indomethacin failed to have any effect on the relaxation to CRF. This indicates that prostacyclin may not be an important factor mediating endothelium-dependent relaxation by CRF in pregnant rat uterine artery.

Removal of endothelium did not abolish the relaxation of uterine artery by CRF. Similar results were obtained in the rat aorta (Jain *et al.*, 1997). These finding are only in partial agreement with the observations that in small mesenteric arteries from male Wistar rats (contracted with arginine vasopressin) the responses to CRF are strictly endothelium-independent (Lei *et al.*, 1993). Since CRF exhibits regional differences in its effects (Gardiner, Compton and Bennet, 1988), the mechanism characterized in the present study may differ from the one examined in the mesenteric artery of the male rat (Lei *et al.*, 1993). Differences in the agonist used to contract the blood vessel and in the gender and strain of rats used are other factors which may account for the discrepancy. Thus, CRF may also cause relaxation by acting directly on the vascular smooth muscle. However, this is not a predominant mechanism in the rat aorta or the uterine artery.

41 Endothelium-derived Hyperpolarizing Factor in Omental Arteries from Normotensive Pregnant Women

Y.P. Vedernikov, M.A. Belfort*, G.R. Saade and R.E. Garfield

Departments of Obstetrics and Gynecology, University of Texas Medical Branch, Galveston, Texas, and University of Utah Medical Center, Salt Lake City, UT, USA Address for correspondence: Yuri P. Vedernikov, Reproductive Sciences, Department of Obstetrics and Gynecology, 301 University Boulevard, Galveston Texas, 77555-1062, USA. Telephone: (409) 747-0488; Fax: (409) 772-2261; E-mail vederni@utmb.edu Michael A. Belfort, Department of Perinatology, Utah Valleuy Regional Medical Center, 1034 North, 500 West Provo, Utah 84604, USA. Telephone: (801) 357-7565; Fax: (801) 370-9061 George R. Saade, Reproductive Sciences, Department of Obstetrics and Gynecology, 301 University Boulevard, Galveston Texas, 77555-1062, USA. Telephone: (409) 747-0482; Fax: (409) 772-2261; E-mail gsaade@utmb.edu Robert E. Garfield, Reproductive Sciences, Department of Obstetrics and Gynecology, 301 University Boulevard, Galveston Texas, 77555-1062, USA. Telephone: (409) 772-7590; Fax: (409) 772-2261; E-mail rgarfiiel@utmb.edu*

In rings of omental artery from five normotensive pregnant women substance P reduced the contraction induced by potassium chloride. Repeated application of bradykinin (alone or in the presence of N^{ω}-nitro-L-arginine and indomethacin) demonstrated no significant change in the effect. Bradykinin failed to relax rings contracted with U 46619 when the concentration of potassium chloride was increased. In rings contracted with endothelin-1 in the presence of indomethacin, N^{ω}-nitro-L-arginine, tetraethylammonium or glibenclamide did not significantly influence the response to bradykinin. In the presence of indomethacin and N^{ω}-nitro-L-arginine, tetraethylammonium and glibenclamide significantly increased the area under the contraction curve, but did not affect the pD_2, while glibenclamide also reduced the maximal relaxation. These findings suggest that endothelium-derived hyperpolarizing factor participates in the relaxation of human omental artery rings via activation of potassium channels (presumably K_{ATP} rather than K_{Ca}).

KEYWORDS: endothelial hyperpolarizing factor, human omental artery, pregnancy, relaxation

INTRODUCTION

Endothelium-dependent relaxation is mediated by three main factors, prostacyclin (Moncada *et al.*, 1977), endothelium-derived relaxing factor (Furchgott and Zawadzki, 1980) and endothelium-derived hyperpolarizing factor, (Taylor and Weston, 1988; Furchgott and Vanhoutte, 1989; Suzuki, Chen and Yamamoto, 1992). The latter two remain operative after inhibition of cyclooxygenase. The contribution of endothelium-derived hyperpolarizing factor to the regulation of tone prevails in resistance as opposed to conduit vessels (Garland *et al.*, 1995). The aim of the present study was to investigate the role of endothelium-

363

derived hyperpolarizing factor (EDHF) in omental arteries from normotensive pregnant women.

METHODS

Five normotensive pregnant women (21 to 39 years old, gestational age 38.8 ± 0.4 weeks and systolic/diastolic blood pressure 103.8 ± 9.4 mm Hg/57.2 ± 1.96 mm Hg) were enrolled in the study. A piece of omentum was isolated and removed following completion of cesarean section. The specimen was immediately transported to the laboratory in cold modified Krebs-bicarbonate buffer of the following composition (in mM): NaCl – 119; KCl – 4.7; $MgSO_4$ – 1.2; KH_2PO_4 – 1.2; $CaCl_2$ – 2.5; $NaHCO_3$ – 25; dextrose – 11.1; sodium EDTA – 0.03. The omental artery was dissected free of the surrounding tissue and the external diameter ($515 + 4.8$ μm) was measured under a dissecting microscope (Olympus SZ30, Olympus Corporation, New York, NY). The blood vessel was then cut into rings (3 mm in width) which were suspended on tungsten wire stirrups (75 μ diameter) in organ chambers containing 10 ml of Krebs-bicarbonate buffer maintained at 37°C, aerated with a gas mixture of 5% CO_2 in air (pH about 7.4) for isometric tension recording (Harvard Apparatus, South Natick, MA). Passive tension was applied gradually to the rings and adjusted to 1 gram, reflecting the optimal length-tension relationship, during an equilibration period of 60 to 90 minutes. The Krebs-bicarbonate buffer bathing the blood vessels was changed every 15 to 20 minutes during the equilibration period. The rings were then contracted with 60 mM KCl repeatedly, at 30 min intervals, until a stable contraction was established. The integrity of the endothelium was checked by exposing rings contracted with 60 mM KCl to a single concentration of substance P (10^{-9} M) and observing the relaxation. The organ chambers were then washed out and the arterial rings were allowed to equilibrate for 60 minutes.

Experimental Protocol

Eight rings were prepared from each omental artery. In two rings the endothelium was removed mechanically. After the equilibration period and checking the integrity of the endothelium, all rings were incubated with indomethacin (10^{-5} M) for 30 min. Three of the rings were then incubated with N^ω-nitro-L-arginine (10^{-4} M), tetraethylammonium (10^{-3} M) or glibenclamide (3×10^{-6} M), respectively for an additional 30 min. A pair of rings, with and without endothelium, served as controls for the bradykinin concentration-relaxation relationship study and an additional ring was used as a time-solvent control. A pair of rings, with and without endothelium, were used to study relaxations to 3-morpholinosydnonimine (SIN-1). Rings were contracted with endothelin-1 (10^{-9} to 5×10^{-9} M) to match tension. After tension stabilized, bradykinin was added to the organ chambers cumulatively in 0.5 log unit increments. The rings were washed out thoroughly and allowed to rest for 90 min. To test the effect of nitric oxide synthase inhibition, rings were then incubated with N^ω-nitro-L-arginine in addition to either tetraethylammonium or glibenclamide, contracted with endothelin-1 and the concentration-response experiments with bradykinin were repeated.

Drugs Used

Bradykinin, glibenclamide, N^{ω}-nitro-L-arginin, substance P, tetraethylammonium, U-46619 (Sigma Chemical Co., St. Louis, MO); 3-morpholinosyndonimine-hydrochloride, SIN-1 (Cassella, Germany). Stock solutions of bradykinin and substance P (10^{-3} M) were made using deionized water, aliquoted and kept at $-20°$ until used. Indomethacin was dissolved in deionized water and an equal molar concentration of Na_2CO_3. A stock solution of glibenclamide (10^{-2} M) was prepared in dimethyl sulfoxide. All other drugs were dissolved in distilled water on the day of the experiment. The required dilutions were made using distilled water and kept on ice. The drugs were added directly to the organ chambers in volumes ranging from 10 μl to 23 μl.

Data Analysis

The negative logarithm of the molar concentration of bradykinin causing 50% inhibition of the tension induced by endothelin-1 (pD_2), the area under the cumulative concentration-relaxation curve and maximal effect were calculated and compared. All data sets were subjected to a test for normalcy (Wilk-Shapiro test) and parametric or non-parametric tests were then used as appropriate. A two-tailed P-value less then 0.05 was considered to be statistically significant.

RESULTS

Substance P (10^{-9} M) relaxed intact omental artery rings contracted with 60 mM potassium chloride (in the absence of indomethacin) by $72.04 \pm 1.4\%$. Removal of the endothelium attenuated the response to $1.2 \pm 2.7\%$. Bradykinin did not relax rings contracted with 60 mM potassium chloride (data not shown). The contraction induced by the thromboxane A_2 mimetic U46619 was not well maintained but bradykinin relaxed the rings contracted with U46619 in an endothelium-dependent manner. An increase in the concentration of potassium chloride in the medium maintained tension, but progressively attenuated endothelium-dependent relaxations induced by bradykinin (Figure 41-1).

Endothelium-dependent relaxations induced by bradykinin in intact omental arteries did not significantly differ between the first and second experimental series in terms of either pD_2 or area under the curve. N^{ω}-nitro-L arginine had no significant effect on the response to bradykinin neither in terms of pD_2 nor area under the curve. The results in the first and second experimental series were also similar (Figure 41-2, Table 41-1). Tetraethyl-ammonium (10^{-3} M) did not significantly influence either the pD_2 or the area under the curve for bradykinin (Figure 41-3, Table 41-1). Incubation with both tetraethylammonium and N^{ω}-nitro-L arginine shifted the concentration-relaxation curve to bradykinin to the right, with significant change in area under the curve, while changes in pD_2 or maximal relaxation did not reach the level of significance (Figure 41-3, Table 41-1).

Glibenclamide attenuated the relaxing effect of low concentrations of bradykinin but did not significantly effect the pD_2, the area under the curve or the maximal effect. In rings incubated with glibenclamide and N^{ω}-nitro-L arginine the inhibition of bradykinin-induced relaxation was more pronounced, both in terms of the area under the curve and maximal

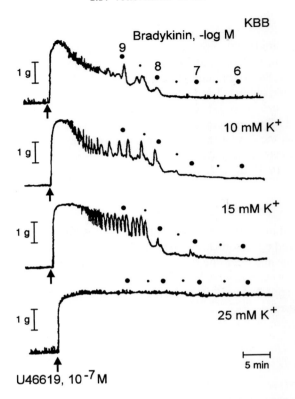

Figure 41-1. Representative tracings of the effect of bradykinin in omental artery rings with endothelium contracted with the thromboxane A$_2$ mimetic U 46619. Note that the contraction induced by U 46619 was not maintained. Increasing concentrations of potassium chloride stabilized the contraction, but inhibited the relaxation.

     KBB-Krebs-bicarbonate buffer. Concentrations of U 46619 and bradykinin are shown in log M.

Table 41-1   Bradykinin (BK) induced changes in pD$_2$, area under the curve and maximal relaxation (% of endothelin-1 (ET)-contraction) after repeated application, alone or in the presence of N$^\omega$-nitro-L-arginine (10^{-4} M) (NLA), and effect of tetraethylammonium (10^{-3} M) (TEA), glibenclamide (3 × 10^{-6} M) (Gli), alone or in combination with N$^\omega$-nitro-L-arginine in omental artery rings from normotensive pregnant women

First trial (n = 5)	BK-control	NLA	TEA	GLI	TEA + NLA	Gli + NLA
PD$_2$	8.87 ± 0.41	9.06 ± 0.26	9.13 ± 0.37	8.49 ± 0.19		
Area	434 ± 69	399 ± 35	391 ± 53	507 ± 32		
% of ET contraction	5.2 ± 2.3	8.7 ± 1.4	4.0 ± 1.2	6.8 ± 1.8		
Second trial (n = 4)						
PD$_2$	9.38 ± 0.49	9.26 ± 0.6			8.17 ± 0.12	8.1 ± 0.3
Area	331 ± 90	353 ± 90			543 ± 35 *	688 ± 58 *
% ET contraction	1.5 ± 1.53	13.4 ± 7.5			20.3 ± 5.9	61 ± 16 *

The asterisk indicates statistically significant difference (P < 0.05) from the value in the absence of N$^\omega$-nitro-L-arginine.

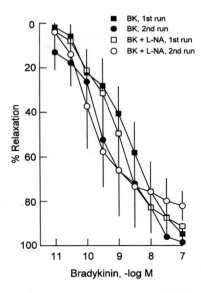

Figure 41-2. Repeated concentration-relaxation curves to bradykinin (BK) in omental artery rings contracted with endothelin-1 in the presence of indomethacin (10^{-5} M). Incubation with N^{ω}-nitro-L-arginine (10^{-4} M) (NLA) did not influence concentration-relaxation relationship to bradykinin. The 100% contractions (in g) were: 3.6 ± 0.2 and 3.7 ± 0.3 in the absence and 3.9 ± 0.7 and 4.2 ± 0.7 in the presence of N^{ω}-nitro-L-arginine for the first and second run, respectively. Data shown as mean $\pm$ SEM.

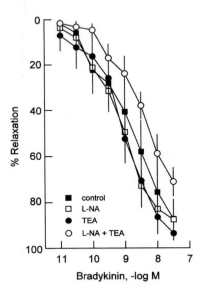

Figure 41-3. Tetraethylammonium (10^{-3} M) had no significant effect on the concentration-relaxation relationship to bradykinin in the presence of indomethacin (10^{-5} M). Incubation of rings with tetraethylammonium and N^{ω}-nitro-L-arginine (10^{-4} M) shifted the concentration-relaxation relationship to bradykinin to the right. The 100% contraction (in g) were: 3.6 ± 0.4, 4.2 ± 0.7 and 3.6 ± 0.2 in tetraethylammonium, N^{ω}-nitro-L-arginine and tetraethylammonium + N^{ω}-nitro-L-arginine groups, respectively, compared to 3.6 ± 0.2 in control group. Data shown as mean $\pm$ SEM.

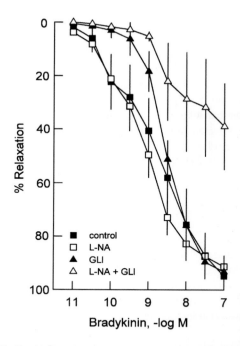

Figure 41-4. Glibenclamide (3×10^{-6} M) (Gli) in the presence of indomethacin (10^{-5} M) attenuated bradykinin-induced relaxation in the low concentrations range. Incubation of rings with glibenclamide and N^{ω}-nitro-L-arginine (10^{-4} M) shifted the concentration-relaxation relationship to bradykinin to the right ($P < 0.05$). The 100% contraction (in g) were: 3.9 ± 0.1, 4.2 ± 0.7 and 4.0 ± 0.2 in glibenclamide, N^{ω}-nitro-L-arginine and glibenclamide + N^{ω}-nitro-L-arginine groups, respectively compared to 3.6 ± 0.2 in control. Data shown as mean $\pm$ SEM (n = 5).

effect, although the change in the pD_2 did not reach a significant level (Figure 41-4, Table 41-1).

There was no significant differences between first and second concentration-relaxation curves to SIN-1 in rings with endothelium, neither in terms of the pD_2 nor the area under the curves (7.12 ± 0.25 and 355 ± 49 and 7.64 ± 0.4 and 265 ± 66 at first and second trials, respectively). In rings without endothelium, the concentration-relaxation curves to SIN-1 were similar on the first and second trials (pD_2 and areas under the curve were: 7.15 ± 0.3 and 358 ± 36 and 7.19 ± 0.32 and 346 ± 54, respectively).

DISCUSSION

The data suggest that endothelium-mediated relaxations of omental artery rings from normotensive pregnant women involves the endothelial cell products NO and EDHF.

Depolarization of the rings with high potassium chloride inhibits relaxation by EDHF (Nagao and Vanhoutte, 1991, 1992a). Therefore, the ability of substance P to relax the intact rings depolarized with potassium chloride demonstrates that another factor(s) is operating. Rings contracted with U46619 failed to relax with bradykinin when the concentration of potassium chloride was raised to 30 mM, supporting a role for EDHF in the response to this peptide. Indomethacin and N^{ω}-nitro-L-arginine did not significantly influence the concentration-relaxation relationship to bradykinin. This finding confirms that EDHF, and not prostacyclin or EDRF is the major endothelial factor responsible for the relaxation to bradykinin in omental artery from pregnant women. In resistance arteries, EDHF apears to be a major determinant of vascular caliber and may be of primary importance in the regulation of vascular resistance (Garland *et al.*, 1995).

Inhibition of K_{Ca}-channels with 10^{-3} M of tetraethylammonium did not influence the concentration-relaxation relationship to bradykinin in the presence of a functional EDRF/NO system, but shifted the relationship to the right in the presence of N^{ω}-nitro-L-arginine. Inhibition of nitric oxide synthase, however, did not affect the sensitivity. Glibenclamide, an inhibitor of K_{ATP} channels, did not significantly influence the concentration-relaxation relationship to bradykinin in rings incubated with indomethacin, although it attenuated the relaxation in the lower concentration range. In the presence of N^{ω}-nitro-L-arginine, however, it shifted the curve to the right and decreased the maximal relaxation induced by bradykinin. These data are consistent with at least partial activation of K_{ATP} channels by EDHF in human omental arteries in pregnancy. Endothelium-dependent hyperpolarizations have been documented in human blood vessels (Nakashima *et al.*, 1993; Peterson *et al.*, 1995), thus confirming the existance of relaxation by EDHF. In those studies, K_{ATP} channels were not involved in EDHF-induced relaxation. The type of potassium channels involved in endothelium-dependent hyperpolarization differs among blood vessels and/or species (Brayden, 1990; Standen *et al.*, 1989; Garland *et al.*, 1995). It is still possible that NO, which could escape inhibition by NO synthase inhibitors, may have induced hyperpolarization through a cyclic GMP independent effect on K_{Ca} channels (Bolotina *et al.*, 1994) or a cyclic GMP-mediated effect on voltage-dependent Ca^{2+}-channels (Blatter and Wier, 1994).

Relaxation by substance P of omental arteries from non pregnant patients contracted with U46619 are mediated both by NO and endothelium-dependent hyperpolarization involving large conductance K_{Ca} channels (Wallerstedt and Bodelsson, 1997). On the other hand, both nitric oxide and cyclooxygenase derivative(s) may be involved and could substitute for each other in relaxing the omental artery (Ohlmann *et al.*, 1997). In human omental microvessels (normalized internal diameter about 200 μm), which are smaller then those used in the present study tetraethylammonium in concentration of 10^{-2} M, but not 10^{-3} M, as well as gliburide abolished relaxation in rings from non pregnant, but not in those from pregnant, women (Pascoal and Umans, 1996).

Acetylcholine- and bradykinin-induced endothelium-dependent relaxation of omental arteries from normotensive pregnant women is uneffected by inhibitors of nitric oxide synthesis (Pascoal *et al.*, 1998). The response to acetylcholine, but not that to bradykinin, was abolished in omental vessels from preeclamptic patients. In the present experiments, bradykinin and substance P were the only agents which produced endothelium-dependent relaxations in human omental artery. Indeed, relaxation to acetylcholine could not be obtained in omental arteries from either nonpregnant, normotensive pregnant or preeclamptic patients (data not shown). Another unusual finding was that inhibitors of either nitric oxide synthase or cyclooxygenase had no effect on basal tension. Inhibition of cyclooxygenase, but not of nitric oxide synthase, potentiated contractions induced by U 46619 and counteracted inhibition of the response by 17β-estradiol (Belfort *et al.*, 1995), indirectly supporting the involvement of cyclooxygenase product(s) in the regulation of the contractions of the human omental artery.

42 Mechanisms of Vasodilatation in Human Conduit Arteries

C.A. Hamilton, G. Berg, K. McArthur, A. Faichney, E. Jardine, J.L. Reid and A.F. Dominiczak

Department of Medicine and Therapeutics, Western Infirmary, Glasgow, G11 6NT.
Tel. No. 0141 211 2042; Fax No. 0141 339 2800

Endothelium-dependent relaxation mediated via activation of potassium channels and hyperpolorization has been documented in a range of animal tissues. Less is known about this pathway in human blood vessels. The aim of the present studies was to investigate the contribution of potassium channel activation to endothelium-dependent relaxations of human conduit vessels. The effect of blockers of potassium channels and inhibitors of nitric oxide synthase on relaxation to carbachol and bradykinin was studied in isolated rings with endothelium of internal thoracic and radial arteries and of saphenous vein taken from patients undergoing coronary artery bypass graft surgery.

In the radial artery nitric oxide synthase inhibition only caused a small attenuation of relaxation. Charybdotoxin and to a greater extent charybdotoxin plus apamin also attenuated it. The effects of N^{ω}-nitro-L-arginine methyl ester and charybdotoxin were additive. In the internal thoracic artery no significant relaxation to carbachol or bradykinin was observed in the presence of nitric oxide synthase inhibitors. However, tetraethylammonium, iberiotoxin and charybdotoxin also significantly attenuated relaxation. In the saphenous vein, nitric oxide synthase inhibitors abolished relaxations to bradykinin and carbachol but none of the potassium channel blockers modified the responses.

These results suggest that in addition to the L-arginine nitric oxide pathway, potassium channel activation may contribute to endothelium-dependent relaxations of human conduit arteries. The nature of the potassium channel differed between the two arteries studied, although in both cases calcium-dependent potassium channels appeared to be involved.

The existence of alternative pathways for endothelium-dependent relaxation may help maintain normal vascular tone in disease. Knowledge of the mechanisms regulating endothelium-dependent dilatation in individual blood vessels may be important in optimising treatment of cardiovascular disease.

KEYWORDS: radial artery, internal thoracic artery, potassium channels, hyperpolarization, vasorelaxation

INTRODUCTION

Activation of potassium channels and hyperpolarization of vascular smooth muscle cells contribute to endothelium-dependent relaxations in a wide range of animal tissues (for review see Garland *et al.*, 1995), and may be the primary mechanism of relaxation in small resistance arteries (Shimokawa *et al.*, 1996). Potassium channel activation and hyperpolarization of vascular smooth muscle may also contribute to agonist-stimulated

371

endothelium-dependent responses in humans (Nakashima *et al.*, 1993; Wallerstedt and Bodelsson 1997; Urakami-Harasawa *et al.*, 1997). How generalised these mechanisms are and whether they are restricted to human blood vessels of smaller diameters is still to be elucidated.

The aim of the present studies was to examine the contribution of potassium channels to relaxation of human conduit blood vessels used for revascularisation in coronary artery bypass graft surgery.

METHODS

Preparation

Human internal thoracic artery, radial artery and saphenous vein were obtained from patients undergoing coronary artery bypass graft surgery. Approval to use discarded tissues was granted by the Western Infirmary Ethical Committee, Glasgow, UK. Human internal thoracic arteries were dissected as a pedicle with the venae commitantes from the thoracic wall by a no-touch technique leaving the vessels surrounded by internal thoracic fascia in the usual manner. The saphenous vein was dissected free (Barker *et al.*, 1994).

Radial arteries were harvested from the non-dominant arm. A pedicle of radial artery with two venae commitantes was dissected without touching the artery.

Segments of the blood vessels were placed in physiological salt solution and transferred to the laboratory. The vessels were cleaned of connective tissue and cut into rings (2–3 mm width). Rings were suspended on wires in organ chambers filled with 10 ml of physiological salt solution, maintained at 37°C, and aerated with a mixture of 95% O_2–5% CO_2. The rings were connected to force transducers and changes in isometric tension were recorded. The physiological salt solution (pH = 7.4 ± 0.1) had the following composition in (mM): 130 NaCl, 4.7 KCl, 14.9 $NaHCO_3$, 1.18 KH_2PO_4, 5.5 glucose, 1.17 $MgSO_4$. $7H_2O$, 1.6 $CaCl_2.2H_2O$, 0.03 $CaNa_2EDTA$. Indomethacin (2×10^{-5} M) dissolved in dimethyl sulphoxide was added to inhibit prostanoid synthesis.

Inhibitors of Nitric Oxide Synthase and Potassium Channels and Relaxations to Carbachol and Bradykinin

Cumulative concentration-response curves to phenylephrine ($10^{-8} - 3 \times 10^{-5}$ M) were obtained. Organ chambers were washed out and tissues allowed to relax. Rings were constricted to their individual EC_{50} values for phenylephrine and relaxation to carbachol ($10^{-8} - 10^{-5}$ M) or bradykinin ($10^{-10} - 3 \times 10^{-7}$ M) studied. The organ chambers were again washed out, inhibitors of nitric oxide synthase or potassium channel blockers added. Twenty minutes later the rings were constricted again to there EC_{50} for phenylephrine. If necessary, the dose of phenylephrine was adjusted so that the tone was comparable to that achieved previously. A second concentration-response curve to carbachol or bradykinin was then obtained. Some rings, which acted as controls, had neither nitric oxide synthase nor potassium channel blockers added before the second concentration-response curve.

The nitric oxide synthase inhibitors used were N^{ω}-nitro-L-arginine methyl ester 2×10^{-4} M, 5×10^{-4} M and 2×10^{-3} M and N^G monomethyl-L-arginine – 10^{-3} M.

The following potassium channel blockers were used: tetraethylammonium (3×10^{-3} M) which is a relatively non-specific blocker of calcium activated potassium channels. Charybdotoxin (5×10^{-8} M) and iberiotoxin (10^{-7} M) which preferentially block large conductance K_{Ca} channels (BK_{Ca}), apamin (5×10^{-7} M) which blocks small conductance channels (SK_{Ca}) and glibenclamide (10^{-5} M) which blocks ATP regulated channels.

Inhibitors of Nitric Oxide Synthase and Potassium Channels and Response to Phenylephrine

Relaxations to carbachol were studied to check the integrity of the endothelium. Blood vessels were then constricted to their EC_{20} to phenylephrine. Either N^{ω}-nitro-L-arginine methyl ester (2×10^{-4} M) or charybdotoxin (5×10^{-8} M) or a combination of the two were then added and any change in tension recorded. After approximately 20 minutes, by which time tension had stabilised, the organ chambers were washed out and contraction to 10^{-1} M potassium examined.

Histological Examination

Internal thoracic artery, radial artery and saphenous vein segments from patients were fixed in 10% phosphate buffered formalin followed by dehydration and embedding in paraffin wax. 4 μM thick sections were stained using haematoxylin and eosin. These sections were assessed for wall structure, integrity, and endothelial continuity.

Materials

N^{ω} mono methyl-L-arginine was supplied by Calbiochem Ltd, Nottingham, UK and glibenclamide by Alexis Corporation, Nottingham, UK. All other chemicals were obtained from Sigma-Aldrich Company, Poole, Dorset, UK.

Statistical Analysis

The maximal relaxation to carbachol or bradykinin (E_{max}) and the concentration of agonist required to produce 50% relaxation (EC_{50}) were calculated for each individual ring using Microsoft Excel (R). Groups contained 5 to 12 rings. Individual groups contained no more than two rings from any one patient.

To allow for differences in the initial relaxation-curves in rings from different subjects statistical analysis was only undertaken when it was possible to compare pre- and post-treatment values in the same rings. For comparison of one treatment with the control pre-treatment values Student's t-test for paired observations was used. All results are expressed as mean $\pm$ SE, when P was less than 0.05 differences were considered to be statistically significant.

RESULTS

Histology

One of more risk factors for atherosclerotic disease were present in most patients from whom blood vessels were obtained. The average plasma cholesterol was 6.4 ± 0.4 mM, 13% had diabetics 26% hypertension and 26% were current smokers. Despite this overt atheroma was not observed in any of the blood vessels taken for histological examination.

In the blood vessels studied the relaxation to carbachol was dependent on the presence of the endothelial layer. Radial and internal thoracic arteries were muscular arteries with average internal diameters of 2.1 and 1.9 mm respectively. The average internal diameter of the saphenous vein was 2.4 mm but the range was wider than for the arteries (1.5 to 4.6 mm).

Effect of Vehicle Treatment on Relaxations to Carbachol and Bradykinin

In none of the blood vessels studied was relaxation to carbachol or bradykinin affected by vehicle treatment. In internal thoracic artery, saphenous vein and radial artery maximal relaxations to carbachol were 62 ± 7, 38 ± 7 and $87 \pm 9\%$ before and 63 ± 7, 36 ± 4 and $86 \pm 8\%$ after-treatment, respectively. The corresponding values for bradykinin were 29 ± 6, 71 ± 4 and $53 \pm 8\%$ before and 30 ± 6, 73 ± 6 and $51 \pm 7\%$ after treatment.

Inhibitors of Nitric Oxide Synthase and Raised Extracellular Potassium and Relaxation to Carbachol

In the internal thoracic artery, the relaxation to carbachol was reduced from $59 \pm 12\%$ to $10 \pm 5\%$ and that to bradykinin from $26 \pm 3\%$ to $9 \pm 1\%$ in the presence of N^{ω}-nitro-L-arginine methyl ester (2×10^{-4} M). In the saphenous vein, the relaxation to carbachol and bradykinin were reduced from $26 \pm 3\%$ to $10 \pm 4\%$ and from $57 \pm 7\%$ to $34 \pm 4\%$, respectively. In the internal thoracic artery, increasing the dose of N^{ω}-nitro-L-arginine methyl ester to 5×10^{-4} M had no additional effect whereas in the saphenous vein the higher dose of N^{ω}-nitro-L-arginine methyl ester caused an additional attenuation of relaxation from $31 \pm 5\%$ to $3 \pm 4\%$ for carbachol, and from $50 \pm 7\%$ to $18 \pm 3\%$ for bradykinin. In contrast in the radial artery, over 50% of relaxation to both agonists was resistant to nitric oxide synthase blockade. In the radial artery, in the presence of N^{ω}-nitro-L-arginine methyl ester (2×10^{-4} M) the relaxation to carbachol was reduced from 97 ± 8 to $74 \pm 7\%$ and that to bradykinin from $52 \pm 3\%$ to $33 \pm 1\%$. Increasing the concentration of N^{ω}-nitro-L-arginine methyl ester had no further effect on the relaxation, and it was not possible to abolish the response even in the presence of supermaximal concentrations of either N^{ω}-nitro-L-arginine methyl ester or N^{ω} monomethyl-L-arginine (2×10^{-3} M or greater) in this blood vessel. In the radial artery, relaxations to carbachol and bradykinin were also attenuated in the presence of raised extracellular potassium (Figure 42-1). Relaxation to carbachol was reduced from $101 \pm 10\%$ to $72 \pm 8\%$ and relaxation to bradykinin was reduced from $53 \pm 3\%$ to $33 \pm 4\%$. The combination of raised potassium and N^{ω}-nitro-L-arginine methyl ester (2×10^{-4} M) had an additive inhibitory effect on the relaxation. In the presence of N^{ω}-nitro-L-arginine methyl ester plus 2×10^{-2} M potassium, the relaxation to carbachol

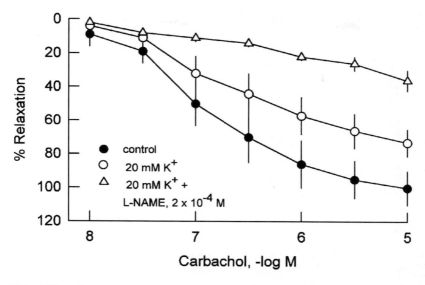

Figure 42-1. Effect of raised extracellular potassium (2×10^{-2} M) on relaxation to carbachol in radial artery.

Closed circles	Control response
Open circles	Response in the presence of 2×10^{-2} M potassium
Open triangles	Response in the presence of 2×10^{-2} M potassium and 2×10^{-4} M N^{ω}-nitro-L-arginine methyl ester

Results are expressed as mean ± SE, n = 7.

The maximal relaxation to carbachol was significantly attenuated in the presence of 2×10^{-2} M potassium ($p < 0.05$). A further significant fall was observed in the presence of 2×10^{-2} M potassium + 2×10^{-4} M N^{ω}-nitro-L-arginine methyl ester ($p < 0.05$). EC_{50} values were not significantly altered.

was reduced to 35 ± 6% and that to bradykinin to 16 ± 3%. In the internal thoracic artery, despite the abolition of relaxation by the inhibitors nitric oxide synthase, increasing the concentration of extracellular potassium to 20 mM also attenuated relaxation to carbachol and bradykinin (Figure 42-2). The relaxation to carbachol was reduced from 64 ± 6% to 36 ± 6% and to bradykinin from 24 ± 6% to 11 ± 2%. In contrast, in the saphenous vein, these agonist-mediated relaxations were not modified significantly by increasing the extracellular potassium concentration, the relaxation to carbachol and bradykinin being 31 ± 6% and 46 ± 6% in normal potassium and 29 ± 5% and 47 ± 8% in the presence of raised extracellular potassium, respectively.

Potassium Channel Blockers and Relaxations to Carbachol in the Internal Thoracic Artery and the Radial Artery

In the internal thoracic artery, tetraethylammonium, iberiotoxin and charybdotoxin all caused significant attenuation of relaxation to carbachol (Table 42-1). However, the EC_{50} value to carbachol was not significantly altered. In the internal thoracic artery, neither apamin nor glibenclamide caused decreases in the relaxation to carbachol.

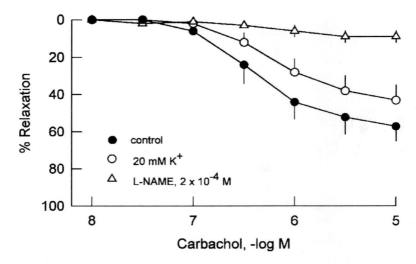

Figure 42-2.   Effect of non-nitro-L-arginine methyl ester and raised extracellular potassium on relaxation to carbachol in internal thoracic artery.

Closed circles   Control response

Open circles   Response in the presence of 2×10^{-2} M potassium

Open triangles   Response in the presence of 2×10^{-4} M N$^\omega$-nitro-L-arginine methyl ester

Results are expressed as mean $\pm$ SE n = 8.

The maximal relaxation to carbachol was significantly attenuated in the presence of 2×10^{-2} M potassium ($p < 0.05$) and also in the presence of N$^\omega$-nitro-L-arginine methyl ester 2×10^{-4} M ($p < 0.01$).

Table 42-1   Relaxation to carbachol in internal thoracic and radial arteries

| | Maximal relaxation to carbachol % | | | |
| | Radial artery | | Internal thoracic artery | |
	Pre treatment	Post treatment	Pre treatment	Post treatment
Treatment				
Vehicle	84 ± 9	87 ± 9	62 ± 7	63 ± 7
2×10^{-2} M K$^+$	101 ± 10	$73 \pm 8^*$	61 ± 5	$41 \pm 6^*$
Charybdotoxin	102 ± 9	$78 \pm 8^*$	57 ± 7	$39 \pm 3^*$
Cx-Apamin	87 ± 6	$47 \pm 11^{*\Delta}$	57 ± 6	$37 \pm 3^*$
Iberiotoxin	88 ± 9	84 ± 8	52 ± 2	$36 \pm 3^*$
TEA	92 ± 10	97 ± 5	63 ± 9	$56 \pm 8^*$
Apamin	93 ± 5	90 ± 3	51 ± 3	59 ± 3
Glibenclamide	89 ± 9	85 ± 10	41 ± 3	42 ± 5

* significantly less than pre treatment response ($p < 0.05$)

Δ significantly less relaxation in the presence of Cx plus Apamin than Cx alone ($p < 0.05$)

Cx   = charybdotoxin
TEA = tetraethylamonium

Blood vessels were constricted to their individual EC$_{50}$ values for phenylephrine and relaxation to carbachol $10^{-8} - 10^{-5}$ M examined. Maximal relaxation was calculated from fitted curves for each blood vessel.

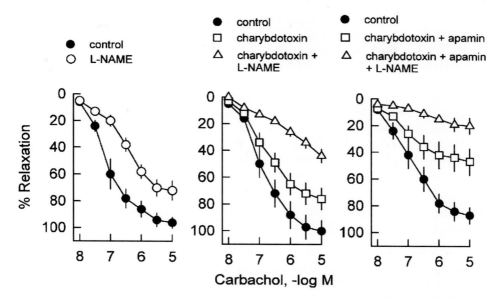

Figure 42-3. Effect of N^{ω}-nitro-L-arginine methyl ester and charybdotoxin on relaxation to carbachol in radial artery.

Left hand panel

Closed circles	Control response
Open circles	Response in presence of N^{ω}-nitro-L-arginine methyl ester 2×10^{-4} M

Middle panel

Closed circles	Control response
Open squares	Response in the presence of charybdotoxin 5×10^{-8} M
Open triangles	Response in the presence of charybdotoxin plus N^{ω}-nitro-L-arginine methyl ester

Right hand panel

Closed circles	Control response
Open circles	Response in the presence of charybdotoxin 5×10^{-8} M plus apamin 3×10^{-3} M
Open triangles	Response in the presence of charybdotoxin plus apamin plus N^{ω}-nitro-L-arginine methyl ester

Results are shown as mean $\pm$ SE n = 7

N^{ω}-nitro-L-arginine methyl ester alone and charybdotoxin alone caused a significant attenuation of the maximal vasorelaxation to carbachol. The combination of charybdotoxin plus apamin caused a significantly greater attenuation of relaxation to carbachol than charybdotoxin alone. N^{ω}-nitro-L-arginine methyl ester plus charybdotoxin and N^{ω}-nitro-L-arginine methyl ester plus charybdotoxin and apamin caused significantly greater attenuation of relaxation than charybdotoxin alone or charybdotoxin plus apamin $p < 0.05$ take as significant.

In the radial artery, the only potassium channel blocker to cause a significant attenuation of relaxation to carbachol was charybdotoxin (Figure 42-3). The attenuation of the relaxation to carbachol by charybdotoxin was significantly greater in the radial artery than the internal thoracic artery (Table 42-1). In addition, in the radial artery although apamin alone had no effect on the relaxation, when combined with charybdotoxin it caused an attenuation of the response that was significantly greater than that observed with charybdotoxin alone.

No other combination of blockers, (for example, iberiotoxin plus apamin or charybdotoxin plus glibenclamide) showed synergistic effects. The synergism between charybdotoxin plus apamin was not observed in the internal thoracic artery.

In the radial artery the effects of charybdotoxin and N^{ω}-nitro-L-arginine methyl ester appeared to be additive (Figure 42-3). N^{ω}-nitro-L-arginine methyl ester $(2 \times 10^{-4}$ M), charybdotoxin, and charybdotoxin plus N^{ω}-nitro-L-arginine caused attenuation of the relaxation by $24 \pm 6\%$, $23 \pm 3\%$ and $60 \pm 3\%$, respectively. Charybdotoxin plus apamin caused a reduction in relaxation of $49 \pm 9\%$, while N^{ω}-nitro-L-arginine methyl ester $(2 \times 10^{-4} M)$, charybdotoxin and apamin together attenuated the relaxation by $78 \pm 4\%$. When the concentration of N^{ω}-nitro-L-arginine methyl ester was increased to 5×10^{-4} M this combination of blockers attenuated the relaxation by $85 \pm 5\%$.

Inhibitors of Nitric Oxide Synthase and Potassium Channel and Response to Phenylephrine

In the internal thoracic artery, the addition of N^{ω}-nitro-L-arginine methyl ester to blood vessels constricted to the EC_{20} for phenylephrine caused a significant increase in tension of $29 \pm 10\%$. In contrast, addition of potassium channel blockers had no effect on the response to phenylephrine, nor did the combination of potassium channel blockade plus nitric oxide synthase inhibition have an effect above that produced by N^{ω}-nitro-L-arginine methyl ester alone (data not shown).

In the radial artery, neither N^{ω}-nitro-L-arginine methyl ester alone nor any of the potassium channel blockers alone had a significant effect on the response to phenylephrine. However, the combination of N^{ω}-nitro-L-arginine methyl ester plus charybdotoxin caused a significant increase in the contractile response to phenylephrine of $39 \pm 8\%$.

In the saphenous vein, potassium channel blockers and N^{ω}-nitro-L-arginine methyl ester, either alone or in combination, did not affect the response to phenylephrine.

DISCUSSION

These results suggest that activation of potassium channels contributes substantially to endothelium-dependent relaxation in human conduit arteries.

In radial arteries nitric oxide synthase inhibitors were unable to inhibit relaxation to bradykinin or carbachol by more than 40%. As these studies were carried out in the presence of indomethacin prostanoids are unlikely to contribute to the relaxation. However, relaxations were attenuated by 30–40% in the presence of raised extracellular potassium suggesting a role for potassium channel activation in mediating relaxation (Adeagbo and Triggle, 1993). The effects of inhibition of potassium channels and nitric oxide synthase were additive suggesting the presence of two independent pathways contributing to relaxation to bradykinin and carbachol in the radial artery.

In contrast, in the internal thoracic artery no significant relaxation to carbachol or bradykinin was observed in the presence of nitric oxide synthase inhibitors. Despite this, raising the extracellular potassium concentration also attenuated relaxations to these agonists. Thus in this blood vessel it would seem that nitric oxide may cause activation of potassium channels.

Endothelium-dependent activation of potassium channels and hyperpolarization of vascular smooth muscle is well documented in a range of animal blood vessels (Murphy and Brayden, 1995b; McCulloch *et al.*, 1997; Petersson, Zygmunt and Hogestatt, 1997) and has also been reported in human coronary arteries (Nakashima *et al.*, 1993; Kemp and Cocks, 1997) gastro-epiploic arteries (Urakami-Harasawa *et al.*, 1997), omental arteries (Wallerstedt and Bodelsson, 1997) and pial arteries (Petersson, Zygmunt and Hogestatt, 1997). In the majority of these studies only small resistance arteries were examined and it has been suggested that endothelium-dependent hyperpolarization is more important in mediating relaxation in such blood vessels whereas the L-arginine nitric oxide pathway predominates in large conduit vessels (Nagao and Vanhoutte, 1992b; Garland and Plane, 1996; Urakami-Harasawa *et al.*, 1997). However, the radial artery and the internal thoracic artery are both conduit vessels of similar diameter yet show very different dependencies on nitric oxide for endothelium-dependent relaxation.

Nitric oxide dependent activation of potassium channels such as that observed in this study in internal thoracic artery has been documented in animal blood vessels (Koh *et al.*, 1995; Archer *et al.*, 1994; Bolatina *et al.*, 1994; Tare *et al.*, 1990). Moreover, in the rabbit carotid artery, nitric oxide can fully account for relaxation to acetylcholine despite the fact that resistance to nitric oxide synthase inhibitors is observed (Cohen *et al.*, 1997). However, nitric oxide-dependent activation of potassium channels has not previously been demonstrated in human vessels.

In the radial artery the prostanoid/nitric oxide-independent relaxations were attenuated in the presence of charybdotoxin. A greater inhibition was observed using the combination of charybdotoxin plus apamin despite apamin itself having no effect on relaxation. Charybdotoxin plus apamin has also been shown to be particularly effective in attenuating endothelium-dependent relaxations and hyperpolarization of vascular smooth muscle cells in the guinea pig carotid artery (Corriu *et al.*, 1996b) and the rat hepatic artery (Zygmunt *et al.*, 1997a).

In the internal thoracic artery, the relaxation to carbachol was attenuated in the presence of tetraethylammonium, iberiotoxin and charybdotoxin. The combination of charybdotoxin plus apamin was no more effective than charybdotoxin alone. This would suggest that in this blood vessel activation of large conductance calcium-dependent potassium channels contributes to relaxation. Thus, the potassium channel(s) involved in carbachol-mediated relaxation differs between the two vessels. Whether this is because different potassium channels are present in radial artery and internal thoracic artery, or whether it is related to the mechanism of activation remains to be elucidated. In the radial artery, the activation of the potassium channel is most likely mediated by an endothelium-derived hyperpolarizing factor whereas in the internal thoracic artery potassium channels may be activated directly by nitric oxide or possibly by cyclic GMP.

In saphenous veins, no attenuation of relaxation to either carbachol or bradykinin was observed when the extracellular potassium concentration was increased to 2×10^{-2} M. Similarly, potassium channel blockers failed to attenuate responses to carbachol and bradykinin in this blood vessel (data not shown). Thus, hyperpolarization appears not to contribute to endothelium-dependent relaxation in this vein. Whether hyperpolarization contributes to relaxation in other venous tissues or whether relaxation dependent on hyperpolarization is restricted to arteries in humans remain to be seen. The role of potassium channel activation and hyperpolarization has been less intensively studied in animal veins than arteries. However, endothelium-dependent relaxation resistant to blockade by N$^\omega$-nitro-L-arginine have been demonstrated in pulmonary veins from sheep with short to medium term hypertension (Kemp *et al.*, 1995).

In vivo the basal release of nitric oxide may be of greater importance than that stimulated by pharmacological concentrations of agonists such as bradykinin and acetylcholine in determining resting tone (Vallance, Collier and Moncada, 1989). Basal release of nitric oxide is higher in arteries than veins (Vallance *et al.*, 1995; Hamilton *et al.*, 1997). This may contribute to the improved patency and survival of arterial versus venous grafts in revascularisation surgery (Lüscher *et al.*, 1988). In the internal thoracic artery, but not the saphenous vein, the addition of N$^\omega$-nitro-L-arginine methyl ester to inhibit nitric oxide synthase causes an increase in tension during the response to phenylephrine (Hamilton *et al.*, 1997). In the radial artery, the addition of N$^\omega$-nitro-L-arginine methyl ester had no effect on the response to phenylephrine but increases in tension were observed in the presence of N$^\omega$-nitro-L-arginine methyl ester plus charybdotoxin although charybdotoxin alone had no effect. The effect of inhibitors of nitric oxide synthase on the pressor response to phenylephrine has previously been used as an indirect measure of the contribution of nitric oxide to the regulation of basal tone (Rees, Ben-Ishay and Moncada, 1996; McIntyre *et al.*, 1997). Thus the present results suggest that, in the radial artery, both nitric oxide and potassium channel activation are involved in the regulation of basal tone.

All the blood vessels used in these studies were taken from patients undergoing coronary artery bypass graft surgery. Most of these patients had risk factors for atherosclerotic disease and were receiving multiple drug therapy. The results obtained in such patients may not be directly comparable to those which would be observed in those blood vessels in healthy younger people. Indeed the presence of multiple endothelium-dependent relaxing factors may have important consequences particularly under pathophysiological conditions. Impaired production of one factor or functioning of one pathway may not greatly affect the overall ability of a vessel to relax. Alternative mechanisms for relaxation may help maintain normal vascular tone in disease. Alterations in the contribution of hyperpolarization to relaxation have been proposed in arteries and veins from sheep with pulmonary hypertension (Kemp *et al.*, 1995) mesenteric arteries from streptozotocin-induced diabetic rats (Fukao *et al.*, 1997a) mesenteric artery of spontaneously hypertensive rats (Fujii *et al.*, 1992) and aortas from cholesterol fed rabbits (Cohen, Najibi and Bolotina, 1996). Thus knowledge of the factors regulating endothelium-dependent vasodilatation in individual blood vessels may be important in optimising treatment of cardiovascular disease.

43 Vasodilatation to Bradykinin is Mediated by an Endothelium-derived Hyperpolarizing Factor in Essential Hypertensive Patients

Stefano Taddei, Lorenzo Ghiadoni, Agostino Virdis and Antonio Salvetti

Department of Internal Medicine, University of Pisa, Italy
Address for correspondence: Dr. Stefano Taddei, Department of Internal Medicine, University of Pisa, Via Roma, 67, 56100 Pisa, Italy. Tel: +39-50-551110; Fax: +39-50-553407; E-mail: s.taddei@int.med.unipi.it

The present study was designed to evaluate whether vasodilatation to bradykinin in humans could be caused by an endothelium-dependent hyperpolarizing factor (EDHF) acting through $Na^+-K^+/ATPase$. In eight normotensive subjects and sixteen hypertensive patients modifications in forearm blood flow (strain-gauge plethysmography) during intrabrachial infusion of bradykinin in the presence of saline, N^{ω}-monomethyl-L-arginine (to inhibit NO-synthase) and ouabain (to inhibit $Na^+-K^+/ATPase$). The response to sodium nitroprusside, an endothelium-independent agonist, was also evaluated. In normotensive subjects, vasodilatation to bradykinin was significantly blunted by N^{ω}-monomethyl-L-arginine and unaffected by ouabain. In essential hypertensive patients, the vasodilatation to bradykinin was reduced as compared to healthy subjects. It was not modified by N^{ω}-monomethyl-L-arginine, but significantly reduced by ouabain. Responses to sodium nitroprusside were not different in essential hypertensive patients compared to normotensive controls. In hypertensive patients the study was repeated six hours after oral administration of the converting enzyme inhibitor lisinopril, able to reduce the breakdown of bradykinin. Lisinopril significantly increased the vasodilatation to bradykinin, an effect which was not altered by N^{ω}-monomethyl-L-arginine, but was abolished by ouabain. These experiments suggest that the vasodilatation to bradykinin is impaired in essential hypertensive patients because of an alteration in the L-arginine-NO pathway. Moreover in essential hypertension, bradykinin-induced vasodilatation seems to be mediated by an alternative pathway involving the production of EDHF.

KEYWORDS: nitric oxide, hyperpolarization, bradykinin, ouabain, forearm blood flow, essential hypertension

INTRODUCTION

After the original report demonstrating that the endothelium releases a vasodilator substance in response to acetylcholine (Furchgott and Zawadzki, 1980), it became clear that various different relaxing and contracting factors contribute in endothelium-dependent responses (Lüscher and Vanhoutte, 1990). The most widely known and probably the most important endothelium-derived relaxing factor is nitric oxide (NO), which is released from endothelial cells in response to shear stress or stimulation of different receptors on the endothelial cell surface. These stimuli increase the activity of a constitutively expressed enzyme, NO synthase, which converts the amino acid L-arginine into NO and citrulline (Moncada *et al.*, 1991). However, this substance does not universally explain endothelium-dependent relaxations. Thus in many studies of isolated blood vessels and the intact circulation, the action of endothelial vasodilators is, at least in part, resistant to NO-inhibitors (Cowan and Cohen, 1991; Mügge *et al.*, 1991b). An alternative mechanism is represented by an endothelial factor that causes hyperpolarization of smooth muscle cells (Cohen and Vanhoutte, 1995). Although the identity of this endothelium-derived hyperpolarizing factor (EDHF) is unknown, its hyperpolarizing activity on smooth muscle cells has been attributed to an increase in conductance to potassium ions (Standen *et al.*, 1989). Nevertheless, other potential explanations exist that could contribute to an understanding of endothelium-dependent hyperpolarization, including activation of the $Na^+K^+/$ATPase or inactivation of chloride channels (Félétou and Vanhoutte, 1988).

In the human circulation *in vivo*, NO plays a major role in modulating vascular tone. In the normal state, several agonists including acetylcholine, bradykinin or substance P cause vasodilatation, when directly injected into the brachial or coronary circulation to avoid reflexogenic hemodynamic modifications induced by systemic effects (Taddei *et al.*, 1998). This vasodilatation is NO- and therefore endothelium-dependent since it can be inhibited by specific NO synthase inhibitors, such as the N^{ω}-monomethyl-L-arginine (Rees *et al.*, 1989b). However several pathological conditions including essential hypertension, diabetes and hypercholesterolemia, are characterized by endothelial dysfunction. In essential hypertensive patients the response to endothelium-dependent agonists, mainly acetylcholine or bradykinin, is blunted in different vascular districts as compared to healthy controls (Taddei *et al.*, 1998). This diminished relaxation to acetylcholine or bradykinin, although widely employed to explore the presence of endothelial dysfunction in hypertension, does not truly identify the mechanism of the pathology. Furthermore, at variance with healthy subjects, in essential hypertensive patients the vasodilator effect of acetylcholine or bradykinin is still present and resistant to N^{ω}-monomethyl-L-arginine, despite its reduction (Panza *et al.*, 1995; Taddei *et al.*, 1998b), indicating that while NO availability is compromised some other mediator(s) account(s) for the vascular relaxing effects of these compounds.

The aim of the present study was to explore the mechanisms responsible for the relaxing response to bradykinin in the peripheral circulation of essential hypertensive patients. Moreover, since converting enzyme inhibitors reduce the degradation of bradykinin and thus increase its vascular effect (Vanhoutte, 1989), this study also evaluated which mediator is responsible for the increased effectiveness of bradykinin when the enzyme is inhibited.

METHODS

Patients

Eight healthy subjects and 16 matched essential hypertensive patients participated in the study (Table 43-1). Subjects with hypercholesterolemia (total cholesterol greater than 5.2 mmol/L), diabetes mellitus, cardiac and/or cerebrovascular ischaemic vascular disease, impaired renal function and other major pathologies were excluded. Subjects or patients smoking more than five cigarettes per day and/or consuming more than 60 g of ethanol (corresponding to half a litre of wine) per day were likewise excluded from the study. In accordance with institutional guidelines, all patients were aware of the investigational nature of the study and gave written consent to it.

Table 43-1 Characteristics of study subjects (data are mean ± S.D.)

Parameter	Normotensive Subjects (n = 8)	Essential Hypertensive Patients (n = 16)
Age (years)	44.3 ± 4.4	45.1 ± 4.1
Sex (male/female)	6/2	11/5
Smoking (yes/no)	No	No
Body Mass Index (Kg/m^2)	22.1 ± 2.4	21.7 ± 2.9
Systolic Blood Pressure (mmHg)	124.3 ± 4.7	157.3 ± 6.1*
Diastolic Blood Pressure (mmHg)	81.2 ± 3.1	101.6 ± 4.1*
Heart Rate (beats/min)	70.8 ± 5.8	71.4 ± 6.2
Plasma Glucose (mg/dL)	83.2 ± 6.1	85.9 ± 5.9
Plasma Total Cholesterol (mg/dL)	186.4 ± 11.2	189.7 ± 10.4
Plasma HDL Cholesterol (mg/dL)	43.8 ± 3.8	41.9 ± 5.3
Plasma LDL Cholesterol (mg/dL)	108.5 ± 7.5	110.4 ± 8.6

HDL: High Density Lipoprotein; LDL: Low Density Lipoprotein
The asterisks indicate statistical, significant ($P < 0.05$) difference between the two groups.

Subjects were defined as normal according to the absence of familial history of essential hypertension and blood pressure values below 140/90 mmHg (Table 43-1). Essential hypertensive patients were recruited from among newly diagnosed cases when they reported the presence of positive family history of essential hypertension and if supine arterial blood pressure (after 10 min of rest), measured by mercury sphygmomanometer three times at one week intervals, was consistently found to be greater than 140/90 mmHg (Table 43-1). Secondary forms of hypertension were excluded by routine diagnostic procedures. Patients were enrolled if never-treated (n = 12) or reporting a history of discontinued or ineffective pharmacological antihypertensive treatment (n = 4).

Experimental Model

Forearm blood flow studies were performed at 08.00 h after overnight fasting, with the subjects lying supine in a quiet air-conditioned room (22–24°C). A polyethylene cannula (21 gauge, Abbot, Sligo, Ireland) was inserted into the brachial artery under local anaesthesia (2% lidocaine). The cannula was connected through stopcocks to a pressure transducer (model MS20, Electromedics, Englewood, Colo.) for the determination of systemic mean arterial blood pressure (1/3 pulse pressure plus diastolic pressure), heart rate (model VSM1, Physiocontrol, Redmond, Wash.) and intra-arterial infusions. Forearm blood flow was measured by strain-gauge venous plethysmography (LOOSCO, GL LOOS, Amsterdam, The Netherlands) (Whitney, 1953). Circulation to the hand was occluded one minute before each measurement of forearm blood flow by inflating a paediatric cuff around the wrist at suprasystolic blood pressure. Forearm volume was determined by the water-displacement method, and the drug infusion rate was adjusted for each subject according to his or her forearm volume. Drug infusion rates were normalized to 100 ml forearm tissue by alteration of the drug concentration in the solvent. Drugs used were infused through three-way stopcocks at concentrations that had no systemic effects.

Study Design

While normotensive subjects were evaluated as outpatients, essential hypertensive patients were evaluated during hospitalization. In ten normotensive subjects and ten essential hypertensive patients, endothelium-dependent forearm vasodilatation was evaluated by obtaining a dose-response curve to intraarterial bradykinin (cumulative increase in infusion rates: 5, 15, 50 ng/100 ml of forearm tissue min for 5 min each dose). Endothelium-independent vasodilatation was assessed with sodium nitroprusside (1, 2 and 4 μg/100 ml of forearm tissue/min for 5 min each dose), that acts directly on the smooth muscle cells (Schultz and Schultz, 1977).

To identify the mediator responsible for the bradykinin-induced vasodilatation, the dose-response curve to the peptide was repeated in the presence of intra-arterial N^{ω}-monomethyl-L-arginine (100 g/100 ml forearm tissue/min, an inhibitor of NO-synthase), and in the presence of intraarterial ouabain (0.7 μg/100 ml forearm tissue/min, to inhibit $Na^{+}K^{+}$/ATPase) (Tonomura, 1986). In essential hypertensive patients, the response to sodium nitroprusside was also repeated in the presence of ouabain.

Finally, in eight further essential hypertensive patients, bradykinin and sodium nitroprusside infusions were performed both under basal conditions (saline infusion at 0.2 ml min) and after acute administration of lisinopril 20 mg (at the peak drug effect, 6 to 8 hours after dosing). To avoid two arterial punctures in the same day, the cannula was kept patent by infusion of heparinized saline (6 (l/min) through a portable minipump (Microject Bolus 2, Miles, Cavenago, Milano, Italy). Both under basal conditions and after lisinopril the administration of bradykinin was repeated during intrabrachial N^{ω}-monomethyl-L-arginine or ouabain.

To evaluate the effectiveness of lisinopril, blood pressure values and heart rate were measured, as well as plasma renin activity [ng angiotensin (ang) I/ml/hr] by radioimmuno-assay (Malvano *et al.*, 1972) and serum converting enzyme activity (nmol/min/ml) by a radioenzymatic method (Lieberman, 1975).

Data Analysis

Data were analyzed in terms of changes in forearm blood flow. Because arterial blood pressure did not change significantly during the forearm blood flow study, increments in forearm blood flow were taken as evidence of local vasodilatation. Results are expressed as mean $\pm$ SD, while in the figures results are described as mean $\pm$ SEM. Differences between two means were compared by the Student's t test for paired or unpaired observations, as appropriate. Responses to bradykinin and sodium nitroprusside were analyzed by ANOVA for repeated measures and Scheffè's test was applied for multiple comparison testing. Differences were considered to be statistically significant when P was less than 0.05.

Drugs

Bradykinin HCl (Clinalfa AG, Läufelfingen, Switzerland), N^{ω}-monomethyl-L-arginine (Clinalfa AG, Läufelfingen, Switzerland), ouabain (Ouabaine Arnaud) and sodium-nitroprusside (Malesci, Milan, Italy) were obtained from commercially available sources

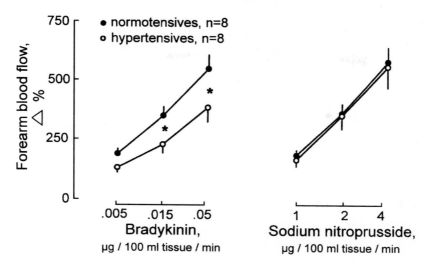

Figure 43-1. Line graphs show forearm blood flow (FBF) increase above basal induced by bradykinin and sodium nitroprusside infused at cumulative increasing rates into the brachial artery of normotensive subjects (●) and essential hypertensive patients (○). Data are shown as means ± SEM (n = 8) and expressed as percent increase above basal. Asterisks denote significant differences (P < 0.05) between infusion in normotensive subjects and essential hypertensive patients. The figure shows that the response to bradykinin, but not to sodium nitroprusside, is blunted in essential hypertensive patients.

and diluted freshly to the desired concentration by adding normal saline. Sodium nitroprusside was dissolved in glucosate solution and protected from light by aluminum foil.

RESULTS

Age, sex, plasma cholesterol, glycemia and smoking history were similar, and within a normal range, between the two study groups, which differed only in blood pressure values (Table 43-1).

The dose-dependent increase in forearm blood flow induced by bradykinin was significantly reduced in essential hypertensive patients (from 3.5 ± 0.3 to a maximum of 17.4 ± 4.6 ml/100 ml/min) as compared to normotensive subjects (from 3.5 ± 0.3 to a maximum of 22.9 ± 4.9 ml/100 ml/min) (Figure 43-1). In contrast, dose-dependent vasodilatation to sodium nitroprusside was similar in normotensive subjects (from 3.4 ± 0.3 to a maximum of 21.4 ± 1.8 ml/100 ml/min) and hypertensive patients (from 3.5 ± 0.3 to a maximum of 20.9 ± 3.9 ml/100 ml/min) (Figure 43-1).

In normotensive subjects, the infusion of N^{ω}-monomethyl-L-arginine, which caused a decrement in basal forearm blood flow (from 3.5 ± 0.4 to 2.0 ± 0.2 ml/100 ml forearm tissue/min), significantly blunted the vasodilator effect of bradykinin (saline: from 3.5 ± 0.5 to 22.9 ± 3.7 ml/100 ml forearm tissue/min; N^{ω}-monomethyl-L-arginine: from 2.0 ± 0.2 to 8.0 ± 1.8 ml/100 ml forearm tissue/min) (Figure 43-2) while ouabain, which caused a decrement in basal forearm blood flow (from 3.5 ± 0.4 to 2.4 ± 0.2 ml/100 ml forearm

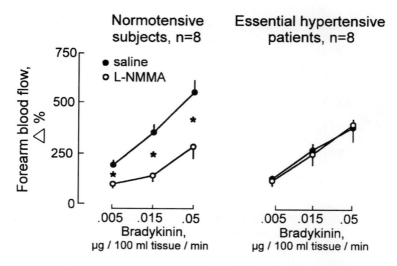

Figure 43-2. Bradykinin-induced increase in forearm blood flow (FBF) under control conditions (saline at 0.2 ml/min) and in the presence of N$^{\omega}$-monomethyl-L-arginine (L-NMMA, 100 μg/100 ml forearm tissue/min) in normotensive subjects and essential hypertensive patients. Data are shown as means ± SEM and, since N$^{\omega}$-monomethyl-L-arginine modifies basal forearm blood flow, expressed as per cent increase above basal. Asterisks denote a significant difference between infusion with and without L-NMMA (P < 0.05). N$^{\omega}$-monomethyl-L-arginine blunts the response to bradykinin in normotensive subjects, but not in essential hypertensive patients, suggesting the presence of impaired availability of NO in the latter.

tissue/min), did not change the response to the agonist (saline: from 3.5 ± 0.5 to 22.9 ± 3.7 ml/100 ml forearm tissue/min; ouabain: from 2.4 ± 0.2 to 13.8 ± 1.8 ml/100 ml forearm tissue/min) (Figure 43-3). In essential hypertensive patients, the infusion of N$^{\omega}$-monomethyl-L-arginine, which caused a lesser decrease in basal forearm blood flow (from 3.5 ± 0.5 to 2.7 ± 0.2 ml/100 ml forearm tissue/min) as compared to normotensive controls (per cent forearm blood flow decrease: 42% vs 23%, respectively), did not change the response to bradykinin (saline: from 3.5 ± 0.5 to 17.4 ± 2.3 ml/100 ml forearm tissue/min; N$^{\omega}$-monomethyl-L-arginine: from 2.7 ± 0.2 to 13.5 ± 1.9 ml/100 ml forearm tissue/min) (Figure 43-2). In contrast, ouabain, which caused a decrease in forearm blood flow comparable to that induced by N$^{\omega}$-monomethyl-L-arginine (from 3.5 ± 0.4 to 2.6 ± 0.2 ml/100 ml forearm tissue/min), significantly blunted the response to bradykinin (from 2.6 ± 0.3 to 9.1 ± 1.6 ml/100 ml forearm tissue/min) (Figure 43-3). Ouabain did not alter significantly the response to sodium nitroprusside (data not shown).

In the hypertensive patients, an acute (6 to 8 hours post dosing) administration of 20 mg lisinopril significantly inhibited serum converting enzyme, increased plasma renin activity and reduced blood pressure (Table 43-2). Lisinopril caused a significant increase in response to bradykinin (saline: from 3.5 ± 0.5 to a max of 16.9 ± 4.1 ml/100 ml/min; lisinopril: from 3.4 ± 0.5 to a max of 22.4 ± 4.9 ml/100 ml/min) (Figure 43-4), while the vasodilatation to sodium nitroprusside was unaffected by lisinopril (saline: from 3.5 ± 0.4 to a maximum of 21.1 ± 4.8 ml/100 ml/min; lisinopril: from 3.4 ± 0.3 to a maximum of

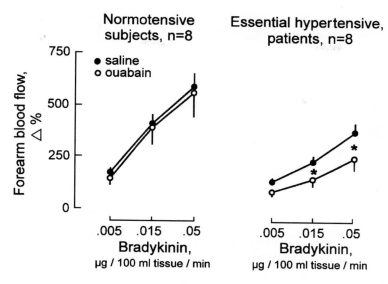

Figure 43-3.   Bradykinin-induced increase in forearm blood flow (forearm blood flow) under control conditions (saline at 0.2 ml/min) and in the presence of ouabain (0.7 μg/100 ml forearm tissue/min) in normotensive subjects and essential hypertensive patients. Data are shown as means ± SEM and, since ouabain modifies basal forearm blood flow, expressed as per cent increase above basal. Asterisks denote a significant difference between infusion with and without ouabain (P < 0.05). Ouabain blunts the response to bradykinin in essential hypertensive patients, but not in healthy controls, suggesting the production of a hyperpolarizing factor as the mechanism responsible for the vasodilatation to bradykinin in the former.

Table 43-2   Hemodynamic and humoral characteristics of essential hypertensive patients (n = 10) before, and after lisinopril 20 mg p.o. single administration (data are mean ± S.D.)

Parameter	basal	Lisinopril
Systolic Blood Pressure (mmHg)	152.6 ± 6.1	133.2 ± 5.1*
Diastolic Blood Pressure (mmHg)	101.3 ± 3.6	88.1 ± 3.2*
Heart Rate (beats/minute)	72.6 ± 4.3	73.4 ± 4.6*
PRA (ng Ang I/ml/hr)	0.84 ± 0.8	1.98 ± 1.3*
ACE (nmol/ml/min)	74.3 ± 26.4	5.9 ± 3.6*

PRA: Plasma Renin Activity; ACE: Angiotensin Converting Enzyme.
The asterisks indicate statistical, significant (P < 0.05) difference between the two groups.

20.8 ± 5.1 ml/100 ml/min). N^{ω}-monomethyl-L-arginine did not change the response to bradykinin either before (from 2.7 ± 0.4 to a maximum of 13.8 ± 3.1 ml/100 ml/min) or after (from 2.6 ± 0.4 to a maximum of 13.4 ± 2.9 ml/100 ml/min) the administration of lisinopril (Figure 43-4). In contrast, ouabain blunted the vasodilatation to bradykinin under basal conditions (from 2.6 ± 0.4 to a maximum of 9.2 ± 1.6 ml/100 ml/min) and completely prevented the facilitating effect of lisinopril (from 2.7 ± 0.4 to a maximum of 9.2 ± 1.5 ml/100 ml/min) (Figure 43-4).

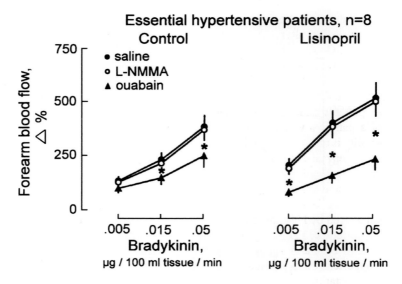

Figure 43-4. Bradykinin-induced increase in forearm blood flow (forearm blood flow) under control conditions (saline at 0.2 ml/min) and in the presence of N^{ω}-monomethyl-L-arginine (L-NMMA, 100 μg/100 ml forearm tissue/min) or ouabain (0.7 μg/100 ml forearm tissue/min) in essential hypertensive patients before and after single oral administration of lisinopril (20 mg). Data are shown as means ± SEM and, since N^{ω}-monomethyl-L-arginine and ouabain modify basal forearm blood flow, expressed as per cent increase above basal. Asterisks denote a significant difference between infusion with and without ouabain (P < 0.05). Ouabain, but not N^{ω}-monomethyl-L-arginine, blunts the response to bradykinin in essential hypertensive patients. Lisinopril increases the bradykinin-induced vasodilatation and this effect is inhibited by ouabain, further suggesting a mechanistic role for hyperpolarization in vascular control in essential hypertension.

DISCUSSION

The present study was designed to investigate the mechanisms involved in the bradykinin-mediated vasodilatation in the peripheral circulation of healthy subjects and essential hypertensive patients. In agreement with previous evidence, the vasodilator effect of bradykinin, an endothelial agonist, was blunted in hypertensive patients compared to controls. This finding is specific for bradykinin, since the vasodilatation to sodium nitroprusside, a dilator acting directly on smooth muscle cell, was similar in the two study groups. Therefore the present results confirm the presence of impaired endothelium-dependent vasodilatation to bradykinin in the peripheral circulation of essential hypertensive patients (Panza *et al.*, 1995) and further reinforce the concept of endothelial dysfunction in essential hypertension (Taddei *et al.*, 1998).

In healthy subjects the vasodilator response to bradykinin was blunted by N^{ω}-monomethyl-L-arginine, an inhibitor of NO-synthase, in confirmation of earlier observation (Panza *et al.*, 1995). This finding indicates that the relaxing activity of this agonist must be predominantly mediated by activation of the L-arginine-NO pathway. This interpretation is reinforced by the present evidence that in the normotensives studied ouabain, an inhibitor of Na^+K^+/ATPase, does not change the response to bradykinin, ruling out any effect related to hyperpolarization. In essential hypertensive patients different results were observed.

Thus the vasodilatation to bradykinin was resistant to N^ω-monomethyl-L-arginine, indicating the presence of an alteration in the L-arginine-NO pathway which leads to impaired availability of NO. On the other hand the response to bradykinin was reduced by ouabain, at a dose that did not change the vasodilator effect of sodium nitroprusside. Taken together these results indicate that while under healthy conditions endothelium-dependent vasodilatation to bradykinin seems to be mainly dependent on NO-production, in essential hypertensive patients it is not dependent on the NO-system, but rather is related to activation of a pathway which leads to hyperpolarization of the smooth muscle cells.

This possibility is reinforced by the study with the converting enzyme inhibitor lisinopril. Since converting enzyme is responsible for the breakdown of bradykinin, lisinopril must increase local concentration of bradykinin and, as a consequence, its vascular effects. In essential hypertensive patients, lisinopril significantly potentiates the vasodilator response to bradykinin (Taddei *et al.*, 1998a). The present study demonstrates that this facilitating effect was not altered by N^ω-monomethyl-L-arginine, but prevented by ouabain. Therefore in essential hypertensive patients, vasodilatation to bradykinin, even at high local concentrations, seems to be mainly dependent on hyperpolarization.

Essential hypertension is characterized by the presence of endothelial dysfunction. This conclusion is substantiated by studies performed with different agonists (acetylcholine, metacholine, bradykinin and substance P) exploring different vascular beds (peripheral cutaneous, subcutaneous and muscle microcirculation, renal circulation and coronary macro and microcirculation) (Taddei *et al.*, 1998). However most of the studies in essential hypertensives, especially in the peripheral circulation, have been performed with acetylcholine or bradykinin. Not only is the agonist-induced vasodilatation curtailed, but also it becomes resistant to N^ω-monomethyl-L-arginine (Panza *et al.*, 1993) (Taddei *et al.*, 1994). This alteration seems to be dependent on oxidative stress, since vitamin C, a scavenger of oxygen-derived free radicals, not only increases the vasodilatation to acetylcholine, but also reverses the inhibitory effect of N^ω-monomethyl-L-arginine, thereby restoring the availability of NO (Taddei *et al.*, 1998b). In the case of bradykinin, the present results suggest that in essential hypertensive patients hyperpolarization is probably a compensatory pathway leading to vasodilatation when the NO system is impaired (Figure 43-5).

This possibility is in line with experimental evidence indicating that in isolated blood vessels and the intact circulation, the action of endothelium-dependent vasodilators is, at least in part, resistant to inhibitors of NO-synthase (Cohen and Vanhoutte, 1995). Therefore the existence of a mediator different from NO has been proposed to account for endothelium-dependent relaxation. This mediator, although still unidentified, acts through hyperpolarization (Vanhoutte *et al.*, 1996). Endothelium-dependent hyperpolarization has been measured only *in vitro*. Its existence *in vivo* and its consequent role in vasodilatation has been deduced from the above mentioned failure of the inhibitors of NO-synthase to completely block the response to endothelium-dependent vasodilators. The existence of an EDHF has been demonstrated in human coronary arteries from transplant hearts (Nakashima *et al.*, 1993). In these blood vessels EDHF production was evoked by bradykinin, was also resistant to NO synthase inhibitors, and was potentiated by the converting enzyme inhibitor perindoprilat.

The present results indicate that in healthy subjects the L-arginine-NO pathway is the main mechanism causing endothelium-dependent vasodilatation to bradykinin, while other

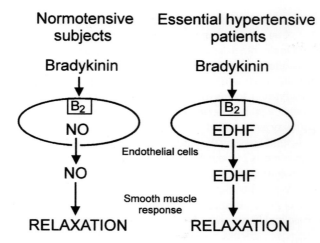

Figure 43-5. In healthy subjects, where the L-arginine-NO pathway is preserved, vasodilatation to bradykinin is mediated by NO. In essential hypertension the NO system is dysfunctional. However bradykinin still exerts a relaxing effect which is mediated by hyperpolarization. B_2: kinin B_2 receptor.

pathways do not seem to play a major role (Figure 43-5). In essential hypertension the NO-system is impaired because of an alteration in the availability of NO. Under the experimental conditions of this study, bradykinin-induced vasodilatation, although clearly impaired, was still present in essential hypertensive patients. The residual response to bradykinin was resistant to N^ω-monomethyl-L-arginine, but could be blocked by ouabain, and this effect was still present when vasodilatation to local bradykinin is potentiated by a converting enzyme inhibitor. Taken together, these results indicate that in essential hypertensive patients the production of a putative EDHF is probably a compensatory mechanism allowing endothelium-dependent vasodilatation in the presence of an impairment of the L-arginine-NO pathway (Figure 43-5).

44 Importance of Endothelium-derived Hyperpolarizing Factor in Human Arteries

Hiroaki Shimokawa, Lemmy Urakami-Harasawa, Mikio Nakashima, Hirofumi Tagawa, Yoshitaka Hirooka and Akira Takeshita

*From the Research Institute of Angiocardiology and Cardiovascular Clinic, Kyushu University School of Medicine, Fukuoka, Japan, and *Surgical Center, Saga Medical School, Saga, Japan*
Address for correspondence: Hiroaki Shimokawa, MD, PhD, Research Institute of Angiocardilogy and Cardiovascular Clinic, Kyushu University School of Medicine, 3-1-1 Maidashi, Higashi-ku, Fukuoka 812-8582, Japan. Tel: +81-92-642-5360; Fax: +81-92-642-5374; e-mail: shimo@cardiol.med.kyushu-u.ac.jp

The endothelium plays an important role in maintaining vascular homeostasis by releasing vasodilator substances, including prostacyclin, nitric oxide (NO), and endothelium-derived hyperpolarizing factor (EDHF). Although the former two substances have been investigated extensively, the importance of EDHF still remains unclear, especially in humans. *In vitro* findings in isolated human gastroepiploic arteries indicate that (a) the contribution of EDHF to endothelium-dependent relaxations is significantly larger in microvessels than in large arteries; (b) the nature of EDHF may not be a product of cytochrome P450 pathway, while EDHF-induced hyperpolarization is partially mediated by calcium-activated K channels; and (c) aging and hypercholesterolemia significantly impair EDHF-mediated relaxations. In the human forearm, the dilator responses to substance P, an endothelium-dependent vasodilating agent, are resistant to the blockade of cyclooxygenase or NO synthases, suggesting an involvement of EDHF. *In vivo* findings with substance P in the human forearm suggest that EDHF-mediated vasodilations are impaired in postmenopausal women and in patients with ischemic heart disease and that estrogen and eicosapentaenoic acid, respectively, are effective in correcting those abnormalities. Taken together, these results indicate that EDHF plays an important role in human arteries, both *in vitro* and *in vivo*.

KEYWORDS: endothelium-derived hyperpolarizing factor, endothelium-derived relaxing factor, estrogen, fish oil, human, nitric oxide

INTRODUCTION

Animal studies have shown that endothelium-dependent relaxation is achieved by combined vasodilator effects of endothelium-derived prostacyclin, nitric oxide (NO), and endothelium-derived hyperpolarizing factor (EDHF) (Lüscher and Vanhoutte, 1990; Shimokawa and Vanhoutte, 1997; Shimokawa, 1999). While the roles of the former two substances have been investigated extensively in both animals and humans, the importance of EDHF (Félétou and Vanhoutte, 1988; Suzuki and Chen, 1990; Vanhoutte, 1997) still remains unclear, especially in humans. The presence of EDHF in human arteries was first documented in coronary arteries excised from failing human hearts (Nakashima *et al.*, 1993). However, since failing hearts were used in this study, the reported results may not necessarily represent the role of EDHF in normal human arteries. In addition, microvessels were not examined in that study. Endothelium-dependent relaxations (to bradykinin and substance P) resistant to the blockade of the production of prostacyclin or NO have been observed in human omental microvessels (Pascoal and Umans, 1996; Wallerstedt and Bodelsson, 1997). However, in these studies no electrophysiological measurements were made and thus it is unknown whether or not such relaxations are associated with endothelium-dependent hyperpolarizations. This chapter briefly summarizes findings illustrating the importance of EDHF in human arteries both *in vitro* and *in vivo*.

IN VITRO FINDINGS

The study included 77 patients (40 men and 34 women; ranging from 30 to 88 yr, mean 59 yr), who underwent a gastrectomy operation. The risk factors examined included hypercholesterolemia, diabetes mellitus, hypertension, smoking, age, gender, and combinations of these risk factors (Urakami-Harasawa *et al.*, 1997). During the operation, right gasteroepiploic arteries and nets were carefully removed, and proximal and distal (100–150 μm in diameter) gastroepiploic arteries were dissected, and cut into rings used for organ chamber experiments. The membrane potentials were also recorded by microelectrode methods.

The relative contribution of prostacyclin, NO, and EDHF to endothelium-dependent relaxations was determined as follows; the contribution of prostacyclin expressed by the relaxation between the responses under control conditions and the responses in the presence of indomethacin (10^{-5} M), the contribution of NO expressed by the relaxation between the responses in the presence of indomethacin and those in the presence of indomethacin and N^{ω}-nitro-L-arginine (10^{-4} M), and the contribution of EDHF expressed by the relaxation between the responses in the presence of prostacyclin and N^{ω}-nitro-L-arginine and those in the presence of the two inhibitors and KCl (20–40 mM) (Shimokawa *et al.*, 1996; Urakami-Harasawa *et al.*, 1997). It is methodologically important to confirm that after inhibition of the production of prostacyclin and NO, EDHF-mediated responses alone were examined. This was achieved by the following points. First, the dose of N^{ω}-nitro-L-arginine used was sufficient to completely suppress the bradykinin-induced increase in the vascular levels of cyclic GMP. Second, the resting membrane potentials and hyperpolarizations were not significantly modified by indomethacin or N^{ω}-nitro-L-arginine, which indicates that endogenous prostacyclin or NO itself does not cause hyperpolarization in human gastroepiploic arteries.

Effect of Vessel Size

The extent of the relaxations to bradykinin was larger than that to acetylcholine in both sizes of human arteries (Figure 44-1). Indomethacin did not affect the bradykinin- or acetylcholine-induced relaxations in either size of human artery. In the presence of indomethacin and N^{ω}-nitro-L-arginine, bradykinin-induced relaxations were suppressed significantly in large arteries but not in microvessels. In the presence of indomethacin, N^{ω}-nitro-L-arginine, and KCl, bradykinin-induced relaxations were largely inhibited in both sizes of human arteries (Figure 44-1). As is the case with bradykinin, the combined

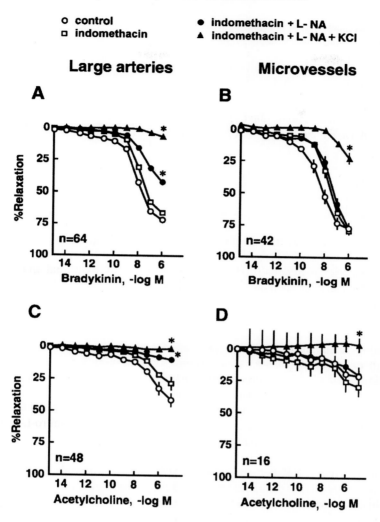

○ control
□ indomethacin
● indomethacin + L- NA
▲ indomethacin + L- NA + KCl

Large arteries

Microvessels

Figure 44-1. Endothelium-dependent relaxations to bradykinin (A and B) and acetylcholine (C and D) in human arteries. Results are presented as mean ± SEM. A and C represent the result with large arteries, and B and D show the result with microvessels.

Open circles, control; *open squares*, indomethacin (10^{-5} M); *filled circles*, indomethacin plus N$^\omega$-nitro-L-arginine (10^{-4} M); *filled triangles*, indomethacin plus N$^\omega$-nitro-L-arginine plus KCl (20–40 mM).

*Denotes $P < 0.05$ in the maximal relaxations compared with those in the control responses. The relaxations to acetylcholine were smaller than those to bradykinin in both large arteries and microvessels. N$^\omega$-nitro-L-arginine in the presence of indomethacin reduced the relaxations in large arteries but not in microvessels. (Adapted from Urakami-Harasawa *et al.*, 1997).

treatment with indomethacin and N$^\omega$-nitro-L-arginine inhibited the acetylcholine-induced relaxations in large arteries but not in microvessels. Those acetylcholine-induced relaxations in microvessels were abolished by KCl in the presence of indomethacin and N$^\omega$-nitro-L-arginine (Figure 44-1).

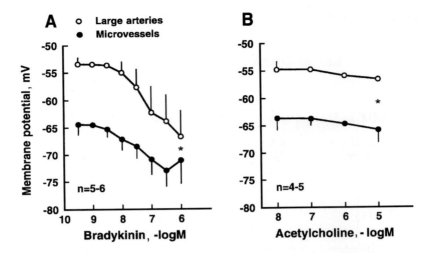

Figure 44-2. Endothelium-dependent hyperpolarizations to bradykinin (A) and to acetylcholine (B) in human arteries. Results are presented as mean ± SEM. The membrane potentials were significantly deeper in microvessels (*filled circles*) than in large arteries (*open circles*). * Denote P < 0.05 between the two whole dose-responses. (Adapted from Urakami-Harasawa *et al.*, 1997).

The resting membrane potentials in rings without endothelium were significantly more negative in microvessels than in large arteries. This was also the case in rings with endothelium, either after the treatment with indomethacin alone or with both indomethacin and N^{ω}-nitro-L-arginine. Bradykinin caused hyperpolarizations in both sizes of arteries, while the acetylcholine-induced hyperpolarizations were not evident (Figure 44-2). The membrane potentials were significantly more negative in microvessels than in arteries in all ranges of concentrations of both agonists (Figure 44-2).

These findings are consistent with those in rat mesenteric arteries in that within large arteries both NO and EDHF equally contribute, while in microvessels most of the relaxation is achieved by EDHF (Shimokawa *et al.*, 1996). The degree of hyperpolarization to acetylcholine was smaller in both sizes of human arteries, together with smaller relaxations to the muscarinic agonist. Several mechanisms explain the larger contribution of EDHF in microvessels when compared with large arteries in humans (Shimokawa *et al.*, 1996). First, microvessels may release more EDHF. Secondly, the vascular smooth muscle in microvessels may be more sensitive to EDHF. Indeed, the difference in resting membrane potentials between large arteries and microvessels was not affected by endothelium removal, or inhibition of the production of either prostacyclin or NO, which suggests that the difference in resting membrane potentials was determined primarily by the difference in smooth muscle properties. Third, EDHF-induced vascular relaxations may be determined by the combined effect of the resting membrane potentials and agonist-induced hyperpolarizations. Indeed, the difference in the membrane potentials in response to bradykinin resulted primarily from the difference in the resting membrane potentials. Finally, EDHF may diffuse more easily to the underlying vascular smooth muscle in microvessels than in large arteries, or EDHF may have too short a half life to diffuse through the smooth muscle layers of large arteries.

The larger contribution of EDHF to the endothelium-dependent relaxations in microvessels than in large arteries is noted not only in rat mesenteric arteries or human gastroepiploic arteries, but also in porcine and rabbit coronary arteries (Urakami *et al.*, 1997). Thus, the greater importance of EDHF in microvessels may be a generalized rather than species specific property of vascular beds.

Nature of EDHF in Human Arteries and Its Vasodilator Mechanism

The endothelium-dependent relaxations to bradykinin of large human gastroepiploic arteries were not suppressed by SKF 525a (an inhibitor of cytochrome P450 monooxygenase), iberiotoxin (an inhibitor of large conductance of calcium-activated K channels), or apamin (an inhibitor of small conductance calcium-activated K channels), while charybdotoxin (an inhibitor of both large and intermediate conductance calcium-activated K channels) partially reduced and tetrabutylammonium (a nonselective inhibitor of K channels) almost abolished the relaxations (Figure 44-3). Tetrabutylammonium also markedly inhibited both the resting membrane potentials and hyperpolarizations to bradykinin, while charybdotoxin significantly inhibited the hyperpolarizations to bradykinin (Figure 44-3). Other inhibitors did not show any inhibitory effect (Figure 44-3).

EDHF may be an arachidonic acid metabolite of cytochrome P450 monooxygenase at least in bovine arteries (Rosolowsky and Campbell, 1993; Hecker *et al.*, 1994; Bauersachs, Hecker and Busse, 1994; Campbell *et al.*, 1996). EDHF probably is not be a metabolite of cyclooxygenase or cytochrome P450 monooxygenase in the porcine coronary artery (Weintraub *et al.*, 1994), the guinea pig carotid artery (Corriu *et al.*, 1996a), or the rat mesenteric artery (Fukao *et al.*, 1997b). The present study demonstrated that in human gastroepiploic arteries also, EDHF is not likely to be a metabolite of cytochrome P450 monooxygenase. Thus, there may be more than one EDHF, depending on the species examined, the blood vessels tested, and the agonists used (Vanhoutte, 1997; Shimokawa, 1999).

Endothelium-dependent relaxations and hyperpolarizations in the presence of inhibitors of prostacyclin and NO synthesis are markedly suppressed by K channel inhibitors (Rosolowsky and Campbell, 1993; Hecker *et al.*, 1994; Bauersachs, Hecker and Busse, 1994; Najibi *et al.*, 1994; Murphy and Brayden, 1995b; Zygmunt, 1996; Corriu *et al.*, 1996a). However, there is no consensus regarding the subtype of K channel that mediates EDHF-induced hyperpolarization. The present study demonstrated that nonspecific blockade of K channels by tetrabutylammonium and blockade of large and intermediate conductance calcium-activated K channels by charybdotoxin exerted an inhibitory effect, while blockade of large conductance calcium-activated K channels showed no inhibitroy effect on the EDHF-induced hyperpolarizations. These results suggest that intermediate conductance calcium-activated K channels may be involved in EDHF-mediated relaxations in human gastroepiploic arteries. However, the exact subtype of calcium-activated K channel mediating EDHF-mediated relaxations in human arteries remains to be fully clarified.

Influences of Cardiovascular Risk Factors

Univariate analysis revealed that age, hypercholesterolemia, and the number of risk factors

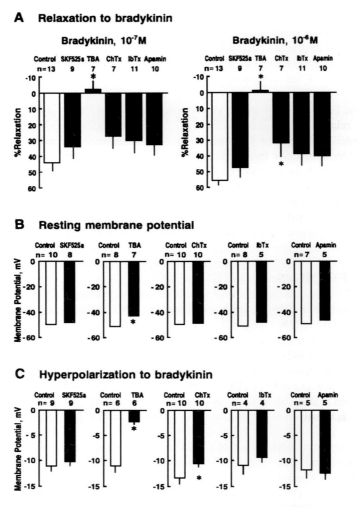

Figure 44-3. Effects of inhibitors on indomethacin/N^{ω}-nitro-L-arginine-resistant endothelium-dependent relaxations to bradykinin (*upper panel*), resting membrane potential (*middle panel*), and endothelium-dependent hyperpolarizations to bradykinin (10^{-6} M) (*lower panel*) in large human arteries. All responses were examined in the presence of 10^{-5} M indomethacin and 10^{-4} M N^{ω}-nitro-L-arginine. Results are presented as mean ± SEM. TBA, tetrabutylammonium; *ChTx*, charybdotoxin; *IbTx*, iberiotoxin. *Denotes P < 0.05 vs. control responses. Tetrabutylammonium significantly inhibited the endothelium-dependent relaxations to bradykinin, resting membrane potentials, and endothelium-dependent hyperpolarizations to the kinin, while charybdotoxin significantly inhibited the relaxations (at 10^{-6} M) and hyperporarizations in response to bradykinin. Other inhibitors showed no inhibitory effect on the relaxations or electrophysiological responses. (Adapted with permissions from Urakami-Harasawa *et al.*, 1997).

are significant predictors, while other risk factors, such as diabetes mellitus or smoking, were not significant predictors of the impairment of EDHF-mediated relaxations in large human gastroepiploic arteries (Urakami-Harasawa *et al.*, 1997). Hypertension also tended to impair the EDHF-mediated relaxations. Multivariate analysis using these significant predictors further demonstrated that age and hypercholesterolemia were again significant predictors (Urakami-Harasawa *et al.*, 1997). The effect of age was equally important in both men and women. In contrast, none of the risk factors significantly affected the EDHF-mediated responses in microvessels.

Risk factors of atherosclerosis, such as hypercholesterolemia, hypertension, diabetes mellitus, and smoking, impair NO-mediated endothelium-dependent relaxations in both humans and animals (Lüscher and Vanhoutte, 1990; Shimokawa and Vanhoutte, 1997; Shimokawa, 1999). Several animal studies also suggest that EDHF-mediated endothelium-dependent relaxations are reduced by such factors, including hypertension (Fujii *et al.*, 1992; Mantelli, Amerini and Ledda, 1995), hypercholesterolemia (Najibi *et al.*, 1994; Cowan and Steffen, 1995; Eizawa *et al.*, 1995; Najibi and Cohen, 1995), diabetes mellitus (Fukao *et al.*, 1996), and aging (Fujii *et al.*, 1993; Mantelli, Amerini and Ledda, 1995). The present study demonstrated that among these risk factors aging and hypercholesterolemia are the significant predictors for impaired EDHF-mediated relaxations in human arteries. However, it remains to be clarified why EDHF-mediated responses of human gastroepiploic microvessels were not affected by those risk factors.

IN VIVO FINDINGS

Since the nature of EDHF has not been confirmed, a selective blocker for this endothelial factor is not yet available. Thus, the exact role of EDHF *in vivo*, especially in humans, is unknown. However, a clinical study demonstrated that forearm vasodilatation in humans in response to substance P, whose vasodilator effect is totally dependent on the presence of intact endothelium *in vitro*, is resistant to the blockade of the synthesis of prostacyclin or NO (Shiramoto *et al.*, 1997). This finding in the human forearm circulation is somewhat different from that in the human coronary circulation where the vasodilator response to substance P is sensitive, to some extent, to the blockade of the synthesis of NO (Tagawa, T. *et al.*, 1997). Thus, substance P could be used as a pharmacological tool to examine the possible role and/or alterations of EDHF-mediated responses at least in the forearm circulation in humans.

Estrogen

The effect of short-term administration of conjugated estrogen was examined in the forearm circulation of postmenopausal women by strain-gauge plethysmography (Tagawa, H. *et al.*, 1997). The treatment resulted in a significant increase in serum levels of estrone and estradiol but not of estriol. The forearm blood flow significantly increased in response to both acetylcholine and substance P after intravenous administration of estrogen (Figure 44-4). However, the increase in flow in response to acetylcholine was abolished by N^{ω}-monometyl-L-arginine, while those in response to substance P were resistant to the inhibitor of NO synthases (Figure 44-4). The estrogen-induced forearm vascular responses to acetylcholine and substance P were comparable between patients who received aspirin and those who did not. The endothelium-independent forearm vasodilatation to sodium nitroprusside was comparable before and after the administration of estrogens. These results indicate that estrogens augment both NO-mediated and non NO-mediated (probably EDHF-mediated) endothelium-dependent forearm dilatation in postmenopausal women (Tagawa, H. *et al.*, 1997).

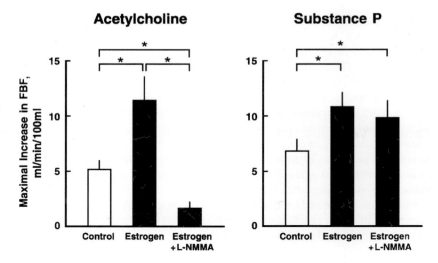

Figure 44-4.   Effect of N$^{\omega}$-monometyl-L-arginine on the forearm blood flow (FBF) responses to acetylcholine (16 μg/min) (left) and substance P (3.2 μg/min) (right) in postmenopausal women. Results are presented as mean ± SEM in six patients. *Denotes P < 0.05 between the two responses. *Control*, under control conditions; *estrogen*, after intravenous administration of conjugated estrogen (10 mg); *estrogen + N$^{\omega}$-monometyl-L-arginine*, after intravenous administration of estrogen and intra-arterial administration of N$^{\omega}$-monometyl-L-arginine (8 μM/min for 5 min). (Adapted from Tagawa, H. *et al.*, 1997).

Estrogen (especially estradiol) upregulates the expression of endothelial NO synthase in human endothelial cells (Hishikawa *et al.*, 1995) and improves endothelium-dependent dilatation in the human coronary (Reis *et al.*, 1994; Gilligan, Quyyumi, Cannon, 1994) and forearm circulation (Gilligan *et al.*, 1994). The present study demonstrates that the beneficial effect of estrogens on endothelial vasodilator functions appears to extend to NO-independent and prostacyclin-independent (presumably EDHF-dependent) mechanism.

Fish Oil

Similar findings to estrogen were obtained in patients with ischemic heart disease after a long-term treatment with eicosapentaenoic acid, a major ω-3 unsaturated fatty acid in fish oil (Tagawa, H. *et al.*, in press). Fish oil improves endothelial vasodilator functions (Shimokawa and Vanhoutte, 1988). The present data further elucidate the beneficial effect of fish oil on the endothelial vasodilator functions as such effect appears to extend to NO-independent and prostacyclin-independent (presumably EDHF-dependent) mechanism.

Taken together all these results in human arteries *in vitro* and *in vivo*, EDHF appears to play an important role, in addition to NO, in maintaining vascular homeostasis in humans as well. However, the exact physiological and pathophysiological roles of EDHF in humans *in vivo* will await the identification of this endothelial factor and the subsequent development of a selective inhibitor for its action.

45 Concluding Remarks: Identified and Unidentified Endothelium-Derived Hyperpolarizing Factors

Paul M. Vanhoutte and Michel Félétou

Inst. Recherches Internationales Servier, 6 Place des Pléiades, 92415, Courbevoie, France; and Dpt. Diabétologie, Inst. Recherches Servier, 11 rue des Moulineaux, 92150 Suresnes, France
Author for correspondence: Prof. Paul M. Vanhoutte, Inst. Recherches Internationales Servier, 6 Place des Pléiades, 92415, Courbevoie, France. Tel: 33. 1. 55.72.61.23; Fax: 33. 1. 55.72.72.76; email: vanhoutt@servier.fr

Despite the increasing number of groups involved in EDHF research and the considerable efforts which have been produced by these teams, the identity of EDHF remains elusive (Figure 45-1). The chapters of this monograph prompt, besides the conviction that EDHF exists, the following conclusions.

Measurements of relaxation of isolated blood vessels, of the membrane potential of the smooth muscle, of cyclic-GMP accumulation and endothelial nitric oxide release show that, in the rabbit carotid artery, NO is EDHF and that an incomplete blockade of the NO-synthase explains the relaxation and the repolarization observed in presence of inhibitors of cyclooxygenase and nitric oxide synthase. In this tissue, the repolarization and the relaxation produced by NO is not strictly dependent on the accumulation of cyclic-GMP and NO can directly activate large conductance calcium-activated potassium channels (BK_{Ca}). Furthermore, NO can be stored in the vascular wall, mainly in the media, in a protein bound dinitrosyl-iron complex (at least when the iNOS is induced). NO can be exchanged with low molecular weight thiols and activates soluble guanylate cyclase and/or potassium channels. These findings indicate that the role of NO originating either from *de novo* NO synthesis or from stored pools, has to be considered when studying EDHF responses. The study of endothelium-dependent responses in transgenic animals subjected to disruption of the endothelial gene for the NO-synthase were designed to address the contribution of NO in EDHF responses. Unfortunately, in the various isolated arteries studied so far taken from either endothelial nitric oxide synthase knock-out mice or wild type controls, endothelium-dependent hyperpolarizations were not observed.

However, the existence of a third endothelial pathway, besides NO-synthase and cyclooxygenase, participating in the endothelium-dependent relaxations can be demonstrated. For example, in the rat hepatic artery, NO does not activate potassium channels and in the guinea-pig carotid artery, it activates ATP-sensitive potassium channels, a population of channels which is not involved in endothelium-dependent hyperpolarizations. In order to avoid contaminating NO production by a NO-synthase which is not fully inhibited, EDHF studies should be performed with a combination of several NO-synthase

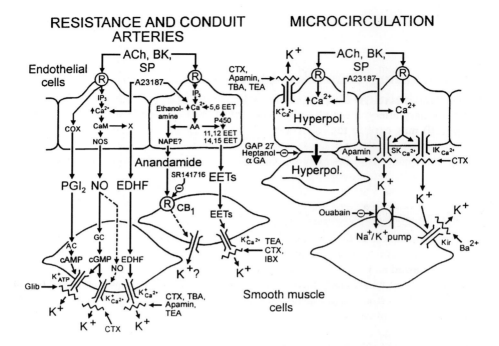

Figure 45-1. Possible mechanisms leading to endothelium-dependent hyperpolarizations.

Substances such as acetylcholine (ACh), bradykinin (BK) and substance P (SP), through the activation of M_3 muscarinic, B_2 bradykinin and NK_1 neurokinin receptor subtypes, respectively, and agents which increase intracellular calcium, such as the calcium ionophore A23187, release endothelium-derived hyperpolarizing factors.

R: receptor; NOS: nitric oxide synthase; COX: cyclooxygenase; X: putative EDHF synthase; P450: cytochrome P450 monooxygenase; CaM: calmodulin; NO: nitric oxide; PGI_2: prostacyclin; EDHF endothelium-derived hyperpolarizing factor; 5,6 EET: 5,6-epoxy-eicosatrienoic acid; 11,12 EET: 11,12-epoxy-eicosatrienoic acid; 14,15 EET: 14,15-epoxy-eicosatrienoic acid; NAPE: N-acylphosphatidylethanolamine; GC: guanylate cyclase, cGMP: cyclic guanosine monophosphate; cAMP: cyclic adenosine monophosphate; ATP: adenosine trisphosphate; IP_3: inositol trisphosphate; Hyperpol. hyperpolarization.

SR 141716 is an antagonist of the cannabinoid CB1 receptor subtype (CB1). Glibenclamide (Glib) is a selective inhibitor of ATP sensitive potassium channels (K^+_{ATP}). Tetraethyl ammonium (TEA) and tetrabutyl ammonium (TBA) are non specific inhibitors of potassium channels when used at high concentrations (> 5 mM) while at lower concentrations (1-3 mM) these drugs are selective for calcium-activated potassium channels (K^+_{Ca2+}). Iberiotoxin (IBX) is a specific inhibitor of large conductance K^+_{Ca2+}. Charybdotoxin (CTX) is an inhibitor of large conductance K^+_{Ca2+}, intermediate conductance K^+_{Ca2+} (IK_{Ca2+}) and voltage-dependent potassium channels. Apamin is a specific inhibitor of small conductance K^+_{Ca2+} (SK_{Ca2+}). Barium (Ba^{2+}) in the micromolar range, is a specific inhibitor of the inward rectifier potassium channel (K_{ir}). Gap27, an eleven amino acid peptide possessing conserved sequence homology to a portion of the second extracellular loop of connexin, 18β-glycyrrhetinic acid (αGA) and heptanol are gap junction uncouplers.

For the sake of clarity, the suggestion that in porcine coronary arteries, EDHF responses could involve the release of arachidonic acid through the activation of cytosolic phospholipase A2 and the subsequent formation of hydroxyeicosatetraenoic acid is not included.

inhibitors or with a combination of a NO-synthase inhibitor (or an inhibitor of soluble guanylate cyclase) and a NO scavenger.

Epoxyeicosatrienoic acids relax blood vessels, hyperpolarize coronary arterial smooth muscle cells and increase the open-state probability of BK_{Ca}. In smooth muscle cells of isolated bovine coronary arteries, 11,12 epoxyeicosatrienoic acid activates BK_{Ca} in a cyclic-AMP and cyclic-GMP-independent manner, through ADP-ribosylation of a guanine nucleotide binding protein, $Gs\alpha$. Muscarinic agonists and bradykinin release epoxyeico-satrienoic acids from isolated bovine and porcine coronary arteries, from bovine coronary arterial endothelial cells as well as from cultures of human umbilical vein endothelial cells. Furthermore, epoxyeicosatrienoic acids can be stored in membrane phospholipids and, in canine coronary artery, these phospholipid-derived epoxyeicosatrienoic acids (not newly synthesized) may explain the relaxation attributed to EDHF. Taken in conjunction these observations support the hypothesis that epoxyeicosatrienoic acids act as EDHF. Interestingly, in the brain, epoxyeicosatrienoic acids are also synthesized by astrocytes (P4502C11) and are released in response to glutamate produced by metabolically active neurons. These epoxyeicosatrienoic acids hyperpolarize vascular smooth muscle cells and relax cerebral arteries, thereby increasing nutritive blood flow to the active brain area. Furthermore, endothelium-derived epoxyeicosatrienoic acids activate smooth muscle cells Erk1/2 suggesting a possible role in the regulation of cellular signaling and possibly gene regulation.

However, epoxyeicosatrienoic acids do not produce relaxation or hyperpolarization in guinea-pig carotid and basilar arteries while in the guinea-pig and porcine coronary artery the mechanism by which epoxyeicosatrienoic acid produces hyperpolarization or relaxation (BK_{Ca}) differs from the mechanism involved in EDHF-mediated responses (sensitive to the combination of charybdotoxin plus apamin). In other studies involving blood vessels from humans, rats, guinea-pigs, dogs and pigs, chemically unrelated inhibitors of cytochrome P450 do not inhibit the EDHF responses or produce a non-specific inhibition.

Activation of cytochrome P450 in endothelial cells increases the intracellular calcium concentration and thus the release of endothelium-derived factors and stimulates endothelial tyrosine kinase. Inhibition of EDHF-mediated responses by cytochrome P450 inhibitors may only be linked to the general requirement for an increase in the endothelial intracellular calcium concentration in order to observe the release of endothelial derived-factors such as NO and EDHF.

The research summarized in this monograph convincingly supports the theory that epoxyeicosatrienoic acid could represent EDHF in vascular beds such as the canine and bovine coronary arteries, but not in other beds such as the guinea-pig and rat arteries. Can these discrepancies be explained by a true species specificity? Does the presence of a pre-stored pool of epoxyeicosatrienoic acids explain the resistance to the inhibitors of cytochrome P450? Does the mandatory contribution of endothelial cytochrome P450 in releasing vasoactive substances complicate the issue? These questions await answers.

In the isolated and perfused mesenteric and coronary arterial bed of the rat, anandamide induces dilatation which mimics responses to EDHF. However, in isolated blood vessels from various species (pig, guinea-pig, rat), the endogeneous cannabinoid does not produce hyperpolarization or if it does so, the underlying mechanism differs from EDHF-mediated responses. In addition, CB1-receptor antagonists do not inhibit endothelium-dependent hyperpolarizations. These observations do not support the suggestion that an endogenous cannabinoid is the ubiquitous mediator of endothelium-dependent hyperpolarizations. However, anandamide or lipoxygenase metabolites such as trihydroxyeicosatrienoic acids, which can be produced and released by endothelial cells and can relax vascular smooth muscle cells by activating potassium channels, may in some vascular beds contribute to endothelium-dependent hyperpolarizations.

In the rat hepatic artery, potassium ion are EDHF. Indeed, endothelial cells hyperpolarize in response to neurohumoral substances which produce the release of vasoactive substances such as NO and EDHF. Potassium ions, flowing through the opening of endothelial small and intermediate conductance potassium channels (sensitive to charybdotoxin and apamin), may accumulate in the intercellular space. This rise in potassium concentration could hyperpolarize the smooth muscle by activating the inward rectifying potassium channels (sensitive to low concentrations of barium) and the sodium/potassium pump (sensitive to ouabain). However, in other blood vessels the addition of potassium does not necessarily produce hyperpolarization, possibly because the inward rectifying potasssium channel is expressed poorly in certain vascular smooth muscle cells especially in those of large blood vessels. Furthermore, in the guinea-pig carotid and porcine coronary arteries, the endothelium-dependent hyperpolarization is not affected by the combination of ouabain and barium (Quignard *et al.*, 1999). Further studies are required to verify the pertinence of this proposal and to dertermine the vascular beds where it applies (Vanhoutte, 1998).

Endothelium-dependent hyperpolarizations could also involve electrical coupling through myo-endothelial junctions. Indeed, substances which produce endothelium-dependent hyperpolarization of vascular smooth muscle cells also hyperpolarize endothelial cells, with the same time course. Gap junctions couple smooth muscle and endothelial cells and conduction of depolarization and hyperpolarization from smooth muscle cells to endothelial cells occurs. However, in most blood vessels, electrical propagation from endothelial to smooth muscle cells does not seem to occur. Transfer of calcium from contracted smooth muscle cells to endothelial cells can produce the release of NO (and possibly EDHF) to act as a negative feedback mechanism to modulate the contraction of the vascular wall. The role of gap junctions between the innermost intimal vascular smooth muscle cells and the deeper layers has to be clarified. Conflicting results have been obtained with the non-specific gap junction uncoupler heptanol. More specific blockers of gap junctions, 18β-glycyrrhetinic acid and Gap27 (a peptide which possesses a conserved sequence homology with a portion of connexin) inhibit EDHF-like responses in rabbit and guinea-pig arteries (Chaytor *et al.*, 1998; Yamamoto *et al.*, 1998; Taylor *et al.*, 1998; Yamamoto *et al.*, 1999; data published after workshop). At present, this mechanism needs to be further explored to better understand its potential contribution to EDHF-mediated responses. Gap junctions could be the site of myo-endothelial electrical cell coupling but could also be the site of transfer of low molecular weight compounds.

In most blood vessels, inhibition of the endothelium-dependent hyperpolarization requires the combination of two toxins, charybdotoxin plus apamin, while each blocker individually is either ineffective (charybdotoxin) or only partially effective (apamin). However, the combination of iberiotoxin (a specific inhibitor of BK_{Ca}) plus apamin does not mimic the effects of charybdotoxin plus apamin, indicating that BK_{Ca} channels are not involved. Attempts to identify, in vascular smooth muscle cells, a potassium conductance specifically sensitive to the combination of these two toxins (without involvement of BK_{Ca}), have so far been unfruitful. In the hepatic artery of the rat, the combination of charybdotoxin plus apamin inhibits the hyperpolarization of endothelial cells produced by acetylcholine suggesting that the targets of these toxins are on the endothelial cells. If so, how does the vascular smooth muscle cell hyperpolarize? 4-Aminopyridine, a blocker of voltage-dependent potassium channels inhibits hyperpolarization of these cells in rat and guinea-pig arteries in the same range of concentrations required to inhibit the delayed rectifier potassium channels in isolated vascular smooth muscle cells. However, the compound is not very selective as it can also inhibit BK_{Ca} and its site of action has not been determined (endothelium-vs smooth muscle). In the rat hepatic artery, the inward rectifying potassium channels and the sodium-potassium pump could represent the target of EDHF on the smooth muscle cells. Alternatively, inhibition of non-selective cation channels can cause profound hyperpolarization. However, endothelium-dependent hyperpolarizations are also associated with an increase in rubidium efflux and a decrease in membrane resistance, suggesting that the hyperpolarization is due to the opening and not to the closing of a conductance. At present the actual target of EDHF on vascular smooth muscle cells also remains elusive. Membrane hyperpolarization produced by EDHF (or by other mechanisms such as potassium channel openers or activation of the sodium/potassium pump) produces relaxation not only by inhibiting voltage-dependent calcium channels but also by reducing the production of inositol triphosphate, the receptor operated calcium influx and possibly the sensitivity of the contractile proteins to calcium.

References

Adams, D.J., Barakeh, J., Laskey, R., Van Breemen, C. (1989) Ion channels and regulation of intracellular calcium in vascular endothelial cells. *FASEB J.*, 3:2389–2400.

Adeagbo, A.S.O., Henzel, M.K. (1998) Calcium-dependent phospholipase A2 mediates the production of endothelium-derived hyperpolarizing factor in perfused rat mesenteric prearteriolar bed. *J. Vasc. Res.*, 35:27–35.

Adeagbo, A.S.O., Triggle, C.R. (1993) Varying extracellular [K$^+$]: A functional approach to separating EDHF- and EDNO-related mechanisms in perfused rat mesenteric arterial bed. *J. Cardiovasc. Pharmacol.*, 21:423–429.

Akaike, T., Yoshida, M., Miyamoto, Y., Sato, K., Kohno, M., Sasamoto, K., Miyazaki, K., Ueda, S., Maeda, H. (1993) Antagonistic action of imidazolineaoxyl N-Oxides against endothelium-derived relaxing factor/ NO through a radical reaction. *Biochem.*, 32:827–832.

Alkayed, N.J., Birks, E.K., Hudetz, A.G., Roman, R.J., Henderson, L., Harder, D.R. (1996a) Inhibition of brain P-450 arachidonic acid epoxygenase decreases baseline cerebral blood flow. *Am. J. Physiol.*, 271:H1541–H1546.

Alkayed, N.J., Birks, E.K., Narayanan, J., Petrie, K.A., Kohler-Cabot, A.E., Harder, D.R. (1997) Role of P-450 arachidonic acid epoxygenase in the response of cerebral blood flow go glutamate in rats. *Stroke*, 28:1066–1072.

Alkayed, N.J., Narayanan, J., Gebremedhin, D., Medhora, M., Roman, R.J., Harder, D.R. (1996b) Molecular characterization of an arachidonic acid epoxygenase in rat brain astrocytes. *Stroke*, 27:971–979.

Alonso-Torre, S.R., Alvarez, J., Montero, M., Sanchez, A., Garcia-Sancho, J. (1993) Control of Ca^{2+} entry into HL60 and U937 human leukemia cells by the filling state of the intracellular Ca^{2+} stores. *Biochem. J.*, 289:761–766.

Alonso, M.J., Salaices, M., Sanchez-Ferrer, C.F., Ponte, A., Lopez-Rico, M., Marin, J. (1993) Nitric-oxide-related and non related mechanisms in the acetylcholine-evoked relaxations in cat femoral arteries. *J. Vasc. Res.*, 30:339–347.

Alonso, M.T., Alvarez, J., Montero, M., Sanchez, A., Garcia-Sancho, J. (1991) Agonist-induced Ca^{2+} influx into human platelets is secondary to the emptying of intracellular Ca^{2+} stores. *Biochem. J.*, 280:783–789.

Alvarez, J., Montero, M., Garcia-Sancho, J. (1991). Cytochrome P-450 may link intracellular Ca^{2+} stores with plasma membrane Ca^{2+} influx. *Biochem. J.*, 274:193–197.

Alvarez, J., Montero, M., Garcia-Sancho, J. (1992a). Cytochrome P450 may regulate plasma membrane Ca^{2+} permeability according to the filling state of the intracellular Ca^{2+} stores. *FASEB J.*, 6:786–792.

Alvarez, J., Montero, M., Garcia-Sancho, J. (1992b) High affinity inhibition of Ca^{2+}-dependent K$^+$ channels by cytochrome P$_{450}$ inhibitors. *J. Biol. Chem.*, 267:11789–11793.

Amagishi, T., Yanagisawa, T., Taira, N. (1992) K$^+$ channel openers, cromakalim and Ki4032, inhibit agonist-induced Ca^{2+} release in canine coronary artery. *Naunyn-Schmiedeberg's Arch Pharmacol.*, 346:691–700.

Andriantsitohaina, R., Okruhlicova, L., Cortes, S.F., Lagaud, G.J.L., Randriamboavonjy, V., Muller, B., Stoclet, J.C. (1996) Role of endothelium nitric oxide in the response to angiotensin II of small mesenteric arteries of the rat. *J. Vasc. Res.*, 33:386–394.

Aptecar, E., Teiger, E., Dupouy, P., Woscoboinik, J., Kern, M.J., Sediame, S., Loisance, D., Rande, J.L.D. (1997) Effects of bradykinin on coronary blood flow and vasomotion in transplant patients. *J. Am. Col. Cardiol.*, 29:142A (abstract).

Archer, S.L., Huang, J.M.C., Hampl, V., Nelson, D.P., Shultz, P.J., Weir, E.K. (1994) Nitric oxide and cGMP cause vasorelaxation by activation of a charybdotoxin-sensitive K channel by cGMP-dependent protein kinase. *Proc. Natl. Acad. Sci. (USA)*, 91:7583–7587.

Archer, S.L., Huang, J.M.C., Reeve, H.L., Hampl, V., Tolarova, S., Michelakis, E., Weir, E.K. (1996) Differential distribution of electrophysiologically distinct myocytes in conduit and resistance arteries determines their response to nitric oxide and hypoxia. *Circ Res.*, 78:431–442.

Asano, M., Aoki, K., Suzuki, Y., Matsuda, T. (1987) Effects of Bay K 8644 and nifedipine on isolated dog cerebral, coronary and mesenteric arteries. *J. Pharmacol. Exp. Ther.*, 234:476–484.

Asano, M., Ito, M.K., Matsuda, T., Suzuki, Y., Oyama, H., Shibuya, M., Sigita, K. (1993) Functional role of charybdotoxin-sensitive K^+ channels inthe resting state of cerebral, coronary and mesenteric arteries of the dog. *J. Pharmacol. Exp. Ther.*, 267:1277–1285.

Asano, M., Masuzawa-Ito, K., Matsuda, T. (1993) Charybdotoxin-sensitive K^+ channels regulate the myogenic tone in the resting state of arteries from spontaneously hypertensive rats. *Br. J. Pharmacol.*, 108:214–222.

Auch-Schwelk, W., Katusic, Z.S., Vanhoutte, P.M. (1992) Nitric oxide inactivates endothelium-derived contracting factor in the rat aorta. *Hypertension*, 19, 442–445.

Badr, K.F., Brenner, B.M., Ichikawa, I. (1987) Effects of leukotriene D4 on glomerular dynamics in the rat. *Am J Physiol*, 253:F239–F243.

Balazy, M. (1994) Peroxynitrite and arachidonic acid. Identification of arachidonate epoxides. *Pol. J. Pharmacol.*, 46:593–600.

Barker, J.E., Anderson, J., Treasure, T., Piper, P.J. (1994) Influence of endothelium and surgical preparation on responses of human saphenous vein and internal thoracic artery to angiotensin II. *Br. J. Clin. Pharmac.*, 38:57–62.

Baron, A., Frieden, M., Bény, J.L. (1997) Epoxyeicosatrienoic acids activate a high-conductance, Ca^{2+}-dependent K^+ channel on pig coronary artery endothelial cells. *J. Physiol.*, 504:537–543.

Bartoli, F., Lin, H.K., Ghomashchi, F., Gelb, M.H., Jain, M.K., Apitz-Castro, R. (1994) Tight binding inhibitors of 85 kDa phospholipase A2 but not 14 kDa phospholipase A2 inhibit release of free arachidonate in thrombin-stimulated human platelets. *J. Biol. Chem.*, 269:15625–15630.

Barton, M., Cosentino, F., Brandes, R.P., Moreau, P., Shaw, S., Lüscher, T.F. (1997) Anatomic heterogeneity of vascular aging. Role of nitric oxide and endothelin. *Hypertension*, 30:817–824.

Bauersachs, J., Hecker, M., Busse, R. (1994) Display of the characteristics of endothelium-derived hyperpolarizing factor by a cytochrome P450-derived arachidonic acid metabolite in the coronary microcirculation. *Br. J. Pharmacol.*, 113:1548–1553.

Bauersachs, J., Lischke, V., Busse, R., Hecker, M. (1996) Endothelium-dependent hyperpolarization in the coronary microcirculation. In *Endothelium-derived hyperpolarizing factor*, edited by P.M. Vanhoutte, Vol. 26, pp. 211–216. The Netherlands: Harwood Academic Publishers.

Bauersachs, J., Popp, R., Hecker, M., Sauer, E., Fleming, I., Busse, R. (1996b) Nitric oxide attenuates the release of endothelium-derived hyperpolarizing factor. *Circulation*, 94:3341–3347.

Bayliss, W.M. (1992) On the local reactions of the arterial wall to changes of internal pressure. *J. Physiol. Lond.*, 28:220–231.

Bazan, E., Campbell, A.K., Rapoport, R.M. (1992) Protein kinase C activity in blood vessels from normotensive and spontaneously hypertensive rats. *Eur. J. Pharmacol.*, 227:343–348.

Beckman, K.B., Ames, B.N. (1998) The free radical theory of aging matures. *Physiological Rev.*, 78:547–581.

Beech, D.J., Bolton TB. (1989) Two components of potassium current activated by depolarization of single smooth muscle cells from the rabbit portal vein. *J. Physiol.*, 418:293–309.

Beetens, J.R., Coene, M.C., Verheyen, A., Zonnekeyn, L., Herman, A.G. (1986) Biphasic response of intimal prostacyclin production during the development of experimental atherosclerosis. *Prostaglandins*, 32:319–334.

Beevers, D.G., Erskine, E., Robertson, M., Beattie, A.D., Campbell, B.C., Goldberg, A., Moore, M.R., Hawthorne, V.M. (1976) Blood-lead and hypertension. *Lancet*, 2:1–3.

Belfort, M., Saade, G., Kramer, W., Suresh, M., Moise, K., Vedernikov, Y.P. (1995) The vasodilator effect of 17 B estradiol on isolated human omental artery is dedominantly mediated by prostacyclin, and not nitric oxide. *Am. J. Obstet. Gynecol.*, 172:319.

Beltramo, M., Stella, N., Calignano, A., Lin, S.Y., Makriyannis, A., Piomelli, D. (1997) Functional role of high-affinity anandamide transport, as revealed by selective inhibition. *Science*, 277:1094–1097.

Bennett, B.M., McDonald, B.J., Nigam, R., Long, P.G., Simon, W.C. (1992) Inhibition of nitrovasodilator- and acetylcholine-induced relaxation and cyclic GMP accumulation by the cytochrome P_{450} substrate, 7-ethoxyresorufin. *Can. J. Physiol. Pharmacol.*, 70:1297–1303.

Bény, J.L. (1990) Endothelial and smooth muscle cells hyperpolarized by bradykinin are not dye coupled. *Am. J. Physiol.*, 258:H836–H841.

Bény, J.L. (1997) Electrical coupling between smooth muscle cells and endothelial cells in pig coronary arteries. *Eur. J. Physiol.*, 433:364–367.

Bény, J.L., Brunet, P.C. (1988) Electrophysiological and mechanical effects of substance P and acetylcholine on rabbit aorta. *J. Physiol.*, 398:277–289.

Bény, J.L., Brunet, P.C. (1988a). Neither nitric oxide nor nitroglycerin accounts for all the characteristics of endothelially mediated vasodilatation of pig coronary arteries. *Blood Vessels*, 25:308–311.

Bény, J.L., Brunet, P.C., Huggel, H. (1987) Interaction of bradykinin and des-Arg[9]-bradykinin with isolated pig coronary arteries: mechanical and electrophysiological events. *Regulatory Peptides*, 17:181–190.

Bény, J.L., Zhu, P., Haefliger, I.O. (1997) Lack of bradykinin-induced smooth muscle hyperpolarization despite heterocellular dye coupling and endothelial cell hyperpolarisation in porcine ciliary artery. *J. Vasc. Res.*, 34:Suppl. 1, 6(abstract 006).

Berridge, M.J. Capacitative calcium entry. *Biochem. J.*, 312:1–11.

Berthiaume, N., Hes, F., Chen, A., Regoli, D., D'Orléans-Juste, P. (1997) Pharmacology of kinins in the arterial and venous mesenteric bed of normal and B2 knockout transgenic mice. *Eur. J. Pharmacol.*, 333:55–61.

Beverelli, F., Bea, M.L., Puybasset, L., Giudicelli, J.F., Berdeaux, A. (1997) Chronic inhibtion of NO synthase enhances the production of prostacyclin in coronary arteries through upregulation of the cyclooxygenase type 1 isoform. *Fundam. Clin. Pharmacol.*, 11:252–259.

Billingham, M.E. (1987) Cardiac transplant atherosclerosis. *Transplant Proc.*, 19(5):19–25.

Blatter, L.A., Wier, W.G. (1994) Nitric oxide decreases $Ca^{2+}i$ in vascular smooth muscle by inhibition of calcium current. *Cell. Calcium*, 15:122–131.

Blomgren, H., Hammarstrom, S., Wasserman, J. (1987) Synergistic enhancement of mitogen responses of human lymphocytes by inhibitors of cyclo-oxygenases and 5, 8, 11-eicostriynoic acid, an inhibitor of 12-lipoxygenase and leukotriene biosynthesis. *Int. Arch. Allergy Appl. Immunol.*, 83(3):247–255.

Bokoch, G.M., Reed, P.W. (1981) Evidence for inhibition of leukotriene A4 synthesis by 5, 8, 11, 14-eicosatetraynoic acid in guinea pig polymorphonuclear leukocytes. *J. Biol. Chem.*, 256:4156–4159.

Bolotina, V.M., Najibi, S., Palacino, J.J., Pagano, P.J., Cohen, R.A. (1994) Nitric oxide directly activates calcium-dependent potassium channels in vascular smooth muscle. *Nature*, 368:850–853.

Bolotina, V.M., Weisbord, R.M., Gerike, M., Cohen, R. (1997) Novel mechanism of nitric oxide-induced relaxation: accelerated refilling of intracellular calcium stores and inhibition of store-operated calcium influx. *Circulation*, 96(8):I-448.

Bolton, T.B., Lang, R.J., Takewaki, T. (1984) Mechanism of action of noradrenaline and carbachol on smooth muscle of guinea-pig anterior mesenteric artery. *J. Physiol.*, 351:549–572.

Bolton, T.B., Lang, R.J., Takewaki, T. (1996) Endothelium-dependent hyperpolarization. In *Endothelium-derived hyperpolarizing factor*, edited by P.M. Vanhoutte, Vol. 3, pp. 19–23. The Netherlands: Harwood Academic Publishers.

Bolzon, T., Xiong, Z., Cheung, D.W. (1993). Membrane rectification in single smooth muscle cells from the rat tail artery. *Pflügers Arch.*, 425:482–490.

Borg-Capra, C., Fournet-Bourguignon MP, Janiak, P., Villeneuve, N., Bidouard, J.P., Vilaine, J.P., Vanhoutte, P.M. (1997) Morphological heterogeneity with normal expression but altered function of G proteins in porcine cultured regenerated coronary endothelial cells. *Br. J. Pharmacol.*, 122:999–1008.

Bornheim, L.M., Kim, K.Y., Chen, B., Correira, M.A. (1993) The effects of cannabidiol on mouse hepatic microsomal P450-dependent anandamide metabolism. *Biochem. Biophys. Res. Comm.*, 197:740–746.

Bornheim, L.M., Kim, K.Y., Chen BL, Correira, M.A. (1995) Microsomal cytochrome P450-mediated liver and brain anandamide metabolism. *Biochem. Pharmacol.*, 50:677–686.

Bowman, A., Gillespie, J.S. (1982) Block of some non-adrenergic inhibitory responses of smooth muscle by a substance from hemolyzed erythrocytes. *J. Physiol.*, 328, 11–25.

Bradford, M.M. (1976) a rapid an sensitive method for the quantitation of microgram quantities of protein utilizing the principle of protein-dye-binding. *Anal. Biochem.*, 72, 248–254.

Brandes, R.P., Behra, A., Lebherz, C., Böger, R.H., Bode-Böger, S., Phivthong-Ngam, L., Mügge, A. (1997) N[ω]-nitro-L-arginine-and indomethacin-resistant endothelium-dependent relaxation in the rabbit renal artery: effect of hypercholesterolemia. *Atherosclerosis*, 135:49–55.

Brayden, J.E. (1993) Membrane hyperpolarization as a mechanism of endothelium-dependent vasodilation. In *Ion Flux in Pulmonary Vascular Control*, edited by E.K. Weir, J.R. Hume and J.T. Reeves, pp. 287–296. New York: Plenum Press.

Brayden, J.E. (1990) Membrane hyperpolarization is a mechanism of endothelium-dependent cerebral vasodilation. *Am. J. Physiol.*, 259:H668–H673.

Brayden, J.E., Murphy, M.E. (1996) Potassium channels activated by endothelium-derived factors in mesenteric and cerebral resistance arteries. In *Endothelium-derived hyperpolarizing factor*, edited by P.M. Vanhoutte, Vol. 18, pp. 137–142. The Netherlands: Harwood Academic Publishers.

Brayden, J.E., Nelson, M.T. (1992) Regulation of arterial tone by activation of calcium-dependent potassium channels. *Science*, 256:532–535.

Brayden, J.E., Wellman, G.C. (1989) Endothelium-dependent dilation of feline cerebral arteries: role of membrane potential and cyclic nucleotides. *J. Cereb. Blood Flow Metab.*, 9:256–263.

Brown, B.L., Albano, J.D.M., Sgherzi, A.M., Tampion, W. (1971) A simple and sensitive saturation assay method for the measurement of adenosine 3',5'-monophosphate. *Biochem. J.*, 121:561–563.

Brugnara, C., Armsby, C.C., De Franceschi, L., Crest, M., Euclaire, M.F., Alper, S.L. (1995) Ca^{2+}-activated K^+ channels of human and rabbit erythrocytes display distinctive pattern of inhibition by venom peptide toxins. *J. Membr. Biol.*, 147:71–82.

Brune, B., Lapetina, E.G. (1989) Activation of a cytosolic ADP-ribosyltransferase by nitric oxide-generating agents. *J. Biol. Chem.*, 264:8455–8458.

Brune, B., Molina y Vedia, L., Lapetina, E.G. (1990) Agonist-induced ADP-ribosylation of a cytosolic protein in human platelets. *Proc. Natl. Acad. Sci. (USA)*, 87:3304–3308.

Brunet, P.C., Bény, J.L. (1989) Substance P and bradykinin hyperpolarize pig coronary artery endothelial cells in primary culture. *Blood Vessels*, 26:228–234.

Brunk, C.F., Jones, K.C., James, T.W. (1979) Assay for nanogram quantities of DNA in cellular homogenates. *Anal. Biochem.*, 92:497–500.

Bryant, R.W., Bailey, J.M. (1981) Rearrangement of 15 hydroperoxyeicosatetraenoic acid (15-HPETE) during incubations with hemoglobin: a model for platelet lipoxygenase metabolism. *Prog. Lipid Res.*, 20, 279–281.

Busse, R., Fichtner, H., Luckhoff, A., Kohlhardt, M. (1988a) Hyperpolarization and increased free calcium in acetylcholine-stimulated endothelial cells. *Am. J. Physiol.*, 255:H965–H969.

Busse, R., Fleming, I. (1996) Endothelial dysfunction in atherosclerosis. *J. Vasc. Res.*, 33:181–194.

Busse, R., Mülsch, A. (1990) Calcium-dependent nitric oxide synthesis in endothelial cytosol is mediated by calmodulin. *FEBS.*, 265:133–136.

Busse, R., Mülsch, A., Fleming, I., Hecker, M. (1993) Mechanisms of nitric oxide release from the vascular endothelium. *Circulation*, 87:V18–V25.

Busse, R., Pohl, U., Luckhoff, A. (1989) Mechanisms controlling the production of endothelial autacoids. (Review). *Zeitschrift fur Kardiologie.*, 78:64–69.

Büsselberg, D., Platt, B., Michael, D., Carpenter, D.O., Haas, H.L. (1994) Mammalian voltage-activated calcium channel currents are blocked by Pb2+, Zn2+, and Al3+. *J. Neurophysiol.*, 71:1491–1497.

Cabot, M., Wesh, C., Cao, H., Chabbott, H. (1988) The phosphatidylcholine pathway of diacylglycerol formation stimulated by phorbol diester occurs via phospholipase D activation. *FEBS Lett.*, 233:153–157.

Cailla, H.L., Vannier, C.J., Delaage, M.A. (1976) Guanosine 3',5' cyclic monophosphate assay at the 10^{-15} mole level. *Anal. Biochem.*, 70:195–202.

Calignano, A., La Rana, G., Beltramo, M., Makriyannis, A., Piomelli, D. (1997) Potentiation of anandamide hypotension by the transport inhibitor, AM404. *Eur. J. Pharmacol.*, 337:R1–R2.

Campbell, E.A., Linton, E.A., Wolfe, C.D.A., Scraggs, P.R., Jones, M.T., Lowry, P.J. (1987) Plasma corticotropin-releasing hormone concentrations during pregnancy and parturition. *J. Clin., Endocrinol. Metab.*, 64:1054–1059.

Campbell, W.B., Gebremedhin, D., Pratt, P.F., Harder, D.R. (1996) Identification of epoxyeicosatrienoic acids as endothelium-derived hyperpolarizing factors. *Circ. Res.*, 78:415–423.

Capdevila, J.H., Chacos, N., Falck, J.R., Manna, S., Negro-Villar, A., Ojeda, S.R. (1983) Novel hypothalmic arachidonic products stimulate somastotatin release from median eminence. *Endocrinology*, 113:421–423.

Capdevila, J.H., Falck, J.R., Dishman, E., Karara, A. (1990) Cytochrome P450 arachidonate oxygenase. In *Methods in Enzymol*, Vol. 187, pp. 385–394. San Diego: Academic Press.

Capdevila, J.H., Gil, L., Orellana, M., Marenett, L.J., Mason, J.I., Yadagiri, P. and Falck, J.R. (1988) Inhibitors of cytochrome P-450-dependent arachidonic acid metabolism. *Arch. Biochem. Biophys.*, 261:257–263.

Carrier, G.O., Fuchs, L.C., Winecoff, A.P., Giulumian, A.D., White, R.E. (1997) Nitrovasodilators relax mesenteric microvessels by cGMP stimulation of Ca-activated K^+-channels. *Am. J. Physiol*, 273:H76–H84.

Catella, F., Lawson, J., Braden, G., Fitzgerald, D.J., Shipp, E., Fitzgerald GA. (1991) Biosynthesis of P450 products of arachidonic acid in humans: increased formation in cardiovascular disease. *Adv. Prostaglandin Thromb. Res.*, 21A:193–196.

Cauvin, C., Loutzenhiser, R., Van Breemen, C. (1983) Mechanisms of calcium antagonist-induced vasodilation. *Annu. Rev. Pharmacol. Toxicol.*, 23:373–396.

Cayatte, A.J., Palacino, J.J., Horten, K., Cohen, R.A. (1994) Chronic inhibition of nitric oxide production accelerates neointima formation and impairs endothelial function in hypercholesterolemic rabbits. *Arterioscler. Thromb.*, 14:753–759.

Chandy, K.G., Gutman, G.A. (1995) Voltage-gated potassium channels genes. In *Ligand- and voltage gated ion channels*, edited by R.A. North, pp. 2–71. Boca Raton: CRC Press.

Chataigneau, T., Félétou, M., Duhault, J., Vanhoutte, P.M. (1998a) Epoxyeicosatrienoic acids, potassium channel blockers and endothelium-dependent hyperpolarization in the guinea-pig carotid artery. *Br. J. Pharmacol.*, 123:574–580.

Chataigneau, T., Félétou, M., Thollon, C., Villeneuve, N., Vilaine, J.P., Duhault, J., Vanhoutte, P.M. (1998b) Cannabinoid CB₁ receptor and endothelium-dependent hyperpolarization in guinea-pig carotid, rat mesenteric and porcine coronary arteries. *Br. J. Pharmacol.*, 123:968–974.

Chataigneau, T., Félétou, M., Huang, P.L., Fishman, M.C., Duhault, J. and Vanhoutte, P.M. (1999) Acetylcholine-induced relaxation in blood vessels from endothelial nitric oxide synthase knockout mice. *Br. J. Pharmacol.*, 126:219–226.

Chaytor, A.T., Evans, W.H., Griffith, T.M. (1998) Central role of heterocellular gap junctional communication in endothelium-dependent relaxations of rabbit arteries. *J. Physiol.*, 508:561–573.

Chen, G., Cheung, D.W. (1992) Characterization of acetylcholine-induced membrane hyperpolarization in endothelial cells. *Circ. Res.*, 70:257–263.

Chen, G., Cheung, D.W. (1996) Modulation of endothelium-dependent hyperpolarization and relaxation to acetylcholine in rat mesenteric artery by cytochrome P450 enzyme activity. *Circ. Res.*, 79:827–833.

Chen, G., Cheung, D.W. (1997) Effect of K⁺-channels blockers on Ach-induced hyperpolarization and relaxation in mesenteric arteries. *Am. J. Physiol.*, 272:H2306–H2312.

Chen, G., Suzuki, H., Weston, A.H. (1988) Acetylcholine releases endothelium-derived hyperpolarizing factor and EDRF from rat blood vessels. *Br. J. Pharmacol.*, 95:1165–1174.

Chen, G., Suzuki, H. (1989) Some electrical properties of the endothelium-dependent hyperpolarization recorded from rat arterial smooth muscle cells. *J. Physiol.*, 410:91–106.

Chen, G., Suzuki, H. (1990) Calcium dependency of the endothelium-dependent hyperpolarization in smooth muscle cells of the rabbit carotid artery. *J. Physiol.*, 421:521–534.

Chen, G., Yamamoto, Y., Miwa, K., Suzuki, H. (1991) Hyperpolarization of arterial smooth muscle induced by endothelial humoral substances. *Am. J. Physiol.*, 260:H1888–H1892.

Chen, H.H., Burnett, Jr. J.C. (1997) C-type natriuretic peptide: a newly identified endothelial factor. *J. Cardiovasc. Pharmacol.*, 30(3):S35–S41.

Chen, J.-K., Falck, J.R., Reddy, K.M., Capdevila, J., Harris, R.C. (1998) Epoxyeicosatrienoic acids and their sulfonimide derivatives stimulate tyrosine phosphorylation and induce mitogenesis in renal epithelial cells. *J. Biol. Chem.*, 273:29254–29261.

Chinkers, M., Lowe, D.G., Bennett, G.L., Minamino, L., Kangawa, K., Matsuo, H., Goeddel, D.U. (1992) Selective activation of the B natriuretic peptide, brain natriuretic peptide and C-type natriuretic peptide. *Endocrinology*, 130:229–239.

Cho, H., Ueda, M., Tamaoka, M., Hamaguchi, M., Aisaka, K., Kiso, Y., Inoue, T., Ogino, R., Tatsuoka, T., Ishihara, T. (1991) Novel caffeic acid derivatives: extremely potent inhibitors of 12-lipoxygenase. *J. Med. Chem.*, 34(4):1503–1505.

Chou, T.C., Yen, M.H., Li, C.Y., Ding, Y.A. (1998) Alterations of nitric oxide synthase expression with aging and hypertension in rats. *Hypertension*, 31:643–648.

Christophersen, P. (1991) Ca²⁺-activated K⁺ channel from human crythrocyte membranes: single channel rectification and selectivity. *J. Membr. Biol.*, 119:75–83.

Chuang, M., Severson, D.L. (1990) Inhibition of diacylglycerol metabolism in isolated cardiac myocytes in U57908(RNC 80267), a diacylglycerol inhibitor. *J. Mol. Cell. Cardiol.*, 22:1009–1016.

Clancy, R.M., Leszczynska-Piziak, J., Abramson, S.B. (1993) Nitric oxide stimulates the ADP-ribosylation of actin in human neutrophils. *Biochem. Biophys. Res. Commun.*, 191:847–852.

Clark, S.G., Fuchs, L.C. (1997) Role of nitric oxide and Ca⁺⁺-dependent K⁺ channels in mediating heterogeneous microvascular responses to acetylcholine in different vascular beds. *J. Pharmacol. Exper. Therap.*, 282:1473–1479.

Clifton, V.L., Read, M.A., Leitch, I.M., Giles, W.B., Boura, A.L.A., Robinson, P.J., Smith, R. (1995) Corticotropin-releasing hormone-induced vasodilatation in the human fetal-placental circulation: involvement of the nitricoxide-cyclic guanosine 3′,5′-monophosphate-mediated pathway. *J. Clin., Endocrinol. Metab.*, 80(10):2888–2893.

Clozel, M., Kuhn, H., Hefti, F. (1990) Effects of angiotensin converting enzyme inhibitors and of hydralazine on endothelial function in hypertensive rats. *Hypertension*, 16:532–540.

Cohen, R.A. (1993) Dysfunction of vascular endothelium in diabetes mellitus. *Circulation*, 87:v67–v76.

Cohen, R.A. (1995) The role of nitric oxide and other endothelium-derived vasoactive substances in vascular disease. *Prog. Cardiovasc. Diseases*, 38:105–128.

Cohen, R.A., Najibi, S., Bolotina, V.M. (1996) The other role of nitric oxide as an endothelium-derived hyperpolarizing factor. In *Endothelium-derived hyperpolarizing factor*, edited by P.M. Vanhoutte, Vol. 21, pp. 19–23. The Netherlands: Harwood Academic Publishers.

Cohen, R.A., Plane, F., Najibi, S., Huk, I., Malinski, T., Garland, C.J. (1997) Nitric oxide is the mediator of both endothelium-dependent relaxation and hyperpolarization of the rabbit carotid artery. *Proc. Natl. Acad. Sci.(USA)* 94:4193–4198.

Cohen, R.A., Vanhoutte, P.M. (1995) Endothelium-dependent hyperpolarization. Beyond nitric oxide and cyclic GMP. *Circulation*, 92:3337–3349.

Cohen, R.A., Zitnay, K.M., Haudenschild, C.C., Cunningham, D. (1988) Loss of selective endothelial cell vasoactive functions caused by hypercholesterolemia in pig coronary arteries. *Circ. Res.*, 63:903–910.

Cornwell, T.L., Pryzwansky, K.B., Wyatt, T.A., Lincoln, T.M. (1991) Regulation of sarcoplasmic reticulum protein phosphorylation by localized cyclic GMP-dependent protein kinase in vascular smooth muscle cells. *Mol. Pharmacol.*, 40:923–931.

Corriu, C., Félétou, M., Canet, E., Vanhoutte, P.M. (1996a) Inhibitors of the cytochrome P450 mono-oxygenase and endothelium-dependent hyperpolarizations in the guinea-pig isolated carotid artery. *Br. J. Pharmacol.*, 117:607–610.

Corriu, C., Félétou, M., Canet, E., Vanhoutte, P.M. (1996b) Endothelium-derived factors and hyperpolarization of the carotid artery of the guinea-pig. *Br. J. Pharmacol.*, 119:959–964.

Corriu, C., Félétou, M., Puybasset, L., Bea, M.L., Berdeaux, A., Vanhoutte, P.M. (1998) Endothelium-dependent hyperpolarization in isolated arteries from animal treated with NO-synthase inhibitors. *J. Cardiovasc. Pharmacol.*, 32:944–950.

Corson, M.A., James, N.L., Latta, S.E., Nerem, R.M., Berk, B.C., Harrison, D.G. (1996) Phosphorylation of endothelial nitric oxide synthase in response to fluid shear stress. *Circ. Res.*, 79:984–991.

Cowan, C.L., Cohen, R.A. (1991) Two mechanisms mediate relaxation by bradykinin of pig coronary artery: NO-dependent and-independent responses. *Am. J. Physiol.*, 261:H830–H835.

Cowan, C.L., Palacino, J.J., Najibi, S., Cohen, R.A. (1993) Potassium channel-mediated relaxation to acetylcholine in rabbit arteries. *J. Pharmacol. Exp. Therap.*, 266(3):1482–1489.

Cowan, C.L., Steffen, R.P. (1995) Lysophosphatidylcholine inhibits relaxation of rabbit abdominal aorta mediated by endothelium-derived nitric oxide and endothelium-derived hyperpolarizing factor independent of protein kinase C activation. *Arterioscler Thromb. Vasc. Biol.*, 15:2290–2297.

Cox, R., Tulenko, T. (1991 Altered excitation-contraction coupling in hypertension: role of plasma membrane phospholipids and ion channels. In *Regulation of smooth muscle contraction*, edited by R. Moreland, pp. 291–302. New York: Plenum Press.

Creager, M.A., Cooke, J.P., Mendelshohn, M.E., Gallagher, S.J., Coleman, S.M., Loscalzo, J., Dzau, V.J. (1990) Impaired vasodilation of forearm resistance vessels in hypercholesterolemic humans. *J. Clin. Invest.*, 86:228–234.

Cui, X.L., Douglas, J.G. (1997) Arachidonic acid activates c-jun N-terminal kinase through NADPH oxidase in rabbit proximal tubular epithelial cells. *Proc. Natl. Acad. (USA)*, 94:3771–3776.

D'Alarcao, M., Corey, F.H., Cunard, C., Ramwell, P., Uotila, P., Vargas, R., Wroblewska, B. (1987) The vasodilation induced by hydroperoxy metabolites of arachidonic acid in the rat mesenteric and pulmonary circulation. *Br. J. Pharmacol.*, 91, 627–632.

D'Angelo, G., Meininger, G.A. (1994) Transduction mechanisms involved in the regulation of myogenic activity. *Hypertension*, 23:1096–1105.

Dashwood, M.R., Andrews, H.E., Wei, E.T. (1987) Binding of [^{125}I] Tyr-corticotropin-releasing factor to rabbit aorta is reduced by removal of the endothelium. *Eur. J. Pharmacol.*, 135:111–112.

Daut, J. (1997) Coming closer: structure and function of calcium-activated K$^+$ channels in coronary arteries. *J. Physiol.*, 502:469.

Davies, P.F., Oleson, S.P., Clapham, D.E., Morel, E.M., Schoen, F.J. (1988) Endothelial communication: state of the art lecture. *Hypertension*, 11:563–572.

Davis, S.F., Yeung, A.C., Meredith, I.T., Charbonneau, F., Ganz, P., Selwyn, A.P., Anderson, T.J. (1996) Early endothelial dysfunction predicts the development of transplant coronary artery disease at 1 year posttransplant. *Circulation*, 93:457–462.

Degawa, M., Arai, H., Kubota, M., Hashimoto, Y. (1994) Ionic lead, an unique metal ion as an inhibitor for cytochrome P450IA2(CYP1A2) expression in the rat liver. *Biochem. Biophys. Res. Commun*, 200:1086–1092.

De Mey, J.G., Claeys, M., Vanhoutte, P.M. (1982) Endothelium-dependent inhibitory effects of acetylcholine, adenosine triphosphate, thrombin and arachidonic acid in the canine femoral artery. *J. Pharmacol. Exp. Ther.*, 222:166–173.

De Mey, J.G., Vanhoutte, P.M. (1983) Anoxia and endothelium-dependent reactivity of the canine femoral artery. *J. Physiol.*, 335:65–74.

Deutsch, D.G., Goligorsky, M.S., Schmid, P.C., Krebsbach, R.J., Schmid, H.H.O., Das, S.K., Dey, S.K., Arreaza, G., Thorup, C., Stefano, G., Moore, L.C. (1997) Production and physiological actions of anandamide in the vasculature of the rat kidney. *J. Clin. Invest.*, 100:1538–1546.

Desarnaud, F., Cadas, H., Piomelli, D. (1995) Anandamide amidohydrolase activity in rat brain microsomes. Identification and partial characterization. *J. Biol. Chem.*, 270:6030–6035.

Devane, W.A., Dysarz, F.A. 3d, Johnson, M.R., Melvin, L.S., Howlett, A.C. (1988) Determination and characterization of a cannabinoid receptor in rat brain. *Mol. Pharmacol.*, 34:605–613.

Devane, W.A., Hanus, L., Breuer, A., Pertwee, R.G., Stevenson, L.A., Griffin, G., Gibson, D., Mandelbaum, A., Etinger, A., Mechoulam, R. (1992) Isolation and structure of a brain constituent that binds to the cannabinoid receptor. *Science*, 258:1946–1949.

Di Marzo, V., Fontana, A., Cadas, H., Schinelli, S., Cimino, G., Schwartz, J.C., Piomelli, D. (1994) Formation and inactivation of endogenous cannabinoid anandamide in central neurons. *Nature*, 372:686–691.

Dinerman, J.L., Dawson, T.M., Schell, M.J., Snowman, A., Snyder, S.H. (1994) Endothelial nitric oxide synthase localized to hippocampal pyramidal cells: implications for synaptic plasticity. *Proc. Nat. Acad. Sci.*, 91:4214–4218.

Dinerman, J.L., Lowenstein, C.J., Snyder, S.H. (1993) Molecular mechanism of nitric oxide regulation. Potential relevance to cardiovascular disease. *Circ. Res.*, 73:217–222.

Dohi, Y., Kojima, M., Sato, K. (1996) Benidipine improves endothelial function in renal resistance arteries of hypertensive rats. *Hypertension*, 28:58–63.

Dohi, Y., Kojima, M., Sato, K., Lüscher, T.F. (1995) Age-related changes in vascular smooth muscle and endohelium. *Drugs Aging*, 7, 278–291.

Dohi, Y., Thiel, M.A., Bühler, F.R., Lüscher, T.F. (1990) Activation of endothelial L-arginine pathway in resistance arteries. Effect of age and hypertension. *Hypertension*, 16:170–179.

Dong, H., Waldron, G.J., Galipeau, D., Cole, W.C., Triggle, C.R. (1997) NO/PGI2-independent vasorelaxation and the cytochrome P450 pathway in rabbit carotid artery. *Br. J. Pharmacol.*, 120:695–701.

Dora, K.A., Doyle, M.P., Duling, B.R. (1997) Elevation of intracellular calcium in smooth muscle causes endothelial cell generation of NO in arterioles. *Proc. Natl. Acad. Sci. (USA)*, 94:6529–6534.

Dorn II, G.W., Becker, M.W. (1993) Thromboxane A$_2$ stimulated signal transduction in vascular smooth muscle. *J. Pharmacol. Exper. Ther.*, 265:447–456.

Dove Pettit, D.A., Harrison, M.P., Olson, J.M., Spencer, R.F., Cabral, G.A. (1998) Immunohistochemical localization of the neural cannabinoid receptor in the rat brain. *J. Neurosci. Res.*, 51:391–402.

Drinkwater, D.C., Rudis, E., Laks, H., Ziv, E., Marino, J., Stein, D., Ardehali, A., Aharon, A., Moriguchi, J., Kobashigawa, J. (1995) University of Wisconsin versus Stanford Cardioplegic solution and the development of cardiac allograft vasculopathy. *J. Heart Lung Transplant.*, 14:891–896.

Drummond, G.R., Cocks, T.M. (1996) Evidence for mediation by endothelium-derived hyperpolarizing factor of relaxation to bradykinin in the bovine isolated coronary artery independently of voltage-operated Ca^{2+} channels. *B.R. J. Pharmacol.*, 117:1035–1040.

Durant, G.J., Ganellin, C.R., Parsons, M.E. (1975) Chemical differentiation of histamine H1 and H2 receptor agonists. *J. Med. Chem.*, 18, 905–909.

Eckman, D.M., Frankovich, J.D., Keef, K.D. (1992) Comparison of the actions of acetylcholine and BRL 38227 in the guinea pig coronary artery. *Br. J. Pharmacol.*, 106:9–16.

Eckman, D.M., Hopkins, N., McBride, C., Keef, K.D. (1998) Endothelium-dependent relaxation and hyperpolarization in guinea-pig coronary artery: role of epoxyeicosatrienoic acid. *Br. J. Pharmacol.*, 124:181–189.

Eckman, D.M., Weinert, J.S., Buxton, I.L.O., Keef, K.D. (1994) Cyclic GMP independent relaxation and hyperpolarization with acetylcholine in guinea-pig coronary artery. *Br. J. Pharmacol.*, 111:1053–1060.

Edwards, F.R., Hirst, G.D.S. (1988) Inward rectification in submucosal arterioles of guinea-pig ileum. *J. Physiol.*, 404:437–454.

Edwards, G., Dora, K.A., Gardener, M.J., Garland, C.J., Weston, A.H. (1998a) K$^+$ is an endothelium-derived hyperpolarizing factor in rat arteries. *Nature*, 396:269–272.

Edwards, G., Gardener, M.J., Walker SD, Weston, A.H. (1998b) Comparison of effects of 1-EBIO and NS1619 on K$^+$ currents in vascular smooth muscle and endothelial cells. *Br. J. Pharmacol.*, 125:106P.

Edwards, G., Weston, A.H. (1997) Recent advances in potassium channel modulation. In *Progress in Drug Research*, edited by E. Jucker, Vol. 49, pp. 93–121. Basel: Birkhauser Verlag.

Edwards, G., Weston, A.H. (1998) Endothelium-derived hyperpolarizing factor — a critical appraisal. In *Progress in Drug Research*, edited by E. Jucker, Vol. 50, pp. 107–133. Basel: Birkhauser Verlag.

Edwards, G., Zygmunt, P.M., Högestätt, E.D., Weston, A.H. (1996) Effects of cytochrome P450 inhibitors on potassium currents and mechanical activity in rat portal vein. *Br. J. Pharmacol.*, 119:691–701.

Eizawa, H., Yui, Y., Inoue, R., Kosuga, K., Hattori, R., Aoyama, T., Sasayama, S. (1995) Lysophosphatidylcholine inhibits endothelium-dependent hyperpolarization and N$^\omega$-nitro-L-arginine/indomethacin-resistant endothelium-dependent relaxation in the porcine coronary artery. *Circ.*, 92:3520–3526.

Ellis, E.F., Moore, S.F., Willoughby, K.A. (1995) Anandamide and Δ9-THC dilation of cerebral arterioles is blocked by indomethacin. *Am. J. Physiol.*, 269:H1859–H1864.

Ellis, E.F., Police, R.J., Yancey, L., Mckinney, J.S., Amruthesh, S.C. (1990) Dilation of cerebral arterioles by cytochrome P-450 metabolites of arachidonic acid. *Am. J. Physiol.*, 259:H1171–H1177.

Elwood, P.C., Yarnell, J.W., Oldham, P.D., Catford, J.C., Nutbeam, D., Davey-Smith, G., Toothill, C. (1988) Blood pressure and blood lead in surveys in Wales. *Am. J. Epidemiol.*, 127:942–945.

England, S.K., Wooldridge, T.A., Stekiel, W.J., Rush, N.J. (1993) Enhanced single channel K$^+$ current in arterial membranes from genetically hypertensive rats. *Am. J. Physiol.*, 264, H1337–1345.

Enoki, T., Miwa, S., Sakamoto, A., Minowa, T., Komuro, T., Kobayashi, S., Ninomiya, H., Masaki, T. (1995) Long-lasting activation of cation current by low concentration of endothelin-1 in mouse fibroblasts and smooth muscle cells of rabbit aorta. *Br. J. Pharmacol.*, 115:479–485.

Erdahl, W.L., Chapman, C.J., Taylor, R.W., Pfeiffer, D.R. (1994) Ca^{2+} transport properties of ionophore A23187, ionomycin, and 4-BrA23187 in a well defined model system. *Biophys. J.*, 66:1678–1693.

Evans, D.H., Weingarten, K. (1990) The effect of cadmium and other metals on vascular smooth muscle of the dogfish shark, Squalus acanthias. *Toxicology*, 61:275–281.

Facci, L., Dal Toso, R., Romanello, S., Buriani, A., Skaper, S.D., Leon, A. (1995) Mast cells express a peripheral cannabinoid receptor with differential sensitivity to anandamine and palmitoylethanolamide. *Proc. Nat. Acad. Sci.*, 92:3376–3380.

Falck, J.R., Manna, S., Siddhanta, A.K., Capdevila, J. (1983) Transformation of 15 HETE to 14, 15-dihydroxyeicosatrienoic acid and 11, 14, 15 and 13, 14, 15-trihydroxyeicosatrienoic acid. *Tetrahedron Letters*, 24:5715–5718.

Falcone, J.C., Bohlen, H.G. (1990) EDRF from rat intestine and skeletal muscle venules causes dilation of arterioles. *Am. J. Physiol.*, 258:H1515–H1523.

Fang, X., Kaduce, T.L., Weintraub, N.L., VanRollins, M., Spector, A.A. (1996) Functional implications of a newly characterized pathway of 11, 12-Epoxyeicosatrienoic acid metabolism in arterial smooth muscle. *Circ. Res.*, 79:784–793.

Fang, X., VanRollins, M., Kaduce, T.L., Spector, A.A. (1995) Epoxyeicosatrienoic acid metabolism in arterial smooth muscle cells. *J. Lipid Res.*, 36:1236–1246.

Faraci, F.M., Brian, J.E. (1994) Nitric oxide and the cerebral circulation. *Stroke*, 25:692–703.

Faraci, F.M., Heistad, D.D. (1998) Regulation of the cerebral circulation: role of endothelium and potassium channel. *Physiological Review*, 78:54–75.

Faraci, F.M., Sigmund, C.D., Shesely, E.G., Maeda, N., Heistad, D.D. (1998) Responses of carotid artery in mice deficient in expression of the gene for endothelial NO synthase. *Am. J. Physiol.*, 274:H564–H570.

Feinmark, S.J., Steel, D.J., Thekkuveettil, A., Abe, M., Li, X.D., Schwartz, J.H. (1992) Aplysia Californica contains a novel 12-lipoxygenase which generates biologically active products from arachidonic acid. In *Neurobiology of Essential Fatty Acids*, edited by N.G. Bazan *et al.*, pp. 159–169. New York: Plenum Press.

Felder, C.C., Joyce, K.E., Briley, E.M., Glass, M., Mackie, K.P., Fahey, K.J., Cullinan, G.J., Hunden, D.C., Johnson, D.W., Chaney, M.O., Koppel, G.A., Brownstein, M. (1998) LY320135, a novel cannabinoid CB receptor antagonist, unmasks coupling of the CB1 receptor to stimulation of cAMP accumulation. *J. Pharmacol. Exp. Ther.*, 284:291–297.

Felder, C.C., Joyce, K.E., Briley, E.M., Mansouri, J., Mackie, K.P., Blond, O., Lai, Y., Ma, A.L., Mitchell, R.L. (1995) Comparison of the pharmacology and signal-transduction of the human cannabinoid CB1 and CB2 receptors. *Mol. Pharmacol.*, 48:443–450.

Félétou, M., Vanhoutte, P.M. (1988) Endothelium-dependent hyperpolarization of canine coronary smooth muscle. *Br. J. Pharmacol.*, 93:515–524.

Félétou, M., Girard, V., Canet, E. (1995) Different involvement of nitric oxide in endothelium-dependent relaxation of porcine pulmonary artery and vein: influcence of hypoxia. *J. Cardiovasc. Pharmacol.*, 25:665–673.

Fenger-Gron, J., Mulvany, M.J., Christensen, K.L. (1995) Mesenteric blood pressure profile of concious freely moving rats. *J. Physiol.*, 488:753–760.

Fitzpatrick, F.A., Murphy, R.C. (1989) Cytochrome P-450 metabolism of arachidonic acid: formation and biological action of « epoxygenase » derived eicosanoids. *Pharmacol. Rev.*, 40:229–241.

Fleming, I., Fisslthaler, B., Busse, R. (1995) Calcium signaling in endothelial cells involves activation of tyrosine kinases and leads to activation of mitogen-activated protein kinase. *Circ. Res.*, 76:522–529.

Fleming, I., Fisslthaler, B., Busse, R. (1996) Interdependence of calcium signaling and protein tyrosine phosphorylation in human endothelial cells. *J. Biol. Chem.*, 271:11009–11015.

Fleming, I., Schermer, B., Popp, R., Busse, R. (1999) Inhibition of the production of the endothelium-derived hyperpolarizing factor by cannabinoid receptor agonists. *Br. J. Pharmacol.*, H126:949–960.

Force, T., Hyman, G., Hajjar, R., Sellmayer, A., Bonventre, J.V. (1991) Noncyclooxygenase metabolites of arachidonic acid amplify the vasopressin-induced Ca^{2+} signal in glomerular mesangial cells by releasing Ca^{2+} from intracellular stores. *J. Biol. Chem.*, 266:4295–305.

Förstermann, U., Closs, E.I., Polluck, J.S., Nakane, M., Schwarz, P., Gath, I., Kleinert, H. (1994) Nitric oxide synthase isozymes: characterization, purification, molecular cloning and functions. *Hypertension*, 23:1121–1131.

Förstermann, U., Mügge, A., Alheid, U., Haverich, A., Frölich, J.C. (1988a) Selective attenuation of endothelium-mediated vasodilation in atherosclerotic human coronary arteries. *Circ. Res.*, 62:185–190.

Fujii, K., Ohmori, S., Tominaga, M., Abe, I., Takata, Y., Ohya, Y., Kobayashi, K., Fujishima, M. (1993) Age-related changes in endothelium-dependent hyperpolarization in the rat mesenteric artery. *Am. J. Physiol.*, 265:H509–H516.

Fujii, K., Tominaga, M., Ohmori, S., Abe, I., Fujishima, M. (1996) Impaired endothelium-dependent hyperpolarization in the mesenteric artery of spontaneously hypertensive rats. In *Endothelium-derived hyperpolarizing factor*, edited by P.M. Vanhoutte, Vol. 30, pp. 247–254. The Netherlands: Harwood Academic Publishers.

Fujii, K., Tominaga, M., Ohmori, S., Kobayashi, K., Koga, T., Takata, Y., Fujishima, M. (1992) Decreased endothelium-dependent hyperpolarization to acetylcholine in smooth muscle of the mesenteric artery of spontaneously hypertensive rats. *Circ. Res.*, 70:660–669.

Fukao, M., Hattori, Y., Kanno, M., Sakuma, I., Kitabatake, A. (1995) Thapsigargin- and cyclopiazonic acid-induced endothelium-dependent hyperpolarization in rat mesenteric artery. *Br. J. Pharmacol.*, 115:987–992.

Fukao, M., Hattori, Y., Kanno, M., Sakuma, I., Kitabatake, A. (1996) Endothelium-dependent hyperpolarizations in arteries from diabetic rats. In *Endothelium-derived hyperpolarizing factor*, edited by P.M. Vanhoutte, Vol. 32, pp. 263–270. The Netherlands: Harwood Academic Publishers.

Fukao, M., Hattori, Y., Kanno, M., Sakuma, I., Kitabatake, A. (1997a) Evidence against a role of cytochrome P450-derived arachidonic acid metabolites in endothelium-dependent hyperpolarization by acetylcholine in rat isolated mesenteric artery. *Br. J. Pharmacol.*, 120:439–446.

Fukao, M., Hattori, Y., Kanno, M., Sakuma, I., Kitabatake, A. (1997b) Sources of Ca^{2+} in relation to generation of acetylcoline-induced endothelium-dependent hyperpolarization in rat mesenteric artery. *Br. J. Pharmacol.*, 120:1328–1334.

Fukao, M., Hattori, Y., Kanno, M., Sakuma, I., Kitabatake, A. (1997c) Alterations in endothelium-dependent hyperpolarization and relaxation in mesenteric arteries from streptozotocin-induced diabetic rats. *Br. J. Pharmacol.*, 121:1383–1391.

Fulton, D., Mahboudi, K., McGiff, J.C., Quilley, J. (1995) Cytochrome P450-dependent effects of bradykinin in the rat heart. *Br. J. Pharmacol.*, 114:99–102.

Fulton, D., McGiff, J.C., Quilley, J. (1992) Contribution of NO and cytochrome P 450 to the vasodilator effect of bradykinin in the rat kidney. *Br. J. Pharmacol.*, 107:722–725.

Fulton, D., McGiff, J.C., Quilley, J. (1994) Role of K^+ channels in the vasodilator response to bradykinin in the rat heart. *Br. J. Pharmacol.*, 113:954–958.

Fulton, D., McGiff, J.C., Quilley, J. (1996) Role of phospholipase C and phospholipase A_2 in the nitric oxide-independent vasodilator effect of bradykinin in the rat perfused heart. *J. Pharmacol. Exper. Ther.*, 278:518–526.

Furchgott, R.F. (1983) Role of endothelium in response of vascular smooth muscle. *Circ. Res.*, 53:557–573.

Furchgott, R.F. (1988) Studies on relaxation of rabbit aorta by sodium nitrite: The basis for the proposal that the acid-activatable inhibitory factor from bovine retractor penis is inorganic nitrite and the endothelium-derived relaxing factor is nitric oxide. In *Vasodilatation: Vascular smooth muscle, peptides, autonomic nerves, and endothelium*, edited by P.M Vanhoutte, Vol. 4, pp. 401–414. New York: Raven Press.

Furchgott, R.F., Vanhoutte, P.M. (1989) Endothelium-derived relaxing and contracting factors. *FASEB J.*, 3:2007–2017.

Furchgott, R.F., Zawadzki, J.V. (1980) The obligatory role of endothelial cells in the relaxation of arterial smooth muscle by acetylcholine. *Nature*, 288:373–376.

Furukawa, K., Ohshima, N., Tawada-Iwata, Y., Shigekawa, M. (1991) Cyclic GMP stimulates Na^+/Ca^+ exchange in vascular smooth muscle cells in primary culture. *J. Biol. Chem.*, 266:12337–12341.

Furukawa, K., Tawada, U., Shigekawa, M. (1988) Regulation of the plasma membrane Ca^{2+} pump by cyclic nucleotides in cultured vascular smooth muscle cell. *J. Biol. Chem.*, 263:8058–8065.

Galle, J., Bauersachs, J., Bassenge, E., Busse, R. (1993) Arterial size determines the enhancement of contractile responses after suppression of endothelium-derived relaxing factor formation. *Pflugers Arch.*, 422:564–569.

Garcia, M.L., Knaus, H.-G., Munujos, P., Slaughter, R.S., Kaczorowski, G.J. (1995) Charybdototoxin and its effects on potassium channels. *Am. J. Physiol.*, 269:C1–C10.

Gardiner, S.M., Compton, A.M., Bennet, T. (1988) Regional haemodynamic effects of depressor neuropeptides in conscious, unrestrained, Long Evans and Brattleboro rats. *Br. J. Pharmacol.*, 95:197–208.

Garland, C.J., McPherson, G.A. (1992) Evidence that nitric oxide does not mediate the hyperpolarization and relaxation to acetylcholine in the rat small mesentery artery. *Br. J. Pharmacol.*, 105:429–435.

Garland, C.J., Plane, F. (1996) Relative importance of endothelium-derived hyperpolarizing factor for the relaxation of vascular smooth muscle in different arterial beds. In *Endothelium-derived hyperpolarizing factor*, edited by P.M. Vanhoutte, Vol. 22, pp. 173–179. The Netherlands: Harwood Academic Publishers.

Garland, C.J., Plane, F., Kemp, B.K., Cocks, T.M. (1995) Endothelium-dependent hyperpolarization: a role in the control of vascular tone. *Trends Pharmacol. Sci.*, 16:23–30.

Gebremedhin, D., Harder, D.R., Pratt, P.F., Campbell, W.B. (1998) Bioassay of an endothelium-derived hyperpolarizing factor from bovine coronary arteries: role of a cytochrome P450 mono-oxygenase. *J. Vasc. Res.*, 35:274–284.

Gebremedhin, D., Kaldunski, M., Jacobs, E.R., Harder, D.R., Roman, R.J. (1996) Coexistence of two types of Ca^{2+} activated K^+ channels in rat renal arterioles. *Am. J. Physiol.*, 270:F69–F81.

Gebremedhin, D., Lange, A.R., Narayanan, J., Harder, D.R. (1998) 20-HETE enhance L-type Ca^{2+} channel current in cat cerebral arterial muscle cells. *FASEB. J.*, 11:A22.

Gebremedhin, D., Ma, Y.H., Falck, J.R., Roman, R.J., VanRollins, M., Harder, D.R. (1992) Mechanism of action of cerebral epoxyeicosatrienoic acids on cerebral arterial smooth muscle. *Am. J. Physiol.*, 263:H519–H525.

Gebremedhin, D., Ma, Y.H., Imig, J.D., Harder, D.R., Roman, R.J. (1993) Role of cytochrome P450 in elevating renal vascular tone in spontaneously hypertensive rats. *J. Vasc. Res.*, 30:53–63.

Gerike, M., Weisbord, R.M., Cohen, R.A., Bolotina, V.M. (1997) Intracellular calcium stores regulate agonist-induced calcium influx in smooth muscle cells from rabbit aorta. *Biophys. J.*, 72:A296.

Gilligan, D.M., Bader, D.M., Panza, J.A., Quyyumi, A.A., Cannon, R.O. (1994) Acute vascular effects of estrogen in postmenopausal women. *Circulation*, 90:786–791.

Gilligan, D.M., Quyyumi, A.A., Cannon, R.O. (1994) Effects of physiological levels of estrogen on coronary vasomotor function in postmenopausal women. *Circulation*, 89:2545–2551.

Gisclard, V., Miller, V.M., Vanhoutte, P.M. (1988) Effect of 17 beta estradiol on endothelial responses in the rabbit. *J. Pharmacol. Exp. Ther.*, 244:19–22.

Godecke, A., Decking, U.K.M., Ding, Z., Hirchenhain, J., Bidmon, H.J., Godecke, S., Schrader, J. (1998) Coronary hemodynamics in endothelial NO synthase knockout mice. *Circ. Res.*, 82:186–194.

Gordon, J.L., Martin, W. (1983) Endothelium-dependent relaxation of the pig aorta: relationship to stimulation of ^{86}Rb efflux from isolated endothelial cells. *Br. J. Pharmacol.*, 79:531–541.

Graier, W.F., Holzmann, S., Hoebel, B.G., Kukovetz, W.R., Kostner, G.M. (1996) Mechanisms of L-N$^\omega$ nitroarginine/indomethacin-resistant relaxation in bovine and porcine coronary arteries. *Br. J. Pharmacol.*, 119:1177–1186.

Graier, W.F., Paltauf-Doburzynska, J., Hill, H., Fleischhacker, B., Hoebel, B.G., Kostner, G.M., Sturek, M. (1998) Submaximal stimulation of porcine endothelial cells cause focal Ca^{2+} elevation beneath the cell membrane. *J. Physiol. Lond.*, 506:109–125.

Graier, W.F., Simecek, S., Sturek, M. (1995) Cytochrome P450 mono-oxygenase-regulated signalling of Ca^{2+} entry in human and bovine endothelial cells. *J. Physiol.*, 482:259–274.

Graier, W.F., Sturek, M., Kukovetz, W.R. (1994). Ca^{2+} regulation and endothelial vascular function. *Endothelium*, 1:223–236.

Grandjean, P., Hollnagel, H., Hedegaard, L., Christensen, J.M., Larsen, S. (1989) Blood lead-blood pressure relations: alcohol intake and hemoglobin as confounders. *Am. J. Epidemiol.*, 129:732–739.

Grigoriadis, D.E., Lovenberg, T.W., Chalmers, D.T., Liaw, C., De Souza, E.B. (1996) Characterization of corticotropin-releasing factor receptor subtypes. *Ann. N.Y. Acad. Sci.*, 780:60–80.

Groschner, K., Graier, W.F., Kukovetz, W.R. (1992) Activation of a small-conductance Ca^{2+}-dependent K$^+$ channel contributes to bradykinin-induced stimulation of nitric oxide synthesis in pig aortic endothelial cells. *Biochim. Biophys. Acta*, 1137:162–170.

Gryglewski, R.J., Palmer, R.M., Moncada, S. (1986) Superoxide anion is involved in the breakdown of endothelium-derived vascular relaxing factor. *Nature*, 320:454–456.

Gunning, M., Ballerman, B.J., Silva, P., Brenner, B., Zeibel, M. (1988) Characterization of atrial natriuretic peptide receptors in rabbit inner medullary collecting duct cells. *Am. J. Physiol.*, 255, F324–F330.

Gunning, M., Silva, P., Brenner, B., Zeibel, M., (1989) Characterization of ANP sensitive guanylate cyclase in inner medullary collecting duct cells. *Am. J. Physiol.*, 255, F766–F775.

Gupta, S., McArthur, C., Grady, C., Ruderman, N.B. (1994) Stimulation of vascular Na$^+$-K$^+$-ATPase activity by nitric oxide: a cGMP-independent effect. *Am. J. Physiol.*, 266:H2146–H2151.

Halpern, W., Osol, G., Coy, G.S. (1984) Mechanical behavior of pressurized in vitro pre-arteriolar vessels determined with a video system. *Ann. Biomed. Eng.*, 12:463–479.

Hamilton, C.A., Berg, G., McIntyre, M., McPhaden, A.R., Reid, J.L., Dominiczak, A.F. (1997) Effects of nitric oxide and superoxide on relaxation in human artery and vein. *Atherosclerosis*, 133:77–86.

Hampson, A.J., Hill, W.A.G., Zanphillips, M., Makriyannis, A., Leung, E., Eglen, R.M., Bornheim, L.M. (1995) Anandamide hydroxylation by brain lipoxygenase: metabolites, structures and potencies at the cannabinoid receptor. *Biochim. Biophys. Acta-Lipids Lipid Metabol.*, 1259:173–179.

Hansen, P.R., Olesen, S.-P. (1997) Relaxation of rat resistance arteries by acetylcholine involves a dual mechanism: activation of K$^+$ channels and formation of nitric oxide. *Pharmacol. & Toxicol.*, 80:280–285.

Hansson, E., Rönnbäck, L. (1995) Astrocytes in glutamate neurotransmission. *FASEB J.*, 9:343–350.

Hanus, L., Gopher, A., Almog, S., Mechoulam, R. (1993) Two new unsaturated fatty acid ethanolamides in brain that bind to the cannabinoid receptor. *J. Med. Chem.*, 36:3032–3034.

Hara, Y., Kitamura, K., Kuriyama, H. (1980). Actions of 4-aminopyridine on vascular smooth muscle tissues of the guinea-pig. *Br. J. Pharmacol.*, 68:99–106.

Harder, D.R., Campbell, W.B., Roman, R.J. (1995) Role of cytochrome P450 enzymes and metabolites of arachidonic acid in the control of vascular tone. *J. Vasc. Res.*, 32:79–92.

Harder, D.R., Narayanan, J., Briks, E.K., Liard, J.F., Imig, J.D., Lombard, J.H., Lange, A.R., Roman, R.J. (1996) Identification of a putative microvascular oxygen sensor. *Circ. Res.*, 79:54–61.

Harder, D.R., Waters, A. (1984) Electrical activation of arterial smooth muscle. *Int. Rev. Cyt.*, 89:137–149.

Harlan, W.R., Landis, J.R., Schmouder, R.L., Goldstein, N.G., Harlan, L.C. (1985) Blood lead and blood pressure. Relationship in the adolescent and adult US population. *JAMA*, 253:530–534.

Harris, D., Kendall, D.A., Randall, M.D. (1998a) Characterization of cannabinoid receptors mediating EDHF responses in the rat isolated mesentery. *Br. J. Pharmacol.*, 125:95P.

Harris, D., Kendall, D.A., Randall, M.D. (1998b) Effects of AM404, a cannabinoid reuptake inhibitor, on EDHF-mediated relaxations in the rat isolated mesentery. *Br. J. Pharmacol.*, 125:15P.

Harrison, D.G. (1996) Endothelial control of vasomotion and nitric oxide production: a potential target for risk factor management. *Cardiol. Clin.*, 14:1–15.

Hashitani, H., Suzuki, H. (1997) K+ channels which contribute to the acetylcholine-induced hyperpolarization in smooth muscle of the guinea-pig submucosal arteriole. *J. Physiol.*, 501(2):319–329.

Hasunuma, K., Terano, T., Tamura, Y., Yoshida, S. (1991) Formation of epoxyeicosatrienoic acids from arachidonic acid by cultured rat aortic smooth muscle cell microsomes. *Prostaglandins Leukot. Essent. Fatty Acids.*, 42:171–175.

Hatake, K., Wakabayashi, I., Hishida, S. (1995) Endothelium-dependent relaxation resistant to N^ω-nitro-L-arginine in rat aorta. *Eur. J. Pharmacol.*, 274:25–32.

Hatton, C.J., Peers, C. (1996) Effects of cytochrome P-450 inhibitors on ionic currents in isolated rat type I carotid body cells. *Am. J. Physiol.*, 271:C85–C92.

Hay, M., Kunze, D.L. (1994) An intermediate conductance calcium-activated potassium channel in rat visceral sensory afferent neurons. *Neurosci. Lett.*, 167:179–182.

Hayakawa, H., Hirata, Y., Suzuki, E., Sugimoto, T., Matsuoka, H., Kikuchi, K., Nagano, T., Hirobe, M., Sugimoto, T. (1993) Mechanisms for altered endothelium-dependent vasorelaxation in isolated kidneys from experimental hypertensive rats. *Am. J. Physiol.*, 264:H1535–H1541.

He, G.W., Yang, C.D., Yang, J.A. (1997) Depolarizing cardiac arrest and endothelium-derived hyperpolarizing factor-mediated hyperpolarization and relaxation in coronary arteries. *J. Thorac. Cardiovasc. Surg.*, 113:932–941.

Hecker, M., Bara, A.T., Bauersachs, J., Busse, R. (1994) Characterization of endothelium-derived hyperpolarizing factor as a cytochrome P_{450}-derived arachidonic acid metabolite in mammals. *J. Physiol.*, 481:407–414.

Hecker, M., Fleming, I., Busse, R. (1995) Kinin-mediated activation of endothelial nitric oxide formation: Possible role during myocardial ischemia. In *Mediators in the Cardiovascular System: Regional Ischemia*, edited by K. Schrör and C.R. Pace-Asciak, pp. 119–127. Basel: Birkhäuser.

Heistad, D.D., Kontos, H.A. (1983) Handbook of physiology, Section 2: The Cardiovascular System, Volume III, Cerebral Circulation. Baltimore, M.D., *Am. Physiol. Society*, 137–182.

Hendrickx, H., Casteels, R. (1974) Electrogenic sodium pump in arterial smooth muscle cells. *Pfügers Arch.*, 346:299–306.

Henricksson, P., Hamberg, M., Diczfalusy, U. (1985) Formation of 15 HETE as a major hydroxyeicosaetraenoic acid in atherosclerotic vessel wall. *Biochem. Biophys. Acta*, 834:272–274.

Henrion, D., Dechaux, E., Dowell, F.J., Maclouf, J., Samuel, J.L., Levy, B.I., Michel, J.B. (1997) Alteration of flow-induced dilatation in mesenteric resistance arteries of L-NAME treated rats and its partial assocation with induction of cyclo-oxygenase-2. *Br. J. Pharmacol.*, 121:83–90.

Henry, Y., Lepoivre, M., Drapier, J.C., Ducrocq, C., Boncher, J.L., Guissani, A. (1993) EPR characterization of molecular targets for NO in mammalian cells and organelles. *FASEB J.*, 7:1124–1134.

Hermus, A.R., Pieters, G.F., Willemsen, J.J., Ross, H.A., Smals, A.G., Benraad, T.J., Kloppenborg, P.W. (1987) Hypotensive effects of ovine and human corticotropin-releasing factors in man. *Eur. J. Clin. Pharmacol.*, 31:531–534.

Heublein, D.M., Clavell, A., Lerman, A., Wold, L., Burnett, J.C. (1992) C-type natriuretic peptide inmunoreactivity in human breast vascular endothelial cells. *Peptides*, 13, 1017–1019.

Hewitt, N., Plane, F., Garland, C.J. (1997) Bioassay of EDHF in the rabbit isolated femoral artery. *Br. J. Pharmacol.*, 122:122P.

Higuchi, Y., Nishimura, J., Kobayashi, S., Kanaide, H. (1996) CPA induces a sustained increase in $[Ca^{2+}]_i$ of endothelial cells in situ and relaxes porcine coronary artery. *Am. J. Physiol.*, 270:H2038–H2049.

Hirata, M., Kohse, K.P., Chang, C.H., Ikebe, T., Murad, F. (1990) Mechanism of cyclic GMP inhibition of inositol phosphate formation in rat aorta segments and cultured bovine aortic smooth muscle cells. *J. Biol. Chem.*, 265:1268–1273.

Hishikawa, K., Nakati, I., Marumo, T., Suzuki, H., Kato, R., Saruta, T. (1995) Up-regulation of nitric oxide synthase by estradiol in human aortic endothelial cells. *FEBS. Lett.*, 360:291–293.

Hoebel, B.G., Graier, W.F. (1998) 11, 12-Epoxyeicosatrienoic acid stimulates tyrosine kinase activity in porcine aortic endothelial cells. *Eur. J. Pharmacol.*, 346:115–117.

Hoebel, B.G., Kostner, G.M., Graier, W.F. (1997) Activation of microsomal cytochrome P450 mono-oxygenase by Ca^{2+} store depletion and its contribution to Ca^{2+} entry in porcine aortic endothelial cells. *Br. J. Pharmacol.*, 121:1579–1588.

Hoebel, B.G., Steyrer, E., Graier, W.F. (1998) Origin and function of epoxyeicosatrienoic acids in vascular endothelial cells — More than just endothelium-derived hyperpolarizing factor? *J. Clin. Exp. Pharmacol. Physiol.*, 25:826–830.

Högestätt, E.D., Andersson, K.E., Edvinsson, L. (1983) Mechanical properties of rat cerebral arteries as studied by a sensitive device for recording of mechanical activity in isolated small blood vessels. *Acta Physiol. Scand.*, 117:49–61.

Holland, M., Plane, F., Garland, C.J., Boyle, J.P. (1997) Comparison of the vasorelaxant effects of anandamide and EDHF. *Br. J. Pharmacol.*, 122:398P.

Holzmann, S. (1982) Endothelium-induced relaxation by acetylcholine associated with larger rises in cyclic GMP in coronary arterial strips. *J. Cyclic. Nucl. Res.*, 8:409–419.

Hong, S.L., Deykin, D. (1982) Activation of phospholipases A_2 and C in pig aortic endothelial cells synthetizing prostacyclin. *J. Biol. Chem.*, 257:7151–7154.

Hongo, K., Nakagomi, T., Kassell, N.F., Sasaki, T., Lehman, M., Vollmer, D.G., Tsukahara, T., Ogawa, H., Torner, J. (1988) Effects of aging and hypertension on endothelium-dependent vascular relaxation in rat carotid artery. *Stroke*, 19:892–897.

Howlett, A.C. (1995) Pharmacology of cannabinoid receptors. *Annu. Rev. Pharmacol. Toxicol.*, 35:607–634.

Hu, S., Kim, H.S. (1993) Activation of K^+ channels in vascular smooth muscles by cytochrome P450 metabolites of arachidonic acid. *Eur. J. Pharmacol.*, 230:215–221.

Huang, A.H., Busse, R., Bassenge, E. (1988) Endothelium-dependent hyperpolarization of smooth muscle cells in rabbit femoral arteries is not mediated by EDRF (nitric oxide). *Naunyn-Schmiedeberg's Arch. Pharmacol.*, 338:438–442.

Huang, P.L., Fishman, M.C. (1996) Genetic analysis of nitric oxide synthase isoforms. Targeted mutation in mice. *J. Mol. Med.*, 74:415–421.

Huang, P.L., Huang, Z., Mashimo, H., Bloch, K.D., Moskowitz, M.A., Bevan, J.A., Fishman, M.C. (1995) Hypertension in mice lacking the gene for endothelial nitric oxide synthase. *Nature*, 377:239–242.

Hwa, J.J., Ghibaudi, L., Williams, P., Chatterjee, M. (1994) Comparison of acetylcholine-dependent relaxation in large and small arteries of rat mesenteric vascular bed. *Am. J. Physiol.*, 266:H952–H958.

Hyshikawa, K., Nakaki, T., Marumo, T., Suzuki, H., Kato, R., Saruta, T. (1995) Up-regulation of nitric oxide synthase by estradiol in human aortic endothelial cells. *FEBS. Lett.*, 360:291–293.

Iadecola, C., Zhang, F. (1994) Nitric oxide dependent and independent components of cerebrovasodilation elicited by hypercapnia. *Am. J. Physiol.*, 266:R546–R552.

Ignarro, L.J. (1989) Biological action and properties of endothelium-derived nitric oxide formed and released from artery and vein. *Circ. Res.*, 65:1–21.

Ignarro, L.J., Buga, R.E., Wood, K.S. (1997b) Endothelium-derived relaxing factor from pulmonary artery and vein possesses pharmacologic and chemical properties identical to those of nitric oxide radical. *Circ. Res.*, 61:866–879.

Ignarro, L.J., Byrns, R.E., Buga, G.M., Wood, K.S., Chaudhuri, G. (1988a) Pharmacological evidence that endothelium-derived relaxing factor is nitric oxide: use of pyrogallol and superoxide dismutase to study endothelium-dependent and nitric oxide-elicited vascular smooth muscle relaxation. *J. Pharmacol. Exp. Ther.*, 244(1):181–189.

Ignarro, L.J., Byrns, R.E., Wood, K.S. (1988b) Biochemical and pharmacological properties of endothelium-derived relaxing factor and its similarity to nitric oxide radical. In *Mechanism of Vasodilatation*, edited by P.M. Vanhoutte, pp. 427–435. New York: Raven Press.

Illiano, S., Nagao, T., Vanhoutte, P.M. (1992) Calmidazolium, a calmodulin inhibitor, inhibits endothelium-dependent relaxations resistant to nitro-L-arginine in the canine coronary artery. *Br. J. Pharmacol.*, 107:387–392.

Inoue, R., Isenberg, G. (1990) Effect of membrane potential on acetylcholine-induced inward current in guinea-pig ileum. *J. Physiol. (Lond.)*, 424:57–71.

Inoue, R., Kuriyama, H. (1993) Dual regulation of cation-selective channels by muscarinic and $\alpha 1$-adrenergic receptors in the rabbit portal vein. *J. Physiol. (Lond.)*, 465:427–448.

Irizar, A., Loannides, C. (1995) Expression and inducibility of cytochrome P450 proteins belonging to families 2, 3 and 4 in the rabbit aorta. *Biochem. Biophys. Res. Commun.*, 213:916–921.

Isenberg, G. (1993). Nonselective cation channels in cardiac and smooth muscle cells. In *Nonselective Cation Channels: Pharmacology Physiology and Biophysics*, edited by D. Siemen and J. Hescheler, pp. 247–260. Basel: Birkhauser Verlag.

Ishikawa, T., Eckman, D.M., Keef, D.K. (1997) Characterization of delayed rectifier K^+ currents in rabbit coronary artery cells near resting membrane potential. *Can. J. Physiol. Pharmacol.*, 75(9):1116–1122.

Ishii, T.M., Maylie, J., Adelman, J.P. (1997) Determinants of apamin and d-tubocurarine block in SK potassium channels. *J. Biol. Chem.*, 272:23195–23200.

Ishii, T.M., Silvia, C.C., Hirschberg, B., Bond, C.T., Adelman, J.P., Maylie, J. (1997) A human intermediate conductance calcium-activated potassium channel. *Proc. Natl. Acad. Sci. USA*, 94:11651–11656.

Ito, A., Egashira, K., Kadokami, T., Fukumoto, Y., Takayanagi, T., Nakaike, R., Kuga, T., Sueishi, D., Shimokawa, H., Takeshita A., (1995). Chronic inhibition of endothelium-derived nitric oxide synthesis causes coronary microvascular structural changes and hyperreactivity to serotonin in pigs. *Circulation*, 92:2636–2644.

Itoh, T., Ito, S., Shafiq, J., Suzuki, H. (1994) Effects of a newly synthesized K^+-channel opener, Y-26763, on noradrenaline-induced Ca^{2+} mobilization in smooth muscle of the rabbit mesenteric artery. *Br. J. Pharmacol.*, 111:165–172.

Itoh, T., Seki, N., Suzuki, S., Ito, S., Kajikuri, J., Kuriyama, H. (1992) Membrane hyperpolarization inhibits agonist-induced synthesis of inositol 1,4,5-trisphosphate in rabbit mesenteric artery. *J. Physiol.*, 451:307–328.

Iwasawa, K., Nakajima, T., Hazama, H., Goto, A., Shin, W.S., Toyo-oka, T., Omata, M. (1997) Effects of extracellular pH on receptor-mediated Ca^{2+} influx in A7r5 rat smooth muscle cells: involvement of two different types of channel. *J. Physiol. (Lond.)*, 503:237–251.

Jackson, W.F., Konig, A., Dambacher, T., Busse, R. (1993) Prostacyclin-induced vasodilation in rabbit heart is mediated by ATP-sensitive potassium channels. *Am. J. Physiol.*, 264:H238–H243.

Jacquemin, C., Thibout, H., Lambert, B., Correze, C. (1986) Endogenous ADP-ribosylation of G_s subunit and autonomous regulation of adenylate cyclase. *Nature.*, 323:182–184.

Jain, V., Ali, M., Purcell, T., Saade, G.R., Chwalisz, K., Garfield, R.E. (1998a) Expression of receptors for corticotropin-releasing factor in vasculature of pregnant rats. *J. Soc. Gynecol. Investig.*, 5(1):147A.

Jain, V., Shi, S.Q., Verdernikov, Y.P., Saade, G.R., Chwalisz, K., Garfield, R.E. (1998b) In vivo effects of corticotropin-releasing factor in pregnant rats. *Am. J. Obstet. Gynecol.*, 178:186–191.

Jain, V., Vedernikov, Y.P., Saade, G.R., Chwalisz, K., Garfield, R.E. (1997) The relaxation responses to corticotropin-releasing factor in rat aorta are endothelium dependent and gestationally regulated. *Am. J. Obstet. Gynecol.*, 176:234–240.

Johns, R.A., Linden, J.M., Peach, M.J. (1989) Endothelium-dependent relaxation and cyclic GMP accumulation in rabbit pulmonary artery are selectively impaired by moderate hypoxia. *Circ. Res.*, 65:1508–1515.

Jones, D.P. (1986) Intracellular diffusion gradients of O2 and ATP. *Am. J. Physiol.*, 250:C663–C675.

Kähönen, M., Mäkynen, H., Wu, X., Arvola, P., Pörsti, I. (1995) Endothelial function in spontaneously hypertensive rats: influence of quinapril treatment. *Br. J. Pharmacol.*, 115:859–867.

Kalsner, S. (1985) Coronary artery reactivity in human vessels: some questions and some answers. *Federation Proc.*, 44:321–325.

Karara, A., Breyer, M., Falck, J.R., Capdevilla, J.H. (1991). Epoxyeicosatrienoic acids (EETs) elevates cytosolic calcium in isolated rat hepatocytes. *FASEB J.*, 6:A1053.

Karara, A., Dishman, E., Blair, I., Falck, J.R., Capdevilla, J.H. (1989) Endogenous epoxyeicosatrienoyl-phospholipids: a novel class of cellular glycerolipids containing epoxidized arachidonate moities. *J. Biol. Chem.*, 266:7561–7569.

Kauser, K., Rubanyi, G.M. (1992) Bradykinin-induced, nitro-L-arginine-insensitive endothelium-dependent relaxation of porcine coronary artery is not mediated by bioassayable substances. *J. Cardiovasc. Pharmacol.*, 20(Suppl. 12):S101–104.

Kauser, K., Stekiel, W.J., Rubanyi, G.M., Harder, D.R. (1989) Mechanism of action of EDRF on pressurized arteries: Effect on K^+ conductance. *Circ. Res.*, 65:199–204.

Keef, K.D., Bowen SM. (1989) Effect of ACh on electrical and mechanical activity in guinea pig coronary arteries. *Am. J. Physiol.*, 257:H1096–H1103.

Kemp, B.K., Cocks, T.M. (1997) Evidence that mechanisms dependent and independent of nitric oxide mediate endothelium-dependent relaxation to bradykinin in human small resistance-like coronary arteries. *Br. J. Pharmacol.*, 120:757–762.

Kemp, B.K., Smolich, J.J., Ritchie, B.C., Cocks, T.M. (1995) Endothelium-dependent relaxations in sheep pulmonary arteries and veins: resistance to block by N^{ω}-nitro-L-arginine in pulmonary hypertension. *Br. J. Pharmacol.*, 116:2457–2467.

Kessler, P., Popp, R., Busse, R., Schini-Kerth, V.B. (1997) Proinflammatory mediators impair the synthesis of endothelium-derived hyperpolarizing factor (EDHF) in arteries through a cGMP-dependent mechanism. *J. Vasc. Res.*, 34(1):23.

Köhler, M., hirschberg, B., Bond, C.T., Kinzie, J.M., Marrion, N.V., Maylie, J., Adelman, J.P. (1996) Small-conductance, calcium-activated potassium channels from mammalian brain. *Science*, 273:1709–1714.

Khalil, M.F., Gonick, H.C., Weiler, E.W., Prins, B., Weber, M.A., Purdy, R.E. (1993) Lead-induced hypertension: possible role of endothelial factors. *Am. J. Hypertens.*, 6:723–729.

Khatsenko, O.G., Gross, S.S., Rifkind, A.B., Vane, J.R. (1993) Nitric oxide is a mediator of the decrease in cytochrome P450-dependent metabolism caused by immunostimulants. *Proc. Natl. Acad. Sci. USA*, 90:11147–11151.

Kiang, J.G. (1994) Corticotropin-releasing factor increases $[Ca^{2+}]_i$ via receptor-mediated Ca^{2+} channels in human epidermoid A-431 cells. *Eur. J. Pharmacol.*, 267:135–142.

Kilpatrick, E.V., Cocks, T.M. (1994) Evidence for differential roles of nitric oxide (NO) and hyperpolarization in endothelium-dependent relaxation of pig isolated coronary artery. *Br. J. Pharmacol.*, 112:557–565.

Kitagawa, S., Yamaguchi, Y., Kunitomo, M., Sameshima, E., Fujiwara, M. (1994) N^{ω}-nitro-L-arginine-resistant endothelium-dependent relaxation induced by acetylcholine in the rabbit renal artery. *Life Sci.*, 55:491–498.

Kleschyov, A.L., Mordvintcev, P.I., Vanin, A.F. (1985) Role of nitric oxide and iron in hypotensive action of nitrosyl iron complexes with various anion ligands. *Studia Biophys.*, 105:93–102.

Kleschyov, A.L., Muller, B., Schott, C., Stoclet, J.C. (1998a) Role of adventitia-derived nitric oxide in vascular hyporeactivity induced by lipopolysaccharide in rat aorta. *Br. J. Pharmacol.*, 124:623–626.

Kleschyov, A.L., Muller, B., Stoclet, J.C. (1998b) Localization of dinitrosyl nonheam iron complexes and their relation to ionic channels in lipopolysaccharide-treated vessels. In *The Biology of nitric oxide*, edited by S. Moncada, N. Toda, H. Maeda and E.A. Higgs, part 6, p. 300. London: Porland Press.

Knot, H.J., Zimmermann, P.A., Nelson, M.T. (1996) Extracellular K⁺-induced hyperpolarizations and dilatations of rat coronary and cerebral arteries involve inward rectifier K⁺ channels. *J. Physiol.*, 492:419–430.

Kockx, M.M., De Meyer, G.R.Y., Jacob, W.A., Bult, H., Herman, A.G. (1992) Triphasic sequence of neointimal formation in the cuffed carotid artery of the rabbit. *Arterioscler. Thromb.*, 12:1447–1457.

Koga, T., Takata, Y., Kobayashi, K., Takishita, S., Yamashita, Y., Fujishima, M. (1989) Age and hypertension promote endothelium-dependent contractions to acetylcholine in the aorta of the rats. *Hypertension*, 14:542–548.

Koga, T., Yoshida, Y., Cai, J.Q., Islam, M.O., Imai, S. (1994) Purification and characterization of 240-kDa cGMP-dependent protein kinase substrate of vascular smooth muscle: close resemblance to inositol 1,4,5-triphosphate receptor. *J. Biol Chem.*, 269:11640–11647.

Koh, S.D., Campbell, J.D., Carl, A., Sanders, K.M. (1995) Nitric oxide activates multiple potassium channels in canine colonic smooth muscle. *J. Physiol.*, 489.3:735–743.

Komalavilas, P., Lincoln, T.M. (1994) Phosphorylation of the inositol 1,4,5-trisphosphate receptor by cyclic GMP-dependent protein kinase. *J. Biol. Chem.*, 269:8701–8707.

Komori, K., Lorenz, R.R., Vanhoutte, P.M. (1988) Nitric oxide, ACh, and electrical and mechanical properties of canine arterial smooth muscle. *Am. J. Physiol.*, 255:H207–H212.

Konishi, M., Su, C. (1983) Role of endothelium in dilator responses of spontaneously hypertensive rats. *Am. J. Physiol*, 260:H1888–H1992.

Kristof, A.S., Noorhosseini, H., Hussain, S.N.A. (1997) Attenuation of endothelium-dependent hyperpolarizing factor by bacterial lipopolysaccharides. *Europ. J. Pharmacol.*, 328:69–73.

Ku, D., Guo, L., Dai, J., Acuff, C.G., Steinhelper, M.E. (1996) Coronary vascular and endothelial reactivity changes in transgenic mice overexpressing atrial natriuretic factor. *Am. J. Physiol.*, 271:H2368–H2376.

Kühberger, E., Groschner, K., Kukovetz, W.R., Brunner, F. (1994). The role of myoendothelial cell contact in non-nitric oxide-, non-prostanoid-mediated endothelium-dependent relaxation of porcine coronary artery. *Br. J. Pharmacol.*, 113:1289–1294.

Kurachi, Y., Ito, H., Sugimoto, T., Shimizu, T., Miki, I., Ui, M. (1989) Arachidonic acid metabolites as intracellular modulators of the G protein-gated cardiac K⁺ channel. *Nature.*, 337:555–557.

Kuriyama, H., Kitamura, K., Nabata, H. (1995) Pharmacological and physiological significance of ion channels and factors that modulate then in vascular tissues. *Pharmacol. Reviews*, 47:387–573.

Kuschinsy, W. (1987) Coupling of function, metabolism, and blood flow in the brain. *News Physiol. Sci.*, 2:217–220.

Lagaud, G.J.L., Stoclet, J.C., Andriantsitohaina, R. (1996) Calcium handling and purinoceptor subtypes involved in ATP-induced contraction in rat small mesenteric arteries. *J. Physiol.*, 492:689–703.

Laher, I., Vorkapic, P., Dowd, A.L., Bevan, J.A. (1989) Protein kinase C potentiates strech-induced cerebral artery tone by increasing intracellular sensitivity to Ca²⁺. *Biochem. Biophys. Res. Comm.*, 165:312–318.

Lake, K.D., Compton, D.R., Varga, K., Martin, B.R., Kunos, G. (1997) Cannabinoid-induced hypotension and bradycardia in rats is mediated by CB₁-like cannabinoid receptors. *J. Pharmacol. Experiment. Therap.*, 281:1030–1037.

Lake, K.D., Martin, B.R., Kunos, G., Varga, K. (1997) Cardiovascular effects of anandamide in anesthetized and conscious normotensive and hypertensive rats. *Hypertension*, 29:1204–1210.

Lange, A., Gebremedhin, D., Narayanan, J., Harder, D.R. (1997) 20-Hydroxyeicosatetraenoic acid-induced vasoconstriction and inhibition of potassium current in cerebral vascular smooth muscle is dependent on activation of protein kinase, C. *J. Biol. Chem.*, 272:27345–27352.

Large, W.A., Wang, Q. (1996) Characteristics and physiological role of the Ca²⁺-activated Cl-conductance in smooth muscle. *Am. J. Physiol.*, 271:C435–C454.

Latorre, R., Oberhauser, A., Labarca, P., Alvarez, O. (1989) Varietes of calcium-activated potassium channels. *Annual Review of Physiol.*, 51:385–399.

Lawrence, R.N., Clelland, C., Beggs, D., Salama, F.D., Dunn, W.R., Wilson, V.G. (1998) Differential role of vasoactive prostanoids in the porcine and human isolated pulmonary arteries in response to endothelium-dependent relaxants. *Br. J. Pharmacol.*, 125:1128–1137.

Lawrence, R.N., Dunn, W.R., Wilson, V.G. (1997) Evidence for different mechanisms of relaxation by ethanol in isolated pulmonary and coronary arteries from the pig. *Br. J. Pharmacol.*, 122:140P.

Lee, M.W., Severson, D.L. (1994) Signal transduction in vascular smooth muscle: diacylglycerol second messengers and PKC action. *Am. J. Physiol.*, 267:C659–C678.

Lei, S., Richter, R., Beinert, M., Mulvany, M.J. (1993) Relaxing actions of corticotropin-releasing factor on rat resistance arteries. *Br. J. Pharmacol.*, 108:941–947.

Li, P.L., Campbell, W.B. (1997) Epoxyeicosatrienoic acids activate K⁺ channels in coronary smooth muscle through a guanine nucleotide binding protein. *Circ. Res.*, 80:877–884.

Li, P.L., Zou, A.P., Al-Kayed, N.J., Rusch, N.J., Harder, D.R. (1994) Guanine nucleotide-binding proteins in aortic smooth muscle from hypertensive rats. *Hypertension.*, 23:914–918.

Li, P.L., Zou, A.P., Campbell, W.B. (1997) Regulation of potassium channels in coronary arterial smooth muscle by endothelium-derived vasodilators. *Hypertension*, 29:262–267.

Lieberman, J. (1975) Elevation of serum angiotesin-converting enzyme (ACE) level in sarcoidosis. *Am. J. Med.*, 59:365–372.

Lin, L.L., Lin, A.Y., Knopf, J.L. (1992) Cytosolic phospholipase A_2 is coupled to hormonally regulated release on arachidonic acid. *Proceedings of the National Academy of Sciences of United States of Amercia.*, 89:6147–6151.

Lincoln, T.M., Cornwell, T.L., Komalavilas, P., MacMillan-Crow, L., Boerth, N. (1996) The nitric oxide-cyclic GMP signaling system. In *Biochemistry of smooth muscle contraction*, pp. 257–268. Academic Press Inc.

Linder, L., Kiowski, W., Bühler, F.R., Lüscher, T.F. (1990) Indirect evidence for release of endothelium-derived relaxing factor in human forearm circulation *in vivo*: Blunted response in essential hypertension. *Circulation*, 81:1762–1767.

Lischke, V., Busse, R., Hecker, M. (1995a) Selective inhibition by barbiturates of the synthesis of endothelium-derived hyperpolarizing factor in the rabbit carotid artery. *Br. J. Pharmacol.*, 115:969–974.

Lischke, V., Busse, R., Hecker, M. (1995b) Inhalation anesthetics inhibit the release of endothelium-derived hyperpolarizing factor in the rabbit carotid artery. *Anesthesiology*, 83:574–582.

Lockette, W., Otsuka, Y., Carretero, O. (1986) The loss of endothelium-dependent vascular relaxation in hypertension. *Hypertension*, 8:II61–II66.

Loesch, A., Burnstock, G. (1996) Immunocytochemistry of vasoactive agents and nitric oxide synthase in vascular endothelial cells with emphasis on the cerebral blood vessels. *Cell. Vision*, 3:346–357.

Logsdon, N.J., Kang, J., Togo, J.A., Christian, E.P., Aiyar, J. (1997) A novel gene, hKCa4, encodes the calcium-activated potassium channel in human T lymphocytes. *J. Biol. Chem.*, 272:32723–32776.

Lonigro, A.J., Weintraub, N.L., Branch, C.A., Stephenson, A.H., McMurdo, L., Sprague, R.S. (1994) Endothelium-dependent relaxation to arachidonic acid in porcine coronary artery: Is there a fourth pathway? *Pol. J. Pharmacol.*, 46, 567–577.

López-Redondo, F., Lees, G.M., Pertwee, R.G. (1997) Effects of cannabinoid receptor ligands on electrophysiological properties of myenteric neurones of the guinea-pig ileum. *Br. J. Pharmacol.*, 122:330–334.

Loscalzo, J., Welch, G. (1995) Nitric oxide and its role in the cardiovascular system. *Prog. Cardiovasc. Dis.*, 38:87–104.

Lot, T.L., Starke, G., Wilson, V.G. (1993) Endothelium-dependent contractions to N^ω-nitro-L-arginine methyl ester in the porcine isolated splenic artery are sensitive to cyclooxygenase and lipoxygenase inhibitors. *Naunyn-Schmiedeberg's Arch. Pharmacol.*, 347:115–118.

Lückhoff, A., Busse, R. (1986) Increased free calcium in endothelial cells under stimulation with adenine nucleotides. *J. Cell. Physiol.*, 126:414–420.

Lückhoff, A., Busse, R. (1990) Calcium influx into endothelial cells and formation of EDRF is controlled by the membrane potential. *Pflügers Arch.*, 416:305–311.

Lückhoff, A., Pohl, U., Mülsch, A., Busse, R. (1988) Differential role of extra- and intracellular calcium in the release of EDRF and prostacyclin from cultured endothelial cells. *Br. J. Pharmacol.*, 95:189–196.

Ludbrook, J. (1994) Repeated measurements and multiple comparisons in cardiovascular research. *Cardiovasc. Res.*, 28:303–311.

Lüscher, T.F., Boulanger, C.M., Dohi, Y., Yang, Z. (1992) Endothelium-derived contracting factors. *Hypertension*, 19:117–130.

Lüscher, T.F., Boulanger, C.M., Yang, Z., Noll, G., Dohi, Y. (1993) Interactions between endothelium-derived relaxing and contracting factors in health and cardiovascular disease. *Circulation*, 87:V36–V44.

Lüscher, T.F., Diederich, D., Siebenmann, R., Lehmann, K., Stutlz, P., von Segesser, L., Yang, Z., Turina, M., Gradel, E., Weber, E., Buhler, F.R. (1988) Difference between endothelium-dependent relaxation in arterial and in venous coronary bypass graft. *New Engl. J. Med.*, 319:462–467.

Lüscher, T.F., Vanhoutte, P.M., Raij, L. (1987) Antihypertensive treatment normalizes decreased endothelium-dependent relaxations in salt-induced hypertension of the rat. *Hypertension*, 9:193–197.

Lüscher, T.F., Vanhoutte, P.M. (1986a) Endothelium-dependent contractions to acetylcholine in the aorta of the spontaneously hypertensive rat. *Hypertension*, 8:344–348.

Lüscher, T.F., Vanhoutte, P.M. (1986b) Endothelium-dependent responses to platelets and serotonin in spontaneously hypertensive rats. *Hypertension*, 8:II55–II60.

Lüscher, T.F., Vanhoutte, P.M. (1990) The endothelium modulator of cardiovascular function, pp. 1–228. Boca Raton, Fla: CRC Press, Inc.

Ma, Y.H., Harder, D.R., Clark, J.E., Roman, R.J. (1991) Effects of 12 HETE on isolated dog renal arcuate arteries. *Am. J. Physiol.*, 261:H451–H456.

Mackie, K., Hille, B. (1992) Cannabinoids inhibit n-type calcium channels in neuroblastoma glioma-cells. *Proc. Nat. Acad. Sci. (USA)*, 89:3825–3829.

Madeja, M., Binding, N., Mulhoff, U., Pongs, O., Witting, U., Speckmann, E.J. (1995) Effects of lead on cloned voltage-operated neuronal potassium channels. *Naunyn-Schmiedeberg's Arch. Pharmacol.*, 351:320–327.

Madhun ZT, Goldthwait, D.A., McKay, D., Hopfer, U., Douglas, J.G. (1991). An epoxygenase metabolite of arachidonic acid mediates angiotensin II-induced rises in cytosolic calcium in rabbit proximal tubule epithelial cells. *J. Clin. Invest.*, 88:456–461.

Malvano, R., Zucchelli, G.C., Rosa, U., Salvetti, A. (1972) Measurement of plasma renin activity by angiotensin I radioimmunoassay (I): An assessment of some methodological aspects. *J. Nucl. Biol. Med.*, 16:24–31.

Mantelli, L., Amerini, S., Ledda, F. (1995) Role of nitric oxide and endothelium-derived hyperpolarizing factor in vasorelaxant effect of acetylcholine as influenced by aging and hypertension. *J. Cardiovasc. Pharmacol.*, 25:595–602.

Marchenko, S.M., Sage SO. (1996) Calcium-activated potassium channels in the endothelium of intact rat aorta. *J. Physiol.*, 492:53–60.

Martin, W., Villani, G.M., Jothianandan, D., Furchgott, R.F. (1985) Selective blockade of endothelium-dependent and glyceryl trinitrate-induced relaxation by hemoglobin and by methylene blue in rabbit aorta. *J. Pharmacol. Exp. Ther.*, 232:708–716.

Mathews, J.M., Dostal, L.A., Bend, J.R. (1985) Inactivation of rabbit pulmonary cytochrome P-450 in microsomes and isolated perfused lungs by the suicide substrate 1-aminobenzotriazole. *J. Pharmacol. Exp. Ther.*, 235:186–190.

Matsuda, L.A., Lolait, S.J., Brownstein, M.J., Young, A.C., Bonner TI. (1990) Structure of a cannabinoid receptor and functional expression of the cloned cDNA. *Nature*, 346:561–564.

Matthys, K.E., Bult, H. (1997) Nitric oxide function in atherosclerosis. *Mediators of Inflammation*, 6:3–21.

Matthys, K.E., Van Hove, C.E., Kockx, M.M., Andries, L.J., Van Ossselaer, N., Herman, A.G., Bult, H. (1998) Exposure to oxidized low-density lipoprotein in vivo enhances intimal thickening and selectively impairs endothelium-dependent dilation in the rabbit. *Cardiovasc. Res.*, 37:239–246.

Mayer, R.J., Marshall, L.A. (1993) New insights on mammalian phospholipase A2(s); comparison of arachidonoyl-selective and -nonselective enzymes. *FASEB J.*, 7:339–348.

McCarron, J.G., Halpern, W. (1990) Potassium dilates rat cerebral arteries by two independent mechanisms. *Am. J. Physiol.*, 259:H902–H908.

McCulloch, A.I., Bottril, F.E., Randall, M.D., Robin Hiley, C. (1997) Characterization and modulation of EDHF-mediated relaxations in the rat isolated superior mesenteric arterial bed. *Br. J. Pharmacol.*, 120:1431–1438.

McGiff, J.C. (1991) Cytochrome P-450 metabolism of arachidonic acid. *Annu. Rev. Pharmacol. Toxicol.*, 31:339–369.

McIntyre, M., Hamilton, C.A., Rees, D.D., Reid, J.L., Dominiczak, A.F. (1997) Sex differences in the abundance of endothelial nitric oxide in a model of genetic hypertension. *Hypertension*, 30:1517–1524.

Mechoulam, R., Ben-Shabat, S., Hanus, L., Ligumsky, M., Kaminski, N.E., Schatz, A.R., Gopher, A., Almog, S., Martin, B.R., Compton, D.R. (1995) Identification of an endogenous 2-monoglyceride, present in canine gut, that binds to cannabinoid receptors. *Biochem. Pharmacol.*, 50:83–90.

Mery, P.F., Pavoine, C., Belhassen, L., Pecker, F., Fischmeister, R. (1993) Nitric oxide regulates cardiac Ca^{2+} current: involvement of cGMP-inhibited and cGMP-stimulated phosphodiesterases through guanylyl cyclase activation. *J. Biol. Chem.*, 268:26286–26295.

Minami, K., Fukuzawa, K., Nakaya, Y. (1993) Protein kinase C inhibits Ca^{2+}-activated K+ channel of cultured porcine coronary artery smooth muscle cells. *Biochem. Biophys. Res. Commun.*, 190:263–269.

Minova, T., Miwa, S., Kobayashi, S., Enoki, T., Zhang, X.F., Komuro, T., Iwamuro, Y., Masaki, T. (1997) Inhibitory effect of nitrovasodilators and cyclic GMP on ET-1-activated Ca^{2+}-permeable nonselective cation channel in rat aortic smooth muscle cells. *Br. J. Pharmacol.*, 120:1536–1544.

Mistry, D.K., Garland, C.J. (1998) Characteristics of single, large-conductance calcium-dependent potassium channels (BKCa) from smooth muscle cells isolated from the rabbit mesenteric artery. *J. Membrane Biol.*, 164:125–138.

Miyoshi, H., Nakaya, Y., Moritoki, H. (1994) Nonendothelial-derived nitric oxide activates the ATP-sensitive K channel of vascular smooth muscle cells. *FEBS*, 345:47–49.

Molina, Y., Vedia, L., Nolan, R.D., Lapetina, E.G. (1989) The effect of iloprost on the ADP-ribosylation of $G_s\alpha$ (the α-subunit of Gs). *Biochem. J.*, 261:841–845.

Mombouli, J.V., Bissiriou, I., Agboton, V.D., Alvear, J., Chelly, F., Kilbourn, R., Vanhoutte, P.M. (1996) Endotoxemia reduces the contribution of endothelium-derived hyperpolarizing factor, but augments that of prostanoids in canine coronary arteries. In *Endothelium-derived hyperpolarizing factor*, edited by P.M. Vanhoutte, Vol. 33, 271–278. The Netherlands: Harwood Academic Publishers.

Mombouli, J.V., Illiano, S., Nagao, T., Scott-Burden, T., Vanhoutte, P.M. (1992b) Potentiation of endothelium-dependent relaxations to bradykinin by angiotensin I converting enzyme inhibitors in canine coronary artery involves both endothelium-derived relaxing and hyperpolarizing factors. *Circ. Res.*, 71:137–144.

Mombouli, J.V., Vanhoutte, P.M. (1993) Purinergic endothelium dependent and independent contractions in rat aorta. *Hypertension*, 22:577–583.

Mombouli, J.V., Vanhoutte, P.M. (1995) Endothelium-derived hyperpolarizing factor(s) and the potentiation of kinins by converting enzyme inhibitors. *Am. J. Hypert.*, 8:19S–27S.

Mombouli, J.V., Vanhoutte, P.M. (1997) Endothelium-derived hyperpolarizing factor(s): updating the unknown. *Trends Pharmacol. Sci.*, 18:252–256.

Moncada, S., Flower, R.J., Vane, J.R. (1985) Prostaglandins, prostacyclin, thromboxane A_2 and leukotrienes. In *The Pharmacological Basis of Therapeutics*, edited by A.G. Gilman, L.S. Goodman, T.W. Rall and F. Murad, pp. 660–673. New-York: Macmillan.

Moncada, S., Herman, A.G., Higgs, E.A., Vane, J.R. (1977) Differential formation of prostacyclin (PGX or PGI_2) by layers of the arterial wall. An explanation for the anti-thrombotic properties of vascular endothelium. *Thromb. Res.*, 11:323–344.

Moncada, S., Higgs, A. (1993) The L-arginine-nitric oxide pathway. *N. Engl. J. Med.*, 329:2002–2012.

Moncada, S., Palmer, R.M.J. (1990) The L-arginine: nitric oxide pathway in the vessel wall. In *Nitric Oxide from L-arginine: A Bioregulatory System*, edited by S. Moncada and E.A. Higgs, pp. 19–33. Amsterdam: Elsevier.

Moncada, S., Palmer, R.M.J., Higgs, E.A. (1991) Nitric oxide: Physiology, pathophysiology, and Pharmacology. *Pharmacol. Rev.*, 43:109–142.

Moncada, S., Vane, J.R. (1979) Pharmacology and endogenous roles of prostaglandin endoperoxydes, thromboxane A2 and prostacyclin. *Pharmacol. Rev.*, 30:293–331.

Montero, M., Alvarez, J., Garcia-Sancho, J. (1992) Control of plasma-membrane Ca^{2+} entry by the intracellular Ca^{2+} stores-kinetic evidence for a short-lived mediator. *Biochem. J.*, 288:519–525.

Montero, M., Garcia-Sancho, J., Alvarez, J. (1993) Inhibition of the calcium store-operated calcium entry pathway by chemotactic peptide and by phorbole ester develops gradually and independently along differentiation of HL60 cells. *J. Biol. Chem.*, 268:26911–26919.

Moore, S.A., Spector, A.A., Hart, M.N. (1988) Eicosanoid metabolism in cerebromicrovascular endothelium. *Am. J. Physiol.*, 254:C37–C44.

Morel, N., Ghisdal, P., Godfraind, T. (1996) Hyperpolarization to acetylcholine in the mesenteric artery from NaCl-loaded hypertensive rats. In *Endothelium-derived hyperpolarizing factor*, edited by P.M. Vanhoutte, Vol. 31, pp. 255–261. The Netherlands: Harwood Academic Publishers.

Morgan, K.G. (1993) Ca^{2+}_i versus $[Ca^{2+}]_i$. *Biophys. J.*, 65:561–562.

Mügge, A., Lopez, J.A.G., Piegors, D.J., Breese KR, Heistad, D.D. (1991) Acetylcholine-induced vasodilatation in rabbit hindlimb in vivo is not inhibited by analogues of L-arginine. *Am. J. Physiol.*, 260:H242–H247.

Muller, B., Kleschyov, A.L., Malblanc, S., Stoclet, J.C. (1998) Nitric oxide-related cyclic GMP-independent effect of N-acetylcysteine in lipopolysaccharide treated rat aorta. *Br. J. Pharmacol.*, 123:1221–1229.

Muller, B., Kleschyov, A.L., Stoclet, J.C. (1996) Evidence for N-acetylcysteine-sensitive nitric oxide stores as dinitrosyl-iron complexes in lipopolysaccharide-treated rat aorta. *Br. J. Pharmacol.*, 119:1281–1285.

Mülsch, A., Mordvintcev, P., Vanin, A.F., Busse, R. (1991) The potent and guanylyl cyclase activating dinitrosyl-iron (II) complex is stored in a protein-bound form in vascular tissue and is released by thiols. *FEBS Lett.*, 294:252–256.

Mülsch, A. (1994) Nitrogen monoxide transport mechanism. *Arzneimittelforschung.*, 44:408–411.

Mulvany, M.J., Aalkjaer, C. (1990) Structure and function of small arteries. *Physiological Reviews*, 70:921–961.

Mulvany, M.J., Halpern, W. (1977). Contractile properties of small arterial resistance vessels in spontaneously hypertensive and normotensive rats. *Circ. Res.*, 41:19–26.

Muraki, K., Imaizumi, Y., Ohya, S., Sato, K., Takii, T., Onozaki, K., Watanabe, M. (1997) Apamin-sensitive Ca^{2+}-dependent K^+ current and hyperpolarization in human endothelial cells. *Biochem. Biophys. Res. Commun.*, 236:340–343.

Murphy, M.E., Brayden, J.E. (1995a) Nitric oxide hyperpolarization of rabbit mesenteric arteries via ATP-sensitive potassium channels. *J. Physiol.*, 486:47–58.

Murphy, M.E., Brayden, J.E. (1995b) Apamin-sensitive K^+ channels mediate an endothelium-dependent hyperpolarization in rabbit mesenteric arteries. *J. Physiol.*, 489:723–734.

Murray, M., Reidy, G.F. (1990) Selectivity in the inhibition of mammalian cytochromes P-450 by chemical agents. *Pharmacol. Rev.*, 42:85–101.

Myers, P.R., Muller, J.M., Tanner, M.A. (1991) Effects of oxygen tension on endothelium dependent responses in canine coronary microvessels. *Cardiovasc. Res.*, 25:885–894.

Nagao, T., Illiano, S.C., Vanhoutte, P.M. (1992a) Heterogeneous distribution of endothelium-dependent relaxations resistant to N^ω-nitro-L-arginine in rats. *Am. J. Physiol.*, 263:H1090–H1094.

Nagao, T., Illiano, S.C., Vanhoutte, P.M. (1992b) Calmodulin antagonists inhibit endothelium-dependent hyperpolarization in the canine coronary artery. *Br. J. Pharmacol.*, 107:382–386.

Nagao, T., Vanhoutte, P.M. (1991). Hyperpolarization contributes to endothelium-dependent relaxations to acetylcholine in femoral veins of rats. *Am. J. Physiol.*, 261:H1034–H1037.

Nagao, T., Vanhoutte, P.M. (1992a) Hyperpolarization as a mechanism for endothelium-dependent relaxations in the porcine coronary artery. *J. Physiol.*, 445:355–367.

Nagao, T., Vanhoutte, P.M. (1992b) Characterization of endothelium-dependent relaxations resistant to nitro-L-arginine in the porcine coronary artery. *Br. J. Pharmacol.*, 107:1102–1107.

Nagao, T., Vanhoutte, P.M. (1993) Endothelium-derived hyperpolarizing factor and endothelium-dependent relaxations. *Am. J. Respir. Cell. Mol. Biol.*, 8:1–6.

Najibi, S., Cohen, R.A. (1995) Enhanced role of potassium channels in relaxation of hypercholesterolemic rabbit carotid artery to nitric oxide and sodium nitroprusside. *Am. J. Physiol.*, 269:H805–H811.

Najibi, S., Cowan, C.L., Palacino, J.J., Cohen, R.A. (1994) Enhanced role of potassium channels in relaxations to acetylcholine in hypercholesterolemic rabbit carotid artery. *Am. J. Physiol.*, 266:H2061–H2067.

Nakashima, M., Mombouli, J.V., Taylor, A.A., Vanhoutte, P.M. (1993) Endothelium-dependent hyperpolarization caused by bradykinin in human coronary arteries. *J. Clin. Invest.*, 92:2867–2871.

Nakashima, M., Vanhoutte, P.M. (1993a) Age-dependent decrease in endothelium-dependent hyperpolarizations to endothelin-3 in the rat mesenteric artery. *J. Cardiovasc. Pharmacol.*, 22:S352–S354.

Nakashima, M., Vanhoutte, P.M. (1993b) Endothelin-1 and -3 cause endothelin-dependent hyperpolarization in the rat mesenteric artery. *Am. J. Physiol.*, 265:H2137–H2141.

Narayanan, J., Imig, M., Roman, R.J., Harder, D.R. (1994) Pressurization of isolated renal arteries increases inositol triphosphate and diacylglycerol. *Am. J. Physiol.*, 266:H1840–H1845.

Narumiya, S., Salmon, J.A., Cottee, F.H., Weatherley, B.C., Flower, R.J. (1981) Arachidonic acid 15-lipoxygenase from rabbit peritoneal polymorphonuclear leukocytes. *J. Biol. Chem.*, 256:9583–9592.

Needleman, P., Turk, J., Jakschik, B.A., Morrison, A.R., Lefkowith, J.R. (1986) Arachidonic acid metabolism. *Ann Rev. Biochem.*, 55:69–102.

Nelson, M.T., Cheng, H., Rubart, M., Santana LF, Bonev, A.D., Knot, H.J., Lederer, W.J. (1995) Relaxation of arterial smooth muscle by calcium sparks. *Science*, 270:633–637.

Nelson, M.T., Quayle, J.M. (1995) Physiological roles and properties of potassium channels in arterial smooth muscle. *Am. J. Physiol.*, 268:C799–C822.

Nelson, P.R., Yamamura, S., Mureebe, L., Itoh, H., Kent, K.C. (1998) Smooth muscle cell migration and proliferation are mediated by distinct phases of activation of the intracellular messenger mitogen-activated protein kinase. *J. Vasc. Surg.*, 27:117–125.

Neri, L.C., Hewitt, D., Orser, B. (1988) Blood lead and blood pressure: analysis of cross-sectional and longitudinal data from Canada. *Environ Health Perspect.*, 78:123–126.

Ngai, A.C., Winn, R.H. (1995) Modulation of cerebral arteriolar diameter by intraluminal flow and pressure. *Circ. Res.*, 77:832–840.

Nishimura, J., Kolber, M., Van Breemen, C. (1988) Norepinephrine and GTP(s increase myofilament Ca $^{2+}$ sensitivity in α-toxin permeabilized arterial smooth muscle. *Biochem. Biophys. Res. Comm.*, 157:677–683.

Nishiye, E., Nakao, K., Itoh, T., Kuriyama, H. (1989) Factors inducing endothelium-dependent relaxation in the guinea-pig basilar artery as estimated from the actions of haemoglobin. *Br. J. Pharmacol.*, 96:645–655.

Ohgushi, M., Kugiyama, K., Fukunaga, K., Murohara, T., Sugiyama, S., Miyamoto, E., Yasue H., (1993) Protein kinase C inhibitors prevent impairment of endothelium-dependent relaxation by oxidatively modified *LDL*. *Arterioscler. Thromb.*, 13:1525–32.

Ohlmann, P., Martinez, M.C., Schneider, F., Stoclet, J.C., Andriantsitohaina, R. (1997) Characterization of endothelium-derived relaxing factors released by bradykinin in human resistance arteries. *Br. J. Pharmacol.*, 121:657–664.

Oishi, H., Nomiyama, H., Nomiyama, K., Tomokuni, K. (1999) Fluorometric HPLC determination of delta-aminolevulinic acid (ALA) in plasma and urine of lead workers; biological indicators for lead exposure. *J. Anal. Toxicol. (in press)*.

Okada, Y., Yanagisawa, T., Taira, N. (1993) BRL 38227(levcromakalim)-induced hyperpolarization reduces the sensitivity to Ca^{2+} of contractile elements in canine coronary artery. *Naunyn-Schmiedeberg's Arch. Pharmacol.*, 347:438–444.

Okey AB. (1990) Enzyme induction in the cytochrome P450 system. *Pharmac. Ther.*, 45:241–298.

Oliveira, M., Antunes, E., De Nucci, G., Lovisolo, S.M., Zatz, R. (1992) Chronic inhibition of nitric oxide synthesis. A new model of arterial hypertension. *Hypertension*, 20:298–303.

Olmos, L., Mombouli, J.V., Illiano, S., Vanhoutte, P.M. (1995) cGMP mediates the desensitization to bradydinin in isolated canine coronary arteries. *Am. J. Physiol.*, 268:H865–H870.

Olson, L.J., Knych, J.R., Herzig, T.C., Drewett, J.G. (1997) Selective guanylyl cyclase inhibitor reverses nitric oxide-induced vasorelaxation. *Hypertension*, 29:254–261.

Onaka, U., Fujii, K., Abe, I., Fujishima, M. (1998) Antihypertensive treatment improves endothelium-dependent hyperpolarization in the mesenteric artery of spontaneously hypertensive rats. *Circulation*, 98:175–182.

Onaka, U., Fujii, K., Takemoto, M., Egashira, K., Abe, I., Takeshita, A., Fujishima, M. (1996) Effects of chronic NO synthase inhibition on endothelium-dependent hyperpolarization in rat artery *Hypert. Res.*, 19:317 (Abstract).

Osol, G., Cipolla, M., Knutson, S. (1989) A new method for mechanically denuding the endothelium of small (50–150 microns) arteries with a human hair. *Blood Vessels*, 26:320–324.

Osol, G., Laher, I., Kelly, M. (1993) Myogenic tone is coupled to phospholipase C and G protein activation in small cerebral arteries. *Am. J. Physiol.*, 265:H415–H420.

Oyekan AO, McGiff, J.C., Rosencrantz-Weiss, P., Quilley, J. (1994) Relaxant responses of rabbit aorta: influence of cytochrome P450 inhibitors. *J. Pharmacol. Exp. Ther.*, 268:262–269.

Pace-Asciak, C.R., Asotra, S. (1989) Biosynthesis, catabolism, and biological properties of HPETEs, hydroperoxide derivatives of arachidonic acid. *Free Radical Biology and Medecine*, 7:409–433.

Pace-Asciak, C.R., Granstrom, E., Samuelsson, B. (1983) Arachidonic acid epoxides: Isolation and Structure of two hydroxy epoxide intermediates in the formation of 8, 11, 12 and 10, 11, 12-trihydroxyeicosatrienoic acids. *J. Biol. Chem.*, 258, 6835–6840.

Pacicca, C., von der Weid, P., Beny, J.L. (1992) Effect of nitro-L-arginine on endothelium-dependent hyper-polarizations and relaxations of pig coronary arteries. *J. Physiol.*, 457:247–256.

Palmer RMJ, Ferridge, A.G., Moncada, S. (1987) Nitric oxide release accounts for the biological activity of endothelium-derived relaxing factor. *Nature*, 327:524–526.

Paltauf-Dobruzynska, J., Graier, W.F. (1997) Temperature dependence of agonist-stimulated Ca^{2+} signaling in cultured endothelial cells. *Cell. Calcium*, 21:43–51.

Panza, J.A., Casino, P.R., Kilcoyne, C.M., Quyyumi, A.A. (1993) Role of endothelium-derived nitric oxide in the abnormal endothelium-dependent vascular relaxation of patients with essential hypertension. *Circulation*, 87:1468–1474.

Panza, J.A., Garcia, C.E., Kilcoyne, C.M., Quyyumi, A., Cannon III, R.O. (1995) Impaired endothelium-dependent vasodilatation in patients with essential hypertension. Evidence that nitric oxide abnormality is not localised to a single signal transduction pathway. *Circulation*, 91:1732–1738.

Panza, J.A., Quyyumi, A.A., Brush, J.E. Jr., Epstein, S.E. (1990) Abnormal endothelium-dependent vascular relaxation in patients with essential hypertension. *N. Engl. J. Med.*, 323:22–27.

Parkington, H.C., Tare, M., Tonta, M.A., Coleman, H.A. (1993) Stretch revealed three components in the hyperpolarization of guinea-pig coronary artery in response to acetylcholine. *J Physiol*, 465:459–476.

Parkington, H.C., Tonta, M., Coleman, H., Tare, M. (1995). Role of membrane potential in endothelium-dependent relaxation of guinea-pig coronary arterial smooth muscle. *J. Physiol.*, 484:469–480.

Parsaee, H., Mcewan, J.R., Joseph, S., Macdermot, J. (1992) Differential sensitivities of the prostacyclin and nitric oxide biosynthetic pathways to cytosolic calcium in bovine aortic endothelial cells. *Br. J. Pharmacol.*, 107:1013–1019.

Pascoal, I.F., Lindheimer, M.D., Nalbantian-Brandt, C., Umans, J.G. (1998) Preeclampsia selectively impairs endothelium-dependent relaxation and leads to oscillatory activity in small omental arteries. *J. Clin. Invest.*, 101:464–470.

Pascoal, I.F., Umans, J.G. (1996) Effect of pregnancy on mechanisms of relaxation in human omental microvessels. *Hypertension*, 28:183–187.

Pearce, W.J., Harder, D.R. (1996) Cerebrovascular smooth muscle and endothelium. In *Neurophysiological basis of cerebral blood flow control. An introduction*, edited by S. Marovitch and R. Sercombe, pp. 145–175. London: John Libbey & Co. Ltd.

Pearson, P.J., Schaff, H.V., Vanhoutte, P.M. (1990) Acute impairment of endothelium-dependent relaxations to aggregating platelets following reperfusion injury in canine coronary arteries. *Circ. Res.*, 67:385–393.

Perrault, L.P. (1997) Coronary endothelial dysfunction after porcine heart transplantation. ScD Thesis. Strasbourg, France: Universite Louis Pasteur.

Perrault, L.P., Bidouard, J.P., Jacquemin, C., Petit, C., Villeneuve, N., Vilaine, J.P., Vanhoutte, P.M. (1997a) Cyclic GMP is decreased after heterotopic heart transplantation. *FASEB J.*, II:A246(abstract).

Perrault, L.P., Bidouard, J.P., Janiak, P., Bruneval, P., Villeneuve, N., Vilaine, J.P., Vanhoutte, P.M. (1997b) Time course and mechanisms of coronary endothelial dysfunction during acute untreated rejection after heart transplantation. *J. Heart Lung Transplant.*, 16:643–657.

Perrault, L.P., Bidouard, J.P., Janiak, P., Villeneuve, N., Vilaine, J.P., Vanhoutte, P.M. (1996) Impairment of G-protein function and endothelial dysfunction after heart transplantation. *FASEB J.*, 10:A572.

Perkins, A.V., Linton, E.A. (1995) Placental corticotrophin-releasing hormone: there by accident or design? *J. Endocrinol.*, 147:377–381.

Petersson, J., Zygmunt, P.M., Brandt, L., Högestätt, E.D. (1995) Substance P-induced relaxation and hyperpolarization in human cerebral arteries. *Br. J. Pharmacol.*, 115:889–894.

Petersson, J., Zygmunt, P.M., Brandt, L., Högestätt, E.D. (1996) Role of hyperpolarization in endothelium-dependent relaxations of human cerebral arteries. In *Endothelium-derived hyperpolarizing factor*, edited by P.M. Vanhoutte, Vol. 35, pp. 287–292. The Netherlands: Harwood Academic Publishers.

Petersson, J., Zygmunt, P.M., Högestätt, E.D. (1996) Charybdotoxin and apamin inhibit relaxations mediated by EDHF in the guinea-pig basilar artery. *Br. J. Pharmacol. Proc. Suppl.*, 117:119P.

Petersson, J., Zygmunt, P.M., Högestätt, E.D. (1997a) Characterization of the potassium channels involved in EDHF-mediated relaxation in cerebral arteries. *Br. J. Pharmacol.*, 120:1344–1350.

Petersson, J., Zygmunt, P.M., Jönsson, P., Högestätt, E.D. (1997b) Involvement of derivatives of arachidonic acid in endothelium-dependent relaxations mediated by EDHF in the guinea-pig basilar artery. *Br. J. Pharmacol. Proc. Suppl.*, 122:401P.

Petersson, J., Zygmunt, P.M., Jönsson, P., Högestätt, E.D. (1998) Characterization of endothelium-dependent relaxation in guinea pig basilar artery — effect of hypoxia and role of cytochrome P450 mono-oxygenase. *J. Vasc. Res.*, 35:274–284.

Pfister, S.L., Campbell, W.B. (1992) Arachidonic acid-and acetylcholine-induced relaxations of rabbit aorta. *Hypertension*, 20:682–689.

Pfister, S.L., Falck, J.R., Campbell, W.B. (1991) Enhanced synthesis of epoxyeicosatrienoic acids by cholesterol-fed rabbit aorta. *Am. J. Physiol.*, 261:H843–H852.

Pfister, S.L., Schmitz, J.M., Willerson, J.T., Campbell, W.B. (1988) Characterization of arachidonic acid metabolism in Watanabe Heritable Hyperlipidemic (WHHL) and New Zealand White (NZW) rabbit aortas. *Prostaglandins*, 36:515–531.

Pfister, S.L., Spitzbarth, N., Edgemond, W., Campbell, W.B. (1996) Vasorelaxation by an endothelium-derived metabolite of arachidonic acid. *Am. J. Physiol.*, 270:H1021–H1030.

Piccinini, F., Favalli, L., Chiari, M.C. (1977) Experimental investigations on the contraction induced by lead in arterial smooth muscle. *Toxicology*, 8:43–51.

Pinto, A., Abraham, N.G., Mullane, K.M. (1986) Cytochrome P450-dependent monooxygenase activity and endothelial-dependent relaxations induced by arachidonic acid. *J. Pharmacol. Exp. Ther.*, 236:445–451.

Pinto, A., Abraham, N.G., Mullane, K.M. (1987) Arachidonic acid-induced endothelial-dependent relaxations of canine coronary arteries: Contribution of a cytochrome P-450 dependent pathway. *J. Pharmacol. Exp. Ther.*, 241:763–770.

Plane, F., Garland, C.J. (1993) Differential effects of acetylcholine, nitric oxide and levcromakalim on smooth muscle membrane potential and tone in the rabbit basilar artery. *Br. J. Pharmacol.*, 110:651–656.

Plane, F., Holland, N., Waldron, G.J., Garland, C.J., Boyle, J.P. (1997) Evidence that anandamide and EDHF act via different mechanisms in rat isolated mesenteric arteries. *Br. J. Pharmacol.*, 121:1509–1511.

Plane, F., Hurrell, A., Jeremy, J.Y., Garland, C.J., (1996) Evidence that potassium channels make a major contribution to SIN-1-evoked relaxation of rat isolated mesenteric artery. *Br. J. Pharmacol.*, 119:1557–1562.

Plane, F., Pearson, T., Garland, C.J. (1995) Multiple pathways underlying endothelium-dependent relaxation in the rabbit isolated femoral artery. *Br. J. Pharmacol.*, 115:31–38.

Plane, F., Wiley, K.E., Jeremy, J.Y., Cohen, R.A., Garland, C.J. (1998) Evidence that different mechanisms underlie smooth muscle relaxation to nitric oxide and nitric oxide donors in the rabbit isolated carotid artery. *Br. J. Pharmacol.*, 123:1351–1358.

Pollock, J.S., Nakane, M., Buttery, L.D.K., Martinez, A., Springal, D., Polak, J.M., Forstermann, U., Murad, F. (1993) Characterization and localization of endothelial nitric oxide synthase using specific monoclonal antibodies. *Am. J. Physiol.*, 265:C1379–C1387.

Popp, R., Bauersachs, J., Sauer, E., Hecker, M., Fleming, I., Busse, R. (1996a) The cytochrome P450 monooxygenase pathway and nitric oxide-independent relaxations. In *Endothelium-derived hyperpolarizing factor*, edited by P.M. Vanhoutte, Vol. 9, pp. 65–72. The Netherlands: Harwood Academic Publishers.

Popp, R., Bauersachs, J., Sauer, E., Hecker, M., Fleming, I., Busse, R. (1996b) A transferable, β-naphthoflavone-inducible, hyperpolarizing factor is synthesized by native and cultured porcine coronary endothelial cells. *J. Physiol.*, 497:699–709.

Popp, R., Fleming, I., Busse, R. (1998) Pulsatile stretch in coronary arteries elicits release of endothelium-derived hyperpolarizing factor: a modulator of arterial compliance. *Circ. Res.*, 82:696–703.

Pratt, P.F., Hillard, C.J., Edgemond, W.S., Campbell, W.B. (1998) N-arachidonylethanolamide relaxation of bovine coronary artery is not mediated by CB1 cannabinoid receptor. *Am. J. Physiol.*, 274:H375–H381.

Pritchard, J.R., K.A., Wong, P.Y.K., Sternerman, M.B. (1990) Atherogenic concentrations of low-density lipoprotein enhance endothelial cell generation of epoxyeicosatrienoic acid products. *Am. J. Pathol.*, 136:1383–1391.

Putney, J.W. (1986) A model for receptor-regulated calcium-entry. *Cell. Calcium*, 7:1–12.

Puybasset, L., Bea, M.L., Ghaleh, B., Giudicelli, J.F., Berdeaux, A. (1996) Coronary and systemic hemodynamics effects of sustained inhibition of nitric oxide synthesis in conscious dogs. *Circ. Res.*, 79:343–357.

Quayle, J.M., Nelson, M.T., Standen, N.B. (1997) ATP-sensitive and inwardly rectifying potassium channels in smooth muscle. *Physiological Reviews.*, 77:1165–1232.

Quast, U., Baumlin, Y. (1991) Cromakalim inhibits contractions of the rat isolated mesenteric bed induced by noradrenaline but not caffeine in Ca 2+-free medium: evidence for interference with receptor-mediated Ca^{2+} mobilization. *Eur. J. Pharmacol.*, 200:239–249.

Quayle, J.M., Dart, C., Standen, N.B., (1996) The properties and distribution of inward rectifier potassium currents in pig coronary arterial smooth muscle. *J. Physiol.*, 494:715–726.

Quignard, J.F., Frapier, J.M., Harricane, M.C., Albat, B., Nargeot, J., Richard, S. (1997) Voltage-gated calcium current in human coronary myocytes: regulation by cyclic GMP and nitric oxide. *J. Clin. Invest.*, 99(2):185–193.

Quignard, J.F., Félétou, M., Thollon, C., Vilaine J-P, Duhault, J., Vanhoutte, P.M. (1999) Potassium ions and endothelium-derived hyperpolarizing factor in guinea-pig carotid and porcine coronary arteries. *Br. J. Pharmacol. (in press)*.

Quilley, J., Fulton, D., McGiff, J.C. (1997) Hyperpolarizing factors. *Biomedical Pharmacol.*, 54:1059–1070.

Raeymaekers, L., Hofmann, F., Casteels, R. (1988) Cyclic GMP-dependent protein kinase phosphorylates phospholamban in isolated sarcoplasmic reticulum from cardiac and smooth muscle. *Biochem. J.*, 252:269–273.

Raeymakers, L., Wuytack F., (1993) Ca^{2+} pumps in smooth muscle. *J. Muscle Res. Cell Motil.*, 14:141–157.

Randall, M.D., Alexander, S.P.H., Bennett, T., Boyd, E.A., Fry, J.R., Gardiner, S.M., Kemp, P.A., McCulloch, A.I., Kendall, D.A. (1996) An endogenous cannabinoid as an endothelium-derived vasorelaxant. *Biochem. Biophys. Res. Commun.*, 229:114–120.

Randall, M.D., Kendall, D.A. (1997) Involvement of a cannabinoid in endothelium-derived hyperpolarizing factor-mediated coronary vasorelaxation. *Eur. J. Pharmacol.*, 335:205–209.

Randall, M.D., Kendall, D.A. (1998a) Anandamide and endothelium-derived hyperpolarizing factor act via a common vasorelaxant mechanism in rat mesentery. *Eur. J. Pharmacol.*, 346:51–53.

Randall, M.D., Kendall, D.A. (1998b) Endocannabinoids: a new class of vasoactive substances. *TIPS*, 19:55–58.

Randall, M.D., McCulloch, A.I., Kendall, D.A. (1997) Comparative pharmacology of endothelium-derived hyperpolarizing factor and anandamide in rat isolated mesentery. *Eur. J. Pharmacol.*, 333:191–197.

Rao, G.N., Baas, A.S., Glasgow, W.C., Eling, T.E., Runge, M.S., Alexander, R.W. (1994) Activation of mitogen-activated protein kinases by arachidonic acid and its metabolites in vascular smooth muscle cells. *J. Biol. Chem.*, 269:32586–32591.

Rapoport, R.M., Murad, F. (1983) Agonist induced endothelium-dependent relaxation in rat thoracic aorta may be mediated through cyclic GMP. *Circ. Res.*, 52:352–357.

Rapoport, R.M. (1986) Cyclic guanosine monophosphate inhibition of contraction may be mediated through inhibition of phosphatidylinositol hydrolysis in rat aorta. *Circ. Res.*, 58:407–410.

Rapp, J.P. (1982) Dahl salt-susceptible and salt-resistant rats, a review. *Hypertension*, 4:753–763.

Rees, D.D., Ben-Ishay, D., Moncada, S. (1996) Nitric oxide and the regulation of blood pressure in the hypertensive-prone and hypertension-resistant sabra rat. *Hypertension*, 28:367–371.

Rees, D.D., Palmer, R.M.J., Hodson, H.F., Moncada, S. (1989) A specific inhibitor of nitric oxide formation from L-arginine attenuates endothelium-dependent relaxation. *Br. J. Pharmacol.*, 96:418–424.

Reilly, T.M., Beckner, S., Mchugh, E.M., Blecker, M. (1981) Isoproterenol-induced ADP-ribosylation of a single plasma membrane protein of cultured differentiated RL-PR-C hepatocytes. *Biochem. Biophys. Res. Commun.*, 98:1115–1120.

Reis, S.E., Gloth, S.T., Blumenthal, R.S., Resar, J.R., Zacur, H.A., Gerstenblith, G., Brinker, J.A. (1994) Ethinyl estradiol acutely attenuates abnormal coronary vasomotor responses to acetylcholine in postmenopausal women. *Circulation*, 89:52–60.

Rengasamy, A., Johns, R.A. (1996) Synthesis of hydroxyeixosatetranoic (HETEs) and epoxyeicosatrienoic acids (EETs) by cultured bovine coronary artery endothelial cells. *Biochim. Biophys. Acta*, 1299:267–277.

Rengasamy, A., Johns, R.A. (1996) Determination of Km for oxygen of nitric oxide synthase isoforms. *J. Pharmacol. Exp. Ther.*, 276:30–33.

Revtyak, G.E., Johnson, A.R., Campbell, W.B. (1988) Cultured bovine coronary arterial endothelial cells synthesize HETEs and prostacyclin. *Am. J. Physiol.*, 254:C8–C19.

Richard, V., Tanner FC, Tschudi, M.R., Lüscher, T.F. (1990) Different activation of L-arginine pathway by bradykinin, serotonin, and clonidine in coronary arteries. *Am. J. Physiol.*, 259:H1433–H1439.

Riley, S.C., Walton, J.C., Herlick, J.M., Challis, J.R. (1991) The localization and distribution of corticotropin-releasing hormone in the human placenta and fetal membranes throughout gestation. *J. Clin. Endocrin. Metab.*, 72:1001–1007.

Rinaldi-Carmona, M., Barth, F., Héaulene, M., Shire, D., Calandra, B., Congy, C., Martinez, S., Maruani, J., Néliat, G., Caput, D., Ferrara, P., Soubrié, P., Breliere, J.C., Le Fur, G. (1994) SR141716A, a potent and selective antagonist of the brain cannabinoid receptor. *FEBS Lett.*, 350:240–244.

Rinaldi-Carmona, M., Calandra, B., Shire, D., Bouaboula, M., Oustric, D., Barth, F., Casellas, P., Ferrara, P., Lefur, G. (1996) Characterization of 2 cloned human CB cannabinoid receptor isoforms. *J. Pharmacol. Exp. Ther.*, 278:871–878.

Robertson, B.E., Bonev, A.D., Nelson, M.T. (1996) Inward rectifier K^+ currents in smooth muscle cells from rat coronary arteries: block by Mg^{2+}, Ca^{2+}, and Ba^{2+}. *Am. J. Physiol.*, 271:H696–H705.

Robitaille, R., Garcia, M.L., Kaczorowski, G.J., Charlton, M.P. (1993) Functional colocalization of calcium and calcium-gated potassium channels in control of transmitter release. *Neuron.*, 11:645–655.

Rosolowsky, M., Campbell, W.B. (1993) Role of PGI_2 and epoxyeicosatrienoic acids in relaxation of bovine coronary arteries to arachidonic acid. *Am. J. Physiol.*, 264:H327–H335.

Rosolowsky, M., Campbell, W.B. (1996) Synthesis of hydroxyeicosatetraenoic (HETEs) and epoxyeicosatrienoic acids (EETs) by cultured bovine coronary artery endothelial cells. *Biochim. Biophys. Acta*, 1299:267–277.

Rosolowsky, M., Falck, J.R., Campbell, W.B. (1990a) Synthesis and biological activity of epoxyeicosatrienoic acids (EETs) by cultured bovine coronary artery endothelial cells. *Adv. Prostaglandin Thromb. Res.*, 21A:213–216.

Rosolowsky, M., Falck, J.R., Willerson, J.T., Campbell, W.B. (1990b) Synthesis of lipoxygenase and epoxygenase products of arachidonic acid by normal and stenosed canine coronary arteries. *Circ. Res.*, 66:608–621.

Ross, G., Stinson, E., Schroeder, J., Ginsburg, R. (1980) Spontaneous phasic activity of isolated human coronary arteries. *Cardiovasc. Res.*, 14:613–618.

Rossouw, J.E. (1996) Estrogens for prevention of coronary heart disease: putting the brakes on the bandwagon. *Circulation*, 94:2982–2985.

Rowe, D.T.F., Garland, C.J., Plane, F. (1998) Multiple pathways inderlie NO-independent relaxation to the calcium ionophore A23187 in the rabbit isolated femoral arteries. *Br. J. Pharmacol.*, 123:1P.

Rubanyi, G.M. (1993) The role of endothelium in cardiovascular homeostasis and diseases. *J. Cardiovasc. Pharmacol.*, 22:S1–S14.

Rubanyi, G.M., Vanhoutte, P.M. (1985b) Ouabain inhibits endothelium-dependent relaxations to arachidonic acid in canine coronary arteries. *J. Pharmacol. Exp. Ther.*, 235:81–86.

Rubanyi, G.M., Vanhoutte, P.M. (1986) Superoxide anions and hyperoxia inactivate EDRF. *Am. J. Physiol.*, 250:H822–H8227.

Rubanyi, G.M., Vanhoutte, P.M. (1987) Nature of endothelium-derived relaxing factor: Are there two relaxing mediators? *Circ Res.*, 61:II61–II67.

Ruiz, E., Tejerina, T. (1998) Relaxant effect of L-citrulline in rabbit vascular smooth muscle relaxation. *Br. J. Pharmacol.*, 125:186–192.

Saez, J.M., Dazord, A., Morera, A.M., Bataille, P. (1975) Interactions of adrenocorticotropic hormone with its adrenal receptors. *J. Biol Chem.*, 250, 1683–1689.

Saito, F., Hori, M.T., Ideguchi, Y., Berger, M., Golub, M., Stern, N., Tuck ML. (1992) 12-Lypoxygenase products modulate calcium signals in vascular smooth muscle cells. *Hypertension*, 20:138–143.

Salari, H., Braquet, P., Borgeat, P. (1984) Comparative effects of indomethacin, acetylenic acids, 15-HETE, nordihydroguaiaretic acid and BW755C on the metabolism of arachidonic acid in human leukocytes and platelets. *Prostaglandins Leukot. Med.*, 13, 53–60.

Salvemini, D., Misko, T.P., Masferrer, J.L., Seibert, K., Currie, M.G., Needlman, P. (1993) Nitric oxide activates cyclooxygenase enzymes. *Proc. Natl. Acad. Sci. (USA)*, 90:7240–7244.

Sargeant, P., Farndale, R.W., Sage, S.O. (1994) The imidazole antimycotics econazole and miconazole reduce agonist-evoked protein-tyrosine phosphorylation and evoke membrane depolarisation in human platelets: cautions for their use in studying Ca^{2+} signalling pathways. *Cell. Calcium*, 16:413–418.

Sautebin, L., Ialenti, A., Ianarro, A., Di Rosa, M. (1995) Modulation by nitric oxide of prostaglandin biosynthesis in the rat. *Br. J. Pharmacol.*, 114:323–328.

Schmidt, H.H.H.W., Gagne, G.D., Nakane, M., Pollock, J.S., Miller, M.F., Murad, F. (1992) Mappings of neural nitric oxide synthase in the rat suggests frequent co-localization with NADPH diaphorase but not with soluble guanylyl cyclase, and novel paraneural function for nitrinergic signal transduction. *J. Histochem. Cytochem.*, 40:1439–1456.

Schultz, K.D., Schultz, K., Schultz, G. (1977) Sodium nitroprusside and other smooth muscle relaxants increase cyclic GMP Level in rat ductus deferens. *Nature*, 265:750–751.

Schwartz, S.M., Benditt, E.P. (1977) Aortic endothelial cell replication. Effects of age and hypertension in the rat. *Circ. Res.*, 41:248–255.

Schwartz, S.M., deBlois, D., O'Brien, E.R.M. (1995) The intima Soil for atherosclerosis and restenosis. *Circ. Res.*, 77:445–465.

Scott-Burden, T., Engler, D.A., Tock, C.L., Schwarz, J.J., Casscells, S.W., (1997) Liposomal induction of NO synthase expression in cultured vascular smooth muscle cells. *Biochem. Biophys. Res. Commun.*, 231:780–783.

Scott-Burden, T., Tock, C.L., Schwarz, J.J., Casscells, S.W., Engler, D.A. (1996) Genetically engineered smooth muscle cells as linings to improve the biocompatibility of cardiovascular prostheses. *Circ.*, 94:235–238.

Sedhev, J., Garland, C.J., Plane, F. (1998) Further characterization of the mediator of NO-independent dilatation to acetylcholine in the rat isolated perfused mesenteric bed. *Br. J. Pharmacol.*, 123:2P.

Seidler, N.W., Jona, L., Vegh, M., Martonosi, A. (1989) Cyclopiazonic acid is a specific inhibitor of the Ca^{2+}-ATPase of sarcoplasmic reticulum. *J. Biol. Chem.*, 264:17816–17823.

Seki, K., Hirai, A., Noda, M., Tamura, Y., Kato, I., Yoshida, S. (1992) Epoxyeicosatrienoic acid stimulates ADP-ribosylation of a 52-kDa protein in rat liver cytosol. *Biochem. J.*, 281:185–190.

Serhan, C.N. (1994) Lipoxin biosynthesis and its impact in inflammatory and vascular events. *Biochim. Biophys. Acta*, 1212:1–25.

Sewer, M.B., Koop, D.P., Morgan, E.T. (1996) Endotoxemia in rats is associated with induction of the P-450 4 subfamily and suppression of several other forms of cytochrome P-450. *Drug Metab. Dispo.*, 24:401–407.

Sewer, M.B., Koop, D.P., Morgan, E.T. (1997) Differential inductive and suppressive effects of endotoxin and particulate irritants on hepatic and renal cytochrome P-450 expression. *J. Pharmacol. Exp. Therap.*, 280:1445–1454.

Sharma, N.R., Davis, M.J., (1995) Substance P-induced calcium entry in endothelial cells is secondary to depletion of intracellular stores. *Am. J. Physiol.*, 268:H962–H973.

Sharp, D.S., Osterloh, J., Becker, C.E., Bernard, B., Smith, A.H., Fisher, J.M., Syme, S.L., Holman, B.L., Johnston, T. (1988) Blood pressure and blood lead concentration in bus drivers. *Environ. Health Perspect.*, 78:131–137.

Sharrocks, A.D., Brown, A.L., Ling, Y., Yates, P.R. (1997) The ETS-domain transcription factor family. *Int. J. Biochem. Cell. Biol.*, 29:1371–1387.

Sheu, H.L., Omata, K., Utsumi, Y., Tsutsumi, E., Sato, T., Shimizu, T., Abe, K. (1995) Epoxyeicosatrienoic acids stimulate the growth of vascular smooth muscle cells. *Adv. Prostagl. Trombox. Leuk. Res.*, 23:199–201.

Shimizu, S., Paul, R.J. (1997) The endothelium-dependent, substance P relaxation of porcine coronary arteries resistant to nitric oxide synthesis inhibition is partially mediated by 4-aminopyridine-sensitive voltage-dependent K^+ channels. *Endothelium*, 5:287–295.

Shimokawa, H. (1999) Primary endothelial dysfunction: atherosclerosis. *J. Mol. Cell. Cardiol.*, 31:23–37.

Shimokawa, H., Aarhus, L.L., Vanhoutte, P.M. (1987) Porcine coronary arteries with regenerated endothelium have a reduced endothelium-dependent responsiveness to aggregating platelets and serotonin. *Circ. Res.*, 61:256–270.

Shimokawa, H., Flavahan, N.A., Vanhoutte, P.M. (1989) Natural course of the impairment of endothelium-dependent relaxations after balloon endothelium removal in porcine coronary arteries. *Circ. Res.*, 65:740–753.

Shimokawa, H., Vanhoutte, P.M. (1988) Dietary cod-liver oil improves endothelium-dependent responses in hypercholesterolemic and atherosclerotic porcine coronary arteries. *Circulation*, 78:1421–1430.

Shimokawa, H., Vanhoutte, P.M. (1997) Endothelium and vascular injury in hypertension and atherosclerosis. In *Handbook of Hypertension*, edited by A. Zanchetti and G. Mancia, pp. 1007–1068. New York: Elsevier Science.

Shimokawa, H., Yasutake, H., Fujii, K., Owada, M.K., Nakaike, R., Fukumoto, Y., Takayanagi, T., Nagao, T., Egashira, K., Fujishima, M., Takeshita, A. (1996) The importance of the hyperpolarizing mechanism increases as the vessel size decreases in endothelium-dependent relaxations in rat mesenteric circulation. *J. Cardiol. Pharmacol.*, 28:703–711.

Shin, J.H., Chung, S., Park, E.J., Uhm, D.Y., Suh, C.K. (1997) Nitric oxide directly activates calcium-activated potassium channels from rat brain reconstituted into planar lipid bilayer. *Febs Lett.*, 415:299–302.

Shiramoto, M., Imaizumi, T., Hirooka, Y., Endo, T., Namba, T., Oyama, J., Hironaga, K., Takeshita A (1997) Role of nitric oxide toward vasodilator effects of substance P and adenosine 5-triphosphate in human forearm vessels. *Clin. Sci.*, 92:123–131.

Shire, D., Calandra, B., Rinaldi-Carmona, M., Oustric, D., Pessegue, B., Bonnincabanne, O., Lefur, G., Caput, D., Ferrara P., (1996) Molecular-cloning, expression and function of the murine cb2 peripheral cannabinoid receptor. *Biochim. Biophys. Acta*, 1307:132–136.

Shire, D., Carillon, C., Kaghad, M., Calandra, B., Rinaldi-Carmona, M., Lefur, G., Caput, D., Ferrara, P. (1995) An amino-terminal variant of the central cannabinoid receptor resulting from alternative splicing. *J. Biol. Chem.*, 270:3726–3731.

Siegel, G., Schnalke, F., Stock, F., Grote, J. (1989b) Prostacyclin, endothelium-derived relaxing factor, and vasodilation. In *Adv. Prostagl. Thrombox. and Leukotric. Res.*, B. Samuelson, P.Y.K. Wong and F.F. Sun, Vol. 19, pp. 267–270.

Singer, H.A., Peach, M.J. (1982) Calcium- and endothelial-mediated vascular smooth muscle relaxation in rabbit aorta. *Hypertension*, 4:19–25.

Singer, H.A., Peach, M.J. (1983a) Endothelium-dependent relaxation of rabbit aorta., I. Relaxation stimulated by arachidonic acid. *J. Pharmacol. Exp. Ther.*, 226:790–795.

Singer, H.A., Peach, M.J. (1983b) Endothelium-dependent relaxation of rabbit aorta. II. Inhibition of relaxation stimulated by metacholine and A23187 with antagonists of arachidonic acid metabolism. *J. Pharmacol. Exp. Ther.*, 226:796–801.

Singer, H.A., Saye, J.A., Peach, M.J. (1984) Effects of cytochrome P-450 inhibitors on endothelium-dependent relaxation in rabbit aorta. *Blood Vessels*, 21:223–230.

Skarsgard, P., Van Breemen, C., Laher, I. (1997) Estrogen regulates myogenic tone in pressurized cerebral arteries by enhanced basal release of nitric oxide. *Am. J. Physiol.*, 273:H2248–H2256.

Slipetz, D.M., Oneill, G.P., Favreau, L., Dufresne, C., Gallant, M., Gareau, Y., Guay, D., Labelle, M., Metters, K.M. (1995) Activation of the human peripheral cannabinoid receptor results in inhibition of adenylyl-cyclase. *Mol. Pharmacol.*, 48:352–361.

Sokol, P.T., Hu, W., Yi, L., Toral, J., Chandra, M., Ziai, M.R. (1994) Cloning of an apamin binding protein of vascular smooth muscle. *J. Protein Chemistry*, 13:117–128.

Soloviev, A., Bershtein, S. (1992) The contractile apparatus in vascular smooth muscle cells of spontaneously hypertensive rats possess increased calcium sensitivity: the possible role of protein kinase, C. *J. Hypertens.*, 10:131–136.

Soloviev, A., Stefanov, A., Bazilyuk, O., Sagach, V. (1993) Phospolopid vesicles (liposomes) restore endothelium-dependent cholinergic relaxation in thoracic aorta from spontaneously hypertensive rats. *J. Hypertens.*, 11:623–627.

Somlyo, A.P., Somlyo, A.V. (1994) Signal transduction and regulation in smooth muscle cells. *Nature*, 372:231–236.

Standen, N.B., Quayle, J.M., Davies, N.W., Brayden, J.E., Huang, Y., Nelson, M.T. (1989) Hyperpolarizing vasodilators activate ATP-sensitive K^+ channels in arterial smooth muscle. *Science*, 245:177–180.

Stein, E.A., Fuller, S.A., Edgemond, W.S., Campbell, W.B. (1996) Physiological and behavioural effects of the endogenous cannabinoid, arachidonylethanolamine (anandamide), in the rat. *Br. J. Pharmacol.*, 119:107–114.

Stella, N., Schweitzer, P., Piomelli, D. (1997) A second endogenous cannabinoid that modulates long-term potentiation. *Nature*, 388:773–778.

Steudel, W., Ichinose, F., Huang, P.L., Hurford WE, Jones, R.C., Bevan, J.A., Fishman, M.C., Zapol, W.M. (1997) Pulmonary vasoconstriction and hypertension in mice with targeted disruption of the endothelal nitric oxide synthase (NOS 3) gene. *Circ. Res.*, 81:34–4.

Stingo, A., Clavell, A., Heublein, D., Wei, C., Pittelkow, M., Burnett, J.C. (1992) Presence of C-type natriuretic peptide in cultured human endothelial cells and plasma. *Am. J. Physiol.*, 63:H1318–H1321.

Stoclet, J.C., Andriantsitohaina, R., Kleschyov, A.L., Muller, B. (1998) Nitric oxide and cGMP in regulation of arterial tone. *Trends Cardiovasc. Med.*, 8:12–17.

Stoclet, J.C., Fleming, I., Gray, G.A., Julou-Schaeffer, G., Schneider, F., Schott, C., Schott, C., Parratt, J.R. (1993) Nitric oxide and endotoxemia. *Circulation*, 87:V77–80.

Street, I.P., Lin, H.K., Laliberte, F., Ghomashchi, F., Wang, Z., Perrier, H., Tremblay, N.M., Huang, Z., Weech, P.K., Gelb, M.H. (1993) Slow and tight-binding inhibitors of the 85-kDa human phospholipase A_2. *Biochem.*, 32:5935–5940.

Suga, S.I., Nakao, K., Itoh, H., Komatsu, Y., Ogawa, Y., Hama, N. and Imura, H. (1992a) Endothelial production of C-type natriuretic peptide and its marked augmentation by transforming growth factor B. *J. Clin. Invest.*, 90, 1145–1149.

Suga, S.I., Saito, Y., Kambayashi, K., Inouye, K., Imura, H. (1992b) Receptor selectively of natriuretic peptide family, atrial natriuretic peptide, brain natriuretic peptide and C-type natriuretic peptide. *Endocrinology*, 130:229–239.

Sugiura, T., Kodaka, T., Nakane, S., Kishimoto, S., Kondo, S., Waku, K. (1998) Detection of an endogenous cannabimimetic molecule, 2-arachidonoylglycerol, and cannabinoid CB1 receptor mRNA in human vascular cells: is 2-arachidonoylglycerol a possible vasomodulator. *Biochem. Biophys. Res. Commun.*, 243: 838–843.

Sun, D., Messina, E.J., Kaley, G., Koller, A. (1992) Characteristics and origin of myogenic response in isolated mesenteric arterioles. *Am. J. Physiol.*, 26:H1483–H1491.

Suzuki, H., Chen, G., Yamamoto, Y. (1992) Endothelium-derived hyperpolarizing factor (EDHF). *Jpn. Circ. J.*, 56:170–173.

Suzuki, H., Chen, G. (1990) Endothelium-derived hyperpolarizing factor (EDHF): an endogenous potassium-channel activator. *News Physiol. Sc.*, 5:212–215.

Taddei, S., Virdis, A., Mattei, P., Ghiadoni, L., Gennari, A., Fasolo, C.B., Sudano, I., Salvetti, A. (1995) Aging and endothelial function in normotensive subjects and patients with essential hypertension. *Circulation*, 91:1981–1987.

Taddei, S., Mattei, P., Virdis, A., Sudano, I., Ghiadoni, L., Salvetti, A. (1994) Effect of potassium on vasodilation to acetylcholine in essential hypertension. *Hypertension*, 23:485–490.

Taddei, S., Virdis, A., Ghiadoni, L., Mattei, P., Salvetti, A. (1998a) Effects of ACE-inhibition on endothelium-dependent vasodilatation in essential hypertensive patients. *J. Hypertens.*, 16:447–456.

Taddei, S., Virdis, A., Ghiadoni, L., Magagna, A., Salvetti, A. (1998b) Vitamin C improves endothelium-dependent vasodilatation by restoring nitric oxide activity in essential hypertension. *Circulation*, 97:2222–2229.

Taddei, S., Virdis, A., Ghiadoni, L., Salvetti, A. (1998c) The role of endothelium in human hypertension. *Curr. Opin. Nephrol. Hypertens.*, 7:203–209.

Tagawa, T., Mohri, M., Tagawa, H., Egashira, K., Shimokawa, H., Kuga, T., Hirooka, Y., Takeshita, A. (1997) Role of nitric oxide in substance P-induced vasodilation differs between the coronary and forearm circulation in humans. *J. Cardiovasc. Pharmacol.*, 29:546–553.

Tagawa, H., Shimokawa, H., Tagawa, T., Kuroiwa-Matsumoto, M., Horooka, T., Takeshita, A. (1997) Short-term estrogen augments both nitric oxide-mediated and non-nitric oxide-mediated endothelium-dependent vasodilation in postmenopausal women. *J. Cardiovasc. Pharmacol.*, 30:481–488.

Tagawa, H., Shimokawa, H., Tagawa, T., Kuroiwa-Matsumoto, M., Hirooka, Y., Takeshita, A. (1999) Long-term treatment with eicosapentaenoic acid augments both nitric-mediated and non-nitric oxide-mediated endothelium-dependent forearm vasodilatation in patients with coronary artery disease. *J. Cardiovasc. Pharmacol.* (in press).

Takase, H., Moreau, P., Küng, C.F., Nava, E., Lüscher, T.F. (1996) Antihypertensive therapy prevents endothelial dysfunction in chronic nitric oxide deficiency: Effect of verapamil and trandolapril. *Hypertension*, 27:25–31.

Tare, M., Parkington, H.C., Coleman, H.A., Neild, T.O., Dusting, G.J. (1990) Hyperpolarization and relaxation of arterial smooth muscle caused by nitric oxide derived from the endothelium. *Nature*, 346:69–71.

Taylor, H.J., Chaytor, A.T., Evans, W.H., Griffith, T.M. (1998) Inhibition of the gap junctional component of endothelium-dependent relaxations in rabbit iliac artery by 18β-glycyrrhetinic acid. *Br. J. Pharmacol.*, 125:1–3.

Taylor, S.G., Southerton, J.S., Weston, A.H., Baker, J.R. (1988) Endothelium-dependent effects of acetylcholine in rat aorta: a comparison with sodium nitroprusside and cromakalim. *Br. J. Pharmacol.*, 94:853–863.

Taylor, S.G., Weston, A.H. (1988) Endothelium-derived hyperpolarizing factor: a new endogenous inhibitor from the vascular endothelium. *Trends Pharmacol. Sci.*, 9:272–274.

Tejerina, T., Sesin, J., Delgado, C., Tamargo, J. (1988) Effect of milrinone contractility and $^{45}Ca^{2+}$ movements in the isolated rabbit aorta. *Eur. J. Pharmacol.*, 148:239–244.

Terashima, M., Michima, K., Yamada, K., Tsuchiya, M., Wakutani, T. (1992) ADP-ribosylation of actins by arginine-specific ADP-ribosyltransferase purified from chicken heterophils. *Eur. J. Biochem.*, 204:305–311.

Thibonnier, M., Bayer, A.L., Laethem, C.L., Koop, D.R., Simonson, M.S. (1993) Role of eicosanoids in vasopressin-induced calcium mobilization in A7r5 vascular smooth muscle cells. *Am. J. Physiol.*, 265:E108–E114.

Thiermermann, C. (1991) Biosynthesis and interaction of endothelium-derived vasoactive mediators. *Eicosanoids*, 4:187–202.

Thollon, C., Bidouard, J.P., Cambarrat, C., Delescluse, I., Vanhoutte, P.M., Vilaine, J.P. (1997) Endothelium-dependent hyperpolarization in porcine coronary arteries with regenerated endothelium: decreased response to serotonin and increased hyperpolarization to bradykinin. *J. Vasc. Res.*, 34:40(abstract).

Thurston, H. (1994) Goldblatt, coarctation and page experimental models of renovascular hypertension. In *Textbook of Hypertension*, edited by J.D. Swales, pp. 477–493. Oxford: Blackwell Scientific Publications.

Tobias, L.D., Hamilton, J.G. (1979) The effect of 5,8,11,14-eicosatetraynoic acid on lipid metabolism. *Lipids*, 14:181–193.

Toda, N. (1974) Responsiveness to potassium and calcium ions of isolated cerebral arteries. *Am. J. Physiol.*, 227:1206–1211.

Tombes, R.M., Auer, K.L., Mikkelsen, R., Valerie, K., Wymann, M.P., Marshall, C.J., Mahon, M.M., Dent, P. (1998) The mitogen-activated protein (MAP) kinase cascade can either stimulate or inhibit DNA synthesis in primary cultures of rat hepatocytes dependending upon whether its activation is acute/phasic or chronic. *Biochem. J.*, 330:1451–1460.

Tomobe, Y., Ishikawa, T., Yanagisawa, M., Kimura, S., Masaki, T., Goto, K. (1991) Mechanisms of altered sensitivity to endothelin-1 between aortic smooth muscles of spontaneously hypertensive and Wistar-Kyoto rats. *J. Pharmacol. Exp. Ther.*, 257:555–561.

Tschudi, M.R., Barton, M., Bersinger, N.A., Moreau, P., Noll, G., Cosentino, F., Malinski, T., Lüscher, T.F. (1996) Effect of age on kinetics of nitric oxide release in rat aorta and pulmonary arteries. *J. Clin. Invest.*, 98:899–905.

Tschudi, M.R., Criscione, L., Novosel, D., Pfeiffer, K. and Lüscher, T.F. (1994) Antihypertensive therapy augments endothelium-dependent relaxations in coronary arteries of spontaneously hypertensive rats. *Circulation*, 89:2212–2218.

Tsou, K., Brown, S., Sanuda-Pena, M.C., Mackie, K., Walker, J.M. (1998) Immunohistochemical distribution of cannabinoid CB receptors in the rat central nervous system. *Neuroscience*, 83:393–411.

Tsuchiya, M., Hara, N., Yamada, K., Osago, H., Shimoyama, M. (1994) Cloning and expression of cDNA for arginine-specific ADP-ribosyltransferase from chicken bone marrow cells. *J. Biol. Chem.*, 269:27451–27457.

Twitchell, W.A., Pena, T.L., Rane, S.G. (1997) Ca^{2+}-dependent K^+ channels in bovine adrenal chromaffin cells are modulated by lipoxygenase metabolites of arachidonic acid. *J. of Membrane Biology*, 158:69–75.

Urakami, L., Shimokawa, H., Nakashima, M., Matsumoto, N., Owada, M.K., Takeshita, A. (1997) Relative importance of endothelium-derived hyperpolarizing factor and nitric oxide in coronary arteries in pigs and rabbits. *Jpn. J. Pharmacol.*, 75:49P (abstract).

Urakami-Harasawa, L., Shimokawa, H., Nakashima, M., Egashira, K., Takeshita, A. (1997) Importance of endothelium-derived hyperpolarizing factor in human arteries. *J. Clin. Invest.*, 100:2793–2799.

Uddin, M.R., Muthalif, M.M., Karzoun, N.A., Bentner, I.F., Malik, K.U. (1998) Cytochrome P-450 metabolites mediate norepinephrine-induced mitogenic signalling. *Hypertension*, 31:242–247.

Vallance, P., Collier, J., Moncada, S. (1989) Effects of endothelium-derived nitric oxide on peripheral arteriolar tune in man. *Lancet*, 2:197–199.

Vallance, P., Patton, S., Bhagat, R., MacAllister, R., Radomski, M., Moncada, S., Malinski, T. (1995) Direct measurement of nitric oxide in human beings. *Lancet*, 346:153–154.

Van Breemen, C., Chen, Q., Leher, I. (1995) The superficial buffer barrier in smooth muscle. *Trends Pharmacol. Sciences*, 16:98–105.

Van de Voorde, J., Vanheel, B. (1997) Influence of cytochrome P-450 inhibitors on endothelium-dependent nitro-L-arginine-resistant relaxation and cromakalim-induced relaxation in rat mesenteric arteries. *J. Cardiovasc. Pharmacol.*, 29:827–832.

Van de Voorde, J., Vanheel, B., Leusen, I. (1992) Endothelium-dependent relaxation and hyperpolarization in aorta from control and renal hypertensive rats. *Circ. Res.*, 70:1–8.

Van Diest, M.J., Verbeuren, T.J., Herman, A.G. (1991) 15-lipoxygenase metabolites of arachidonic acid evoke contractions and relaxations in isolated canine arteries: role of thromboxane receptors, endothelial cells and cyclooxygenase. *J. Pharmacol. Exp. Therm.*, 256:194–203.

Van Put, D.J.M., Van Hove, C.E., De Meyer, G.R.Y., Wuyts, F., Herman, A.G., Bult, H. (1995) Dexamethasone influences intimal thickening and vascular reactivity in the rabbit collared carotid artery. *Eur. J. Pharmacol.*, 294:753–761.

Vandlen, R., Arcur, K., Napier, M. (1985) Identification of a receptor for atrial natriuretic factor in rabbit aorta membranes by affinity crosslinking. *J. Biol. Chem.*, 260:10889–10892.

Vandlen, R., Arcur, K., Hupe, L., Keegan, M., Napier, M. (1986) Molecular characteristics of receptor for atrial natriuretic factor. *Fred. Proc.*, 45:2366–2370.

Vane, J.R., Bunting, S., Moncada, S. (1982) Prostacyclin in physiology and patophysiology. *Int Rev. Exp. Pathol.*, 23:161–207.

Vanheel, B., Van de Voorde, J., Leusen, I. (1994) Contribution of nitric oxide to the endothelium-dependent hyperpolarization in rat aorta. *J. Physiol.*, 475:277–284.

Vanheel, B., Van de Voorde, J. (1996) Impaired endothelium-dependent hyperpolarization and relaxation in the aorta from the renal hypertensive rat. In *Endothelium-derived hyperpolarizing factor*, edited by P.M. Vanhoutte, Vol. 29, pp. 235–245. The Netherlands: Harwood Academic Publishers.

Vanheel, B., Van de Voorde, J. (1997) Evidence against the involvement of cytochrome P450 metabolites in endothelium-dependent hyperpolarization of the rat main mesenteric artery. *J. Physiol.*, 501:331–341.

Vanhoutte, P.M. (1989) State of the Art Lecture: Endothelium and control of vascular function. *Hypertension*, 13:658–667.

Vanhoutte, P.M. (1997) Endothelial dysfuntion in atherosclerosis. *Eur. Heart. J.*, 18:E19–E29.

Vanhoutte, P.M. (1998) An old-timer makes a comeback. *Nature*, 396:213–215.

Vanhoutte, P.M., Auch-Schwelk, W., Biondi, M.L., Lorenz, R.R., Schini, V.B., Vidal, M.J. (1989) Why are converting enzyme inhibitors vasodilators? *Br. J. Clin. Pharmacol.*, 28:95S–103S.

Vanhoutte, P.M., Cohen, R.A. (1996) Endothelium-dependent hyperpolarization: beyond nitric oxide and cyclic GMP. *Circulation*, 2:3337–3349.

Vanhoutte, P.M., Félétou, M. (1996) Conclusion: Existence of Multiple endothelium-derived hyperpolarizing factor(s)? In *Endothelium-derived hyperpolarizing factor*, edited by P.M. Vanhoutte, Vol. 37, pp. 303–305. The Netherlands: Harwood Academic Publishers.

Vanhoutte, P.M., Félétou, M., Boulanger, C.M., Höffner, U., Rubanyi, G.M. (1996) Existence of multiple endothelium-derived relaxing factors. In *Endothelium-derived hyperpolarizing factor*, edited by P.M. Vanhoutte, Vol. 1, pp. 1–10. The Netherlands: Harwood Academic Publishers.

Vanin, A.F., Kleschyov, A.L. (1998) EPR detection and biological implications of nitrosyl nonheme iron complexes. In *Nitric oxide in transplant rejection and Anti-tumor defense*, edited by S. Lukiewicz and J.L. Zweier, Chapter 3, pp. 49–82. Norwell, MA: Kluwer Academic Publishers.

Vanin, A.F., Stukan, R.A., Manukhina, E.B. (1996) Physical properties of dinitrosyl iron complexes with thiol-containing ligands in relation with their vasodilator activity. *Biochim. Biophys. Acta*, 1295:5–12.

VanRollins, M., Kaduce, T.L., Knapp HR, Spector, A.A. (1993) 14, 15-epoxyeicosatrienoic acid metabolism in endothelial cells. *J. Lipid Res.*, 34:1931–1942.

Varga, K., Lake, K.D., Huangfu, D., Guyenet, P.G., Kunos, G. (1996) Mechanism of the hypotensive action of anandamide in anesthetized rats. *Hypertension*, 28:682–686.

Varga, K., Lake, K.D., Martin, B.R., Kunos, G. (1995) Novel antagonist implicates the CB1 cannabinoid receptor in the hypotensive action of anandamide. *Eur. J. Pharmacol.*, 278:279–283.

Vazquez, B., Rios, A., Escalante B., (1995) Arachidonic acid metabolism modulates vasopressin-induced renal vasoconstriction. *Life Sci.*, 56:1455–1466.

Vedernikov, Y.P., Mordvintcev, P.I., Malenkova IV, Vanin, A.F. (1992) Similarity between the vasorelaxing activity of dinitrosyl iron cysteine complexes and endothelium-derived relaxing factor. *Eur. J. Pharmacol.*, 211:313–317.

Venance, L., Sagan, S., Giaume, C. (1997) (R)-methanandamide inhibits receptor-induced calcium responses by depleting internal calcium stores in cultured astrocytes. *Pflugers Archiv.*, 434:147–149.

Verbeuren, T.J., Jordaens, F.H., Zonnekeyn, L.L., Van Hove, C.E., Coene, M.C., Herman, A.G. (1986) Effect of hypercholesterolemia on vascular reactivity in the rabbit. 1. Endothelium-dependent and endothelium-independent contractions and relaxations in isolated arteries of control and hypercholesterolemic rabbits. *Circ. Res.*, 58:552–564.

Vidrio, H., Sanchez-Salvatori, M.A., Medina, M. (1996) Cardiovascular effects of (–)-11Oh-delta (8)-Tetrahydrocannabinol-dimethylheptyl in rats. *J. Cardiovasc. Pharmacol.*, 28:332–336.

Volk, K.A., Shibata, E.F. (1993) Single delayed rectifier potassium channels from rabbit coronary artery myocytes. *Am. J. Physiol.*, 264:H1146–H1153.

Volpi, M., Sharafi, P.M., Epstein, P.M., Andrenyak, D.M., Feinstein, M.B. (1981) Local anesthetics, mepacrine and propanolol are antagonists of calmodulin. *Proc. Natl. Acad. Sci. USA*, 78:795–801.

Wadsworth, R.M. (1994) Vasoconstrictor and vasodilator effects of hypoxia. *Trends Pharmacol. Sci.*, 15:47–53.

Wagner, J.A., Varga, K., Ellis, E.F., Rzigalinski, B.A., Martin, B.R., Kunos, G. (1997) Activation of peripheral CB1 cannabinoid receptors in haemorrhagic shock. *Nature*, 390:518–521.

Wahler, G.M., Dollinger, S.J. (1995) Nitric oxide donor SIN-1 inhibits mammalian cardiac calcium current through cGMP-dependent protein kinase. *Am. J. Physiol.*, 268:C45–C54.

Waldron, G.J., Garland, C.J. (1994a). Contribution of both nitric oxide and a change in membrane potential to acetylcholine-induced relaxation in the rat small mesenteric artery. *Br. J. Pharmacol.*, 112:831–836.

Waldron, G.J., Garland, C.J. (1994b) Effect of potassium channel blockers on L-NAME insensitive relaxations in rat small mesenteric artery. *Can. J. Physiol. Pharmacol.*, 72:115.

Wallerstedt, S., Bodelsson, M. (1997) Endothelium-dependent relaxation by substance P in human isolated omental arteries and veins: relative contribution of prostanoids, nitric oxide and hyperpolarization. *Br. J. Pharmacol.*, 120:25–30.

Wang, M.H., Brand-Schieber, E., Zand, B.A., Nguyen, X., Falck, J.R., Balu, N., Schwartzman, M.L. (1998) Cytochrome P450-derived arachidonic acid metabolism in the rat kidney: characterization of selective inhibitors. *J. Pharmacol. Exp. Therap.*, 284:966–973.

Wang, X., Chu, W., van Breemen, C. (1996) Potentiation of acetylcholine-induced responses in freshly isolated rabbit aortic endothelial cells. *J. Vasc. Res.*, 33:414–424.

Watts, S.W., Traub, O., Lamb, F.S., Myers, J.H., Webb, R.C. (1993) Effect of ramipril on alpha-adrenoceptor-mediated oscillatory contractions in tail artery of hypertensive rats. *Eur. J. Pharmacol.*, 242:245–253.

Webb, R.C., Winquist, R.J., Victery, W., Vander, A.J. (1981) In vivo and in vitro effects of lead on vascular reactivity in rats. *Am. J. Physiol.*, 241:H211–216.

Weidelt, T., Boldt, W., Markwardt, F. (1997) Acetylcholine-induced K+ currents in smooth muscle cells of intact rat small arteries. *J. Physiol.*, 500:617–630.

Weiler, E., Khalil, M.F., Gonick, H. (1988) Effects of lead and natriuretic hormone on kinetics of sodium-potassium-activated adenosine triphosphatase: possible relevance to hypertension. *Environ. Health. Perspect.*, 78:113–117.

Weiner CP, Lizasoain, I., Baylis, S.A., Knowles, R.G., Charles, I.G., Moncada, S. (1994) Induction of calcium-dependent nitric oxide synthase by sex hormones. *Proc. Natl. Acad. Sci. (USA)*, 91:5212–5216.

Weintraub, N.L., Fang, X., Kaduce, T.L., VanRollins, M., Chatterjee, P., Spector, A.A. (1997) Potentiation of endothelium-dependent relaxation by epoxyeicosatrienoic acids. *Circ. Res.*, 81:258–267.

Weintraub, N.L., Joshi, S.N., Branch, C.A., Stephenson, A.H., Sprague, R.S., Lonigro, A.J. (1994) Relaxation of porcine coronary artery to bradykinin. Role of arachidonic acid. *Hypertension*, 23:976–981.

Weintraub, N.L., Stephenson, A.H., Sprague, R.S., McMurdo, L., Lonigro, A.J. (1995) Relationship of arachidonic acid release to porcine coronary artery relaxation. *Hypertension*, 26:684–690.

Weiss, R.H., Arnold, J.L., Estabrook, R.W. (1987) Transformation of an arachidonic acid hydroperoxide into epoxyhydroxy and trihydroxy fatty acids by liver microsomal cytochrome P-450. *Arch. Biochem. Biophys.*, 252:334–338.

White, R., Hiley, C.R. (1997) A comparison of EDHF-mediated and anandamide-induced relaxations in the rat isolated mesenteric artery. *Br. J. Pharmacol.*, 122:1573–1584.

White, R., Hiley, C.R. (1998) The actions of some cannabinoid receptor ligands in the rat isolated mesenteric artery. *Br. J. Pharmacol.*, 125:533–541.

Whitney, R.J. (1953) The measurement of volume changes in human limbs. *J. Physiol. (Lond)*, 121:1–27.

Williams, S.P., Shackelford, D.P., Iams, S.G., Mustafa, S.J. (1988) Endothelium-dependent relaxation in estrogen-treated spontaneous hypertensive rats. *Eur. J. Pharmacol.*, 145:205–207.

Wink, D.A., Osawa, Y., Darbyshire, J.F., Jones, C.R., Eschenauer, S.C., Nims, R.W. (1993) Inhibition of cytochrome P450 by nitric oxide and a nitric oxide-releasing agent. *Arch. Biochem. Biophys.*, 300:115–123.

Winquist, R.J., Bunting, P.B., Schofield, T.L. (1985) Blockade of endothelium-dependent relaxation by the amiloride analog dichlorobenzamil: Possible role of Na+/Ca++ exchange in the release of endothelium-derived relaxing factor. *J. Pharmacol. Exp. Ther.*, 235:644–650.

Wong, W.S., Roman, C.R., Fleisch, J.H. (1995) Differential relaxant responses of guinea-pig lung strips and bronchial rings to sodium nitroprusside: a mechanism independent of cGMP formation. *J. Pharm. Pharmacol.*, 47:757–761.

Worral, N.K., Chang, K., Suau, G.M., Allison, W.S., Misko, T.P., Sullivan, P.M., Tilton, R.G., Williamson, J.R., Ferguson, T.B. (1996) Inhibition of inducible nitric oxide synthase prevents myocardial and systemic vascular barrier dysfunction during early cardiac allograft rejection. *Circ. Res.*, 78:769–779.

Wu, C.-C., Chen, S.-J., Yen, M.-H. (1997) Loss of acetylcholine-induced relaxation by M_3-receptor activation in mesenteric arteries of spontaneously hypertensive rats. *J. Cardiovasc. Pharmacol.*, 30:245–252.

Wu, K.K. (1995) Inducible cyclooxygenase and nitric oxide synthase. *Adv. Pharmacol.*, 33:179–207.

Xiong, Z., Cheung, D.W. (1994) Neuropeptide Y inhibits Ca^{2+}-activated K+ channels in vascular smooth muscle cells from the rat tail artery. *Pflugers Arch.*, 429:280–284.

Yamamoto, Y., Fukuta, H., Nakahira, Y., Suzuki, H. (1998) Blockade by 18β-glycyrrhetinic acid of intercellular electrical coupling in guinea-pig arterioles. *J. Physiol.*, 511:501–508.

Yamamoto, Y., Imaeda, K., Suzuki, H. (1999) Endothelium-dependent hyperpolarization and intercellular electrical coupling in guinea-pig mesenteric arterioles. *J. Physiol.*, 514:505–513.

Yamamoto, Y., Tomoike, H., Egashira, K., Nakamura, M. (1987) Attenuation of endothelium-related relaxation and enhanced responsiveness of vascular smooth muscle to histamine in spastic coronary arterial segments from miniature pigs. *Circ. Res.*, 61:772–778.

Yamanaka, A., Ishikawa, T., Goto, K. (1998) Characterization of endothelium-dependent relaxation independent of NO and prostaglandins in guinea pig coronary artery. *J. Pharmacol. Exp. Ther.*, 285:480–489.

Yamori, Y., Swales, J.D. (1994) The spontaneously hypertensive rats. In *Texbook of Hypertension*, edited by J.D. Swales, pp. 445–455. Oxford: Blackwell Scientific Publications.

Yanagisawa, T., Teshigawara, T., Taira, N. (1990) Cytoplasmic calcium and the relaxation of canine coronary arterial smooth muscle produced by cromakalim, pinacidil and nicorandil. *Br. J. Pharmacol.*, 101:157–165.

Yao, Y., Tsien, R.Y. (1997) Calcium current activated by depletion of calcium stores in xenopus oocytes. *J. Gen. Physiol.*, 109:703–715.

Yi-ju, Z.J., Wang, L.J., Rubin, L.J., Yuan, X.J. (1997) Inhibition of voltage-gated and calcium-activated potassium channels antagonizes nitric oxide induced relaxation in pulmonary artery. *Am. J. Physiol.*, 271(40):H904–H912.

Yuan, X.J., Tod, M.L., Rubin, L.J., Blaustein, M.P. (1995) Inhibition of cytochrome P450 reduces voltage-gated K^+ currents in pulmonary arterial myocytes. *Am. J. Physiol.*, 268:C259–C270.

Zakharov, S.I., Cohen, R.A., Bolotina, V.M. (1998) Monovalent cation current inhibited by extracellular divalent cations in smooth muscle cells. *Biophys. J.*, 74:A103.

Zakharov, S.I., Mongayt, D.A., Cohen, R.A., Bolotina, V.M. (1999) Monovalent cation and L-type Ca^{2+} channels participate in calcium paradox-like phenomenon in rabbit aortic smooth muscle channels. *J. Physiol.*, 514:71–81.

Zeldin, D.C., Foley, J., M.A.J., Boyle, J.E., Pascual, J.M.S., Moomaw, C.R., Tomer, K.B., Steenbergen, C., Wu, S. (1996) CYP2J subfamily P450s in the lung: expression, localization, and potential functional significance. *Molec. Pharmacol.*, 50:1111–1117.

Zhu, P., Bény, J.L., Flammer, J., Lüscher, T.F., Haefliger, I.O. (1997) Relaxation by bradykinin in porcine ciliary artery. *Invest. Ophthalmol. Vis. Sci.*, 38:1761–1767.

Zimmermann, P.A., Knot, H.J., Stevenson, A.S., Nelson, M.T. (1997) Increased myogenic tone and diminished responsiveness to ATP-sensitive K^+ channels openers in cerebral arteries from diabetic rats. *Circ. Res.*, 81:996–1004.

Zolkiewska, A., Nightingale, M.S., Moss, J. (1992) Molecular characterization of NAD: arginine ADP-ribosyltransferase from rabbit skeletal muscle. *Proc. Natl. Acad. Sci. (USA)*, 89:11352–11356.

Zolkiewska, A., Okazaki, I.J., Moss, J. (1994) Vertebrate mono-ADP-ribosyltransferases. *Mol. Cell. Biochem.*, 138:107–112.

Zou, A.P., Fleming, J.T., Falck, J.R., Jacobs, E.R., Gebremedhin, D., Harder, D.R., Roman, R.J. (1996) Stereospecific effects of epoxyeicosatrienoic acids on renal vascular tone and K^+ channel activity. *Am. J. Physiol.*, 270:F822–F832.

Zou, A.P., Ma, Y.H., Sui, Z.H., Ortiz de Montellano, P.R., Clark, J.E., Masters, B.S., Roman, R.J. (1994) Effects of 17-octadecynoic acid, a suicide substrate inhibitor of cytochrome P450 fatty acid ω-hydroxylase, on renal function in rats. *J. Pharmacol. Exp. Ther.*, 268:474–481.

Zweifach, A., Lewis, R.S. (1993) Mitogen-regulated Ca^{2+} current of T lymphocytes is activated by depletion of intracellular Ca^{2+} stores. *Proc. Natl. Acad. Sci. (USA)*, 90:6295–6299.

Zygmunt, P.M. (1995) Endothelium-dependent hyperpolarization and relaxation of vascular smooth muscle. Lund, Sweden: ScD Thesis, Department of Clinical Pharmacology, Lund University.

Zygmunt, P.M., Edwards, G., Weston, A.H., Davis, S.C., Högestätt, E.D. (1996) Effects of cytochrome P450 inhibitors on EDHF-mediated relaxation in the rat hepatic artery. *Br. J. Pharmacol.*, 118:1147–1152.

Zygmunt, P.M., Edwards, G., Weston, A.H., Larsson, B., Högestätt, E.D. (1997a) Involvement of voltage-dependent potassium channels in the EDHF-mediated relaxation of rat hepatic artery. *Br. J. Pharmacol.*, 121:141–149.

Zygmunt, P.M., Grundemar, L., Högestätt, E.D. (1994a) Endothelium-dependent relaxation resistant to N^ω-nitro-L-arginine in the rat hepatic artery and aorta. *Acta Physiol. Scand.*, 152:107–114.

Zygmunt, P.M., Högestätt, E.D. (1996a) Endothelium-dependent hyperpolarization and relaxation in the hepatic artery of the rat. In *Endothelium-derived hyperpolarizing factor*, edited by P.M. Vanhoutte, Vol. 24, pp. 191–201. The Netherlands: Harwood Academic Publishers.

Zygmunt, P.M., Högestätt, E.D. (1996b) Role of potassium channels in endothelium-dependent relaxation resistant to nitroarginine in the rat hepatic artery. *Br. J. Pharmacol.*, 117:1600–1606.

Zygmunt, P.M., Högestätt, E.D., Grundemar, L. (1994b) Light-dependent effects of zinc protoporphyrin IX on endothelium-dependent relaxation resistant to N^ω-nitro-L-arginine. *Acta Physiol. Scand.*, 152:137–143.

Zygmunt, P.M., Högestätt, E.D., Waldeck, K., Edwards, G., Kirkup, A.J., Weston, A.H. (1997b) Studies on the effects of anandamide in rat hepatic artery. *Br. J. Pharmacol.*, 122:1679–1686.

Zygmunt, P.M., Plane, F., Paulsson, M., Garland, C.J., Högestätt, E.D. (1998) Interactions between endothelium-derived relaxing factors in the rat hepatic artery: focus on regulation of EDHF. *Br. J. Pharmacol.*, 124:992–1000.

Zygmunt, P.M., Ryman, T., Högestätt, E.D. (1995) Regional Differences in endothelium-dependent relaxation in the rat-contribution of nitric oxide-independent mechanisms. *Acta Physiol. Scand.*, 155:257–266.

Zygmunt, P.M., Waldeck, K., Högestätt, E.D. (1994c). The endothelium mediates a nitric oxide-independent hyperpolarization and relaxation in the rat hepatic artery. *Acta. Physiol. Scand.*, 152:375–384.

Index